D0948347

*The Physics and Circuit Properties of*
# TRANSISTORS

# The Physics and
# Circuit Properties of
# TRANSISTORS

James M. Feldman
*Professor of Electrical Engineering*
*Northeastern University*

John Wiley & Sons, Inc. New York · London · Sydney · Toronto

**Library of Congress Cataloging in Publication Data**

Feldman, James M.
The physics and circuit properties of transistors.

1. Transistors.   I. Title.

TK7871.9.F45          621.3815′28          71-39722
ISBN 0-471-25706-0

Printed in the United States of America

10  9  8  7  6  5  4  3  2  1

# PREFACE

Solid-state device theory is both a profitable and an enjoyable subject. It is a required discipline for those who would do rational design of integrated circuits, and it is fun—at least for engineers—because of the remarkably clever things that its proper application can achieve. Unfortunately, the polyglot character of the discipline makes it seem forbidding and difficult and sometimes rather arbitrary. Satisfaction in the application of device theory comes when one develops sufficient physical insight to see purpose and reason in the analytical steps and approximations by which one models a device. This insight is obtained only partly by studying fundamental principles and their application. Although this groundwork is obviously needed, the insight comes when one can make a connection between abstract principle and those data in which one has experience and in which one believes *emotionally*. This book is an attempt to form such a bridge for electrical engineering students at the senior year of undergraduate study or first-year graduate level.

The book is divided almost equally between solid-state principles (the first five chapters) and device analysis (the last five chapters). The whole book requires about two semesters to cover. The last five chapters form a sufficiently discrete unit to be effective in a one-semester device course for students who have had an adequate background in solid-state physics.

As a model of the student who might read this book, I have selected the electrical engineering major who is beginning or has just finished his senior year. Such students are normally well versed in circuit theory and are familiar with the elements of field theory, control theory, and some communication theory. Generally, their motivation for taking the course is an interest in the design and realization of integrated circuits and interesting electronic devices. Such students are willing and able to learn the requisite physics, but their goal is self-satisfying insight into the operation and design of electronic devices.

To satisfy such a student, I have shown the connection between physics and device theory even when devices are still many pages away. The text and the problems are both aimed at making the connection between experience (in which one believes) and the theory (in which one professes to believe). I have made the text more interactive than traditional textbooks. The reader is frequently asked to try the next step, to test his insight, before proceeding to read the development within the text. All too often a smooth presentation beguiles the reader into confusing his ability to follow the individual steps with the understanding of the whole procedure. If the next step is not within the reader's grasp, I have assumed that it helps the reader to know this. The result is a style that at times becomes quite informal. I hope the presentation is effective for self-instruction.

In order to keep the book to a reasonable size, I have limited the material to solid-state physics and circuit models applicable to the several types of transistor. Even with this delimitation, many of the details of interest in particular applications had to be excluded. My rationale has been that for an introductory book, the principal task is the development of physical insight. This book has been written for textbook use, and it makes no pretense at being a general reference work. It is intended to take the reader from an undergraduate background to a degree of sophistication in device theory sufficient for reading the current literature with understanding and appreciation.

It is certainly a pleasant duty to acknowledge the many people who have contributed to this book. I thank the many students who have read the preliminary versions of these chapters and have suggested ways to make them clearer. I also thank Peter Salmon, who helped with the revision of the first three chapters; Michael Riccio, who read the entire final version; and Professors Paul E. Gray and Basil Cochrun, whose suggestions and critical comments contributed substantially to this book.

I wish the reader good reading, and I hope that his efforts and mine acheive the goals he has set for this course.

Haifa, Israel                                                          James M. Feldman
August 1971

# CONTENTS

*The Physics and Circuit Properties of*

# TRANSISTORS

## CHAPTER 1

# A CAREFUL REINTRODUCTION
# TO THE OBVIOUS

*Common sense is prejudice developed before the age of eighteen.*
A. EINSTEIN

The quote above is a very good place to begin the discussion of the behavior of semiconductor devices. As the text proceeds I will ask many questions such as: "What is a...?" or "What would happen if...?" You will probably be tempted to give the obvious answer, but more caution is in order. The problem is that the "obvious" answer is often based on common misperceptions or extrapolations. For example, the statement that "What goes up must come down" sounds pretty idiotic to today's youth, familiar as they are with the fine sport of throwing Our Air Force Captains further than theirs. But 20 years ago, the adage seemed good. It fitted common experience within the limits of casual observation. Counter examples were available—what happened to the light from a flashlight when pointed up?—but they were unnecessary to dealing with everyday choices. Thus, they were mostly ignored.

There are two ways of shooting down "common-sense" fallacies. One is to perform an experiment that violates the "rule" in a clear and obvious way. The second is the rather elegant technique called the gedanken experiment. In this method one proposes a simple experiment to be performed in two ways, both of which have straightforward solutions from the rule under test. If we are clever, we will propose experiments that yield very obvious logical inconsistencies for false rules. The name derives from the fact that the experiments are normally performed in the head rather than physically. One of the prettiest examples of the gedanken technique was employed by Galileo to demolish once and for all the "obvious" law that heavy things fell faster than light. (No, not the leaning-tower-of-Pisa experiment. Aerodynamics could have been used in that experiment to make light things fall faster, slower, or at the same speed as the heavy object. That people accept the Pisa experiment as definitive proves them gullible indeed.) Galileo pro-

1

poses that we drop a heavy and a light object. According to the "obvious" rule, the heavy one will fall faster. "Ah!" says Galileo. "Now tie the two objects together. The heavy one will pull the lighter one down and the lighter one will pull the heavier one up. Thus they will fall at a speed intermediate to their two free velocities. But tied together, the two weigh more than either of them individually and so must fall faster than the heavier of the two." Thus the obvious rule is demonstrably false. (The only difficulty with this technique is contained in the phrase above "If we are clever . . . ." Being the first to be clever earns Nobel prizes today. Fortunately being second, third or $n$'th to be clever takes much less cleverness. All through this book, we will be coasting on other people's cleverness. All that is required is an open mind to new ideas and a willingness to try to use the novel concepts until they become familiar enough to be useful to you.)

## WHAT IS AN AMPLIFIER?

A large portion of the effort in the industrial revolution of the last two centuries has been devoted to power control. Progress in control can be described as getting better amplifiers. It is very important to understand the definition of an *amplifier*, because the object of this course is the understanding of one of our best small amplifiers, the transistor. Try a definition yourself. If you described it as a device to make something bigger, you have missed the point entirely. An amplifier is a device through which a large amount of power is controlled by a small amount of power. Levers, transformers, or hydraulic jacks are NOT amplifiers, since the power transfer ratio is, at best, one. On the other hand, the simple light switch on the wall is an amplifier and a most useful and ubiquitous one. The small effort in controlling the light switch can call forth a substantial effort on the part of the local power company.

There are only two elements necessary to make up an amplifier: a source of energy and a device for controlling its flow. Consider a simple and rather useful example. Connect a centrifugal water pump to a motor. If you dump some water into the inlet (a small amount of work) you can get the same amount of water at the outlet with the ability to do a great deal more work than your original efforts could have achieved. The energy source was the motorized pump. The control of the output power was achieved by regulating the amount of water dumped in at the input.

A conventional transistor operates on essentially the same plan. By expending a very small amount of work, we can inject a substantial number of electrons into one of the leads (the emitter) of a transistor; they appear at another lead (the collector) but are now being pushed along by a battery that serves as the pump. A power gain of 1000 is easily achieved in this process. The operation of the hydraulic amplifier above is conceptually easy because it fits our everyday experience. Let us begin now to develop some of the experience that makes the operation of the transistor as clear.

## WHAT'S AN ELECTRON?

You should be able to give a self-satisfying definition of the electron. Had you been really up on your constants, you might have stated that the electron is the lightest elementary charged particle, with a mass of about $10^{-27}$ grams and a charge of $-1.6 \times 10^{-19}$ coulombs. From your classical-physics instruction you should have a pretty good operational (i.e., experimental) definition for "charge," "mass," and even possibly "elementary." But what about "particle"? How do you know whether something is a particle? It is very tempting to define "particle" in some off-hand fashion, since it *seems obvious* that everyone knows what a particle is. Actually, however, it took physicists 300 years to come up with a good understanding of those special properties that constitute being a particle. To see what difficulties arise, it is necessary that you first try to define particle by some physical test that gives a yes-no answer to the question, "Is this a particle?" Try one.

Some of the definitions that are usually volunteered (or dragged kicking and screaming into the open) center about the sharp boundaries that we associate with *large* particles. However, note well what experiments define these "sharp boundaries": To observe sharp boundaries we either look at the object, which is an optical experiment, or we strike the object with another object, which is a measurement of the forces between objects as a function of their separation. Consider the second experiment first. If the forces we measure do not start abruptly—for example, the gravitational forces fall off smoothly as $1/r^2$ and extend all the way out to infinity—then we do not observe a sharp boundary. On the other hand, the forces that bind solids together invariably extend only a few atomic diameters from the outermost atoms, so these forces seem very abrupt indeed when we look at objects that are $10^7$ atoms in diameter. A definition that was so strongly dependent on the type of forces involved in the experiment would be poor indeed. We are looking for an operational definition that gives the same answer for ALL ways of testing it.

Looking at the optical experiment involved with "seeing" the boundaries, we immediately run into similar difficulties. The optical waves that our eyes can detect have wavelengths of 0.4 to 0.7 microns (millionths of a meter). If instead of micron wavelengths we employed meter wavelengths, most of our familiar objects become either very diffuse or lose their boundaries altogether and become points. This is the reason why we cannot see atoms even though they certainly ought to be included in a good definition of particles. Since the sharpness or fuzziness of a given boundary observed by "seeing" depends on the ratio of wavelength of the light to the size of the object, we cannot accept the visual sharp boundary test as our criteria.

Well, what's left? How about the property of discreteness or countability? This is certainly one of the properties of particles; you can count the grains of sand in a bucket or the atoms in hunk of germanium. But is this all? Almost. To be complete, we must specify the physical parameter to be counted. After all, the days of the year are denumerable, but they hardly fit into the class of things particlelike. The thing to be counted is energy. Although there are many ways of measuring energy, our definition of particle suggests for any of them, that if an energy flow is slowed sufficiently, the energy can change only by discrete amounts if what we are observing is particles. Consider the sand-in-a-bucket example above. If you dump sand quickly from one bucket to another, it is very difficult to perceive that the sand is arriving in discrete grains. However, if the sand is poured more and more slowly, eventually the sand will arrive one grain at a time. We then observe single "quanta" of sand. Slowing the rate still further does not change the size of the discrete grains—only their rate of arrival. This is all right for discreteness, but what about energy? Well, Einstein showed that Newton's laws contained in them the fact that the total energy $E = mc^2$ (where $m =$ = mass and $c =$ speed of light). We could observe the change in the total energy of the bucket, since it will change by a discrete amount as each single grain arrives. Thus we have satisfied our definition. There are, of course, much easier ways of observing sand grains (e.g., the other side of the equation says we can just as easily measure $m$), but that does not detract from the essential simplicity and reasonableness of our definition. It will turn out later, when the chips are down, that some particles can be discovered only by observing the abrupt arrival of a discrete amount of energy. And in the meanwhile, be thankful when there is an easier experiment.

When we observe the arrival of electrons by this energy technique, we find that each time an electron arrives, we obtain $9 \times 10^{-14}$ joules or, in the units we will use for energy throughout this book, about 500,000 electron volts (eV).† The electron fits our definition because it always arrives abruptly in the sense that either we do or do not have another 500,000 eV. However, you already believed that electrons were particles. The surprises come when we apply this definition to less obvious energy flows such as light or sound waves in solids. But more of that shortly.

---

† This unit is of great usefulness in discussing electrons because an electron volt is the energy that an electron will have if it falls through a potential of 1 volt. An electron volt is equal to $1.6 \times 10^{-19}$ joules. You will note that the number eV/joule is exactly the same as the number of coulombs per electron. That makes exceedingly good sense, since the energy that one coulomb of charge has after moving through one volt is one joule.

## WHAT'S A WAVE?

Let us pursue our electron a bit further first. In 1925, Louis de Broglie annoyed his professors and started a most incredible revolution in physics by suggesting that electrons were waves. To the people of his day, such a statement conflicted very strongly with the particle properties that had already been demonstrated for the electron. The conflict was artificial since there was no a priori reason to assume that electrons were not waves. But to see why these two properties—wave character and particle character— are not mutually exclusive, we must have a very good operational definition of what a wave is. Once again, you will get the sharpest insight into the problem if you begin with a definition of your own. A thing is a wave if....

Your physics and math background is sufficient for a reasonable first try if you can organize the concepts you want. You probably want to include two properties: periodic motion and the propagation or traveling of the wave from one place to another. The simplest kind of one-dimensional wave then looks something like $e^{j(\omega t - kx)}$. This is the sort of wave that we find on vibrating strings or on the ocean. This definition has in it much of the kernel of the definition we want. But it lacks one very important property that our particle definition had—it contains no universal physical test of wave character.

Let us assume that when we are talking about wave motion we are discussing one of the modes of energy propagation. Energy propagates by other methods, but most waves of physical interest involve energy flow. We will test the energy flow for wave properties using Young's *two-slit experiment*. We will place two screens at some distance from the source of the energy as shown in Figure 1.1. To illustrate the experiment, let us say that the source

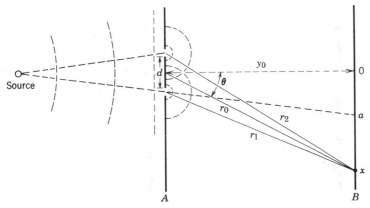

**Figure 1.1** Two-slit experiment.

of energy is a sodium vapor lamp running in air. The lamp puts out energy in two ways. It heats the air around it and it radiates a great deal of energy as yellow light. We make two parallel slits in screen $A$ and examine the energy distribution on screen $B$. Let us first open only one slit. We would observe at $B$ a similar distribution for both the radiation and thermal fields. It would show that the energy tended to spread out from the slit in all directions, although the optical energy might have a much higher density in the region of the point marked $a$. If we now open the second slit, the two fields act very differently. The energy density in the thermal field simply doubles. The optical field, however, develops some very distinct structure. Although the average energy density does indeed go up by two, the local energy density is given by

$$I = 2I_0 (1 + \cos 2\alpha x) \tag{1.1}$$

That is, you see a series of bright and dark bands (see Figure 1.2). The distance between bright bands is given by $\Delta x = \pi/\alpha$. We interpret our test as indicating that light is wavelike but that thermal energy is not.

What has the test really shown? It has shown that both thermal energy and the radiation field obey a linear differential equation, since the sum of the solutions for each slit is the solution for both slits. It has shown that the light is periodic in time and the thermal energy is not. And it has shown that both fields propagate through appropriate media. Looking for equations of motion that have these wave properties, we have a truly infinite set to choose from. However, *the simplest one* that has all of these properties is a second-order linear differential equation. Incorporating both the propagation and the periodicity, the simplest solution we could have for the electric field is $\mathbf{E}(x, t) = \mathbf{E}_0 e^{j(\omega t - kx)}$. This solution satisfies

$$\frac{\partial^2 \mathbf{E}}{\partial x^2} + \frac{k^2}{\omega^2} \frac{\partial^2 \mathbf{E}}{\partial t^2} = \frac{\partial^2 \mathbf{E}_0}{\partial x^2} + k^2 \mathbf{E}_0 = 0 \tag{1.2}$$

In the somewhat more complicated geometry of the two-slit experiment, we would expect a slightly more complicated solution of the form

$$\mathbf{E}(r, t) = \mathbf{E}_0(r) e^{j(\omega t - \mathbf{k} \cdot \mathbf{r})} \tag{1.3}$$

In our experiment, the slits are much closer together than $r_0$ so we may make several simplifying steps. With reference to Figure 1.1 it is easy to see that $r_1 \approx r_2$. As our single-slit experiment showed, the function $\mathbf{E}_0(r)$ does not change very rapidly with $r$, so $\mathbf{E}_0(r_1)$ and $\mathbf{E}_0(r_2)$ must be equal. Accordingly, we may write

$$\mathbf{E}(r, t) = \left[ \mathbf{E}_0(r_1) e^{-j\mathbf{k} \cdot \mathbf{r}_1} + \mathbf{E}_0(r_2) e^{-j\mathbf{k} \cdot \mathbf{r}_2} \right] e^{j\omega t}$$

$$= \left[ e^{-jkr_1} + e^{-jkr_2} \right] \mathbf{E}_0(r_0) e^{j\omega t}$$

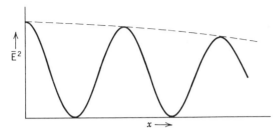

**Figure 1.2**  Young's two-slit fringes.

Since $r_{\frac{1}{2}} = r_0 \pm \Delta r$:

$$\mathbf{E}(r, t) = \left[ e^{jk\Delta r} + e^{-jk\Delta r} \right] \mathbf{E}_0(r_0) e^{j(\omega t - kr_0)}$$

$$= 2\mathbf{E}_0(r_0) \cos(k\Delta r) e^{j(\omega t - kr_0)}$$

We do not observe $E(r, t)$, however, but the intensity which is $E^2(r, t)$. In just the same way we deal with complex notation in electric circuits, we find the average square of a complex sinor by taking the product of the sinor with its complex conjugate. Accordingly, if we stay in the plane $z = 0$:

$$\bar{I}(r) = \mathbf{E}(r, t)\mathbf{E}^*(r, t) = 4\mathbf{E}_0^2(r) \cos^2(k\Delta r)$$

$$= 2\mathbf{E}_0^2(r) \left[ 1 + \cos(2k\Delta r) \right]$$

Simple trigonometry gives us $\Delta r$ as $\Delta r = (d/2) \sin \theta \cong xd/2y_0$. From this we may write

$$\bar{I}(r) = 2\mathbf{E}_0^2(r) \left[ 1 + \cos\left(\frac{kxd}{y_0}\right) \right] \tag{1.4}$$

From (1.4) it is easy to identify $kd/2y_0$ as $\alpha$ in (1.1) above.

It is worthwhile to ask here what the last few pages have accomplished. The answer is that we have developed a technique of analyzing energy flows to see whether they have wave properties. Our test is unambiguous, but it is not necessarily trivial to perform. A great deal of the problems that would be encountered depend on the constant, $k$. The fringe spacing, $\Delta x$, which we can measure with a ruler, gives us this constant as

$$k = \frac{2\pi y_0}{d\Delta x}$$

Note that for a given $k$ and $y_0$, if $d$ is large, $\Delta x$ will be small and vice versa. But both must be physically realizable for our experiment to work. If $k$ gets very large, we will have difficulty doing the experiment simply because we can make slits only so close together and can observe fringes only if they are at least so big. What does a large value of $k$ imply? Going back to (1.3) you can see that if $k$ is large, the sinor is rotating rapidly as the distance changes. In other words, the phase of the sine wave changes rapidly with distance. We normally refer to the distance between identical phase points as the wavelength. That means the constant $k$ gives us the wavelength:

$$\lambda = 2\pi/k \tag{1.5}$$

It is important to notice that in our fundamental definition of wave nature, our universal test yields the wavelength. It does not give the value of $\omega$. Nor is there any propagation velocity contained in our definition. All we insist is that the wave does propagate.

We might have one other difficulty in performing our experiment in a nice clean way. If we get two waves with different values of $k$, their patterns will be superimposed on screen $B$, since by definition, waves are linear and their solutions additive. Two values of $k$ wouldn't be too bad, but if we had 100, the pattern would be a real mess. There are experimental ways of avoiding this problem, and they will become clearer as we go on. But it is important to observe that if we want sharply defined fringes, $k$ must be limited to a very narrow range. This is a fundamental limitation that arises immediately from our operational definition.

Is there anything easy about the experiment? Yes, indeed. You do not have to go through any rigmarole to find the fringes. That is, it did not matter much where you put screens $A$ and $B$. The formal way of saying this is to state that the fringes are not localized.

## CAN PARTICLES BE WAVES?

If you reread our operational definitions again, you will see that there is nothing in them that would prohibit us from answering "Why not?" However, in our discussion of particles, the whole crux of the definition centered about the fundamental indivisibility of fundamental particles. The interference pattern arises from the interaction of the light coming from both slits simultaneously. What if we let only one photon at a time through our apparatus? We still observe the pattern. As was pointed out in the discussion of the two-slit experiment, the interference pattern that characterizes waves disappears if the second slit is closed. The obvious question, normally conveyed in tones of slight indignation, is how can ONE thing go through TWO slits simultaneously? The answer is: ONE thing can go through TWO slits simultaneously, and that's life. The only emotional comfort that can be offered is that for big particles, such as the ones we deal with in everyday life, a slit spacing that would allow us to observe the interference pattern we are looking for would be so small that we could guarantee that the particle would go through NEITHER slit. It would bounce. It would, in fact, bounce in a most satisfying, everyday way. But as I cautioned at the beginning of this chapter, it is not necessarily reasonable to extrapolate everyday experience further than the everyday world. Our definitions are simple and operational. To find out whether light is particulate, we experimentally determine how small a piece of it we can get. If we cannot indefinitely subdivide the energy (or equivalently, the mass), then light is particulate. Similarly, if we wish to know whether light is wavelike, we perform our two-slit experiment and look for fringes. Interestingly enough, the result of both of these experiments is positive. Light is made up of particles that happen to be wavelike.

(As an essentially trivial exercise just to see that you know what to do with the data, compute the appropriate value of $k$ and for $\lambda$ for the two sets of data below. Note that the dimensions of the slit in the second experiment are ridiculously small. The actual experiment used to determine the wavelength of the electron is closely related to our fundamental one and will appear in Chapter 2.)

**1.** Some near-infrared light is examined for its particle character. It is found to come in lumps of 1 eV of energy or multiples thereof. In Young's two-slit experiment, the slit separation was 0.1 mm and the distance between screens $A$ and $B$ was 10 cm. The fringe spacing is found to be $1.240 \times 10^{-2}$ cm. What are $k$ and $\lambda$ for such 1 eV photons? (Answer: $k = 5.08 \times 10^4$ cm$^{-1}$; $\lambda = 1.240$ $\mu$.)

**2.** Some electrons are examined for particle character. They are found to have a mass of $9.1 \times 10^{-28}$ gm ($5 \times 10^5$ eV $= m_0 c^2$). After acceleration between the cathode and screen $A$, the electrons pass through a pair of slits with a *kinetic energy* of 1 eV. The slit spacing $A$ was 100 Å (ha!) and

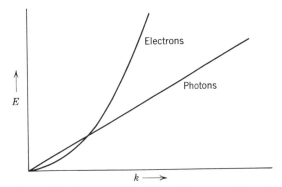

**Figure 1.3** $E$ versus $k$ for free photons and free electrons.

the distance between screens $A$ and $B$ was 1 cm. The fringe spacing was found to be .125 cm. What are $k$ and $\lambda$ for 1 eV electrons? (Answer: $k = 5.04 \times 10^{-9}$ cm$^{-1}$; $\lambda = 12.5$ Å.)

A rather interesting plot can be made for $E$ versus $k$ for photons and electrons. This is presented in Figure 1.3. These are the results obtained for *free* electrons and free photons. Electrons or photons that are constrained will frequently yield different results. The graphs show that for light, which is a particle that moves, oddly enough, with the speed of light, the wavelength is linearly proportional to the energy. For low-energy electrons, on the other hand, the relationship is quadratic: $E \alpha k^2$.

## NEWTON'S LAWS REVISITED

Newton stated quite correctly and with incredible perception that:

1. A body continues in uniform motion unless acted on by an external force.

2. The rate of change of momentum is equal to the net force acting on a body.

3. The sum of the static and dynamic forces acting on a body is zero.

After a number of years of solving classical physics problems with these three laws, you begin to feel comfortable with them and start developing the prejudice of common sense referred to in the quote at the chapter's beginning. How is it possible to have particles with wave characteristics if they obey Newton's laws? After all, waves obey a wave equation such as

$$\frac{\partial^2 \psi}{\partial x^2} + k^2 \psi = 0 \tag{1.2}$$

That certainly does not look very much like $\boldsymbol{F} = d\boldsymbol{p}/dt$. To see that they are equivalent we need two brilliant insights, fortunately not our own. The first that we need was developed by Louis de Broglie, who in 1923 showed that *if* there is to be a wavelength associated with a particle of momentum $p$, then it must be $p = h/\lambda = \hbar k$ ($\hbar = h/2\pi$). This did not appear out of the blue. What de Broglie demonstrated was that any other formulation would not fit in with the principal of relativity. The constant, $h$, would have to be Planck's constant if the formulation was to be correct for photons, so dè Broglie guessed, quite correctly, that the constant would be universal—the same for all particles. Let us see where this leads us. Using the wave equation (1.2) but substituting $p/\hbar$ for $k$ gives us

$$\frac{\partial^2 \psi}{\partial x^2} + \frac{p^2}{\hbar^2} \psi = 0 \tag{1.6}$$

Now in classical physics, such as that which applies to nonrelativistic particles,

$$\frac{p^2}{2m} = (E - V)$$

That is, the kinetic energy equals the total energy minus the potential energy. Putting this in (1.6) above gives

$$\frac{\partial^2 \psi}{\partial x^2} + \frac{2m}{\hbar^2}(E - V)\psi = 0 \qquad (1.7)$$

This is the time-independent (i.e., steady-state) form of Schroedinger's equation. Note carefully that we have included Newton's second and third laws when we invoked conservation of momentum and energy. To see that solutions to this equation obey the first law, notice what happens to the equation if we have no external forces ($V$ equal to zero); according to Newton, the motion must be in a straight line. The solution to (1.7) is just $\exp\left[j(2mE/\hbar^2)^{1/2}x\right]$.

This is the formula for a plane wave propagating in a *constant* direction and at a *constant* velocity. Thus we have the best of both worlds: Newton's friendly old formulas and a nice modern wave formulation.

It is important to understand what is meant by a *time-independent* wave equation. You already have seen that waves are fundamentally and inevitably time varying. So how can the equation describing the motion be time independent? Well consider the simple RC circuit shown in Figure 1.4. If it is driven by an arbitrary voltage, we must write the equation describing the charge, $q$, as

$$R\frac{\partial q}{\partial t} + \frac{q}{C} = V(t)$$

But if the generator puts out a sine wave, $V_0 e^{j\omega t}$, then we have

$$R\frac{\partial}{\partial t}(q_0 e^{j\omega t}) + \frac{q_0 e^{j\omega t}}{C} = V_0 e^{j\omega t} = \left[Rj\omega q_0 + \frac{q_0}{C}\right]e^{j\omega t}$$

We can cancel out the common exponential term and obtain an equation in which time does not appear explicitly:

$$\left(j\omega R + \frac{1}{C}\right)q_0 = V_0$$

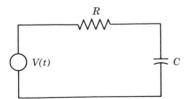

**Figure 1.4** RC circuit.

It must be remembered that the solution includes $e^{j\omega t}$, for indeed it does. But we do not have to carry that term around, nor does our differential equation have to contain the derivatives or integrals with respect to time that the time-dependent equation must. Each of the derivatives or integrals is replaced by $j\omega$ to an appropriate power.

How do we get the time-dependent formula for the Schroedinger equation? We now must make use of the other brilliant insight, this one developed by Max Planck. Classical physics including Maxwell's equations had failed to describe the laws of radiation that were observed experimentally. As a matter of fact, the only classically correct formulation was so obviously wrong that the derivers of the law (Rayleigh and Jeans) referred to it as the "ultraviolet catastrophe." The problem was that classical physics treated light as if it were indefinitely divisible—that is, it assumed that light was NOT particulate. In 1900, Planck showed that he could derive the correct radiation law if he assumed that light comprised particles of energy (photons) with the relationship between energy and frequency given by

$$E = h\nu = \hbar\omega \qquad \text{(where } \nu \text{ is the frequency)} \qquad (1.8)$$

So startling was this photon concept that even Planck did not easily accept it. But Einstein, who apparently lacked common sense or at least the prejudice centered therein, grasped the full significance of Planck's hypothesis. He predicted the low-frequency cutoff for the photoelectric effect that, when demonstrated, proved the particulate nature of light exactly as our definition would suggest.

Using Planck's hypothesis to get the time-dependent formulation of Schroedinger's equation, we simply substitute (1.8) into (1.7). This yields

$$\frac{\hbar^2}{2m} \frac{\partial^2 \psi}{\partial x^2} - V\psi = -E\psi = -\hbar\omega\psi$$

The only way this equation can be extracted from a time-dependent form is if we write it as

$$-\frac{\hbar^2}{2m} \frac{\partial^2 \psi}{\partial x^2} + V\psi = -jh\frac{\partial \psi}{\partial t} \qquad \text{where} \qquad E\psi = -jh\frac{\partial \psi}{\partial t} \qquad (1.9)$$

Note that if, and only if, the *total* energy is a constant, we get the time-independent equation.

## WHAT IS THIS THING CALLED $\psi$?

Establishing experimentally that electrons obey a wave equation does not really tell us what $\psi$ is. Let me ask as related question to arrive at a proper point of view. When I wrote some pages back (equation 1.3) that the electric field was given by $\mathbf{E}(r, t) = \mathbf{E}_0(r)e^{j(\omega t - k \cdot r)}$, what did it mean? You should answer that the electric field could be represented by a sinor that gave its amplitude and phase as a function of time and position. Presumably, you also know that the electric field is the *measured* force on a unit charge at some point in space and time. So far so good. Now, what happens if I perform the two-slit experiment using a piece of film for screen $B$ and letting only one photon at a time through the slits? Think this out carefully, remembering that the photon essentially goes through both slits as we discussed. Being quite specific, what would I find on the film if I developed it after only a few dozen photons had gone through the slits? The answer is that I will find a few dozen exposed spots of very small size. (The photon arrives with barely enough energy to cause one chemical reaction—a single bond is broken. Thus only one silver grain will precipitate for each photon absorbed.[1] That photon really is a particle.) What happened to the interference pattern? It is still there. None of the exposed spots are at the center of a dark fringe. Most of them lie near the center of one or another light fringe. If I run $10^{15}$ photons through the slit, even one at a time, the interference pattern will be recorded as expected. It will simply be the sum of all the spots. To see the fringes on a ground-glass screen, you have to let enough photons through to construct the whole pattern for your eye in a relatively short time (about 1/10th of a second). Our wave solution gives us the value that will be measured if a large number of photons are coming through the slit in a measurement time. (It is important to remember that the film [and the eye] measures the energy density. Thus, we are measuring $EE^*$ rather than $E$.) The fact that the light is particulate says that each photon makes only one very small spot. How can these two concepts be reconciled?

Fist, let us dispose, once and for all of the argument that such things CAN'T happen (or the question usually phrased "How can that be?"). If you find something that does not fit your preconceived ideas, the fact must displace the erroneous concept. What we want here is a proper method of treating the wave and particle concepts so that we can use them to predict physical phenomena correctly.

In discussing photographing the two-slit pattern one photon at a time, I made the statement that the exposed spots NEVER occur at the center of a dark fringe. I further stated that the spots are MOST LIKELY to be found near the center of a light fringe. *Never* and *most likely* are words that are used to describe the probability that an event will take place. It is tempting to identify $E^*(x)E(x)$ with the probability of finding photons at that place $x$. That is certainly consistent with results that one gets taking one photon at

---

[1] Photographic emulsions aren't really quite that simple, but the only property of the emulsion that we are using is its ability to store information on the spatial distribution of photons.

a time or $10^{18}$ at a time. If $EE^*$ is large at some spot, the probability that you will find a photon at that spot per photon put into the system is quite large. Similarly, if at some position $EE^*$ is zero, then you will NEVER find a photon there.

If you now ask whether that interpretation makes sense, I must respond that of course it makes sense. It makes sense precisely because it works. It has been found to be the best interpretation ever since Max Born suggested it in the 1920's. If you do not like this interpretation—this idea of knowing only the likelihood of events rather than whether they will happen or not— you can join a very unexclusive club of people who have looked for alternative interpretations. But keep in mind that unattractive as this uncertainty is, the Born interpretation has never yet failed to yield a proper answer. Once you have accepted the probabilistic interpretation of the wave function and the experimental basis of the wave equation, you are in a position to learn how to predict what a particular wave or particle experiment will yield. Since we will be concerned with the behavior of electrons in a periodic lattice (the crystal), it is very nice to be in a position to interpret electron-in-a-periodic-lattice experiments. This will form the central subject of the next chapter. But meanwhile, you still have some studying to do on how to use the wave function. After all, we have only interpreted what the square of its absolute value is.

Let us proceed from the meaning of $\psi$, the solution to equation 1.9. If we are talking about electrons, for example, $\psi(x, t)\psi^*(x, t) = P(x, t)$ is the probability of finding the electron at $x$ at time $t$. If $\psi$ happens to be $\psi_0(x)e^{j\omega t}$, we can re-write equation 1.9 as equation 1.7, since for that particular type of function, the total energy $E = \hbar\omega$ is a constant. However, if $V(x)$ is not a function of time, the left side, and, of necessity then, both sides of equation 1.9 cannot be functions of time. But how can it be true if we cannot write $\psi(x, t)$ as $\psi_0(x)e^{j\omega t}$? To see what we are trying to do we need an example. Let us put an electron in a perfectly reflecting box of length $A$ (or we could put a photon between a pair of conventional mirrors—it's much the same thing). Let's first look at equation 1.7, the time-independent form. This is a second-order linear differential equation, so we need two boundary conditions to do the problem. Figure 1.5 shows the problem. Inside the box, the potential, $V$, is zero. The box walls are perfectly reflecting. From your knowledge of violin strings or other standing waves at perfectly reflecting walls, the boundary condition that seems most obvious is that $\psi$ must vanish at $x = 0$ and $x = a$. Our problem then is to find $\psi$ such that

$$\frac{\hbar^2}{2m}\frac{\partial^2\psi}{\partial x^2} + E\psi = 0 \qquad \psi(a) = 0 \qquad \psi(0) = 0 \qquad (1.10)$$

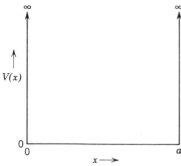

**Figure 1.5a** Potential well for first sample problem.

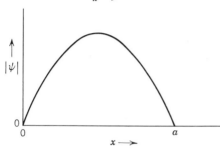

**Figure 1.5b** Solution $\psi_1$.

The solution is $\psi = A \sin kx$ where $k$ can take on any value of the sequence $k = \pm n\pi/a$. Actually, of course, any sum of such solutions is also a solution. This was one of the principal experimental results that led us to use a wave equation in the first place.

Let us consider two solutions. Specifically

$$\psi_1 = A_1 \sin \frac{\pi x}{a}$$

$$\psi_2 = B_1 \left[ \sin \frac{\pi x}{a} + \sin \frac{2\pi x}{a} \right]$$

First, for wave function $\psi$, find the probability of finding the particle at any point $x$ within the box. By our rule, the answer must be $A_1^2 \sin^2 (\pi x/a)$. But we certainly want to find the particle SOMEWHERE in the box, so the probability of finding it anywhere must be 1. This is just the SUM of the probabilities of finding it in each interval $dx$. Accordingly

$$1 = \int_0^a A_1^2 \sin^2 \left( \frac{\pi x}{a} \right) dx = A_1^2 \frac{a}{2}$$

$$A_1 = \left( \frac{2}{a} \right)^{1/2}$$

Such a step is called *normalizing the wave function.*

Now what is the total wave function for $\psi_1$? In other words, how do you write the wave function including both the time and position variation? [*Hint:* look at the equation just before (1.9).] The answer is

$$\psi_1 = \frac{2}{a} \sin \frac{\pi x}{a} e^{j\omega_1 t}; \qquad \text{where} \qquad \omega_1 = \frac{\hbar \pi^2}{2ma^2}$$

You should notice that I have written $\psi_1$ both with and without the time-dependent factor $e^{j\omega_1 t}$. This is not quite legitimate, but it is common practice. The writer assumes the reader knows that the exponential factor belongs there. (You will note that I left "that" out of the previous sentence. It makes sense only because you know that a "that" belongs there.) You should *always* add the exponential factor in this book and any other—it always belongs.

The next problem is to do the same steps for $\psi_2$. First, normalize the wave function. You should get $B_1 = (1/a)^{1/2}$. If you recognized that the solutions of (1.10) are mutually orthogonal

$$\int_0^a \psi_i \psi_j^* dx = \begin{cases} 0 & i \neq j \\ \int_0^a |\psi_i|^2 dx & i = j \end{cases}$$

then the answer to "What is $B_1$?" can be written down as easily as $A_1$. (It can be proved without much difficulty that the solutions to the wave equation always form a complete and orthogonal set.) The next question is a bit tougher. How do you write the total wave function for $\psi_2$?

*Hint:* $\psi_2$ is obviously constructed by adding two of the solutions to (1.10).

The answer is

$$\psi_2 = \left(\frac{1}{a}\right)^{1/2} \left[ \sin \frac{\pi x}{a} e^{j\omega_1 t} + \sin \frac{2\pi x}{a} e^{j\omega_2 t} \right] \qquad \omega_2 = 4\omega_1$$

This solution represents two sinors rotating at different frequencies. Just as in the case of adding the first and second harmonics on a violin string (E.E. notation; a musician would call them the fundamental and first harmonic.), the observed pattern does not remain stationary. Figure 1.6 shows the amplitude of $\psi_2$ and the probability of finding the electron ($\psi_2 \psi_2^*$) as a function of $x$ for two different times. This is the first wave-solution example you have of a particle in nonuniform motion. This one is bouncing back and forth in the sense that the pattern repeats at twice the frequency of the first harmonic. From your knowledge of Fourier series, you should be able to see that ANY nonpathological curve that you can draw between 0 and $a$ can be represented by our Fourier sine series. That would mean that you could distribute your particle any way you wanted within the box. However, the wave equation

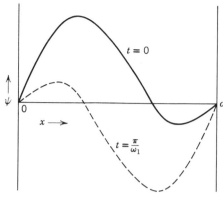

**Figure 1.6a** $\psi_2$ at $t = 0$ and $t = \pi/\omega_1$.

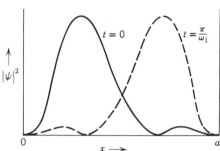

**Figure 1.6b** $|\psi|^2$ at $t = 0$ and $t = \pi/\omega_1$. Observe that the particle has moved from left to right between 0 and $\pi/\omega_1$.

puts some penalties on your arbitrariness. It states, as we shall now see, that the more harmonics you require, the higher the energy it takes to get the particle in that configuration. An extreme example of the penalty imposed by being too arbitrary in choosing a wave function is shown at the end of this chapter. As a routine exercise, answer the following questions:

1. Is $\psi_1$ a solution of (1.7)?

Yes, by direct substitution if $E_1$ is given by $E_1 = \hbar\omega_1$.

2. Is $\psi_2$ a solution to (1.9)?

Yes, by direct substitution

3. Is $\psi_2$ a solution to (1.7)?

No! Since $j\hbar(\partial\psi/\partial t) \neq$ const. $\psi$ we cannot replace (1.9) with an equivalent equation such as (1.7).

We are now back at the problem of what to do when $j\hbar(\partial\psi/\partial t) \neq$ const. $\psi = E\psi$. I suggested above that if $V \neq V(t)$, there was a way to get the energy for any solution to (1.9). This is the last really new procedure that we must include in our wave mechanics; the remainder of our work with wave mechanics will be interpretive.

We have interpreted the square of the wave-function's amplitude as the probability of finding the particle at that place and time. If I wanted to find the *average* position I must perform the following step:

$$\bar{x} = \int_0^a x\psi\psi^* \, dx \tag{1.11}$$

There are several ways of interpreting (1.11). It is certainly the way to find the center of mass of the particle, since $\psi\psi^*$ is the particle density. Another point of view would label $\bar{x}$ as the *expectation value* in the statistical sense of being the average of a large number of independent measurements of $x$. The operation is the same one that you would employ if you wanted to answer the question: "How many batters is a particular pitcher likely to strike out today?" The method of (1.11) says that you should first take the number of times he has obtained some $n$ strikeouts and divide it by the total number of games he has pitched. This operation yields the relative probability that he will get exactly $n$ strikeouts. It is equivalent to $\psi^*(n)\psi(n)$. To get the expectation value of $n$, we must sum $(n\psi^*\psi)$ over all the possible values of $n$ (or integrate if $n$ were continuous). We must be prepared for the possible result that the expectation value may never occur. (For example, the most likely value of the number of strikeouts might well be noninteger, but by the rules of the game, a half-struck-out batter is not struck out).

Now we apply the same operation to equation 1.9. We derived equation 1.7 from the familiar rule that the sum of the potential energy and the kinetic energy was the total energy. We extended (1.7) to (1.9) in a way that did not conflict with that interpretation. Accordingly, we can write

$$\frac{p^2}{2m} + V = E \Leftrightarrow -\frac{\hbar^2}{2m}\frac{\partial^2\psi}{\partial x^2} + V\psi = j\frac{\hbar\partial}{\partial t}\psi$$

As stated before, any solution to (1.9) can be constructed by a Fourier series of solutions to (1.7) as long as $V \neq V(t)$. If we take any solution to (1.9), $\psi$, and Fourier-analyze it, we get

$$\psi = a_1\phi_1 + a_2\phi_2 + a_3\phi_3 + \ldots$$

Each $\phi_i$ is the solution to (1.7) with $E = E_1 = \hbar\omega_1$ $[\omega_i = E_i/\hbar = \hbar^2\pi^2i^2/2ma^2$ for our example of $V = O]$. Let us put $\psi$ into (1.9) and multiply through from the left with $\psi^*$. The result is

$$\psi^* \frac{\hbar^2}{2m}\frac{\partial^2\psi}{\partial x^2} + \psi^*V\psi = \psi^*jh\frac{\partial\psi}{\partial t} = \left(\sum a_i^* \, \phi_i^*\right)\left(\sum E_i a_i\phi_i\right)$$

At the last equal sign, we have taken advantage of the fact that for solutions of (1.7):

$$-jh\frac{\partial\phi_i}{\partial t} = E_i\phi_i$$

Now notice what happens if we integrate the equation from 0 to $a$. Since the $\phi$'s are orthogonal over that interval, all the cross-products $(\phi_i^* \phi_j)$ disappear, leaving us with

$$\int_0^a -\psi^* jh \, \frac{\partial \psi}{\partial t} \, dx = \sum |a_i|^2 E_i = \bar{E} = \int_0^a -\psi^* \frac{h^2}{2m} \frac{\partial^2 \psi}{\partial x^2} \, dx +$$

$$+ \int_0^a \psi^* V \psi \, dx$$

Now if we are going to interpret (1.11) as giving the expectation value of $x$, then it makes sense to interpret the results above as giving the expectation value of the energy. Furthermore, the individual terms on the right are interpreted as giving the expectation values of the kinetic and potential energy terms separately. Since any solution to (1.9) with $V \neq V(t)$ can always be constructed from solutions to (1.7), we can find these expectation values, and our interpretation is at least mathematically consistent.

How consistent is it with physical experimentation? Consider the sensation you observe when looking at a mixture of yellow and red light. You see orange. Let us say that the mixture is 30% yellow and 70% red photons. From the wave viewpoint, the field can be written as $I = I_{red} + I_{yellow}$.

$$E(t) = \sqrt{0.7} e^{j\omega_1 t} + \sqrt{0.3} e^{j\omega_2 t} \quad \omega_1 = 2.7 \times 10^{15}/\text{sec} \quad \omega_2 = 3.2 \times 10^{15}/\text{sec}$$

If we reduce the intensity until we see only one photon at a time, it is either a red or a yellow photon. It is never an orange photon. From the field expressions above, we can say that the expectation value of the energy is "orange," and that if the photons are received one at a time, the odds are 7/3 that a given photon will be red and 3/7 that it will be yellow. Thus, the interpretation of

$$\int_0^a \psi^* jh \, \frac{\partial}{\partial t} \psi \, dx$$

as giving the expectation value is quite consistent with what we observe. If you don't see this step try performing the operation $\psi^* ih(\partial/\partial t)\psi$ and remember that $E = h\omega$. The integral is a bit ambiguous because we didn't put our photon in a box. However, the average value of the cross-product terms is obviously never far from zero.

We differentiate between solutions to equations 1.7 and 1.9 by saying that those that satisfy both equations are *eigenfunctions* of energy. The name arises because to satisfy both equations we must have $jh(\partial \phi_i/\partial t) = E_i \phi_i = (\text{const.})(\phi_i)$. We can thus characterize the solution $\phi_i$ by giving the constant $E_i$. The German word *eigen* means characteristic, hence the name. We call $\phi_i$ the eigenfunction and $E_i$ the eigenvalue. Energy is not the only interesting

**4.** We decided that any definable operator that was equivalent to a physical parameter $[-jh(\partial/\partial x)$ for momentum, as an example] could be treated in the same fashion as the energy operator. Thus we could get eigenfunctions of other physical quantities, we could find expectation values for these quantities, and we could predict what values of the quantities we could measure and with what relative frequency.

This is all the wave mechanics we will need to obtain a pretty good understanding of how electrons behave in crystals. It is only an introduction to wave mechanics, but it is remarkable just how far you can go on this small amount. What is needed now is a problem or two to find out whether you see how to use the rules we have established and to make sure you understand the relationship between the essentially mathematical operations and the real world to which they correspond.

## AN EXERCISE

Figure 1.7 shows a particle uniformly filling half a box. Presumably, this would not seem hard to arrange physically. Note that the wave function is entirely real at this moment, so its square is easily taken. What must the amplitude of the wave function be to establish that there is only one particle in the box? Answer: $\sqrt{2/a}$ so that

$$\int_0^a \psi\psi^* \, dx = 1$$

What values of energy might I find if I measured the energy of this particle? Answer: All the values of $n^2\pi^2\hbar^2/2ma^2$, where $n$ is an integer, except those values for which $n$ is divisible by 4. To find this answer you must Fourier-analyze $\psi$. The eigenfunctions of energy are obviously the same as in our example before. That is:

$$\phi_n = \left(\frac{2}{a}\right)^{1/2} \sin\frac{n\pi x}{a};$$

the coefficients of the eigenfunctions are
$$\begin{cases} = 2/n\pi & \text{for } n \text{ odd} \\ = 0 & n/2 \text{ even} \\ = 4/n\pi & n/2 \text{ odd} \end{cases}$$

Now comes a most amusing result. Find the expectation value of the energy. (Answer: INFINITE! The energy goes up as $n^2$ but the amplitude of the eigenfunctions go down only as $n^{-1}$. Thus, the infinite series [of energy eigenvalues times the square of the amplitudes] does NOT converge. Doesn't look like we could push our particle into this shape, does it?)

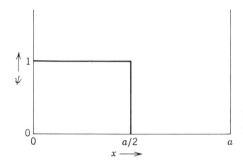

**Figure 1.7** A particle uniformly filling half a box.

## A SECOND EXERCISE WITH A HAPPIER ENDING

The mathematical difficulty that lead to our problem above was the discontinuity in the wave function. As a general rule, we can get away with discontinuous derivatives, but the functions themselves are *always* continuous. The wave equation shows that real potentials (no infinite potentials allowed) always yield bounded second spatial derivatives. Thus we can state that the first derivative of $\psi$ is continuous.

Let us try a problem with a finite potential step. Consider the potential well shown in Figure 1.8. In classical billiard-ball mechanics, a particle in the well ($V$ less than $V_0$) would have to be found between $a/2$ and $-a/2$. All we can insist on here in the real world is that the solution to our wave equation be bounded and continuous with continuous derivatives. Let us see what this implies.

*First*, we divide the $V - x$ plane into three regions: $x < -a/2$, $-a/2 < < x < a/2$, $a/2 < x$. We get two different wave equations ($V$ equals 0 or it doesn't equal 0) depending on which region we are in. If we solve the equation in each region and match the functions and their first derivatives at the common edges we will satisfy the wave equation and the boundary condi-

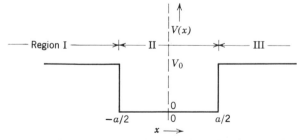

**Figure 1.8** Potential well for exercise 2. Region I and III are spatial regions of high potential. Region II is a region of zero potential.

tions everywhere. That would mean that we were done.

*First*, is the constant $E_i$ in equation 1.7 the same in all three regions? (Yes. A particle has one total energy, not several. We are seeking the distribution of the particle in the three regions, but it is all the same particle.)

*Second*, what sort of boundary conditions do we have on $\psi$?

*Hint:* How many boundary conditions may you impose on a second-order linear differential equation whether or not the coefficients of equation are a function of position?

We have two boundary conditions. The solutions must vanish for large positive and negative values of $x$ in such a way that the integral of $\psi^*\psi$ converges. (The total probability of finding the particle must be 1.)

*Third*, what is the form of the solutions in each of the regions?

In I:  $\phi = Ae^{+\alpha x}$    $\alpha = \sqrt{(V_0 - E_i)\dfrac{2m}{\hbar^2}}$

(vanishes for large negative values)

In II:  $\phi = B \sin \beta x + C \cos \beta x$    $\beta = \sqrt{E_i \dfrac{2m}{\hbar^2}}$

In III:  $\phi = De^{-\alpha x}$    $\alpha = \sqrt{\left(V_0 - E_i\right)\dfrac{2m}{\hbar^2}}$

(vanishes for large positive values).

*Fourth*, what restrictions do we have on $A$, $B$, $C$, and $D$ in order that we can match the solutions and their derivatives across the interfaces? We get four equations. By subtracting one equation from another we obtain

$$(A - D)e^{-\alpha(a/2)} = -2B \sin \left(\beta \frac{a}{2}\right) \quad \alpha(A + D)e^{-\alpha(a/2)} = 2\beta C \sin \left(\beta \frac{a}{2}\right)$$

$$(A + D)e^{-\alpha(a/2)} = 2C \cos \left(\beta \frac{a}{2}\right) \quad \alpha(A - D)e^{-\alpha(a/2)} = 2\beta B \cos \left(\beta \frac{a}{2}\right)$$

Now, if none of the coefficients vanish, we may take ratios to eliminate the exponentials and coefficients. We thus obtain

$$+ \alpha = -\beta \cot \left(\beta \frac{a}{2}\right) \quad \text{and} \quad \alpha = \beta \tan \left(\beta \frac{a}{2}\right)$$

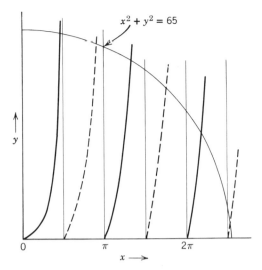

**Figure 1.9** Solution of the two equations which give the energy eigenvalues for the second exercise. The plot is in terms of the normalized coordinates $x = \beta a/2$ and $y = \alpha a/2$. This normalization takes:

$$\beta^2 + \alpha^2 = \frac{2mV_0}{\hbar^2} \qquad \text{to} \qquad x^2 + y^2 = \left(\frac{2mV_0}{\hbar^2}\right)\left(\frac{a^2}{4}\right) \qquad [\text{the circle}]$$

$$\beta \tan \beta a/2 = \alpha \qquad \text{to} \qquad x \tan x = y \qquad [\text{the solid curves}]$$

and

$$\beta \cot \beta a/2 = -\alpha \qquad \text{to} \qquad x \cot x = -y. \qquad [\text{the dashed curves}]$$

For the values of $V$ and $a$ given, six intersections are found between the circle and the tangent or cotangent curves. There are, thus, six and only six solutions for a particle bound in this potential well. The ground state is found at $x = 1.4$, giving a ground state energy of 0.61 eV.

We cannot simultaneously satisfy both of these equations. However, if $A = D$ and $B = 0$, two of the equations above vanish, leaving us with

$$+ \alpha = \beta \tan \left(\beta \frac{a}{2}\right)$$

Similarly, if $A = -D$ and $C = 0$, the other two equations vanish, leaving us with

$$- \alpha = \beta \cot \left(\beta \frac{a}{2}\right)$$

We can solve both of these last two equations graphically to obtain the values of $\alpha$ and $\beta$. (See Figure 1.9.) To solve these equations we must use the fact that

$$\beta_i = \sqrt{E_i \frac{2m}{\hbar^2}} \qquad \alpha_i = \sqrt{(V_0 - E_i) \frac{2m}{\hbar^2}} \qquad \text{or} \qquad \beta^2 + \alpha^2 = V_0 \frac{2m}{\hbar^2}$$

*Fifth,* if $a$ is 1000 Å and $V_0 = 10$ eV, what is the lowest energy that an electron can have? Answer: 0.61 eV. (Remember that $mc^2 = 5 \times 10^5$ eV. Thus, $m = 5/9 \times 10^{-15}$ eV sec$^2$/cm$^2$.)

*Sixth,* sketch the wave-function amplitude as function of position in the ground state (Figure 1.10).

*Seventh,* for the lowest energy level, what is the total probability of finding the particle outside the box? You must first normalize the wave function, giving:

$$\phi_0 = \begin{cases} C_0 \cos \beta_0 x & |x| \leq a/2 \\ A_0 e^{\pm \alpha_0 x} & |x| > a/2 \end{cases}$$

Then you must integrate $\phi_0 \phi_0^*$ over the interval outside the box, giving $P = .0036$. In other words, the electron will be found outside the box about 0.36% of the time if it is in its ground state. For higher energies, the particle would spend more of its time outside the box.

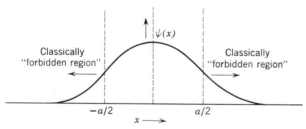

**Figure 1.10** Solution to exercise 2. The function and the derivatives must match at the boundaries.

## Problems

1. A plane wave of electrons is incident on the barrier shown below. Plot the percentage that pass through the barrier as a function of $E$ for values of $E$ both larger and smaller than $V_0$. Are there any values for which all of them pass through? Show that the sum of the reflected and transmitted percentage is 100%.

*Note: This problem is exactly equivalent to several very interesting electrical problems. For example, if two pieces of waveguide are connected by a third of some different impedance and higher cutoff frequency, the description of their transmission as a function of frequency is the same as the above problem. The solution also has some bearing on the operation of the tunnel diode.*

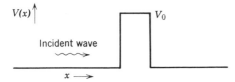

2. A particle finds itself at the bottom (ground state) of an infinitely high, rectangular potential well. Its wave function is thus:

$$\psi_1 = \frac{1}{\sqrt{2}} \sin(kx) \qquad k = \frac{\pi}{a} \qquad a = \text{width of well}$$

Let us say that at $t = 0$ it was possible to instantly double the width of the well. It follows from conservation of energy that the particle could not move instantly, so that

$$|\psi_1(0^-)| = |\psi_1(0^+)|$$

However, $\psi_1$ is no longer an eigenstate of the new well. Therefore, it is possible that work was done on the particle or that the particle did work on the moving wall. Find the energy change, if any, in the transformation from one system to the other. (Note: a possible answer is zero.)

**Comment:** A wave is a wave, for all that. Consider, if you prefer, the precisely analogous problem of abruptly doubling the length of a violin string initially vibrating it its fundamental mode.

3. A particle is an infinitely high potential well of width $a$. At $t = 0$ its wave function is

$$\psi = (a)^{-1/2} \left[ \sin \frac{\pi x}{a} + \sin \left( \frac{2\pi x}{a} \right) \right]$$

Where is the center of mass of the particle (i.e., the expectation value of its position) and what is the velocity of the center of mass?

4. Consider an electron constrained to remain within a cube of volume $a^3$ where $a = 1 \text{ Å} = 10^{-8}$ cm. This is roughly the "size" of a hydrogen atom. Within the cube, the potential is 0. The cube walls are impenetrable. What is the energy of the ground state? What is the energy of first excited state? What is the *degeneracy* of the first excited state (i.e., how many times

does the same energy occur for states with different arrangements of the mode numbers)? What are the mode numbers for the first state in which the electron is never found at the center of the cube?

5. If you wanted to determine the distribution of the electron in problem 4, you might probe it with a very energetic (i.e., fast) small test charge. The test charge would scatter from the electrostatic potential of the electron. By examining the scattering pattern, you might then hope to reconstruct the charge distribution. Assuming that the charge distribution is the same as the mass distribution, determine the electrostatic potential in the ground state for an arbitrary point, $r$, lying within the box. Let the center of the box represent zero potential.

6. One of the few potential-energy functions that leads to a readily soluble wave equation is the one associated with the harmonic oscillator—the weight on a spring. (You will find the solution easily enough if you have had some background in the series-solution of differential equations. If you don't have that background, try another problem.)

   The harmonic oscillator is characterized by a potential function, $V(x) = -Kx$. Find the eigenvalues of energy and the ground state eigenfunction for a particle of mass $m$ subject to this potential.

*Hints.* If you examine the wave equation for large values of $x$, you will find that the leading term suggests a trial function of the form $\phi = u(x)e^{-ax^2}$ where $u(x)$ is a polynomial in $x$. Substitution into the wave equation yields a new equation in $u$ that is quite handy for a series solution. Note that you get two completely independent series, one even and one odd. Under certain circumstances, one of the two series will be truncated to form a *finite* series. Only under these special circumstances will the integral of $|\phi(x)|^2$ from $-\infty$ to $+\infty$ converge. Since this convergence is required, the "special circumstances" give you the discrete and readily determined characteristic values and functions.

*Solution.*

$$E_n = \left(n + \frac{1}{2}\right)\sqrt{K/m}; \quad \phi_0 = \left(\frac{Km}{h^2\sqrt{\pi}}\right)^{1/2} \exp\left[-\frac{\sqrt{Km}}{h^2}\frac{x^2}{2}\right]$$

## CHAPTER 2

# ELECTRONS AND X RAYS IN CRYSTALS

In Chapter 1 some careful definitions and two fundamental hypotheses led to a single equation (1.9) that was supposed to describe the behavior of electrons in *all* situations. Of course, $F = ma$ and the fundamental theorem of the calculus purport to do the same thing in their respective spheres of influence. As you undoubtedly have noticed, there is a substantial practical difference between having the equation and having the answer. It is worthwhile to ask just how well we can do in getting answers to the problems that really count. As the Cheshire Cat might suggest, how well you will do depends on just where you want to get.

The central object for the first half of this text is to develop enough physics to understand the behavior of semiconductor devices. Although semiconductor devices form an amazingly diverse family, they all derive from the behavior of electrons in a crystal subjected to externally applied thermal or electromagnetic or mechanical stimulation. If we could solve equation 1.9 for each case of interest, we could launch immediately into a discussion of device behavior. Unfortunately, the number of exact solutions of even the time-independent wave equation are pitiably few in number. They include the particle in a box (samples in the last chapter), the harmonic oscillator, the hydrogen atom and very little else. However, before you chuck the wave equation into the nearest wastebasket, think how much you can achieve with $F = ma$ even though only one- or two-body problems can be solved exactly. In the first place, a very large portion of the real world can be described fairly accurately in terms of cavities and variations on the hydrogen atom. Quite frequently we are forced into making approximations, but almost as often, we can carry the approximation to any degree of accuracy that we require. Often a set of rules can be developed that describe the general character of the solutions to a problem even if the particular solutions themselves are essentially unobtainable. Then, a moderate amount of laboratory measurement to fit numbers to the rules is all that is required to achieve a really operational understanding. This is the case for electrons in crystals.

In this chapter, we will examine the behavior of electrons and X rays in crystals. An accurate definition of a crystal must be obtained and from this two very important general rules. The first of these rules is that the energy eigenvalues of solutions to (1.7) are grouped into bands. These bands are frequently separated by rather large intervals of energy in which no solutions are found. This rule is quite sufficient to allow us to predict which crystalline solids could be insulators. (Not all of them will be.) The second rule is that *if* electrons are free to move about in a crystal, the incredibly complicated interaction of the moving electron and the crystal can be described simply by $F = ma$ if for $m$ we use a measurable parameter, $m^*$, known as the effective mass.

Those are our goals; along the way we will obtain some techniques for measuring the wavelengths of light or particles and for using known wavelengths to measure crystal structure. So much for the forest; now to the trees.

## THE CRYSTALLINE STATE

How can you tell whether something is a crystal? The answers usually given by students relate to gem crystals: the properties of hardness and the very regular geometric facets that characterize a cut diamond. The first property, hardness, is irrelevant, since some viscous liquids (e.g., window glass) show much greater hardness than some perfectly good crystals (e.g., lead). The facets, however, are very close to the heart of the matter. Consider a diamond, for example. Let us say that you owned an uncut diamond about the size of a ping-pong ball. One day, in a moment of unrestrained scientific curiosity, you drop it on a cement floor. The immediate observation would be that diamonds do not absorb shocks very well. Close examination of the small diamonds you just obtained would show that they were all faceted—that is, they were solids bounded by planes. Most of the planes would intersect in such a way that the faces would be bounded by either four or six straight lines. (See Figure 2.1.)

The regularity of the facets indicates the inherent atomic pattern. Although noncrystalline solids can be carved to any such shape, crystal facets occur naturally and are reproduced in the fracture or cleavage patterns of many crystals.

Not being content with a half-done experiment, you smash one of the larger samples with a hammer, obtaining VERY small diamonds. Looking at these small chips under a microscope would show precisely the same

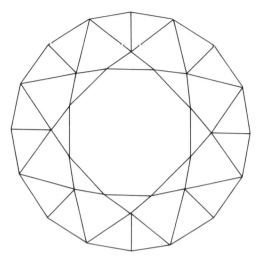

**Figure 2.1** "Cut" diamond—an example of the geometric patterns of single-crystal facets.

"diamondlike" facet pattern. Can this continue indefinitely? Obviously not. Continual subdivision would eventually produce carbon atoms which are, for practical purposes, indivisible. Carbon atoms and diamond *are not the same*. But a diamond is nothing more than atoms of carbon. How many do we need before the atoms begin to act like diamond? The answer depends to some extent on what property you are measuring. To the lady wearing a gem, your diamond dust has already passed beyond the pale. By our facet test, however, they will continue to look like diamonds even if we go to the limits of optical observation (a few thousand angstroms).

What property is it that gives us the regular facets? The only possible explanation is that the atoms are arranged in a regular three-dimensional array. By regular I mean that the array can be extended indefinitely in any direction and still be the same. That is, after all, the only way that we could find the same facet patterns on both large and small diamonds. One atom cannot be arranged at all. It would be hard to tell whether four or five were a "complete" pattern. By the time you have, say, a thousand atoms in each direction, any pattern that is going to show up had better be pretty evident. How big would this $10^9$ atom clump be? [Answer: about $0.5 \times 10^{-11}$ cm$^3$ or roughly a piece about 2 microns on a side. This was obtained from the density of diamond (3.51 gm/cc), the atomic weight of carbon (12), and Avogadro's number ($6.023 \times 10^{23}$/mole).] One can buy submicron diamond grinding grit. Since the grit acts like diamond and not plain carbon, the critical size is considerably less than 1000 atoms on a side.

An interesting question arises. How many shapes are there that can be extended indefinitely in any direction? Shapes such as standard building bricks, cubes, prisms, and such regular geometric forms as tetrahedrons and octahedrons certainly suggest themselves. Actually, if you consider the molecule or atom as a point, there are only 32 distinct arrangements (point lattices) that can be extended indefinitely in three dimensions. But although we can guess at the underlying symmetry behind a particular gem's shape, can we really tell how the atoms are distributed? The exact answer lies in an extension of the two slit diffraction experiments developed below.

## PERIODICITY AND WAVE PROPAGATION

The major portion of this book will concern itself with the motion of electrons in crystals of germanium or silicon. Since the electrons exhibit wave character and the crystals are regular arrays of atoms, we must first establish how waves propagate through a periodic array. Although our purpose in developing these rules is to describe electronic conduction, in the course of the development it will also become clear how we could measure the wavelength of an electron or determine the locations of the atoms in a crystal.

In Chapter 1, a definition was developed that purported to discriminate between things that were and were not waves. This is a simple, binary class distinction. However, within the class of things wavelike there are many subclasses—shear waves, compressional waves, vector waves, scalar waves, and the like. Thus, all waves are not the same even though they share the common properties of propagation and constructive interference. If we want to develop a technique or rule with one kind of wave and then use it with another, we must proceed with reasonable caution to make sure that we have used only those wave properties common to the two types of waves under consideration. What I plan to do now is to develop wave-propagation rules using surface water waves and arrays of posts. Then, with a little bit of caution, I will carry the same technique over to the more complicated case of the three-dimensional crystal with light or electron waves propagating through it. I will be on the most solid footing when I employ only the propagation and interference properties common to *all* waves.

## THE HUYGENS CONSTRUCTION

Consider the simple apparatus shown in Figures 2.2 and 2.3. At one end of a large water tank is a plane-wave generator. This is a board driven up and down in a sinusoidal fashion. Traveling away from the board will be plane waves—that is, waves whose phase is constant over a "plane," which in our two-dimensional tub example is really a line. In Figure 2.3, the hori-

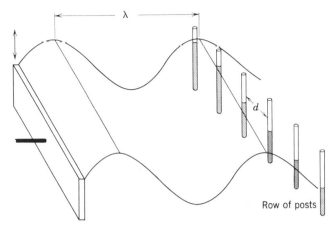

**Figure 2.2** Water-tank wave-scattering example.

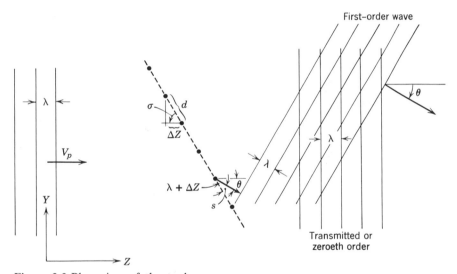

**Figure 2.3** Plan view of the tank.

zontal lines are the loci of points at which the phase is $2\pi$ or an integer multiple thereof. These are spaced a wavelength apart and move down the tank (propagate) with the phase velocity, $v_p$.

Part way down the tank, the waves meet a periodic array of rigid posts set at an angle, $\sigma$, to the lines of constant phase. The first question to answer is how the wave interacts with just one post. Try an answer. You might guess that the post would scatter the part of the wave that hits the post. If you allow a fairly flexible view of just how much of the wave "hits" the post, your guess is plausibly correct.

You probably based your answer on actual observation of the behavior of water waves. In that case, your answer would have been that you would observe approximately circular wave fronts propagating away from the post. This answer is accurate only if the post is reasonably small compared to the wavelength, $\lambda$, a situation that is normally found in water-wave experiments. With that restriction, it is not too far from the correct answer. Since the almost cylindrical waves are observationally correct and the greater precision of the exact solution would not be very useful to this discussion, permit me to proceed with the approximation that the scattered energy appears as cylindrical waves propagating away from the post.

The next question I want to ask is: In which directions from our post array are plane waves propagating? The interest in plane waves is simple enough; wave fronts of shapes other than planes will lead to intensities that decrease as $1/r$ or faster from the array. Thus, if I am at any great distance from the array, I will perceive only the plane waves. The answer to the question can be derived by a simple geometric construction originally developed by Christian Huygens (1625–1695). First, I must have an operational definition of "plane wave." Above, the definition given states that the phase anywhere in a particular plane is constant. It is all very well to make such a definition, but how does one tell the difference between a phase of $\phi$ and $(\phi + 2\pi)$? All I can really insist on is that there be no *detectable* difference in phase from one point in the plane to another. This definition, and my assumption that the scattered wave from each post is a cylindrical wave, is all I need to arrive at the Huygens construction. Two things are clear a priori: (1) each post is identical to all the others, from which (2) the phase of each cylindrical wavelet will have the same phase delay with respect to the time that the incident wave crest arrives at the scattering post in question. Another way of saying the same thing: Whatever phase delays occur in the scattering process at one post will be the same at all posts.

To do the Huygens construction, draw circles of equivalent phase around each post. Any line that is tangent to an equivalent phase circle from *each* and *every* post is a plane of constant phase. The wave propagates in a direction normal to this plane. (Note that the notches in the wave front disappear as the wave gets far away from the scattering posts. These notches represent nonplanar components of the wave and they are not observable far from the posts.)

Since the posts are arranged in a periodic array, the only possible arrangements of circles that will have a common tangent will be orderly monotonic sequences of phases such as 0, $2\pi$, $4\pi$, $6\pi$, etc. or 0, $4\pi$, $8\pi$, $12\pi$, and so on. In performing the construction, you must remember that the incident crest

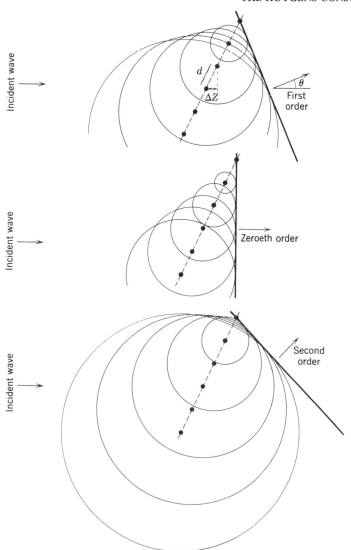

**Figure 2.4** Huygens construction.

gets to post 1 a substantial time before it gets to the last post (six in the figure). Figure 2.4 shows three different plane waves constructed by this technique. In the first example, the phase delay between successive circular phase fronts is $2\pi$. The radius for a given circle is the radius of the circle before it plus $\Delta z$ plus $\lambda$. (Where does the $\lambda$ come from? From the phase difference of $2\pi$ between circles.) Thus, the first circle has a radius of 0, the second has a radius of $(\lambda + \Delta z)$, the third $(2\lambda + 2\Delta z)$, and so on. The second example is the "zeroeth order" construction—that is, all of the circles have the same phase. This results, logically enough, in the same plane wave as the incident wave, although there can be a phase delay between the incident and scattered waves. (The measurement of this phase delay in the zeroeth-order wave yields the index of refraction for the medium. Media consist of atoms, each of which scatters a little bit of the incident wave but with a little delay. The resultant sum of all the little wavelets is a retarded wave. If you have a volume of scatterers [rather than a line], the propagation velocity can be reduced by the continued retardation and Snell's law results. Note that it is not necessary to have the scatterers in a regular array to have the zero-order scattered wave. That one is determined only by the delay in the scattering process, which is characteristic of the scatterer and not the position. But to have the energy scattered into *other* directions grow to a significant magnitude, it is necessary to have the scatterers in a precisely regular pattern.)

There are several direct results of the construction shown in Figure 2.4. Far away from the array of posts, scattered energy is found only at certain very particular and discrete angles. These correspond to the various orders (the order is the number of $2\pi$'s delay between the phases of the wavelets that add to the wave), 0, 1, 2, etc.

Since the angles that are found for a particular incident wave depend on the wavelength of that wave, another wavelength implies a different angle. However, if I were to use $\lambda$ in one case and $\lambda/2$ in another, the first order for the first case is in precisely the same direction as the second order for the second case. Thus, it is evident that in any particular direction, $\theta$, I will find a countable infinity of wavelengths, while for any particular wavelength, $\lambda$, I will find a finite series of angles that satisfy the construction. This is illustrated in Figure 2.5, which is a plot of $\theta$ versus $\lambda$ for the various orders. Note that all orders have all possible values of $\theta \leqq \sigma$ and that the range of wavelengths found in any order is limited. What simple property of the Huygens construction requires that the maximum wavelength in order $n$ be limited to $\lambda_n = (d - \Delta z)/n$? [Answer: In order to have a tangent to the set of circles for order $n$, it is necessary that $n\lambda$ be less than $(d - \Delta z)$. Otherwise, the set of circles forms a "nesting set," in which each successive circle is contained within all of the circles before it.] If what you are measuring is angles or wavelengths, Figure 2.5 is the presentation you want. It was obtained from the Huygens construction in the following way. Referring to Figure 2.3, the angle $s_n$ is given by $(n\lambda + \Delta z)/d = \cos(s_n)$. The angle, $\theta_n$, is related to $s_n$ by $\theta_n = \sigma - s_n$. Using these last two equations and the fact that $\Delta z/d = \cos \sigma$, you can obtain the form of (2.1) below

$$\theta = \sigma - \cos^{-1}\left[\frac{n\lambda}{d} + \cos \sigma\right] \tag{2.1}$$

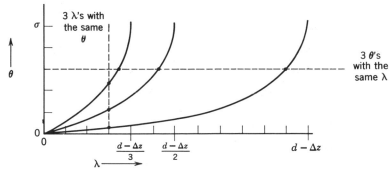

**Figure 2.5** $\theta$ versus $\lambda$ for a particular set of $d$ and $\Delta z$.

Figure 2.5 is a plot of the solutions of equation 2.1 for a particular $\Delta z$ and $d$. By de Broglie's rule,

$$p = \hbar k = \frac{h}{\lambda}$$

Since $p_z/|p| = p_z\lambda/h = \cos\theta$, $p_y/|p| = p_y\lambda/h = \sin\theta$, we can obtain $p_z$ or $p_y$ from $\lambda$ and $\theta$ in Figure 2.5 or equation 2.1.

To get a $k_z$ versus $E$ plot we must have some information on how the total momentum, $k$, relates to $E$. Let us say that we want $k_z$ versus $E$ for electrons. By using the electron-wave equation developed in Chapter 1, the appropriate relationship should be derivable. If the electrons were free ($V = 0$), what is the relationship of $E$ to $k$? [Answer: For the case of free electrons, the potential energy term in (1.9) is zero. Solutions are thus of the form $e^{j(kx - Et/\hbar)}$. Direct application of the energy and momentum operators gives $p = \hbar k$ and $E = \hbar^2 k^2/2m$. Accordingly, $k^2 = 2mE/\hbar^2$ or $\lambda^2 = h^2/2mE$. It is VERY important to note that this simple result comes about only because $V = 0$. Thus, it applies to free electrons but NOT to bound electrons.] Using the free electron $E:k$ relationship and de Broglie's rule gives

$$\hbar k_z = p_z = \sqrt{2mE}\,\cos\theta$$

$$\hbar k_y = p_y = \sqrt{2mE}\,\sin\theta$$

This is plotted in Figure 2.6. It shows the allowed values of $E$ for various values of $k_y$ and $k_z$, or, alternately but equally good, the allowed values of $k_y$ and $k_z$ for various values of $E$. The $k_z$ curves are asymptotic to a parabola, which is the zero-order solution (a result of the fact that $V = \text{constant} = 0$ in the wave equation; also a result of using the electron-wave equation. For light or water waves, $k^2 \propto E^2$). The dotted line drawn across Figure 2.6 intersects the $n = 0$, 1, and 2 curves. Thus, at energy $E_1$, there are three possible solutions with the values of $k_y$ and $k_z$ given by the intersections of the $E = E_1$ and $E$ versus $k$ curves. This is the first example we have had of

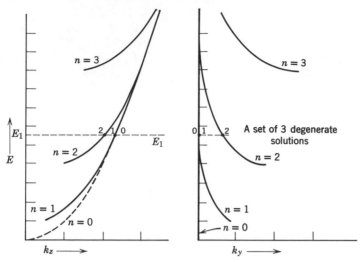

**Figure** 2.6 $E$ versus $k_z$ and $E$ versus $k_y$ for electrons passing through a grating.

a system that has several independent solutions with a common eigenvalue, $E_1$. Such solutions are called *degenerate*. (In solid-state physics, the word "degenerate" is used to mean several almost-unrelated things. When eigenstates are called degenerate, it means they have the same energy. A second meaning of degenerate will be introduced in Chapter 3.) For other potential functions, such as those that are found by electrons in crystals, we should not expect the same form of solution as Figure 2.6. It is entirely possible to have any number of $k$ values for a given $E$. This "any" includes zero.

It is worthwhile to pause for a moment to see how these latest results fit into our objective of finding the eigenvalues of electrons in crystals. What we have found are the solutions for a free electron wave propagating through a regular array of scatterers. Two aspects of this problem are particularly important. First, we have developed some tools for finding and characterizing the solutions to electron waves propagating through a periodic array. Second, the $E:k$ plot has been introduced and with it the idea of degenerate solutions.

What remains to be done is to extend our techniques to two and three dimensions and to examine the consequences of having a periodic $V(x, y, z)$ term in the wave equation.

## A LATTICE OF POSTS

In the example just given, only a few posts were shown. If I add more posts but merely extend the line in so doing, nothing very much happens.

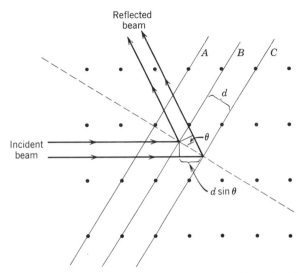

**Figure 2.7** Scattering from planes in a two-dimensional array of posts. The calculation of the Bragg condition for constructive interference uses this figure.

As a matter of fact, my solution assumes that the row of posts continues indefinitely, since I took no special cognizance of the finiteness of the array. However, if I add more *rows* of posts so that the array is now two-dimensional (see Figure 2.7), new restrictions are imposed on the solution. One way of looking at what is happening is to note that the scattered waves are themselves scattered. If the array continues for a very long distance, this scattering can completely attenuate a wave that would otherwise have been generated by a single row. However, conservation of energy requires that the wave go somewhere. What we want to find are the solutions that can propagate without attentuation, because if the lattice is extensive, these are the only solutions we could expect to find. [A problem that frequently proves puzzling at this point is: If I have a lossless array of scatterers and I "illuminate" it with a wave that does *not* meet the lossless propagation requirement, what happens to the energy of that wave? The only possible answer, if the wave is "losslessly attenuated," is that the wave is reflected in toto from the array. That is, the input wave is attenuated by being scattered into directions that are allowed. A good example, which is very relevant to our present subject, is the wave function we found for the particle in a finite potential well in Chapter 1. The solution within the well was of the form $\sin(kx)$, while outside the well it was of the form $e^{-kx}$. The first is an unattenuated wave and the second is attenuated. Now, in a classical billiard-ball sense, the particle cannot propagate through the wall into region of high potential energy, so it is reflected at the wall. The wave point of view is really in general agreement with the billiard-ball approach, but nothing happens abruptly in the wave domain. Accordingly, the wave viewpoint would say that the particle can penetrate *only a small distance* into an unallowed region. Note that the particle is, however, equally well reflected in both cases.]

To find the directions of unattenuated propagation, we must ask the question: "In which directions will the scattered wavelets reinforce each other?" A possible answer is *none*, in which case, the observer far from the scattering crystal would see no scattered *plane* waves. (He would probably observe a variety of nonplanar waves emanating from the illuminated crystal. The difference between this type of scattered wave and the plane wave is something you have observed many times when looking into a dirty mirror. If you shine a flashlight at a dirty mirror and observe what happens from any angle but the angle of reflection, you see the light scattered from the dust on the surface of the mirror. That is, the mirror *surface* looks bright. At the angle of reflection, however, the direct beam is observed, and is many, many times brighter even for a very dirty mirror. You should also note that the scattered light comes from the surface but the reflected light appears to originate well behind the mirror. What we are looking for are the directions in which the crystal "glints" like a mirror.) Looking at Figure 2.7, you will note that there are many ways of selecting the rows of posts to make up the same array. The most obvious are the vertical or horizontal rows, but the diagonal rows are equally good and more interesting for our present discussion. Consider the light scattered by the planes indicated by the heavy lines in the illustration. Let me assume that the radiation is monochromatic. Then, using Figure 2.5, I can find the various angles into which the incident plane wave is scattered. Of course, as soon as one of these plane waves starts off in some particular direction, it is scattered by the planes above it. For example, the radiation scattered from plane $A$ is partially scattered again by planes $B$ and $C$. Now comes the critical question: Are there any combinations of wavelength, $\lambda$, and the angle, $\theta$, that lead to unattenuated propagation? The answer is yes. Note first that if the primary scattered wave ($b$) coming off row $B$ is in phase with the primary wave ($c$) from $C$ as ($c$) reaches $B$, the two waves will reinforce each other. Second, if the angle of reflection is equal to the angle of incidence, then all of the secondary and tertiary scatterings will also add constructively. [I.e., under the conditions of equal angle of incidence and reflection, if wavelet ($c$) is in phase with wavelet ($b$) at plane $B$, then the energy scattered from wavelet ($c$) at plane $B$ will add constructively to the incident wave and not be lost at all.] This second condition of equal angles is met only for the zeroeth-order reflection. The first condition is met if the additional distance the wave travels from plane $B$ to plane $C$ and back ($2a$ in Figure 2.7) is one or more wavelengths. The condition required of $\theta$ and $\lambda$ to have an unattenuated wave is that

$$2a = 2d \sin \theta = n\lambda \qquad (2.2)$$

This is called *Bragg's law*.

A reasonable skeptic might well ask just *how* equal the distance must be to an integer number of wavelengths. This is a most important question, since in answering it you must state one of the many forms of a most important law: the *Heisenberg uncertainty principle*. An appropriate way to phrase the question is: How far from $\theta$ must I go before I find that the intensity falls to $1/2$ its maximum value? Let there be $m$ planes. The maximum amplitude occurs when each of the $m$ scattered wavelets of amplitude $A_0$ adds in phase. That is:

$$A_{\max} = A_0 \sum_{m=0}^{m-1} e^{jn\theta} = mA_0$$

In some other direction, $\theta + \Delta\theta$, each successive wavelet is out of phase with the one preceding it by $\Delta\theta$, and the amplitude is given by

$$A^2 = \left[ A_0 \sum_{m=0}^{m-1} \cos (n\Delta\phi) \right]^2 + \left[ A_0 \sum_{m=0}^{m-1} \sin (n\Delta\phi) \right]^2$$

The answer asked for is the solution to $A^2/A_0^2 = m^2/2$. The solution is readily obtained at the two extremes ($m = 2$ and $m = \infty$). The first case, by direct substitution, gives $\Delta\phi = \pi/2$. The second case may be solved by converting the sums to integrals. [I.e., one finds the average value of $\cos (x)$ and $\sin (x)$ for $x$ varying from 0 to $m\Delta\phi$. The sum of the squares of $m$ times each of these averages must equal $m^2/2$.] The result for very large $m$ is $\Delta\phi = 2.8/m$ radians.

We may now convert from $\Delta\phi$ to $\Delta\theta$ and then solve for the uncertainty in determining $d$. From Figure 2.7 we have

$$\Delta\phi = \frac{4\pi d \cos (\theta)}{\lambda} \Delta\theta$$

so the $3dB$ point that we have just found is at

$$\Delta\theta_{3dB} = \frac{2.8\lambda}{4\pi md \cos (\theta)}$$

From Bragg's law (2.2) we may write

$$\Delta d = -\frac{n\lambda \cos (\theta)}{2 \sin^2 (\theta)} \Delta\theta$$

Evaluating $\Delta d/d$ at the $3dB$ point gives

$$\frac{\Delta d_{3dB}}{d} = \frac{2.8n\lambda^2}{8\pi d^2 m \sin^2 (\theta)} = \frac{2.8}{2\pi nm}$$

The last step uses Bragg's law once again.

We have just obtained a most important result. Regardless of how mono-chromatic our illuminating beam is or how carefully the angles are measured, the reflection angle and the plane separation can be determined with only a limited degree of precision.

The uncertainty in measuring the separation of the planes is inversely proportional to the number of planes. This inherent inexactitude is part of any wave system. Waves do not start and stop abruptly; therefore, no ab-solutely abrupt measurement is possible. Furthermore, as you may well have seen in Fourier transform theory, in any transform pair (e.g., $\omega$ and $t$), the reduction in the spread of one of the variables leads to a proportional increase in the spread of the other. In the case given above, we have another example of the classical optical problem of resolution depending on the aperture (spatial frequency and spatial extent being the Fourier transform pair). It is also a fundamental result of optics that the resolution of a grating spectrometer is proportional to the total number of grooves in the grating.

It is well worthwhile to get a little feel for the size of real measurements and the uncertainties involved. Typical plane separation in a crystal is 5 Å (0.5 nm). Typical X rays used in crystal measurements have energies of the order of $10^4$ eV. With these numbers find the first four angles that satisfy the Bragg condition (counting from $\theta = \pi/2$). Compute the maximum fractional accuracy that could be obtained in the first order if we had 1000 layers (our 2-micron-diameter diamond of a few pages ago). (Answer: 1.41, 1.04, 0.83, and 0.66 radians respectively. $\Delta d/d = 0.446 \times 10^{-3}$.)

Bragg's law is very useful in two ways. If I know $\lambda$, by measuring the various angles, $\theta$, I can determine $d$. Conversely, if I know $d$, I can use the array to determine $\lambda$ or to separate one $\lambda$ from another. It was by using Bragg's law that the wavelengths of electrons (and, indeed, their very wave properties) were first measured by Davison and Germer in 1927; and it is by using Bragg's law with electron, light, or neutron waves that the details of crystal structure have been deduced. With three-dimensional structures, there are many values of $\theta$ that give reflections, and the inversion from angles to the scattering

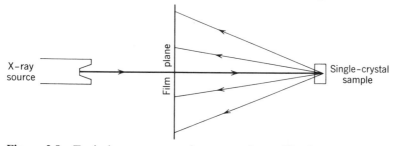

**Figure 2.8a** Typical arrangement for generating a "back-scattered" Laue photo-graph for a single-crystal specimen.

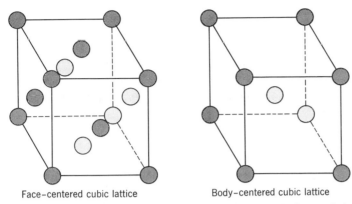

Face-centered cubic lattice          Body-centered cubic lattice

**Figure 2.8***b* The dark spheres are those on the visible faces of the cube. The lighter spheres lie on the hidden faces or within the cube itself.

**Figure 2.8***c* Back-reflection Laue pattern for germanium with the beam normal to the $\langle 111 \rangle$ plane. The hexagonal symmetry is clearly evident in the outer six spots. That the six spots are not quite equal in intensity shows that the beam is not precisely normal to the $\langle 111 \rangle$ plane. Very precise crystal orientations are possible with this technique.

array is frequently far from trivial. However, certain symmetries jump right out in a very satisfactory way. Figure 2.8*c* shows the pattern obtained by illuminating a germanium crystal in a particular direction. The sixfold symmetry is quite evident.

We now proceed to a better way of stating Bragg's law, a simple way of determining the Bragg scattering angles and a more numerical way of discussing crystals, which will stand us in good stead in the second task we face.

## THE "RECIPROCAL LATTICE" AND BASIS VECTORS

A rather useful geometric construction can be used to find all the possible planes that can be drawn through the crystal-lattice points. It will also serve us well when we discuss $E:k$ plots for solids. Consider once again an array of posts such as that shown in Figure 2.9. Picking any point in the lattice as an origin, it is easy to see several ways of defining a pair of vectors that I can use to get to any other point in the lattice. Two such sets are shown ($a$ and $b$ and $a'$ and $b'$). It should be noted that $a'$ and $b$ is an equally good set. Any such pair of vectors is called a *basis set*, since they may be used to construct the entire lattice. (*Construct* is used in the sense that any lattice point [i.e., post] has integer coordinates in terms of the basis vectors. Please observe that there is no reason why the basis vectors should be orthogonal or of equal length. In Figure 2.9, for example, they are not.)

Arbitrarily, I choose $a$ and $b$ as the basis set. I now define two new vectors, $A$ and $B$, that have the following property:

$$A \cdot a = 1, \qquad A \cdot b = 0; \qquad B \cdot b = 1, \qquad B \cdot a = 0 \qquad (2.3)$$

(Note that this means $A \perp b$ and $B \perp a$.) Using a conventional Cartesian coordinate system, the vectors $a$ and $b$ in Figure 2.9 are given as $a = (1,3)$ and $b = (2,0)$. Find $A$ and $B$ to satisfy (2.3) above. [Answer: $A = (0, 1/3)$, $B = (1/2, -1/6)$.] A new lattice can be constructed using $A$ and $B$. This is shown in Figure 2.9. The new lattice is called the *reciprocal lattice*. The first lattice is usually referred to as the *direct lattice*. In three dimensions equation (2.3) become $A = (b \times c)/a \cdot (b \times c)$, $B = (c \times a)/b \cdot (c \times a)$, and so forth.

I now wish to prove that the vector to any point in the reciprocal lattice is perpendicular to (i.e., defines) a complete set of parallel lattice "planes" in the direct lattice. Furthermore, the spacing between successive planes is the inverse of the defining vector's length. Since the direction normal to the planes and the spacing between the planes is the information needed to compute the Bragg scattering angles, there should be very close relationship between the observed scattering pattern (Figure 2.8) and the reciprocal lattice. As a matter of fact, the relationship between the planes in the direct-lattice and the reciprocal-lattice vectors is so direct and fundamental that directions in a crystal are given most easily by the coordinates of the defining reciprocal-lattice points. Now to the proof.

First, it is necessary to get the vector equation of a straight line. With reference to Figure 2.10, let 0 be the origin and $CD$ a line. Then, by inspection it can be seen that

$$r \cdot a = c \qquad (2.4)$$

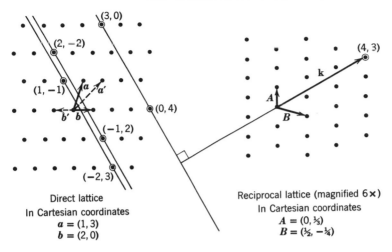

**Figure 2.9** For the planes shown, the defining reciprocal lattice vector expressed in the basis set of vectors is $k = (4, 3)$ (i.e., $k = 4A + 3B$). The coordinates of points in the direct lattice are expressed in terms of $a$ and $b$ so that $r = (n, m) = na + mb$.

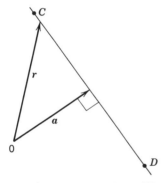

$r$ = Radius vector to any point on $CD$
$a$ = Unit vector perpendicular to $CD$    **Figure 2.10**

a constant. Inspection of the figure shows the constant is $a \cdot a = |a|^2 = c$, and the distance to the line from the origin is $c/a$.

Next, using the symbols of Figure 2.9, I wish to prove that for any vector, $r$,

$$r = (r \cdot a)A + (r \cdot b)B \tag{2.5}$$

where $a$ and $b$ and $A$ and $B$ are the basis vectors for the direct and reciprocal lattices respectively. Certainly if I take the dot product of both sides of (2.5) with, say, $a$, I get the rather satisfying result that $r \cdot a = r \cdot a$. But since $A$ is not perpendicular to $B$, I need something more than that for my proof.

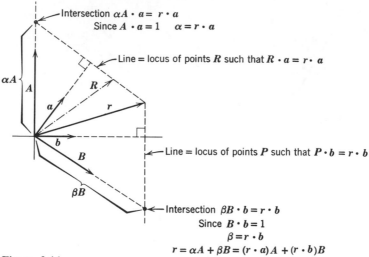

**Figure 2.11**

Figure 2.11 completes the proof by construction. Let $R$ and $P$ be two vectors whose tips travel along the two lines $R \cdot a = r \cdot a$ and $P \cdot b = r \cdot b =$ = const. These lines are drawn out to where they intersect the extensions of $A$ and $B$ respectively. Because $b$ is perpendicular to $A$ and $a$ is perpendicular to $B$, the figure thus formed is a parallelogram. (Note that we would not get a parallelogram if our vectors did not satisfy equation 2.3. By the parallelogram rule for vector addition, $r = \alpha A + \beta B$. From equations 2.3 and 2.4 I get $\alpha = r \cdot a$ and $\beta = r \cdot b$, which proves (2.5). Thus I may express any vector $r$ in either coordinate system by using (2.5).

Now if vector $r$ in (2.5) happens to go from the origin to a lattice point, $r \cdot a$ and $r \cdot b$ will be integers. It is then an easy step to prove that if $k$ is a vector to a lattice point in the reciprocal lattice and $c$ is an integer, $r \cdot k = c$ defines a line passing through a regular sequence of points in the direct lattice. First, if $k$ goes to a reciprocal-lattice point, then $k = k_1 A + k_2 B$ where $k_1$ and $k_2$ are integers. Similarly, if $r$ defines a lattice point in the direct lattice, $r = r_1 a + r_2 b$ where $r_1$ and $r_2$ are integers. Thus $r \cdot k = r_1 k_1 + r_2 k_2$, which is also an integer since it is the sum of a product of integers. Of course, for any given $k$, there are many $r$'s that will give the same constant. For example, let $k = (4, 3)$. Then, if I choose $r = (1, -1)$, I get 1 for the constant. I will also get 1 for $r = (-2, 3), (-5, 7), (-8, 11)$, and so on. In Figure 2.9, the lines given by $k = (4, 3)$ and $c$, the constant, equal to 1, 2 and 12 are drawn. All of these lines are perpendicular to $k$ and the sequence of lines is separated by spacing $1/k$ just as equation 2.3 demands. These lines are referred to as $(4, 3)$ lines, and they are perpendicular to the $(4, 3)$ direction. (Note that the direction is a lattice vector of the *reciprocal lattice*, not of the direct lattice.)

The traditional name for a coordinate of the $k$ vector perpendicular to a set of planes is *Miller index*. That is, the Miller indices of the planes shown in the figure are (4, 3). The older method of finding these indices corresponds to finding the plane closest to the origin that intersected the crystalline coordinate axes at integer values of the direct coordinates. In the direction shown in the figure, this is the "plane" defined by $c = 12$. Direct substitution in the expression $r \cdot k = 12$ gives (3, 0) and (0, 4) as the intersections. To find the Miller indices you then add the vectors, giving (3, 4). Finally, you divide through by the lowest common denominator [yielding (1/4, 1/3)] and take the reciprocal of each of the numbers obtained [yielding (4, 3)]. Of course, after all that fuss and feathers, what you have just obtained is $k$. The advantage of the old method is that if you have the Miller indices (a set of numbers that frequently appears in the literature on semiconductors) and wish to find the plane described by inverting the process above for finding the indices, you can find three points from which to draw the plane in the direct lattice.

## THE EWALD CONSTRUCTION

Now that we have the reciprocal lattice, it turns out that there is a very simple way to compute *all* the scattering directions for a given wavelength and illumination angle. It was devised by P. P. Ewald in 1921 and bears his name.

Dividing equation 2.2 by $\lambda d$ gives a form most suitable for the reciprocal lattice:

$$n/d = (2 \sin \theta)/\lambda \qquad (2.2')$$

The left-hand side of the equation is satisfied every time $d$ is the spacing between a set of planes in the direct lattice. This will occur for each point in the reciprocal lattice, as I just showed. The right-hand side of the equation has two parameters, $\theta$ and $\lambda$, which relate to the illumination. The locus of all possible solutions to the right-hand side of the equation is a circle of radius $1/\lambda$ passing through the origin and with its center on the line connecting the origin to the X-ray source. By examining Figure 2.12 it is fairly easy to see that the length of line $OA$ is $(2 \sin \theta)/\lambda$. To see the relationship to the Bragg condition, note that the vector $OA$ defines a plane that I have labeled the "reflecting plane" in Figure 2.12. Any time that $OA$ happens to fall on a point in the reciprocal lattice, the reflecting plane corresponds to a whole series of parallel, equally spaced planes in the direct lattice. By requiring $OA$ to lie on circle of points satisfying the right side of equation 2.2', the spacing of lattice planes so defined will be correct to have the reflected wavelets re-

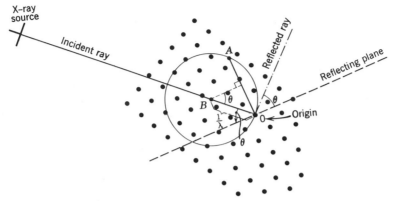

**Figure 2.12** Ewald construction powder technique.

inforce each other. Thus, the intersection of the circle and the reciprocal-lattice points yields all the scattering angles.

It is quite possible to pick an illumination direction or a wavelength, $\lambda$, such that the circle passes through *no* points in the reciprocal lattice. However, in the usual methods of X-ray crystallography, one either uses a discrete angle and a continuum of wavelengths (as in generating Figure 2.8) or a single wavelength and a continuum of angles. In one technique, for example (Debeye-Scherrer), the crystal sample is ground to a randomly oriented powder, thus providing all possible angles of incidence. Figure 2.13 illustrates this method. Since the crystallites that happen to present a proper set of crystal planes could be rotated about the incident beam without changing the angle of incidence, the light scattered by a particular set of planes forms a cone as shown in the figure. By using a single $\lambda_0$, the angles measured correspond to the reciprocal-lattice points lying between 0 and $2/\lambda$ from the origin. (Note that the random orientation rotates the circle in Figure 2.12 about the origin, sweeping out all the points between 0 and $2/\lambda_0$.) If the film is of length $2\pi$, the radius of the cone at the film plane is $2\theta$ in radians. Since $\lambda_0$ is known a priori and $\theta$ is easily measured, this powder technique gives a very quick method of finding $d$, the lattice-plane spacing. The lattice type can also be determined with this technique. First you find the *smallest* angle, $\theta$, that gives a line on the film. From this you find the largest lattice-plane spacing, $d_1$. The lines corresponding to higher orders in (2.2′) are easily identified. Then you find the next angle, $\theta_2$, (not corresponding to $d_1$ in any order) and from this $d_2$. Now if you wanted to differentiate between face-centered cubic and body-centered cubic crystals (defined in the next two paragraphs and Figure 2.8), you have all you need. Straightforward geometry shows that for an f.c.c. direct lattice, the ratio of $d_1/d_2$ is $2/\sqrt{3}$, while for a b.c.c. direct lattice it is $\sqrt{2}$. Other lattice types have other characteristic ratios.

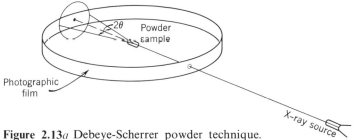

**Figure 2.13**a Debeye-Scherrer powder technique.

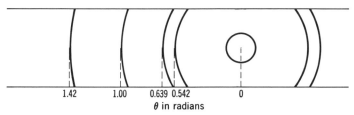

| 1.42 | 1.00 | 0.639 0.542 | 0 |

**θ in radians**

**Figure 2.13**b Debeye-Scherrer diffraction photography for a Ge : Si alloy illuminated by 4.0 Å X rays. The figure shows all of the rings that would be observed with this illumination.

## A PROBLEM FROM THE REAL WORLD

To see how the Ewald construction and the Debeye-Scherrer technique relates to finding the parameters of simple but real crystals, do the following problem: We wish to find the lattice spacing for a GeSi alloy. Both Ge and Si freeze into the face-centered cubic (f.c.c.) point lattice, so it is a good guess that any alloy of Ge and Si will do the same thing. The f.c.c. lattice is a cube with the points located at the corners of the cube and the centers of each of the faces (see Figure 2.8b). If the corners of the cube are given by $xyz$ coordinates, $(1, 0, 0), (0, 1, 0), (1, 1, 1)$ and so on, find a suitable set of direct-lattice basis vectors. [One answer: $(1/2, 1/2, 0) = a$, $(1/2, 0, 1/2) = b$, and $(0, 1/2, 1/2) = c$.] Show that the basis vectors are not orthogonal.

Now find the reciprocal-lattice basis vectors, and show that they generate a body-centered cubic lattice (b.c.c.). (A b.c.c. lattice has points at each of the corners of a cube and one more point at the center of the cube. See Figure 2.8b.) [Answer: For the $a$, $b$, and $c$, selected above, $A = (+ 1, + 1, - 1)$, $B = (+ 1, - 1, + 1)$, and $C = (- 1, + 1, + 1)$.]

If the sample were ground to a fine powder and the Debeye-Scherrer photograph taken with an illuminating wavelength of 4.0 Å, it would look like Figure 2.13b. The angles are marked below each line. Using the data in Figure 2.13b, show that the length of the side of the cube in the direct lattice is 6.71 Å, that the crystal is indeed f.c.c., and that only the four circles shown in the figure should be there. [Solution: For most of the parts of this problem, it is necessary to know to which points in the reciprocal lattice each circle corresponds. Looking at the body-centered cubic lattice of Figure 2.8$b$, it is easy to see that the nearest neighbors to the origin point are the body centers. If the cube is considered to be a unit cube, the nearest neighbors are at a point such as $(1/2, 1/2, 1/2)$ a distance $\sqrt{3}/2$ from the origin. The second-nearest neighbors are at points such as $(1, 0, 0)$; the third-nearest at points such as $(1, 1, 0)$; the fourth-nearest a bit of a surprise at points such as $(3/2, 1/2, 1/2)$; and the fifth-nearest at points such as $(1, 1, 1)$. If Figure 2.13$b$ is correct, the first four should lie inside the Ewald sphere, and the fifth outside.

To show that the lattice is f.c.c., it is necessary to show that $d_1/d_2 = 2/\sqrt{3}$. From (2.2'), $d_1/d_2 = \sin\theta_2/\sin\theta_1$. Figure 2.13$b$ shows that the first two angles satisfy this ratio.

To show that the cube side is 6.71 Å, we must first obtain an appropriate relationship between $d_1$ and the size of the cube in the direct lattice. $d_1$ is the spacing between the planes of widest separation. Its length is proportional to the length of the direct cube side and, as proved several pages ago, equal to $1/|A|$. For the direct lattice of side length, $l$, the value of $d_1 = l/\sqrt{3}$. That gives the proportionality factor. For the angle, $\theta_1$, given in Figure 2.13$b$, $d_1 = \lambda/2\sin\theta_1 = 4.0/2 \times 0.516 = 3.88$ Å. Multiplication by $\sqrt{3}$ then gives the desired result.

The last question is solved by showing that the Ewald radius is greater than $1/d_4$ but less than $1/d_5$. From the calculations above, we have $1/d_1 = = 0.258$. From the list of nearest neighbors still further above, the radii to the sequential near neighbors goes: $\sqrt{3}/2, 1, \sqrt{2}, \sqrt{11}/2, 3$. Thus $1/d_4 = = 0.494$ and $1/d_5 = 0.516$. Since the radius of the Ewald sphere is $2/\lambda = 0.5$, the fourth line is indeed the last one that should appear.]

## A CLOSER LOOK AT THE ELECTRON IN A CRYSTAL

The problem that we have just discussed treated the atoms in a crystal just as we treated the posts in the water. We had to assume that the incident wave was only slightly attenuated by each layer of atoms. In that way, we could add up a large number of almost identical wavelets—one from each plane of atoms—to get the Bragg reflected waves. We would obtain a very different answer if the scattering coefficients were very large. Our observation from the outside would be that the waves bounced off the surface, that they never penetrated the crystal. However, each atom in the crystal went in with its full complement of electrons. Thus, even if the scattering cross-sections are large, it must still be possible to have electron waves propagating in the crystal.

One can imagine three different situations. The first we have just discussed. It is the situation in which each layer of atoms scatters a little of the incident wave. The second is the opposite extreme in which each plane of atoms is almost totally reflecting. Since each atom is surrounded by planes of other atoms, this second approximation implies that the electrons from one atom stay in the immediate vicinity of that atom. Finally, there is the possibility of strong but not total reflection at each plane.

We have a few direct clues that help us choose between the three models. First, the electrons remain within the crystal, so there must be at least as many possible electron states as there are electrons. Since a crystal can sit around indefinitely, these states must be independent of time. Accordingly, we can use the time-independent wave equation (eqn. 1.7).

$$(-\hbar^2/2m)(\partial^2\psi/\partial x^2) + (V - E)\psi = 0 \tag{1.7}$$

## SOME SPECIAL PROPERTIES OF ELECTRON WAVES IN CRYSTALS

The first portion of this chapter was devoted to Bragg reflections and reciprocal lattices, tools useful with all types of waves in crystals. What we need is some results that apply specifically to electrons. These derive from solutions to equation 1.7 with appropriate crystal potential functions for $V$.

Certain things are evident from everyday experience. First, crystals are normally electrically neutral. Thus, each atom or molecule enters the crystal with its full complement of electrons—no more and no less. From this we can conclude that there are at least as many allowed electron states as there are electrons. Second, some crystals are electrical conductors. Accordingly, in those crystals, it must be possible to move electrons. A reasonable conclusion is that, for at least some crystals, the number of allowed states is greater than the number of electrons, and it is possible to make electrons change states with small electrical fields. Of course, some crystals are insulators. For those crystals it must be quite difficult to get electrons to change states. What we are about to do is to show that the electron-wave equation predicts just such behavior. The world of crystals will be clearly segregated into two classes: conductor and insulator. Rather amusingly, the remainder of our work will concern *conduction in insulators*.

The relationship of what we are about to do to the scattering problem we have just done is quite direct. Solutions to equation 1.7 with $E > V$ and $V = $ const. are of the form $e^{jkx}$. If we have two such regions adjacent to one another (for example, the finite potential well problem of Chapter 1), a wave incident on the junction between the two regions will be partially reflected and partially transmitted. By imposing the restraints of continuity of the wave function and its derivative, it is quite straightforward to get the reflection coefficient (the ratio of squares of the reflected and incident wave amplitudes). The result is

$$R = (k_0 - k_1)^2/(k_0 + k_1)^2$$

where the $k$'s are the propagation constants in the two regions. In particular, $k_1^2 = (E - \delta V)(2m/\hbar^2)$, $k_0^2 = E(2m/\hbar^2)$, and $\delta V$ is the potential energy difference between the two regions.

It is most interesting to note that the same result applies to light waves in going from a medium of one index of refraction into a region of another. As a matter of fact, William Rowan Hamilton, the Irish mathematical physicist, pointed out quite clearly in the 1840's that the *same* equation governed the trajectories of light and "material" particles. He showed that the motion of particles was equivalent to the trajectory of light in a medium of continuously variable index of refraction. This tantalizing analogy between the motions of waves and particles stayed around in the advanced physics texts for *80 years* before de Broglie made his breathtaking assumption: it was no analogy; they were both waves.

From the equation for $R$ above, you can see that the largest reflections occur for the largest differences in the propagation constants. Thus, the maximum reflection will occur in regions of the maximum rate of change of $V$ (for electrons) or $\varepsilon$ (for light). This gives us a direct clue to what we are observing when we measure the scattering patterns for X rays, electrons, or neutrons. With X rays or electrons we are observing scattering by the electrons or charge centers respectively. The neutrons observe only the nuclear forces, so their scattering patterns give information on the locations of the nuclei. By using two or three different particles, a great deal of detailed information can be obtained about each site in the lattice.

But the wave equation has much to tell in its own right. Since the atoms in a perfect crystal form a completely ordered array, we can represent the ideal crystal in the wave equation by the periodic potential shown in Figure 2.14. This figure represents a one-dimensional crystal with every site occupied by a unit positive charge. At the ends of the crystal, the potential rises rather quickly to 0. Within the crystal, however, the potential is completely periodic. (We will deal with real, imperfect crystals later.)

For this simplest of all crystals, three essentially different problems arise. For electrons with energy, $E_1$ (see Figure 2.14), the $V(x)$ term is only a small perturbation in equation 1.7. Solutions will be essentially those of the equation:

$$-(\hbar^2/2m)(\partial^2\psi/\partial x^2) - E\psi = 0$$

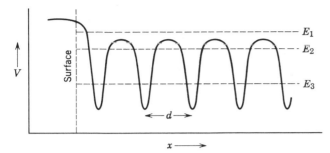

**Figure 2.14** Electron potential in a one-dimensional crystal.

That is, $\psi = Ae^{jkx}$, $k = \sqrt{2mE/\hbar^2}$, or, in other words, plane-wave solutions. This is the case we just examined. The other two cases are represented by $E_2$ and $E_3$ for the total electron energy. It is much more difficult to write down the appropriate approximate solutions for these two cases. However, the basic character of the solution is the same as the second exercise in Chapter 1. That is, in the region where $E \geqq V$, the solution is an undamped wave (e.g., $e^{jkx}$). In regions where $E < V$, the wave function is exponentially attenuated. Since the exponential factor is proportional to $(V - E)^{1/2}$, the wave functions for electrons with total energies $E \simeq E_3$ will be almost entirely located in the vicinity of nucleus. In a sense, the electron at one site hardly knows that any other electrons are around. One would expect, therefore, to find wave functions appropriate to individual atoms rather than those appropriate to periodic arrays. This is the so-called *tight binding approximation*. To the lowest order of approximation, the solutions to (1.7) with $E \simeq E_3$ are single-atom wave functions. (These can be found in any good book on atomic physics, such as Pauling and Wilson, *Introduction to Quantum Mechanics*.)

The case that will turn out to be the most interesting to us is for electrons with energies such as $E_2$, which are reasonably close to the tops of the potential wells. The reason is that the outermost electrons on the atoms in solids find themselves in just such a position. Furthermore, essentially all the observed properties of solids result from the behavior of these outer electrons. Solutions to (1.7) with $E \simeq E_2$ are extremely complicated. However, a few things can be ascertained by careful inspection, and the general character of the solutions can be extracted by drastic simplification of $V$. The specific details in each practical case are obtained experimentally, so it is possible to acquire all the information that we will need.

Since $V$ is never very much larger than $E$ for $E \simeq E_2$, the wave functions have considerable amplitude even in the classically forbidden regions $(V > E)$. This means that the electrons in this energy range "see" each other. Since there are about $10^{23}$ valence electrons/cm$^3$ of almost any solid, you might well wonder how one can account for all the interactions. As I show immediately below, it is reasonable to say that instead of being tied to a single nucleus, each electron spreads out *over the whole crystal*. Although the average electron density in all cases is the same as in the tight binding approximation ($\sim 10^{23}/cm^3$), the addition of each electron in the tight binding approximation changes some local site rather drastically; in the case we are here considering, the additional electron is so diffuse that its addition hardly changes the local potential at any site. Accordingly, I can ask what possible energies each electron could have in the same periodic potential, ignoring the fact that the interaction between the electrons must slightly alter the potential as each one is added.

## BLOCH WAVES

In 1928 Felix Bloch rediscovered a mathematical theorem (Floquet's theorem) that states that for a general class of second-order linear differential equations such as equation 1.7, if the equation is periodic in $x$, with period $a$, the solutions can be written as $\psi = u_k(x)e^{jkx}$ with $u_k(x) = u_k(x + a)$. We call such functions Bloch waves. The proof is simple and is based on the fact that for any second-order linear differential equation there exist exactly two linearly independent solutions. The proof for equation 1.7 goes as follows: If $V(x) = V(x + a) = V(x) + 2a)\ldots$—that is, if $V(x)$ is a perfectly periodic potential—then (1.7) is also periodic with period $a$. Let $u_1(x)$ and $u_2(x)$ be a set of linearly independent solutions to our periodic equation. Now it follows that $u_1(x + a)$ and $u_2(x + a)$ must also be solutions, since the substitution of $x + a$ for $x$ does not change the equation. From the fact that only two solutions may be linearly independent, it must be that we can write

$$u_1(x + a) = \alpha u_1(x) + \beta u_2(x) \tag{2.6}$$

$$u_2(x + a) = \gamma u_1(x) + \delta u_2(x) \tag{2.7}$$

Regardless of how I selected the $u$'s, I may form a new linearly independent set, $u_3$ and $u_4$, such that

$$u_3(x + a) = c_3 u_3(x) \quad (c_3 = \text{constant})$$
$$u_4(x + a) = c_4 u_4(x) \quad (c_4 = \text{constant}) \tag{2.8}$$

To show this, multiply (2.7) by a factor, $g$, and add it to (2.6). What you obtain is

$$u_1(x + a) + g u_2(x + a) = (\alpha + g\gamma) u_1(x) + (\beta + g\delta) u_2(x)$$

To achieve the result desired—that is, (2.8)—we must require that

$$u_1(x + a) + g u_2(x + a) = c_3 [u_1(x) + g u_2(x)]$$

By comparing the last two equations, this can be true only if

$$g = (\beta + g\delta)/(\alpha + g\gamma)$$

Solving the resulting quadratic for $g$ yields two solutions:

$$g = \frac{-(\alpha - \delta) \pm \sqrt{(\alpha - \delta)^2 + 4\gamma\beta}}{2\gamma} \tag{2.9}$$

It is thus possible to achieve (2.8) (one equation from each of the roots of (2.9)) in at least one way. It is not difficult to show that there is *only* one way.

Once (2.8) has been established, I may apply a chaining operation to obtain

$$u_3(x + na) = c_3 u_3(x + \{n - 1\}a) = c_3^2 u_3(x + \{n - 2\}a) = \ldots = c_3^n u_3(x)$$

We may proceed either physically or mathematically to the next and final step. Mathematically, we state that if the solutions are bounded and the equation extends over all space, the only possible values for $c_3$ have magnitude 1. Otherwise $c_3^n$ would diverge for either positive or negative $n$.

The physical argument is similar. Although crystals are finite, they can be an arbitrary number of periods long. Accordingly $c_3^n$ would "almost diverge" if $|c_3|$ were not "almost indistinguishable" from 1. Bloch's theorem is now established, since all numbers of magnitude 1 can be written as $e^{jkx}$, and $c_3$ is thus $e^{jka}$.

It is important to reexamine what has just been proved to see what it says physically. Bloch's theorem states that, regardless of the exact form of the periodic potential, if the equation is periodic, the solutions may be written as

$$\psi_k(x) = u_k(x)\, e^{jkx} \qquad \text{with} \qquad u_k(x) = u_k(x + a) \tag{2.10}$$

Since a Bloch wave has equal magnitude at any equivalent point in the crystal (i.e., at $x$ and $x + a$), the electrons in these Bloch states must extend *throughout the crystal*. This result is true, independent of the degree of "binding." It is just as exact for the tightly bound inner electrons as it is for the outer, freer electrons. However, it is only for the case of loosely bound electrons that there is much observable difference between the Bloch solution and the atomic-wave function.

Our next task is to find out what forms the periodic kernel of the Bloch waves $[u_k(x)]$ may take on. This we now proceed to do for a particularly simple crystal model.

## THE KRONIG-PENNEY MODEL

An important common property of all crystal potential functions is that there are some regions where the bound electrons are classically allowed

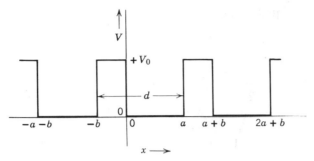

**Figure 2.15** Kronig-Penney model of the one-dimensional crystal potential.

$[V(x) \leq E]$ and some where they are not $[V(x) > E]$. Since the solution of
(1.7) with the authentic $V(x)$ is exceedingly difficult, there is a strong tempta-
tion to simplify the arithmetic while attempting to preserve as much of the
real solution as possible. One such simplification, which preserves only
the periodic allowed and unallowed regions, is shown in Figure 2.15. This
is the very popular Kronig-Penney model of the crystal. It has two endearing
properties: first and probably foremost is that equation 1.7 can be readily
solved for such a potential; second, the solutions give a very good outline
of what to expect in a real crystal. Without much more mathematics, you can
step beyond the Kronig-Penney solutions to solutions that lack only some
obtainable physical data to be quite usefully accurate.

To find the appropriate solutions to Figure 2.15, you should put a Bloch
wave into (1.7). From that operation you obtain

$$(d^2u/dx^2) + [2jk\,(du/dx)] + (2m/\hbar^2)[E - (\hbar^2k^2)/(2m) - V]u = 0 \qquad (2.10)$$

As with all linear, constant-coefficient differential equations, the solution
may be written as a sum of exponentials. [Note that (2.10) would have variable
coefficients if $V$ were a function of $x$.] We must choose one function in the
allowed region (e.g., $0 \leq x \leq a$) and another in the unallowed region, since
$V$ [and hence (2.10)] is not the same in the two regions. What conditions
must we put upon the $u$'s so that they are solutions? The answer was discussed
in Chapter 1. [Answer: The solution and its derivative must be continuous
across the boundaries.] This follows directly from (2.10), since it is evident
from that equation that $d^2u/dx^2$ is everywhere finite. We also require
$u(-b) = u(a)$ for periodicity. Just as in problem 1.2, we will eliminate all the
coefficients to solve for the allowed values of $E$. As stated at the beginning
of the chapter, these eigenvalues of $E$ are one of our primary objectives.
In a manner similar to problem 1.2, we will require that the wave function

be bounded everywhere. Since the wave function is a Bloch wave, this means that we demand that $k$ be real.

The $u$'s in the two regions are given by

$$u_1(x) = Ae^{j(-k+\alpha)x} + Be^{j(-k-\alpha)x} \qquad 0 \leqq x \leqq a \qquad (2.11)$$

$$u_2(x) = -Ce^{j(-k+\beta)x} - De^{j(-k-\beta)x} \qquad -b \leqq x \leqq 0 \qquad (2.12)$$

Putting in the boundary conditions leads to the following set of four equations:

$$A + B + C + D = 0$$

$$(-k+\alpha)A + (-k-\alpha)B + (-k+\beta)C + (-k-\beta)D = 0$$

$$Ae^{-j(-k+\alpha)b} + Be^{-j(-k-\alpha)b} + Ce^{j(-k+\beta)a} + De^{j(-k-\beta)a} = 0$$

$$(-k+\alpha)Ae^{-j(-k+\alpha)b} + (-k-\alpha)Be^{-j(-k-\alpha)b} + (-k+\beta)Ce^{j(-k+\beta)a} +$$
$$+ (-k-\beta)De^{j(-k-\beta)b} = 0$$

By setting the determinant of the coefficients equal to zero, we obtain the nontrivial solution, after much uninspiring algebra:

$$-(\alpha^2 + \beta^2)/2\alpha\beta \sin(\alpha a)\sin(\beta b) + \cos(\alpha a)\cos(\beta b) = \cos[k(a+b)] =$$
$$= \cos(kd)$$

Replacing $\beta$ with $j\gamma$ yields an equation with all real numbers:

$$[(\gamma^2 - \alpha^2)/(2\alpha\gamma)]\sinh(\gamma b)\sin(\alpha a) + \cosh(\alpha b)\cos(\gamma a) = \cos(kd) \qquad (2.13)$$

You should note that the left-hand side of the equation contains only $E$ as a variable while the right-hand side contains only $k$. Under what conditions does (2.13) yield real values for $k$? (Only if the left-hand side of the equation is between $-1$ and $1$.) Since the left-hand side *must* be greater than 1 for some values of $E$, it follows that for some values of $E$ there are no real (i.e., allowed) values of $k$. The allowed and unallowed regions show up very clearly in a graph of the left-hand side of (2.13). This is presented in Figure 2.16. It should also be pointed out that even when a real $k$ exists, *it is not at all unique*. From the right-hand side of (2.13) or directly from the definition of a Bloch wave, it is evident that any Bloch wave has the property that the replacement of $k$ by $k + 2\pi/d$ produces another equally good solution.

A very interesting plot may be made of $E$ versus $k$. This is done in Figure 2.17, a classical plot known as the "unreduced" Bloch diagram. The plot is essentially the same thing as Figure 2.6, but with one VERY important difference: There are some energies for which *no* stable states exist. In the usual nomenclature of the literature, the allowed states are said to form *bands*.

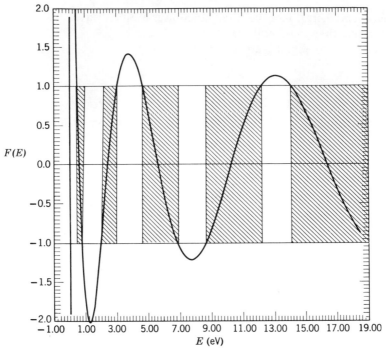

**Figure 2.16** The left side of equation 2.13 versus the electron energy. The lattice spacing is 10 Å; the barrier is 0.5 Å by 20 eV; and the mass is the rest mass of the free electron. The allowed solutions lie in the cross-hatched regions.

Figure 2.16 shows that the bands are wide when $(V_0 - E)$ is small and quite narrow when $(V_0 - E)$ is large. This fits well with what I said about tight binding implying wave functions very close to the free-atom wave functions. For a free atom, a particular state (say, the ground state) has one particular energy. IF the electrons in the solid had the free-atom wave function, they would all have the *same* energy. Figure 2.16 (and also Figure 2.17) shows that for the tightly bound case, the spread in energies is very small indeed. This is not the case for the more loosely bound electrons, where a single free-atom state is spread into a wide band of possible energies.

Take note that we have obtained one of the central results that I set as a goal for this chapter: to show that the allowed states in a crystal form bands. However, you might well ask how well this result holds when *real* potential functions are used. It turns out that this question has been answered quite well by some purely mathematical investigations of the second-order linear, periodic differential equation with quite arbitrary coefficients. The solutions *always* form bands. For a remarkably lucid and readable account

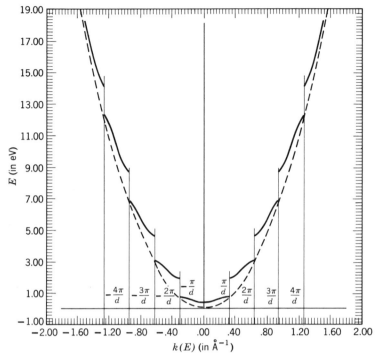

**Figure 2.17** The Bloch diagram for the Kronig-Penney one-dimensional crystal. The conditions are the same as in Figure 2.16. The dashed line is the Bloch diagram for the free electron $(E = \hbar^2 k^2 / 2m)$.

of the general problem of wave propagation in periodic structures, the reader could do no better than Leon Brillouin's book *Wave Propagation in Periodic Structures*, Dover (paperback), 1953.

Several features of the bands do depend rather strongly on the actual potential function. This would certainly make sense. Otherwise all solids would act the same. The most obvious change occurs when we go from our unrealistic one-dimensional model to a three-dimensional one. We now must plot the three components of **k** versus *E*, a four-dimensional plot. Doing this plot on two-dimensional paper gets a bit sticky, but there is nothing very fundamental about that. However, in our one-dimensional model, the bands never crossed. That is, there were no degeneracies except for the rather trivial one of the same *E* for ±*k*. In three dimensions, however, it is not only possible but quite common to find bands crossing. This has quite a dramatic effect on the behavior of certain materials. To show this, I must first answer the question: How many states are there in a given band? The answer is startlingly simple: two for each point in the lattice. This I now must prove.

## THE NUMBER OF STATES IN A BAND

In the derivation of (2.13), no specific attention was paid to the fact that the crystal is of finite length. Two rather heavy-handed approaches are usually employed to add this detail to the solution. Since surfaces of crystals are complicated affairs at best (and not yet really well understood), we propose either a very simple surface, such as an infinite potential step, or a crystal of finite dimensions with no end, such as a donut. The latter approach has always annoyed me, since donuts have surfaces if not ends, and it should be possible to describe solutions for crystals that do not happen to come out as toroids. Accordingly, as a matter of personal preference, let me assume that the crystal is a simple parallelopiped and that the surface represents a very abrupt and large change in potential—essentially an infinite potential step.

Going back to our one-dimensional model, let the length of the crystal be $L$. If the crystal lattice spacing is $d$, it follows that

$$L = md \qquad \text{with } m \text{ an integer} \qquad (2.14)$$

What are our boundary conditions on $\psi(x)$ at 0 and $L$? ($\psi$ must be 0 at both 0 and $L$.) Satisfying this boundary condition requires a proper ratio of the two independent solutions to (1.7), namely

$$\psi(x) = A\left[(\mu(x))/(2j)\right]\left[e^{jkx} - e^{-jkx}\right] = A\mu(x)\sin(kx), \quad k = n\pi/L \qquad (2.15)$$

Equation 2.15 satisfies both the boundary conditions and (1.7). The allowed values of $k$ are now discrete, although with typical $L$s, the distance between $k$'s is very small. Using (2.14) and noting from Figure 2.17 that a single band (or half-band, since there is now only one state for either $\pm k$) is of width $\pi/d$, I obtain the number of states in a given band as

$$n = (\pi/d)/(\pi/md) = m \qquad (2.16)$$

That is, there is one state for each point in the lattice. At this moment you should be frothing mad because I claimed I was going to show that the number of states per point was two. Well, I am. First, I had to prove that it was one. Now I am going to get it to split into two.

## THE SPIN OF THE ELECTRON AND THE PAULI PRINCIPLE

The electron has some properties other than mass and charge. The two that concern us are spin (angular momentum) and magnetic moment. It is an experimental fact that an electron has a magnetic moment. This should cause no more surprise or concern than the statement that it is an experimental fact that electrons have charge. However, most people like to know the "sources" of things. Since magnetic dipole moments usually arise from moving charges, it is tempting to suppose that the electron is "spinning." That would imply that the electron has intrinsic angular momentum. Once again, experiment verifies that electrons have an intrinsic angular momentum. However, it is a most unclassical angular momentum in that it obeys the following two rules:

**1.** The magnitude of the angular momentum is a constant, namely $\hbar/2$. Note that this says that any self-energy associated with the angular momentum never changes and is thus unobservable.

**2.** In any experiment that measures the direction of the spin vector (say, by measuring the direction that its associated magnetic field takes on), the spin is always found in one of only two possible states, usually referred to as "spin up" and "spin down." That is, the spin is either aligned with, or counter to, the defining polar direction.

This second rule is always emotionally difficult to swallow. It produces roughly the same reaction as the statement in Chapter 1 that one particle goes simultaneously through two slits. However, Dirac showed that this two-state property was consistent with the wave equation (1.9). What he demonstrated was that if one wanted to make equation 1.9 relativistically invariant AND have spin, the form that the equations took on predicted that the lowest amount of spin was $\hbar/2$, and it further predicted the observed fact that only two spin states would be observed. This makes spin self-consistent if not emotionally satisfying.

It should be pointed out that not all particles have spin. For example, mesons do not but protons, neutrons, and photons do. The magnetic moment that is derived from the Dirac equation depends on the mass of the particle, so that in spite of the fact that protons, for example, have the same spin as the electron ($\hbar/2$), their magnetic moment is substantially less than the electron's.

Once you have accepted spin, with its associated magnetic moment, it is easy to see that each of the states we found above (with $k$ and $E$ specified) must really be two states. In other words, what was state $(E_1, k_1)$ becomes two states: $(E_1, k_1, \uparrow)$ and $(E_1, k_1, \downarrow)$. The arrows stand for spin up or spin down. In a magnetic field, the energies of the two states would be different so that the degeneracy occurs only with no magnetic field. We now have the two states per band per point in the lattice that I guaranteed to find. Thus

$$n = (\pi/d)/(\pi/2md) = 2m \qquad (2.16')$$

One more extremely important principle, related to spin, is needed to show quite clearly which materials could be insulators. This is the *Pauli exclusion principle*. In a (successful) effort to explain the handsome periodicity of the periodic table of the elements, Pauli suggested a rule that no two electrons in the same system (e.g., atom, crystal, molecule) can have all the same quantum numbers. That is, no two electrons can occupy the same state at the same time. According to this rule, a state is filled if it has one electron in it. It turns out that all particles with spin $\hbar/2$, or odd multiples thereof, obey the Pauli principle. Particles with spin 0 or even multiples of $\hbar/2$ do not. For example, you may put all the photons you want in the same state — e.g. the same mode of a resonant cavity. This may seem plausible enough, but when you consider that the common isotope of helium has an *even* number of protons, neutrons, and electrons you might wonder whether some very peculiar behavior might not be observed. You would not be disappointed. Liquid He, cooled below the so-called lambda point ($2.17°K$), exhibits some of the weirdest phenomena you could imagine. It does indeed act as if a substantial portion of the liquid were in one single state. Macroscopic quantum effects have been observed, the liquid has incredible thermal conductivity and essentially no viscosity, and it runs up the sides of small bore tubing as if gravity were a thing of the past. We shall return to this topic in Chapter 3, but now back to conductors and insulators.

## CONDUCTORS, INSULATORS, AND SEMICONDUCTORS

If a band is entirely empty, can it contribute to the electrical conductivity of a crystal? (Answer: No. What does an empty band have to contribute?) Since the spacing between nonoverlapping bands is usually pretty large, a small electric field cannot pull electrons from a lower band to a higher band. Furthermore, if a band is entirely filled, the Pauli principle prohibits any electron from changing states within the band. Accordingly, entirely full bands contribute nothing to the conductivity. Only crystals with partially filled bands can exhibit electronic conductivity. Now if we wish to know which materials are *inherently* conductors, we must remove such external stimulants as light and heat and then find out whether the crystal still has partially full bands.

Using the Pauli principle and (2.16′), state the rule which guarantees that a given material is a conductor. (Answer: Since the number of states per band is even, there will be at least one partially filled band if the number of electrons per point is odd.) The inverse statement is not necessarily true. Even numbers of electrons do not necessarily imply an insulator, since, in three-dimensional crystals, bands can overlap. Some examples are in order.

Which of the following materials could be insulators, assuming one molecule or atom per lattice point? Ge, C, Pb, Al, Ca, $Al_2O_3$, AgBr. (Answer: All but Al, since all the others have an even number of electrons.) Now, how many *are* insulators? (Answer: Ge, C in its diamond form but not graphite, $Al_2O_3$, and AgBr.) You can conclude immediately that Ca and Pb must have overlapping bands. Note that most compounds result in each atom having the even number of electrons of a noble gas. Thus most compounds are insulators.

Now that I have neatly divided the crystalline world into conductors and insulators, what is a *semiconductor*? As it happens, insulator and semiconductor mean the same thing when these words are applied to crystals. The difference between a conductor and a semiconductor is that a conductor always has at least one partially full band, while a semiconductor is nominally an insulator that has had one or more bands made partially full under the influence of heat, light, or other external stimuli. In general, those materials which are *called* semiconductors can have their room temperature conductivities strongly influenced by impurities. Those that are called insulators remain insulating even with substantial densities of impurities. As you might suspect, such a distinction is not especially sharp.

The essential property that permits both conducting and "insulating" crystals to conduct is the distributed character of the eigenstates. Once an electron is moved into a previously empty band, that electron is reasonably free to meander around the crystal. There are, of course, many useful insulators that are not crystalline (e.g., nylon, silk and transformer oil). Raising an electron to a higher state in one of these materials produces substantially no electrical effect, since the electron is still not free to move.

Another two words that are synonymous when applied to crystals are *conductor* and *metal*. Metals are crystals with partially filled bands, and all of the properties that we usually attribute to the metallic state arise from the free and easy motion of the electrons that lie in the partially full band(s). These include large thermal and electrical conductivities, ductility, and such optical properties as high reflectivity. (There is also a "semimetal," which is a material with just barely overlapping bands.)

Since the motion of the electrons within the partially filled bands is so fundamental to the behavior of crystals, both metallic and semiconducting, that should be the next topic for us to examine. I leave to the end of the chapter the question of how much "external stimulation" is needed to get a given insulator to contribute a few conduction electrons. For the present, I will assume they are already there. What we are going to obtain below is the second rule stated at the beginning of this chapter—that the motion of the electron within the partially full band is generally describable by old, trusty $F = m^*a$, with $m^*$ a measurable crystal parameter.

## THE MOTION OF ELECTRONS UNDER THE INFLUENCE OF AN ELECTRIC FIELD

The most simplistic model of conductivity in a crystal requires three things: electrons free to move under the influence of an electric field; some randomly distributed objects that scatter moving electrons; and a fixed neutralizing charge (i.e., the atomic core) that keeps the electrons from repelling each other. This last condition is necessary to keep the mobile electrons within the volume of the crystal. With this very simple picture, when you apply a voltage across the sample, the electrons accelerate according to $F = -q\mathsf{E} = ma$. Every so often the accelerating electron smashes into one of the scattering centers and dashes off in another direction. On the average, then, the electron accelerates for an average time, $\tau$, and then loses its new-found velocity by scattering. If the average field-induced velocity is small compared to the mean thermal velocity, $\tau$ will be independent of $\mathsf{E}$. The mean thermal velocity will be zero. Under this circumstance, then, what is the average velocity of the electrons? [Answer: $\bar{v} = -q\mathsf{E}\tau/2m$.] What is the conductivity of this "metal"? (Answer: The conductivity is defined as $\sigma = J/\mathsf{E}$. If there are $n$ electrons, the current density is $J = -qn\bar{v}$. Substituting for $\bar{v}$ gives $J = q^2 n\mathsf{E}\tau/2m$ and $\sigma = q^2 n\tau/2m$.] Although this derivation ignores or assumes away all possible difficulties, you are going to find that it is not too far from the proper answer. What must be done is to find what $\tau$ and $m^*$ must be according to the wave equation and the crystal parameters. $m^*$ is discussed below. $\tau$ is undertaken in Chapter 4.

In discussing the effects of the crystal boundaries on the Kronig-Penney solutions, I came up with (2.15) as the proper form for the time-independent solutions. Note that these solutions are *standing* waves. They cannot, by themselves, have any velocity in any direction. However, in studying the solutions of the particle in a box in Chapter 1, I showed that a sum of solutions had the property that the expectation value for position was an oscillatory function of time. In other words, the particle appeared to bounce back and forth between the walls. Our next obvious question is how fast is the particle going? The answer to this question requires that we delve into two of the sundry velocities associated with a wave.

If I take a typical traveling wave—to wit, $e^{(j\omega t - kx)}$—and observe the motion of a point of *constant phase* I get the phase velocity, $v_p$, which is quite directly $v_p = \omega/k$. However, since the observable part of the wave function is the square of the amplitude (i.e., $\psi^*\psi = $ constant for this $\psi$), $v_p$ is not the velocity of anything I can observe. If I take instead the velocity of a point of constant amplitude, I would have a velocity that meant something. However, to have any variation in amplitude, I must have more than one frequency. That is, I need a group of waves. For this reason, the velocity of a point of constant amplitude is called the *group velocity*. You should note immediately that it makes no sense to define a group velocity if there are no points of reasonably constant amplitude in the wave train. But with that caution in mind, let us proceed. If I take the smallest group I can pick, namely two waves, I can write down the traveling wave as

$$\psi = e^{j(\omega t - kx)} + e^{j[(\omega + \Delta\omega)t - (k + \Delta k)x]}$$

$$= [1 + e^{j(\Delta\omega t - \Delta kx)}] e^{j(\omega t - kx)}$$

The second form shows very clearly a traveling wave with traveling modulation. A point of constant amplitude moves with velocity:

$$v_g = \Delta\omega/\Delta k \Rightarrow \partial\omega/\partial k \tag{2.17}$$

To find electrons in motion in the sense that I spoke of above, (2.17) says that I must have an electron in a composite state comprised of a number of eigenstates.

The statement just made brings up two problems that must be examined immediately. The first is that wave functions of the sort above have very obvious momentum associated with them (namely $\hbar k$), but with Bloch waves [e.g., (2.10)], what is the physical meaning of $k$? Does a Bloch wave carry momentum? If so, how much? The second problem concerns the use of Bloch waves of the form $e^{jkx}u(x)$ when I just argued for the use of $\sin(kx)u(x)$. This second problem is more easily disposed of, so let us proceed with it first.

In a transmission line (or violin string) terminated with 0 or infinite impedance, one always obtains a perfect reflection. Accordingly, the solutions are always standing waves of the sort I employed in (2.15). However, if I want a current flow, Kirchhoff's laws require that I put current in at one place on the surface of the crystal and take it out at another. These contacts must not have perfect reflectivity, so the boundary conditions I applied to get (2.15) do not apply to this problem. Some other ratio of the amplitudes of the incident and reflected wave will be appropriate, but this really adds nothing new to our solution technique or our knowledge. Another point could also be raised. In the presence of the applied field, $V(x)$ is no longer completely periodic. You might well wonder whether the use of Bloch waves is proper. Strictly speaking, Bloch waves are *not* proper, but the added term in $V$ is so small compared to the crystal fields that make up the periodic potential that the Bloch waves must certainly be an exceptionally accurate approximation to the exact solution.

We must now return to the thornier question of interpreting $k$. $\hbar k$ is usually called the *crystal momentum*, but a most peculiar momentum it turns out to be. First, show that a Bloch wave is *not* an eigenfunction of momentum. Demonstration: $-jh(\partial/\partial x)[e^{jkx}u(x)] = jh[jk(e^{jkx}u(x)] + e^{jkx}\partial u/\partial x$. Un-

less $u(x)$ is a constant (i.e., the free electron solutions) $- jh(\partial/\partial x)\psi \neq$ (const.) $\cdot$
$\cdot (\psi)$.      Q.E.D. Your next thought might be that $\hbar k$ was the *expectation* value
of the momentum. Without TOO much difficulty, it is possible to show
that this is a viable interpretation. After all, it makes pretty good sense that
$p$ would not be a constant if $V(x)$ were not, because $E$ is a constant and
$E = p^2/2m + V$. But the periodicity certainly guarantees that both $p$ and
$V$ will have an average (expectation) value.

Unfortunately, neat as that interpretation is, it leaves you completely
unprepared for the next interesting detail in our picture of $k$. If you will recall (Figure 2.6), we found several values of $E$ for the same $k$ and several
values of $k$ for each $E$. With Bloch waves, this feature becomes more prominent: for every value of $E$ there are an *infinite* number of $k$'s. This point was
brought out just after (2.13) was developed, when I stated that $k$ and $k + 2n\pi/d$
were equally good solutions. In order not to claim more exactness than I
have really shown, I should draw Figure 2.17 as it appears in Figure 2.18.
NOW what is the meaning of $k$? The answer appears when you properly
consider what you really want to do with momentum. You wish to know
the momentum because the law of conservation of momentum allows you
to solve problems. Let us ask, therefore, what the law of conservation of $k$
is. First, note that $\pi/d$ is a reciprocal-lattice vector for our one-dimensional
crystal. When one asks the question: Can a single particle make a transition
between states with different $k$'s?, the answer is: Yes, if and only if the $k$'s
differ by $2\pi$ times a reciprocal-lattice vector. This may seem like a pretty
peculiar "conservation law," but it is really just another example of the fact
that $\sin \sigma = \sin (\sigma + 2\pi)$. In your normal dealings with angles, you would
automatically replace $(\sigma + 12\pi)$ by $\sigma$ because the angles are *almost* always
precisely *equivalent*. Our rule above for conservation of momentum can be
restored to the familiar Newtonian form if we do the same thing for $k$ that
you would customarily do for $\sigma$—that is, pick a single zone which contains
all the inequivalent $k$'s and use only $k$'s from that zone. The usual scheme
for $\sigma$ is to select a zone that is symmetrical about 0. In other words, $-\pi \leq \sigma \leq \pi$.
Figure 2.18 is divided into zones in an analogous way. The heavy vertical
lines indicate the zone boundaries. The first zone occupies the interval
$- \pi/d \leq k \leq \pi/d$. The second zone occupies the interval from $-2\pi/d$ to
$2\pi/d$, leaving out the first zone; the third from $- 3\pi/d$ to $3\pi/d$, leaving out the
first and second zones; and so forth. Such zones in a $k:E$ plot are called Brillouin
(bree wan') zones. They become much more amusing figures in three dimensions, as you will see later, but the central idea in any number of dimensions is the same as in one: the first Brillouin zone is the region in $k$-space
centered on the origin containing all the points that are less than $2\pi$ times
a reciprocal lattice vector from each other. More of this in a moment, but
now back to the motion of electrons in an applied electric field.

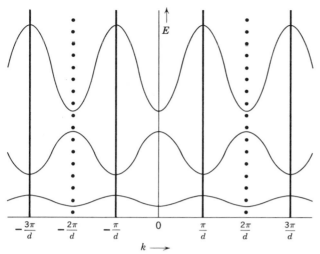

**Figure 2.18** Complete unreduced Bloch diagram. Any interval of $2\pi/d$ along the *k* axis (a *zone*) contains a complete set of solutions. The usual set of zones is indicated by the bold vertical lines.

## THE EFFECTIVE MASS OF AN ELECTRON IN A CRYSTAL

We wish to pursue the motion of a small group of waves subjected to an acceleration by the applied field. The field does some work in accelerating the electron, so I write

$$v_g = (d\omega/dk) = (1/\hbar)\, dE/dk \tag{2.17'}$$

Then:

$$dE = (dE/dk)\,dk = -\,q\,E\,dx = F v_g\,dt = (F/\hbar)\,(dE/dk)\,dt \tag{2.18}$$

From the second and fifth expressions in (2.18), I get

$$dk = (F/\hbar)\,dt \qquad \text{or} \qquad dk/dt = F/\hbar \tag{2.19}$$

Differentiating equation 2.17' with respect to time gives:

$$dv_g/dt = (1/\hbar)\,(d^2E/dk^2)\,dk/dt \tag{2.20}$$

Using (2.19) in (2.20) yields a form very much like Newton's second law, $F/m = a$:

$$a_g = dv_g/dt = (1/\hbar^2)\,(d^2E/dk^2)F \tag{2.21}$$

According to (2.21), the electron behaves as if it were a particle with an *effective* mass, $1/m^* = (1/\hbar^2)\,(d^2E/dk^2)$. It is important to observe what we have finally accomplished. We have found that the very complicated interaction of a periodic lattice and an electron subject to an external force field is describable by an equation quite similar to Newton's second law. Since you are familiar with Newtonian mechanics, it is a great conceptual convenience to think of the ratio of the force to the group acceleration as the particle's effective mass. This we shall do for the remainder of this book. However, before comfortably forgetting that what we are describing by $m^*$ is really a crystal-wave interaction and NOT an inertial effect, take a good look at three very unmasslike properties of $m^*$, namely negative mass, nonconstant mass, and tensor mass.

First let us consider the constancy of $1/m^* = (1/\hbar^2)d^2E/dk^2$. If the second derivative is a constant, the curve is a parabola. Now, a quick glance at Figure 2.18 should convince you that these curves are not parabolas. However, over any sufficiently small interval, any curve looks like a parabola, and for these curves, the important regions at the top and bottom of the bands are very parabolic. Once again, this points out that $m^*$ is a measure of the $E:k$ curvature, not inertia. However, in at least some portions of the curves, we will find reasonably constant curvature, so $F = m^*a$ is useful if not omnipotent.

Now let us examine negative mass. Again, from Figure 2.18, you can see that there are regions where the curvature of the bands is negative. According to (2.21), electrons in these states should behave as if they had negative masses. These negative masses lead to the concept of a hole in the following way. Consider a full band. It can show no conductivity, so the sum of all the electron group velocities must be zero. Accordingly, I can write

$$J = -q\sum_i v_i = 0 \quad \text{where } J \text{ is the current density} \qquad (2.22)$$

If I remove a few of the electrons from the top of the full band—note that the top is where the curvature is negative—equation 2.22 need not add to zero, since some of the terms are now missing. Thus:

$$J = -q\left[\sum_i v_i - \sum_j v_j\right] = q\sum_j v_j \qquad (2.23)$$

where the $i$'s run over all the states in the band and the $j$'s over the empty states. Equation 2.23 is very much the same thing that you would get if you had only a few electrons in an otherwise empty band. That is, for the almost empty band, $J = -q\sum v_k$ with the $k$'s running over the full states. The only difference between (2.23) and the equation above is the sign of the charge. In other words, the electrons in an almost-full band act as if the band were empty except for a few POSITIVELY charged particles with positive effective masses. It should not be difficult to see that a positively charged particle with a positive mass acts like a negatively charged particle with negative mass. This "equivalent" particle is called a *hole*. Unlike the electron, it has no extracrystalline existence. One can generate lots of analogies. Bubbles go "the wrong way" in a gravitational field. Shockley proposed the idea of looking at an almost-full parking lot from a high building. A time sequence of photographs would reveal the motion of the empty spaces. For an almost-empty lot, the sequence would reveal the motion of the cars. Most of the time it is simplest to think of holes as real, positive particles that can recombine with an electron and otherwise behave as a positive particle would be expected to. Since the curvature of the $E:k$ plot determines the hole's effective mass, just as it does the electron's, you must anticipate the possibility of non-constant $m^*$. You should also expect the hole and electron effective masses to be different if they are "traveling" in different bands.

A third property of $m^*$ is its tensor character. If I want to work in three dimensions, I must employ a vector $k$. Group velocity is also a vector given by

$$v_g = (1/\hbar)(\partial E/\partial k_x, \partial E/\partial k_g, \partial E/\partial k_z) = (1/\hbar)\nabla_k E \qquad (2.24)$$

Working through the same steps that lead to (2.21) yields the effective mass in three dimensions:

$$(1/m^*)_{ij} = (1/\hbar^2)\,\partial^2 E/\partial k_i\partial k_j \qquad (2.25)$$

Equation 2.25 may look a bit unlike anything you have considered before, but its interpretation is not too difficult. Since we wish to relate two vector quantities, force and acceleration, we must permit the $i$'th component of one vector to be related to the $j$'th of the other. In a very real sense, you might push an electron in the $x$ direction and find that it accelerates rapidly in the $y$ direction. This was exactly what happened to the X ray waves we sent through the crystal. There is no reason to expect the electron waves to behave noticeably differently. A more mundane example would be pushing a weight along an inclined plane. Pushing in the horizontal direction would result in a vertical component of motion. You might well have attributed this behavior to a tensor mass were it not for your commendable desire to have a mass that did not change with the experimental conditions. In the same interest of simplicity, it is easier to describe the behavior of the electrons in the crystal by a tensor mass such as (2.25). Writing out the three equations relating each of the three components of the acceleration to each of the three components of the force gets unbearably messy. Instead, I take equation 2.25 and use a neat notation originally developed by Einstein. I wish to write

$$a_i = \sum_j (1/m)_{ij}F_j \qquad (2.26)$$

Since we will always want to write such sums and almost never wish to write the individual components, Einstein suggested that the summation be understood. In Einstein's notation, you can write (2.26) as

$$a_i = (1/m)_{ij}F_j \qquad (2.26')$$

This rather complicated relationship between force and acceleration might suggest to you that single-crystal samples of some materials would have conductivities that depended on the direction of application of the field. This does turn out to be the case for *most* materials. The reason why you have not observed tensor conductivity is that (1) the most common conductors (Al, Cu, Ag, etc.) form cubic crystals that may have tensor masses, but the cubic symmetry averages out any anisotropy in the conductivity, and (2) most samples are randomly oriented polycrystalline materials that average out any anisotropies. We will return to tensor masses in a moment when we consider how to measure the effective masses of electrons and holes. First, I must introduce you to techniques for plotting the information in four-dimensional $E:k$ plots. For the moment, try your hand at the following few questions on one-dimensional crystals.

### A FEW QUESTIONS ON $m^*$

Examine Figure 2.19, which is the first Brillouin zone of Figure 2.18. It contains a complete set of unredundant, inequivalent points and thus has all the information that Figure 2.18 has. It is called a "reduced" Bloch diagram. Are the smallest electron masses to be found in bands I, II, or III? (Answer: III, since the curvatures at the bottom and top of the highest band are quite evidently larger than the other two bands. The $m^*$ is inversely proportional to the band curvature.) Are there any points in the three bands where the electrons will not respond to an applied force? [Answer: Yes,

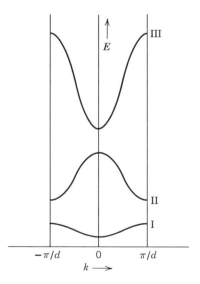

**Figure 2.19** Reduced Bloch diagram in one dimension.

two points in each band where the curvature vanishes. At such a point, (2.21) says that the acceleration vanishes.]

Consider a case where bands I and II were completely filled and III completely empty. Heat the crystal a bit and boil a few electrons from the second band into the third. Does the number of holes equal the number of electrons? (Answer: Yes. Each electron that arrives in band three left behind a hole in band two.) Is the effective mass of the holes (in band II) greater or less than that of the electrons (in band III)? (Answer: By inspection of the figure, the curvature of band II is less than that of band III in the vicinity of their closest approach. This is the region where the holes and electrons will be, so the holes will have the heavier masses.)

Will the interval of energies $\Delta E$ occupied by the holes in band II be greater or smaller than the interval occupied by the electrons in band III? (Answer: Since the interval of energy between states is proportional to the slope, $dE/dk$, the lower effective mass of the electrons implies a greater energy between electron states. [Note, of course, that $dE/dk$ is 0 at $k = 0$ for both bands.] Accordingly, the electrons in band III occupy a larger interval of energy than their equal number of holes in band II). Another very important way of saying the same thing is that *the density of states* is greater in band II. Density in this case means the number of states per unit energy difference per unit volume of crystal. It is quite generally true to state that a large effective mass implies a large density of states and vice versa. As the next chapter will show, the number of electrons and holes that you get in a given material at a given temperature depends directly on the density of states.

At what value of $k$ will the minimum energy for transitions between band II and band III occur? (Answer: At $k = 0$, by inspection of Figure 2.19.) At what value of $k$ will the minimum energy occur for transitions from I to II? (Answer: $k = \pi/d$. Note that $-\pi/d$ is an equivalent point, since it differs from the one I gave by $2\pi/d$.)

## $E:k$ PLOTS IN TWO AND THREE DIMENSIONS

In a two-dimensional crystal, with its two-dimensional reciprocal lattice, the $E:k$ plots for each of the bands are surfaces that must be presented in a three-dimensional representation. A figure such as Figure 2.18 becomes unbearably confusing when extended from two to three dimensions, so it is customary to use reduced Bloch diagrams such as Figure 2.19. To obtain these plots, we must first find the Brillouin zones in the reciprocal lattice. Define the first Brillouin zones in the reciprocal lattice. (It is the locus of all points that lie no further from the origin than $\pi$ times a reciprocal lattice vector.) Note that this definition assures us that no two points within the zone are equivalent.

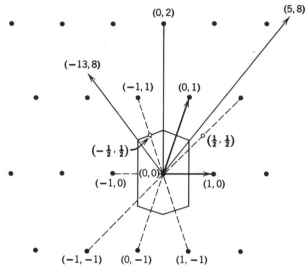

**Figure 2.20** Construction of the first Brillouin zone for a general two-dimensional reciprocal lattice.

The definition of the zone leads directly to the method of constructing or drawing it. Figure 2.20 is used for the demonstration. The scale factor between the reciprocal lattice and the Brillouin zone is $\pi$, so all points that lie within one-half reciprocal-lattice vector of each other will scale to the Brillouin zone on multiplication by $\pi$. Let us therefore work directly in reciprocal space. You start by drawing lines from the origin $(0, 0)$ to all of the reciprocal-lattice points that have only $\pm 1$ or $0$ for their coordinates in terms of the basis vectors [e.g., $(-1, 0)$ or $(-1, 1)$]. These are shown as dashed lines in Figure 2.20. The perpendicular bisectors to these lines are then drawn, but only far enough to intersect. The smallest figure that these intersecting lines can make is the first Brillouin zone. This is the irregular hexagon in the figure. Note that this leaves out the bisectors of $(1, 1)$ and $(-1, -1)$. Except for rectangular lattices, this will always be the case, and it is helpful to observe that points beyond the intersection of the $(1, 1)$ vector and the bisector of the $(1, 0)$ vector are legitimately outside the first zone. To see this, prove that the point $(1/2, 1/2)$ is equivalent to a point at the edge of the Brillouin zone. [Proof: If the reciprocal lattice vector $(1, 0)$ is subtracted from $(1/2, 1/2)$, you obtain the equivalent point $(-1/2, 1/2)$. This is the intersection of the vector $(-1, 1)$ and its own bisector. Thus, $(-1/2, 1/2)$ lies on the Brillouin zone boundary. Q.E.D. For the point picked, an equally good proof follows subtraction of the other basis vector, $(0, 1)$.]

To see that this construction is exactly what was called for, note from the figure that the Brillouin zone so generated has mirror symmetry about the

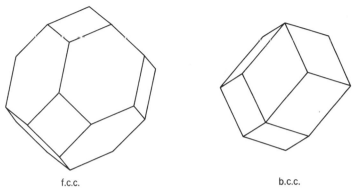

f.c.c.                                    b.c.c.

**Figure 2.21** First Brillouin zones for f.c.c. and b.c.c. lattices.

vertical and horizontal axes. Thus, the entire figure is specified by the portion in the first quadrant. The bisector of (1, 0) defines all the points in the first quadrant that are one half of that reciprocal-lattice vector from the origin. The bisector of (0, 1) defines all the points that are one half of the other basis vector from the origin. Thus, in the first quadrant, all the points within the area defined by those two lines—and no others—fulfill the definition of the first Brillouin zone.

In three dimensions, the construction is much the same, but now the perpendicular bisectors are planes rather than lines. The first Brillouin zone for the f.c.c. and b.c.c. lattices are shown in Figure 2.21. Almost all the commonly used semiconductors are f.c.c. (Si, Ge, GaAs, PbS, AgB), although there are a vast number of noncubic semiconductors.

Now comes the important part of this section—$E:k$ plots. Since $k$ has the dimensionality of the crystal, the $E:k$ plot will have three or four dimensions for two- and three-dimensional crystals respectively. There are a number of acceptable alternatives for presenting three-dimensional figures in two-dimensional representations—plan and section drawings, orthographic projection, perspective drawing, topographic mapping, and the like—but the four-dimensional figure in toto is a bit more of a problem. Two techniques are customarily employed to give essential $E:k$ information for three-dimensional crystals. These derive from the topographic mapping technique and from sections taken along particular crystallographic directions. These will be illustrated for a two-dimensional crystal so you can "see" what is being done. Then some four-dimension $E:k$ plots will be discussed so you can extend your "seeing" to less-familiar representations. Finally, in the last section, two techniques for experimentally obtaining the gross and fine details of the plots will be examined.

Inspection of Figure 2.18 and some thinking about the periodic character of the unreduced Bloch diagram should give you a rule about $dE/dk$ at the Brillouin zone edge (surface in three-dimensional crystals). What is it? (Answer: The normal component of group velocity—i.e., the slope in the direction perpendicular to the zone boundary—is zero at the zone edge.) Do not be too restrictive in your concept of *periodic*. In all but the one-dimensional crystal, the $n$'th zone is VERY different from the $(n + 1)$'th. For example, the first zone of a simple cubic lattice is also a simple cube, while the second zone is the same as the first zone for the b.c.c. lattice (Figure 2.21). However, the volume (area in two dimensions) of each zone is always the same as all the others, and for every point in zone 2, there exists an equivalent point in zone 1. Naturally, two points with different $k$'s would not be equivalent if they did not have the same $E$'s. Since each zone contains all the information on $E$ versus $k$, the $E:k$ plot reproduces itself in each and every zone. In this sense we have a periodic function; it cycles once every zone.

## SIMPLIFICATION OF THE EFFECTIVE MASS TENSOR

Consider the reciprocal lattice and first Brillouin zone shown in Figure 2.20. If I plot the value of energy $E(k)$ as the altitude at point $k$, I obtain Figure 2.22. This orthographic projection represents $E$ versus $k$ for a single (arbitrary) band in the well-known two-dimensional material *phlatnium*. The dotted line around the base shows the edges of the first Brillouin zone. The surface reaches its maximum altitude at $k = (0, 0)$ and its minimum value at the four vertices of the hexagon that are not in the $\pm [-1, 2]$ directions (i.e., $\pm [5, 8]$, $\pm [-13, 8]$).

With this mountainlike model as your guide, try your hand at the following questions. First, is there *inversion* symmetry? The question asks whether $E(k) = E(-k)$. (Answer: By inspection of the figure, yes.) Second, are there any directions about which there is *mirror* symmetry? If there were mirror symmetry about the $x$-axis, which happens to be the $[1, 0]$ direction in this drawing, $E(k_x, k_y) = E(k_x, -k_y)$. (Answer: Again by examination of the figure, mirror symmetry occurs about the $[1, 0]$ and $[-1, 2]$ directions.)

Based on the symmetries in Figure 2.20, are there any axes or special points where Newton's second law is simply $F_i/m_i^* = a_i$? This would come about if $(1/m^*)_{xy} = 0$. [Answer: The mirror symmetry implies the vanishing of $\partial E/\partial x$ or $\partial E/\partial y$ along the $y = (-1, 2)$ or $x = (1, 0)$ directions respectively. Since the order of differentiation is unimportant, the vanishing of either derivative implies $(1/\hbar^2)\, \partial^2 E/\partial x \partial y = (1/m^*)_{xy} = (1/m^*)_{yx} = 0.$]

It would be very helpful analytically if we could get rid of these cross-derivative terms altogether. Although $(1/m^*)$ would still be a tensor, it would be a particularly simple one leading to a Newton's second law of the form:

$$F_x = m_{xx}^* a_x \qquad (2.27)$$

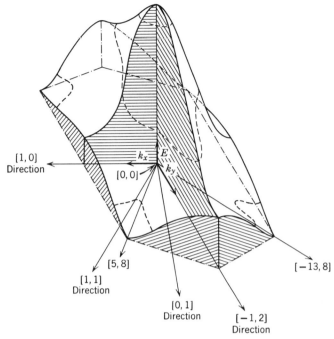

**Figure 2.22** An $E:k$ plot for a single band in the two-dimensional Brillouin zone of Figure 2.20. Three contour lines (lines of constant $E$) are shown, as well as two vertical sections. The directions in the zone are the same ones shown in Figure 2.20.

If $(1/m^*)_{xy} = 0$ everywhere, what is the form of the relationship between $E$ and $k$? That is, what type of terms may occur in an expansion of $E$ in powers of $k_x$ and $k_y$? (Answer: Since $\partial^2 E/\partial k_x \partial k_y = 0$, the only series terms allowed are of the form $k_x^n$ or $k_y^n$.) Thus,

$$E = \sum_0^\infty a_n k_x^n + b_n k_y^n \qquad (2.28)$$

It is important to note that cross-derivative terms will reappear in (2.28) if you rotate the coordinate system. For example, let $k_y' = 1/\sqrt{2}\,(k_x + k_y)$, $k_x' = 1/\sqrt{2}\,(k_x - k_y)$. Then, a term like $k_x^2$ becomes $[1/\sqrt{2}\,(k_x' - k_y')]^2 = 1/\sqrt{2}\,(k_x' - 2k_x'k_y' + k_y'^2)$. From this it is evident that $\partial^2 E/\partial k_x' \partial k_y' \neq 0$ in general.

The interesting question is, given a quite general form of (2.28) including cross-product terms, is it possible to rotate the coordinates to obtain the simple form of (2.28) in which $(1/m^*)_{xy} = 0$? Let me answer the question for the case that most concerns us—the case of parabolic bands. To have parabolic bands, the highest derivatives are the second. Let me locate my origin at the extremum of the $E:k$ plot as I did in Figure 2.22. Then the general form of (2.28) will be

$$E - E_0 = ak_x^2 + bk_xk_y + ck_y^2 \tag{2.29}$$

where $E_0$ is the energy at the extremum. Why are there no first-order terms? (Answer: The presence of first-order terms would indicate that the slope did not vanish at the extremum. If the origin were not at the extremum, the first-order terms would appear.) By a simple rotation of the coordinate system, I wish to convert equation 2.29 into the better form:

$$E - E_0 = a'k_{x'}^2 + c'k_{y'}^2 \tag{2.30}$$

As I do this, you will see that if (2.29) holds, it is *always* possible to obtain (2.30). For any arbitrary rotation, $\phi$, about the $E$-axis I obtain

$$k_x = \alpha k_{x'} - \beta k_{y'} \qquad \alpha = \cos\phi$$
$$\text{where}$$
$$k_y = + \beta k_{x'} + \alpha k_{y'} \qquad \beta = \sin\phi$$

Putting $x$ and $y$, in terms of the transformed coordinates, into (2.29) yields

$$E - E_0 = a\alpha^2 k_{x'}^2 - 2a\alpha\beta k_{x'}k_{y'} + a\beta^2 k_{y'}^2$$
$$+ b\alpha\beta k_{x'}^2 + b(\alpha^2 - \beta^2)k_{x'}k_{y'} - b\alpha\beta k_{y'}^2$$
$$+ c\beta^2 k_{x'}^2 + 2c\alpha\beta k_{x'}k_{y'} + c\alpha^2 k_{y'}^2$$

To obtain (2.30), I must select $\phi$ such that the coefficient of $k_{x'}k_{y'}$ vanishes. Thus, $\alpha^2 + (2/b)(c - a)\alpha\beta - \beta^2 = 0$. Putting all the trigonometric functions back in and converting squares and products to the appropriate first-order expressions in $2\phi$ yields

$$(c - a)/b = \cot 2\phi \tag{2.31}$$

Equation 2.31 has solutions for all possible values of the coefficients $a$, $b$, and $c$, so it is always possible to reduce (2.29) to (2.30) for the case of a quadratic $E:k$ plot. As a check on the result, note that the solution to (2.31) with $b = 0$ is $\phi = 0$.


## TWO-DIMENSIONAL PRESENTATIONS FOR THREE-DIMENSIONAL PLOTS

Now let's look at two two-dimensional presentations, each giving some of the same information, and then some uses for them. The two most obvious ways of taking a slice through the solid figure shown in Figure 2.22 are plan sections and elevations. Symmetry makes half the elevation and all but one quadrant of the plan superfluous. It is customary to avoid the duplication

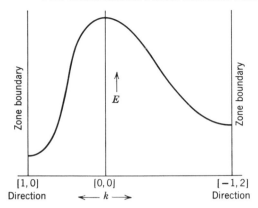

**Figure 2.23** Elevations from Figure 2.22 in two crystalline directions. The two directions shown happen to be at right angles and correspond to the $x$ and $y$ axes in Figure 2.20.

in the elevation but not in the plan. Two elevations are shown in Figure 2.22, one in the $[1, 0]$ direction, the other in the $[- 1, 2]$ direction. A typical $E : k$ plot would present both of them on one graph as shown in Figure 2.23. Since Figure 2.23 is much the same as Figure 2.19, the elevation drawing is the one to use to show the relationships between the various bands. The only difference between the elevations in Figure 2.23 and those one would obtain for three-dimensional crystals is the requirement for another coordinate to specify the direction (e.g., $[1, 0, 0]$ where Figure 2.23 uses $[1, 0]$ ). It is quite important to note that Figure 2.23 gives information of the relationship between energy and crystal momentum in two *specific* directions in the crystal. It does not give any information about other directions in the crystal. You cannot have your cake and eat it too. Figure 2.23, with judicious selection of the directions, gives some of the useful information about the relationships between the various bands in the crystal. For information about the details within the band, you require a topographic display—a display that can show but one band at a time. Examination of Figure 2.23 will show that in both the $[1, 0]$ and $[- 1, 2]$ directions, the $E : k$ plots are reasonably parabolic near $k = 0$. However, it is *most* important to observe that the curvature of the two parabolas shown are quite different. Thus, if we choose the $xy$ coordinate system shown in the figure as the coordinate system to describe forces and accelerations, we get (2.27) as a particularly simple version of (2.26).

The topograph is obtained by plotting a regular sequence of plans— regular in the sense that the contour lines represent outlines of slices spaced at equal *vertical* distances. Each contour is the locus of all points of a given energy. Three contour lines are shown in Figure 2.22 and presented in the standard topographic form in Figure 2.24. This is a very advantageous form for computing effective mass parameters, since the slope is the number of

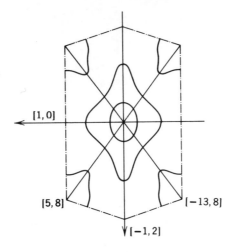

**Figure 2.24** A topographic presentation of Figure 2.22.

contours per unit length in any direction. To obtain useful numbers one would normally need many more contours than appear in Figure 2.22. Nonetheless, even with only three contours a number of facts can be obtained. First, what shape would the contour lines have if the effective mass tensor was constant but $(1/m^*)_{xx} \neq (1/m^*)_{yy}$? [Answer: If $\partial^2 E/\partial k_x^2 = c_1$, $\partial^2 E/\partial k_y^2 = c_2$, and $\partial^2 E/\partial k_y \partial k_x = 0$, the relationship between $E$ and $k$ is essentially equation 2.30:

$$E - E_0 = c_1 k_x^2 + c_2 k_y^2 \qquad (2.30')$$

For any fixed energy, $E$, (2.30') is the formula for an ellipse.] Figure 2.24 shows that in the vicinity of $k = 0$, lines of constant $E$ are elliptical. Accordingly, near $k = 0$ we can expect a constant effective mass tensor. Furthermore, if the energy corresponding to the contour line were known, you could obtain the effective mass parameters directly from the lengths of the major and minor semiaxes.

For the second and third contour lines, you can see that the effective mass tensor is not constant. The relationship between $E$ and $k$ will contain terms of the order $k_i^3$ and higher, and Newton's second law becomes complicated to a point of uselessness.

If you will recall that the allowed values of $k_i$ are discrete and distributed evenly along the axis, you should be able to see that the allowed values of $k$ are the intersections of a rectangular grid, $(n\pi/L_1, m\pi/L_2)$. Thus, for the figure we are considering, the number of states with energy greater than the energy of the contour is proportional to the area enclosed by the contour.

An important parameter that we will have much use for is the *density-of-states function*, $g(E)$, which is defined as the number of states per unit energy interval. In other words, in a crystal of volume 1 cm³, the number of states with energy between $E$ and $E + dE$ is given by $dm = g(E)\,dE$.

$g(E)$ can be obtained from the topographic display. One measures the area, $\Delta A$, between two contours separated by an energy difference $\Delta E$. From (2.15) and (2.16), the area occupied by a single state (not counting spin) is $A_s = = \pi^2/L_1L_2$. If we chose the crystal dimensions, $L_1$ and $L_2$, to be 1 cm, $A_s = \pi^2$. From the discussion accompanying (2.16) it is clear that we want to count the area only in the "first quadrant." (This is to avoid counting the same state twice. Equation 2.15 showed that the state labeled $k$ was the same state as the state labeled $-k$. A similar requirement restricts us to the first quadrant in two or three dimensions.) Thus, the number of states in $\Delta A$, counting spin but not counting any state more than once, is

$$\Delta m = 2\Delta A/4\pi^2$$

From the definition of $g(E)$, we have at once:

$$g(E) = \Delta m/\Delta E = (1/2\pi^2)\,\Delta A/\Delta E$$

For three dimensional crystals (2.32) becomes:

$$g(E) = (1/4\pi^3)\,\Delta V/\Delta E \tag{2.32'}$$

## THREE-DIMENSIONAL REPRESENTATIONS OF FOUR-DIMENSIONAL GRAPHS

A three-dimensional crystal can yield some pretty spectacular topographs. The topograph reduces the dimensionality of the figure represented by one. Thus, for three-dimensional crystals, the topographic $E:k$ display is three-dimensional. It is customary to draw only a single contour surface in a given drawing, the one chosen being best for whatever is under discussion. Figure 2.25 shows such topographic presentations for the first empty band in Ge and in Si. This band is referred to as the *conduction* band. The contour surface chosen is one very close to the minimum energy in the band. Each of the ellipses represents the same energy, so six equivalent minima exist in Si and four in Ge. (Why *four* when you can see eight? Answer: The minimum points in Ge occur at the surface of the Brillouin zone. Thus, the ones on opposite sides are equivalent and should be counted only once.)

## AN ASIDE ON NOTATION

The minima for Si lie on the Cartesian axes of the cube; those of Ge lie on the cube diagonals. If I used the primitive set of basis vectors, some of the rather obviously equivalent directions do not have similar coordinates.

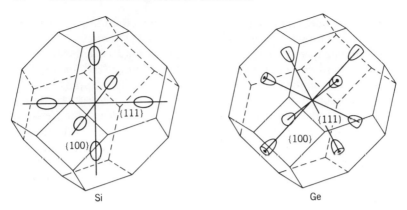

**Figure 2.25** Experimentally determined constant-energy surfaces at the conduction-band minimum in germanium and silicon. The labeling of the reciprocal-lattice planes follows the customary practice with cubic crystals of using the simple cubic lattice notation. Thus, the $x$ axis is $[100]$. If one used a proper basis set, the $x$ axis would be a $\{110\}$ direction. This is discussed further in the text.

For example, some of the hexagonal faces on the Brillouin zone are of the form $[1, 0, 0]$, while others are of the form $[1, 1, 1]$. To bring out the strong cubic symmetry and also because thinking is easier in Cartesian coordinates, it is customary to label the crystal directions as if we were dealing with a simple-cubic cell. The advantages are pretty obvious, the disadvantages less so. For an example of the advantages, the complete reflection and inversion symmetry makes equivalent all directions containing the same integers in their labels, regardless of sign. Thus $[1, 3, 2]$ is equivalent to $[1, 2, 3]$ and $[3, \bar{1}, \bar{2}]$ and $[\bar{2}, \bar{1}, \bar{3}]$. [The bar on the top is the standard way of indicating a negative coordinate. Thus, $(-1, 0, 0) = (\bar{1}, 0, 0)$.] Since all these directions are completely equivalent, it is customary to indicate the entire set of equivalent directions by using pointed or curly brackets. It is also customary to leave out the commas. For example, the Si conduction band minima lie on the $\{100\}$ axes. For Ge, they lie along the $\{111\}$ directions.

The disadvantage of using this system is that it does not retain the simple one-to-one correspondence between the reciprocal-lattice points and the lattice planes in the direct lattice. To use the simple-cubic notation for X-ray crystallography requires the invention of an artifact called the "geometric factor." This factor eliminates the incorrect terms that were inserted by using the wrong lattice in the first place. Messy as this may sound, it is really rather easy to do in practice. Since it is the custom to use the simple-cubic lattice labels in discussing all cubic lattices, I shall follow that custom throughout the remainder of the book.

## BACK TO THE TOPOGRAPHS

A number of rather important features of Si and Ge can be observed in Figure 2.25. Each of the closed surfaces is an ellipsoid of revolution. What does this imply for the relationship between $m_y^*$ and $m_z^*$ along the [100] axis in Si? (Answer: Since the surface is an ellipsoid of *revolution*, $m_y^* = m_z^* \neq m_x^*$.) You can also state that the effective mass tensor is a constant in the vicinity of the minima. Since each of the ellipsoids is the same, only two different mass numbers are needed—the mass along the major axis and the mass perpendicular to it. These are customarily labeled $m_\parallel^*$ and $m_\perp^*$ respectively. If I had one electron in each of the minima, and applied a force in an arbitrary direction, what would the average electron effective mass appear to be? For Si, the answer is simple. I can break my force into three Cartesian components, and since each of the effective masses has the simple form appropriate to (2.27) along the same set of Cartesian axes, I get $F/m_\parallel^* = a_x$ for two electrons and $F/m_\perp^* = a_x$ for the other four. Accordingly,

$$(1/m^*)_{av} = (2/m_\perp^* + 1/m_\parallel^*)/3$$

For Ge, the same result is obtained but not quite so simply. First, note that the cubic symmetry assures you that anything that occurs for forces in the $x$-direction will occur identically for $y$- and $z$-directed forces. In other words, there is complete isotropy. This is a result true for all cubic crystals. Consider a force $F_x$ directed along the [100] axis. Examining Figure 2.25, you can see that, for $F_x$, any acceleration not along the $x$-axis for one minimum will cancel with the equal but opposite acceleration from another of the minima. Accordingly, you need only consider the [100] acceleration. It is also evident from the symmetry that the electrons in each of the minima contribute equally to acceleration in the [100] direction. (N.B. This is true only because the minima lie on the cube diagonals. It is not true for Si, for example, where the minima lie on the Cartesian axes.) That leaves you only the job of finding the $x$-component of the acceleration for any one of the minima. I choose the [111] minimum and proceed as follows. Using $\alpha_{ijk}$ to stand for a unit vector in [$ijk$] direction, the force in the [111] direction is $F_{111} = F_x(\alpha_{100} \cdot \alpha_{111})$. To find the other component of the force, I must find the direction normal to [111] in the same plane as [111] and [100]. That direction is given by

$$\alpha_{111} \times (\alpha_{100} \times \alpha_{111}) = [2\bar{1}\bar{1}]$$

Writing the accelerations along the major and minor axes of the ellipsoid is now easy. I have

$$a_{111} = F_{111}/m_\parallel^*, \qquad a_{2\bar{1}\bar{1}} = F_{2\bar{1}\bar{1}}/m_\perp^*$$

To find the acceleration in the [100] direction:

$$a_{100} = a_{111}(\boldsymbol{\alpha}_{111} \cdot \boldsymbol{\alpha}_{100}) + a_{2\bar{1}\bar{1}}(\boldsymbol{\alpha}_{2\bar{1}\bar{1}} \cdot \boldsymbol{\alpha}_{100})$$

$$= F_x \left[ (\boldsymbol{\alpha}_{111} \cdot \boldsymbol{\alpha}_{100})^2/m_{\parallel}^* + (\boldsymbol{\alpha}_{2\bar{1}1} \cdot \boldsymbol{\alpha}_{100})^2/m_{\perp}^* \right]$$

$$= F_x (1/m_{\parallel}^* + 2/m_{\perp}^*)/3$$

This is the same result that was obtained for Si, and it applies quite generally to cubic crystals with parabolic bands.

For the two-dimensional topograph of $E:k$, $g(E)$ was proportional to the area enclosed between an adjacent pair of contour lines. For the three-dimensional case, the first quadrant scales to $1/8$ of the volume between contour surfaces, and $g(E) = 1/4\pi^3 (\Delta V/\Delta E)$. $\hspace{2cm}$ (2.32′)

Finally, as the piece de resistance, take a look at Figure 2.26. This is the "Fermi surface" for copper and it is certainly as topographic a topograph as one would want to see. This particular surface represents the "level" to which the conduction band is filled in copper. In other words, it is like the surface of water partially filling a bucket; all the states below it are filled, all above are empty. It should be pretty obvious that copper does not have a constant effective mass tensor. It does have cubic symmetry, however, and is therefore isotropic.

As you might imagine, the topographs of metals that have overlapping bands could become quite involved. However, for semiconductors, we normally need data on either the very top or the very bottom of a band, and this is relatively easy to present. Furthermore, it is the nature of extrema to be pretty close to parabolic in shape. Thus, we are spared the rigors of doing calculations with an objet d'art like Figure 2.26.

## MEASUREMENT OF THE EFFECTIVE MASS PARAMETERS

There are only a few pieces of data that we need to know about the Bloch diagrams in order to work most useful problems. These comprise the effective masses of holes and electrons, the width of the forbidden energy region, and the differences in $k$ and $E$ between certain states in the same or adjacent bands. This section discusses the measurement of the effective mass parameters; the measurement of energy and momentum differences will be briefly presented at the end of the chapter. The standard method of measuring effective mass parameters employs a phenomenon known as *cyclotron resonance*. In your sophomore physics you may have obtained the fact that a moving charged particle in a uniform magnetic field circles around the magnetic-field direction with a frequency that is *independent* of the particle's velocity. The frequency is a function only of $B$, $q$, and $m$. If you drive the particle with an oscillating electric field that is tuned to the cyclotron resonance, the

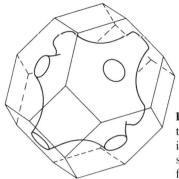

**Figure 2.26** $E:k$ topographic plot for copper. This is the "Fermi surface"—the surface to which the band is filled at absolute zero. The Fermi surface intersects the Brillouin zone near the center of the $\{111\}$ faces.

particle can absorb a great deal of energy from the driving field. The energy is stored in the particle's kinetic energy. With $B$ and $q$ known a priori, the measurement of the frequency of maximum absorption gives the mass quite directly.

Analytically, the argument proceeds as follows: Let the magnetic field point in the $z$-direction $[B = (0, 0, B_0)]$. Then, the Lorentz force equation $(F = qE + qvxB)$ leads to

$$F_x = qE_x + qv_y B_0 = m(dv_x/dt) \tag{2.33}$$

$$F_y = qE_y - qv_x B_0 = m(dv_y/dt) \tag{2.34}$$

$$F_z = qE_z \tag{2.35}$$

Let us consider first the simple case of $E = 0$. Then, differentiating (2.33) and substituting for $dv_y/dt$ from (2.34) leads to

$$m^2(d^2v_x/dt^2) = -q^2v_x B_0^2 \tag{2 35}$$

with the solutions $v_x = Ae^{j\omega t}$, $v_y = Ae^{j(\omega t + \pi/2)}$, and $\omega_0 = qB_0/m$. Thus, the particle circles around the $z$-axis at the frequency, $\omega_0$.

Now let's add an oscillating electric field in the $x$-direction. The only restraint we must impose to keep our solutions simple is that the oscillating magnctic field associated with the electric field be negligible compared to $B_0$. Equation 2.35 now becomes

$$qE_0 e^{j\omega t}/m = d^2v_x/dt^2 + (\omega_0^2 v_x) \tag{2.35'}$$

with the particular solution:

$$v_x = [qE_0/m(\omega_0^2 - \omega^2)](e^{j\omega t}) \tag{2.36}$$

From (2.36) it is evident that the solution "blows up" when the driving frequency equals the cyclotron resonance frequency. In practice, of course, the solution is limited to finite values by collisions with other particles, but if these collisions are minimized, the resonance can be very sharp indeed.

## CYCLOTRON RESONANCE WITH
## A TENSOR EFFECTIVE MASS

Equation 2.36 was derived for a real particle with an isotropic (scalar) effective mass. The solution with $m^*$'s appropriate to Figure 2.25 is similar, but it now contains terms for both $m_\perp^*$ and $m_\parallel^*$, as well as information on the orientation of the crystal in the magnetic field. Since each of the ellipsoids for a given energy has the same shape, we need solve the problem for only one ellipsoid and then "rotate" the notation to the others.

With reference to Figure 2.27, let the major axis of the ellipsoid lie on the $z$-axis and the magnetic field lie on the $x$-$z$ plane. Writing the Lorentz force equation in terms of its components now leads to

$$F_x = q\,(E_x + v_y B_z - v_z B_y) = m_\perp^*\,(dv_x/dt) \tag{2.37'}$$

$$F_y = q\,(E_y + v_z B_x - v_x B_z) = m_\perp^*\,(dv_y/dt) \tag{2.38'}$$

$$F_z = q\,(E_z + v_x B_y - v_y B_x) = m_\parallel^*\,(dv_z/dt) \tag{2.39'}$$

By my choice of axes, $B_y = 0$, $B_x = B_0 \sin\phi$, $B_z = B_0 \cos\phi$. In analogy with the solution to (2.35), define $\omega_\perp = qB_0/m_\perp^*$ and $\omega_\parallel = qB_0/m_\parallel^*$. With a sinusoidal electric field, you will find a resonance at each of the roots of the homogeneous equations (the equations without the electric-field forcing functions). Differentiation of (2.38') and substitution from (2.37') and (2.39') leads to

$$0 = d^2 v_y/dt^2 + v_y[\omega_\perp\,\omega_\parallel \sin^2\phi + \omega_\perp^2 \cos^2\phi] \tag{2.40}$$

with resonant frequency: $[\omega_\perp\,\omega_\parallel \sin^2\phi + \omega_\perp^2 \cos^2\phi]^{1/2} = \omega_{RES}$.

Thus, the cyclotron resonance frequency is a function of the magnetic field, the angle between the field and the major axis of the ellipsoid, and $m_\parallel^*$ and $m_\perp^*$.

The experiment that is usually performed to measure the effective mass must be done at very low temperatures ($\sim 4°K$) to achieve a reasonably sharp resonance. Otherwise, the thermal motion of the electrons and holes would lead to a great deal of scattering and its attendant damping of the cyclotron motion. Damping in this experiment has the same effect as in an LCR circuit; the effect of $R$ is to slightly shift and greatly broaden the resonance, until at critical damping, the resonance disappears altogether.

The economic requirements of low-temperature work demand that the total mass of all the objects to be cooled to liquid-helium temperature be kept as low as possible. To keep the parts small, it is customary to work at relatively high microwave frequencies ($f > 10^{10}$ Hz). The sample is placed in a microwave cavity in a known orientation. The cavity is immersed in liquid He, and the dewar containing the He, cavity, and sample is placed between the poles of an electromagnet. A microwave generator is carefully

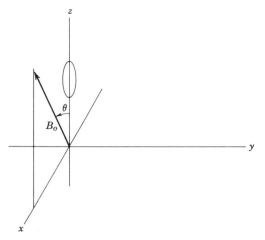

**Figure 2.27** The magnetic field and a constant-energy surface for the cyclotron-resonance calculation.

tuned to resonate with the cavity, and the relative amount of power reflected by the cavity is monitored. (It would be 100% if the cavity had no losses at all.) The magnet is turned on and slowly swept through a range that is typically 0.1 to 1 weber/m$^2$. The microwave power absorbed is recorded as a function of $B$, leading to a plot such as that shown in Figure 2.28.

You might ask how the labels for the various peaks were obtained. The answer has two parts. First, why three peaks for electrons? Note that the conduction band ellipsoids in Ge (Figure 2.25) lie along the {111} directions. For a field along an arbitrary direction, the solution to (2.40) has four values, one for each angle $\phi$ between a {111} direction and the magnetic field. However, Figure 2.28 was taken with $B$ lying in a (110) plane but not along the cube diagonal. Let us say that the plane is the (1$\bar{1}$0) plane; then the [111] and the [11$\bar{1}$] vectors lie in that plane and their angle $\phi$'s are different. The other two cube diagonals are [1$\bar{1}$1] and [$\bar{1}$11]. These lie in the (110) plane, and will make the same angle $\phi$ with $B$. (A unit vector in the direction of $B$ is $\frac{1}{4}(\sqrt{3/2}, -\sqrt{3/2}, 1)$. You can check the angles by taking the dot products of the appropriate lattice vectors with $B$.) There are thus three angles and three solutions to (2.40). If the magnet were rotated so that $B$ moved out of the (110) plane, the two solutions corresponding to [1$\bar{1}$1] and [$\bar{1}$11] would split apart. The other two peaks would also shift.

Interestingly enough, the hole peaks would not shift! This would quickly separate the holes from the electrons, but what does it imply about holes? (Answer: The angular isotropy implies that $m_{\parallel}^* = m_{\perp}^*$. In other words, the constant-energy surfaces are spheres. But a sphere yields only a single cyclo-

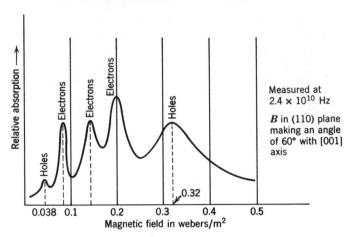

**Figure 2.28***a* Cyclotron-resonance curve for Ge. [From G. Dresselhaus, A. F. Kip, and C. Kittel, *Phys. Rev.* **98**, 368 (1955).]

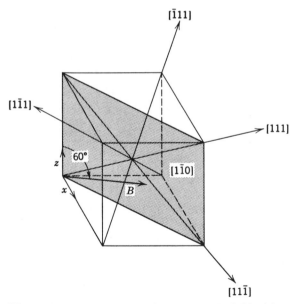

**Figure 2.28***b* Orientation of the magnetic field with respect to the cubic crystal axes.

tron resonance peak, so there must be two spheres of different radii for one energy. The symmetry of the crystal is reproduced in the constant energy surfaces, so the only way to have two spheres with the same energy is to have them centered on $k = 0$. This tells you that there are two very closely spaced bands, both of which are contributing to the conduction process with holes. These bands are called *valence* bands. [Note: The "heavy hole" band is actually not quite spherical—a squashed sphere being a more accurate description. Thus the uppermost resonance peak would actually shift by about 20% from 0 to 60%. The answer given in the problem is correct for the 60° direction.]


## A SHORT PROBLEM

Using the data in Figures 2.27 and 2.28, calculate the two effective masses for holes. A useful technique in solving problems with electrons is to work in units where the electron charge is 1 unit. This requires knowing the electron mass in eV ($E = mc^2$). The electron mass corresponds to $5.1 \times 10^5$ eV. Then, by using the speed of light in mks or cgs units, you may arrive at the units of your choice. For example, at $2.4 \times 10^{10}$ Hz, the field for cyclotron resonance for the free electron is found in two steps:

$$m_0 = E/c^2 = 5.1 \times 10^5/9 \times 10^{16} = 0.566 \times 10^{-11} \text{ eV sec}^2/\text{m}^2$$

$$B_0 = m_0\omega_{\text{RES}}/q = (0.566 \times 10^{-11})(2 \times 2.4 \times 10^{10})/1 = 0.852 \text{ webers/m}^2$$

(Note that a weber equals a volt-second.) The answers in terms of $m_0$ are $m_h^* = 0.375 \, m_0$ and $m_l^* = 0.044 \, m_0$. Had you worked the same problem for the electrons, you would have obtained $m_\parallel^* = 1.64 \, m_0$ and $m_\perp^* = 0.082 \, m_0$. Note that the major to minor axis ratio is 20:1. That is certainly a very skinny ellipse! Silicon's ratio is only 5.2:1.


## MEASURING THE WIDTH OF THE FORBIDDEN ENERGY BAND

Pause for a moment to see where you've gotten to and where you might go next. From the awesome complexity of crystalline solids and the wave characteristics of electrons, some lengthy but uncomplicated algebra has shown that you can use $F/m^* = a$ to describe the motion of the electron. Furthermore, you have found that an excellent description of the motion of electrons near the top of the band is the *hole*. And finally, you have examined the circumstances under which $m^*$ is constant, making $F/m^* = a$ easy to use. Then a few graphical aids were discussed, and the principal experimental technique for measuring $m^*$ was developed. What's left?

Well, if we had only completely filled and completely empty bands, all the previous work would have been pretty useless; none of the electrons could move. In semiconductor problems, one is normally dealing with *almost* full and *almost* empty bands, a situation where holes and electrons have constant $m^*$'s and the possibility of motion. However, even if you know how AN electron will behave, you could not predict the current unless you know how many electrons there were to behave that way. Our next step, therefore, should be to determine the number of electrons in the conduction band and the number of holes in the valence band(s).

There is only one way for electrons to get from the valence to the conduction band in a homogeneous semiconductor; they must be "kicked" across. In other words, they must acquire enough energy to get across the forbidden gap in a very short time. How short? The answer requires some rather sharp perception about the solutions to the wave equation within the forbidden bands. The solutions that we have obtained so far have all been for the time-independent wave equation, equation 1.7. What we are discussing here, however, are solutions to (1.9). Note well that $E$ is not a constant for an electron going from one band to another. However, from the method we first used to examine wave propagation in periodic structures, you can make an estimate of the *kicking* time. An unallowed solution is one that is attenuated as it propagates through the lattice. To move through the unallowed region with no significant attenuation, the transition must be made in a time short compared to the attenuation rate. Certainly a minimum guess would allow us at least one or two lattice constants. A free electron with an eV of kinetic energy will have a velocity of the order of $5 \times 10^7$ cm/sec. A lattice constant is typically of the order of $5 \times 10^{-8}$ cm. Thus, for each lattice constant you allow, you get $\sim 10^{-15}$ seconds to kick in.

There are two conventional ways of exciting electrons across the forbidden gap. In one, you "boil" them (discussed at length in the next chapter) and in the other you bang into them with energetic particles from the outside. Both methods provide data about $E:k$ plots; the energetic-particle technique is more obvious so I will present that first. Thermal techniques are saved until Chapter 3, where the background for them is developed.

The widths of the forbidden regions in all insulators range from 0 to 8 or 9 eV. To do careful probing of the $E:k$ plot, we want a particle that can penetrate into the crystal, interact directly with the electrons, and be readily available in the 0–9 eV range. Furthermore, we want a particle that is easy to measure. The obvious one to use is the photon. With light, it is easy to measure the wavelength by the techniques discussed in these first chapters. What is the relationship between the wavelength and the momentum? (Answer: De Broglie's hypothesis states that $p = h/\lambda$.) What is the relationship between the frequency and the wavelength in air or vacuum? (Answer:

From the definition of phase velocity, $\omega = 2\pi c/\lambda$.) Since $c$ is a well-known constant, we get both the energy ($E = \hbar\omega = hc/\lambda$) and the momentum from a single measurement.

It is important to get some idea of the relationship between the photon's momentum and the crystal momentum of a typical electron. If the lattice spacing in Si is 5.42 Å, what is the crystal momentum for an electron at the Brillouin zone edge along the (100) axis? [Answer: $p = \hbar k = (0.659 \times \times 10^{-15}) \times (0.58 \times 10^8) = 3.82 \times 10^{-8}$ eV sec/cm.] The width of the forbidden zone in Si is 1.07 eV. What is the momentum of a 1.07 eV photon? [Answer: $p = h/\lambda$, $E = hc/\lambda$. Thus, $p = E/c$. For a 1.07 eV photon, that gives $p = 1.07/3 \times 10^{10} = 3.34 \times 10^{-11}$ eV sec/cm.] You can see that a photon carries an insignificant amount of momentum when compared to the electron in a crystal.

If a photon is going to be absorbed, the change it induces must satisfy both the conservation of energy and momentum. If only the photon and an electron are involved, you have just been shown above that the conservation of momentum requires that the electron not change momentum. In other words, to about 0.1 % accuracy, $k_{\text{initial}} = k_{\text{final}}$. Conservation of energy says that $E_{\text{initial}} + \hbar\omega = E_{\text{final}}$. Such a transition is called *direct*.

Is it possible to have a so-called *indirect* transition—that is, one in which $k_i \neq k_f$? Yes, but only if three or more particles interact to satisfy conservation of momentum. What we need are particles that can carry a modicum of momentum, and only two fit the bill. These are electrons and phonons. It is found experimentally that the electron-electron-photon interaction (called an Auger [o'zhay] process) does not take place at these energies. The photon-phonon-electron process, on the other hand, is extremely important.

The *phonon* is the particle associated with lattice vibrations. It fulfills all our criteria for being a particle as well as all the criteria for being a wave. Although you should have no trouble "seeing" that vibrations in a crystal should be wavelike, you probably find *particles of sound* a bit hard to swallow. However, if you accepted particles of light any more easily, it was only because you have been brainwashed. If you perform particle experiments on sound waves in crystals, you would get satisfyingly positive results. That is —and must be—our only criterion for accepting their particle character.

Since de Broglie's and Planck's laws must hold for phonons, what sort of phonon energy would be required to carry a momentum of $7 \times 10^{-8}$ eV sec/cm if the velocity of sound is about 3000 m/sec? [Answer: Once again, $p = E/v_{\text{phase}}$. Thus, $E = (7 \times 10^{-8})(3 \times 10^5) = .021$ eV.] Note that it takes only about 2 % of the photon's energy to create the phonon to carry away all of momentum required to go from the edge to the center of the Brillouin zone in Si.

## OPTICAL ABSORPTION IN Ge WITH DIRECT AND INDIRECT TRANSITIONS

Figure 2.29 shows five bands in Si and Ge. The shapes were derived from a mixture of theoretical analysis and experimental data. Each of the bands is plotted in the same two crystalline directions. The dotted line indicates the uppermost full state at $0°K$. Since Ge contains 32 electrons, how many bands are there below the lowest one shown? (Answer: The three filled bands shown contain 6 electrons per atom. Thus there are 13 bands below the last shown.) The upper two filled bands contain the four valence electrons and thus bear the name *valence bands*. The empty bands above are called conduction bands, the first one being called *the* conduction band.

With reference to the plot for Ge, what is the minimum energy required to excite an electron from the valence to the conduction band? (Answer: 0.66 eV.) Is the transition direct? (Answer: No.) What is the minimum energy for a direct transition? (Answer: 0.8 eV.)

Now consider what the chances are for a photon to get through a volume containing $N$ possible transitions corresponding to that photon's energy. Say the probability is $P(N)$. If the transitions are independent, what is the ratio $P(N)/P(2N)$? (Answer: No, not 1/2, but $1/P(N)$.) The probability that a photon will get through $N$ states is $P(N)$. The probability that it will do it twice in two tries is $[P(N)]^2$. But going through two sets of $N$ states sequentially is the same as going through $2N$ states. Therefore, $P(2N) = = [P(N)]^2$). Such a relationship means that you can write $P(N) = e^{-\gamma N}$. If the photons are propagating through a crystal, $N$ is proportional to the crystal length and you get $I(x) = I_0 e^{-\alpha x}$ for the law relating the intensity to the distance the light has traveled through the medium. $\alpha$ is called the attenuation factor. Such quantities as the thickness of a sample and the intensity of light are easily measured. With some care to subtract the attenuation due to surface reflections, a direct measurement of $\alpha$ versus $E$ is reasonably easy. The result of such a measurement on Ge is shown in Figure 2.30. Don't be misled by the logarithmic plot; the change in absorption with a change in photon energy is quite dramatic. For example, the $300°K$ plot is drawn on a linear scale in Figure 2.31.

Let us take a moment to see how you could discriminate between the direct and indirect transitions on such plots. Consider the direct transition first. The least energy that will cause a direct transition is called the *direct gap*, $E_{go}$. Consider the simplest band structure you could have, namely two parabolic bands with spherical energy surfaces centered at $k = 0$. (See Figure 2.32.) Find the locus of all values of $k$ that define states separated by $E_p = \hbar\omega$. [Solution: Since both bands are parabolic and the effective mass is independent of direction, the conduction band is described by

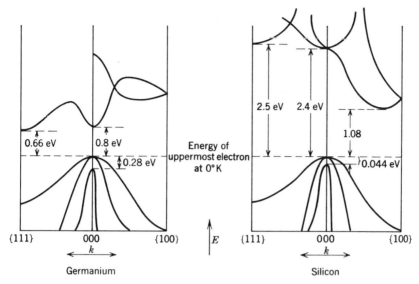

**Figure 2.29** Energy versus wave number along the Cartesian axes and the cube diagonals. Gap energies are for room temperature. (After H. P. R. Frederikse, 9-47 in *The American Institute of Physics Handbook*, second edition, McGraw-Hill, 1963.)

$$E_1 - E_{c0} = ak^2 \tag{2.41}$$

where $E_{c0}$ is the minimum in the conduction band at $k = 0$. Similarly, for the valence band, you have

$$E_{v0} - E_2 = bk^2 \tag{2.42}$$

where $E_{v0}$ is the maximum in the valence band at $k = 0$. To have a direct transition, $E_1$ and $E_2$ must have the same $k$. The problem requires that $E_1 - E_2 = E_p$. Adding (2.41) to (2.42) yields

$$E_p - E_{g0} = (a + b) k^2 \tag{2.43}$$

Equation 2.43 is the equation of a sphere. Thus, the locus required is a sphere of radius $\sqrt{(E_p - E_{g0})/(a + b)}$.] The next problem is: given that the attenuation factor, $\alpha$, is linearly proportional to the number of states, you want to know how many pairs of states interact with photons between $E_p$ and $E_p + dE_p$. [Answer: The number of pairs of states within the sphere found above is $V/\pi^3$ pairs/cm$^3$ of crystal (see equation 2.15). That is, $N = 4k^3/3\pi^2$.] Putting this into (2.43) yields

$$E_p - E_{g0} = (a + b)(3\pi^2/4)^{2/3}N^{2/3} \tag{2.44}$$

Differentiating (2.44) gives

$$dN = \text{(constant)} (E_p - E_{g0})^{1/2}dE \tag{2.45}$$

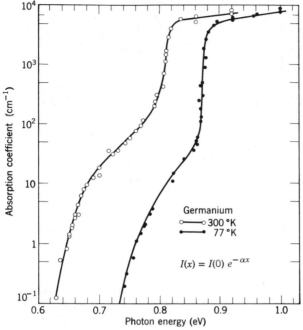

**Figure 2.30** Attenuation factor (in nepers/cm) versus photon energy (eV) for germanium at two temperatures. [After W. C. Dash and R. Newman, *Phys. Rev.* **99**, 1151 (1955).]

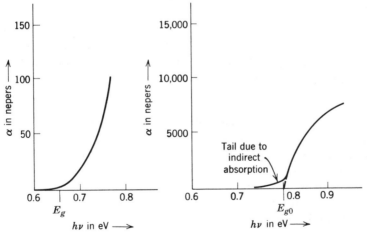

**Figure 2.31** Attenuation factor (nepers/cm) versus photon energy for Ge (on linear scales). Note that for the indirect transition, $\alpha$ is proportional to $(E_p - E_g - E_{phon})^2$, while for the direct transition $\alpha$ is proportional to $(E_p - E_{g0})^{1/2}$. Dash et al., loc. cit. in previous figure.)

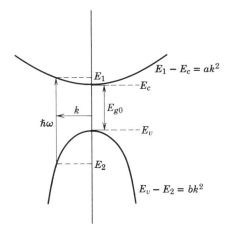

**Figure 2.32** Simplified $E:k$ plot for the calculation of $dN/dE_p$ for direct transitions.

Equation 2.45 is the relationship that we want. It tells us that for parabolic bands, the absorption constant for direct transitions goes as $(E_p - E_{g0})^{1/2}$. If you examine Figures 2.30 and 2.31, you will find that the behavior of the measured absorption constant for photons with $E > E_{g0}$ is exactly as predicted.

In the derivation of the half-power law for $N$ versus $E$, no assumptions were made. But in stating that $\alpha$ is simply proportional to $N$, I have assumed that all direct transitions are equivalent. This should certainly be true over the narrow interval of energy that we are considering, but we have no information from what we have done here that would permit me to extrapolate the half-power law over any really substantial range. Discussion of the dependence of the absoprtion coefficient on the energy difference between states and on the character of the states themselves requires a nontrivial extension of the wave mechanics and field theory that we have been using so far. Since the derivation employed yields the experimentally observed law, we should expect that a quantum-mechanically complete derivation would only support our results. It might well add something that we did not anticipate, but in the interest of stopping somewhere, let us forego that pleasure. (A reasonably thorough introduction to the interaction of light with atomic states can be found in Schiff, *Quantum Mechanics*, McGraw-Hill, 1949.)

## THE ENERGY DEPENDENCE OF ABSORPTION FOR INDIRECT TRANSITIONS

Since our object was to determine the energy gap between bands and whether the transition across the gap is direct or indirect, we must continue the above analysis for the indirect case. You might wonder why you

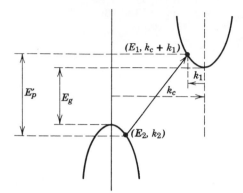

**Figure 2.33** Simplified $E:k$ plot for the calculation of $\alpha$ versus $E_p$ for indirect transitions.

should care whether the transition is direct or not. The answer is that direct materials (materials whose absorption edge begins with direct transitions) make excellent lasers but cannot be used to make conventional bipolar transistors. Conversely, an indirect material can be used to make a bipolar transistor but cannot be used to make a laser. Why will become clear later in the text, but at least you now have a reason why you might want this information.

I am going to do the same thing with the indirect transition that I did with the direct—count the number of possible transitions and assume that the absorption is proportional to the number of transitions—but the arithmetic will be somewhat different. Figure 2.33 defines the various $k$'s and $E$'s that I will use. First, write down the laws of conservation of energy and momentum for the indirect transition:

$$E_p - E_{\text{phon}} = E_1 - E_2 \equiv E'_p \tag{2.46}$$

$$k_{\text{phon}} = k_c + k_1 - k_2 \tag{2.47}$$

where $k_c$ is the momentum difference between the extrema of the valence and conduction bands. (The valence band maximum is assumed to lie at $k = 0$.)

Now if we are dealing with transitions close to the absorption edge and $k_c$ is reasonably large, (2.47) reduces to

$$k_{\text{phon}} \simeq k_c \tag{2.47'}$$

and $E_p$ differs from $E'_p$ by a small constant. We are thus concerned with counting the number of states that satisfy $E_1 - E_2 = E'_p$.

Equations 2.41 and 2.42 now become

$$E_1 - E_c = ak_1^2 \tag{2.48}$$

$$E_v - E_2 = bk_2^2 \tag{2.49}$$

(Notice the definition of $k_1$ in Figure 2.33.) Adding the last two equations gives us the counterpart of (2.43):

$$E'_p - E_g = ak_1^2 + bk_2^2 \tag{2.50}$$

For a particular photon energy and particular $k_2$, equation 2.50 says that the possible transitions lie on a sphere of radius

$$k_1 = [\,(E'_p - E_g)/a - (b/a)k_2^2\,]^{1/2}$$

The number of states that intersect the surface of the sphere is proportional to the area of the sphere. Thus:

$$N(k_1, k_2) = (\text{const.})k_1^2$$

The number of valence band states with the same $|k_2|$ is also proportional to the area of a sphere of radius $k_2$. Since each valence band state is connected to $N(k_1, k_2)$ states in the conduction band, the number of transitions with lattice vector length $k_2$ for a photon of energy $E_p$ is given by

$$N(k_2) = (\text{const.})k_2^2 N(k_1, k_2) = (\text{const.})[\,(E'_p - E_g/a)\,k_2^2 - b/a(k_2^4)\,] \tag{2.51}$$

Equation 2.50 states that the allowed values of $k_2^2$ run from 0 to $k_{2\,\text{max}}^2 = = (E'_p - E_g)/b$. To get the total number of transitions for a given photon, we must integrate (2.51) over the allowed range of $k_2$. That is:

$$N(E_p) = 1/k_{2\text{max}} \int_0^{k_{2\text{max}}} N(k_2)\,dk_2$$

$$= (\text{const.})/k_{2\text{max}}\,[\,(E'_p - E_g/a)\,(k_{2\text{max}}^3/3) - (b/5a)k_{2\text{max}}^5\,] \tag{2.52}$$

Substituting the value of $k_{2\text{max}}$ into (2.52) yields

$$N(E_p) = (\text{const.})[\,(E'_p - E_g)^2/ab\,] \tag{2.53}$$

From (2.53) we have the result we were seeking—that $\alpha$ is proportional to the square of the photon energy for indirect transitions.

Once again, let me stress that I have assumed all the transitions are equivalent. The quantum-mechanical evaluation of the absorption coefficient would not only corroborate this result but would also add information concerning the constants. For example, it is evident by inspection of Figures 2.30 and 2.31 that the coefficient of the square law (indirect) term is several orders of magnitude smaller than the coefficient of the half-power (direct) term. Although nothing I have done here would predict the difference in coefficients, the large difference makes sense when you consider that the direct $\alpha$ is a measure of the probability of a photon being absorbed while the indirect $\alpha$ is a measure of the *joint* probability of a photon being absorbed while a phonon is either absorbed or emitted. Certainly the latter would be less than the former.

It is interesting to note that even with the difference between the coefficients, the square-law term must eventually overwhelm the half-power dependence. Thus, if the direct and indirect terms are sufficiently widely separated, the optical-attenuation experiment would not show two distinct absorption edges. This is the case, for example, in silicon, where the difference is 1.32 eV.

## TWO SHORT EXERCISES

**1.** To get some feeling for what makes a big or small $\alpha$, let us say that you wanted to make an optical filter that had an attenuation of an order of magnitude (about equivalent to a typical pair of sunglasses). What thickness of Ge would you employ for an 0.65 eV photon? What thickness for an 0.9 eV photon? (You see photons with energy between 1.75 and 3.1 eV.) (Answer: From Figure 2.31, I need a thickness of 2.3 cm for the first case and $2.44 \times 10^{-4}$ cm for the second. The last would be considered a "thick film." It is rather interesting to compare this last figure with aluminum, which would require a film only 1/20th as thick to give the same attenuation. [Note that in both cases I have ignored the surface reflectivity, which can easily account for an order of magnitude attenuation in itself.] Even so, you should expect polished Ge to look quite metallic at the energies you can see.)

**2.** Using the data in Figure 2.30, estimate the direct and indirect energy gaps for Ge at 77°K. (Answer: $E_g(77°K) = 0.75$ eV and $E_{g0} = 0.87$ eV.)

## CHAPTER SUMMARY

This chapter has been devoted to the behavior of electrons and photons in perfectly periodic structures. As was the case in the first chapter, we only skimmed the cream, there is a very long drink of milk awaiting those who want to pursue these ideas to their current limits. But limited as we were, the list of rules and techniques that you have been introduced to is rather impressive. You have learned the rudiments of measuring crystal-lattice parameters using X rays, including how to unscramble some of the simpler Laue and Debeye-Scherrer spectra by using the reciprocal lattice. You have studied some of the implications of periodicity on the electron-wave equation, first showing quite generally that the solutions are Bloch waves and then looking at the details of the solution for the most simplified of crystal models—the one-dimensional Kronig-Penney model.

For both X rays and electrons, the solutions were presented in the form of $E:k$ plots to emphasize the periodic character of the solutions. The reduced Bloch diagram was introduced as one very useful way to select a unique set of solutions. On the reduced Bloch diagram, the concept of energy bands is clearly illustrated. With the introduction of the Pauli exclusion principle, it was possible to assess which materials might be insulators. It is important to remember that conductivity requires two properties of a material; first, the material must have electronic states that extend a large distance through the material (all ideal crystal states are of this variety) and second, the states must be closely spaced and only partially filled to permit the electron to change its state under the influence of small electric fields.

We then looked more closely at the motion of electron waves under the influence of an electric field, deriving the extremely important result that the motion of an electron under the combined influence of the electric field and the crystal field obeyed a law very much like Newton's second law: $F/m^* = a$. $m^*$ was named the effective mass, since it acted like an inertial mass, but it was really related to wave propagation in the lattice. A large part of the last half of the chapter was spent exploring the vagaries of $m^*$ and learning how to measure its value(s). Of greatest importance was the discovery that near the maxima of a band, $m^*$ was negative. This led to the invention of the "hole" to describe the behavior of electrons in almost filled bands. You also learned how to average the effective mass where several valleys (minima) contributed, leading to the result for cubic lattices that the average effective mass was isotropic.

The final portion of the chapter was spent on direct and indirect optical transitions and on using them to measure $E_g$ and $E_{g0}$. This was the first process that we have studied that permits us to get an electron from one band to the next. Pervading the whole of the last half of the chapter was the use of the reciprocal lattice, the Brillouin zone, and various plotting techniques to present, predict, and interpret information about the electron's behavior.

This is all the material we will need about the behavior of electrons and photons in perfect crystals. What remains before we can examine device theory is an introduction to thermal influences on the distribution of electrons among the bands (next chapter) and a study of the motion of electrons in real, imperfect crystals (Chapter 4).

### Problems

1. It comes as a surprise to most people to find that plane gratings can have focusing properties. However, when you consider that "focusing" means causing a phase front that originated at some point to converge at another point, it is not unreasonable that the phase selective scattering of a regular grating might lead to focusing for monochromatic illumination. (This property proves most useful in the design of spectrometers and in

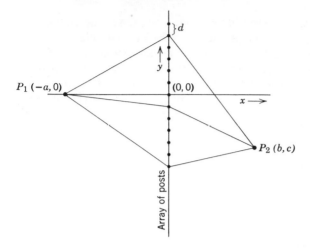

holography.) Consider the "array of posts" example shown below. Show that for given values of $a$, $d$, and $\lambda$, the waves from $P_1$ $(- a, 0)$ are focused at $P_2(b, c)$ where $b = \pm a$ and, for the $m$'th order, $c = \pm m\lambda/d^2$. (*Hint:* Take full advantage of the fact that in any reasonable system, $d \ll a$.)

2. **(a)** Many metals and semiconductors crystallize in a hexagonal close-packed lattice (hcp) whose Bravais lattice is a simple hexagon. (As in the diamond lattice, for hcp there are two inequivalent sites for each "point" in the Bravais lattice.) If the spacing between nearest neighbors in the planes of the hexagons is $a$ and the spacing between the planes is $c$, find an appropriate set of Cartesian basis vectors for the direct and reciprocal lattices.

*Answer:* One such set is

$$\boldsymbol{a} = a\,(1, 0, 0) \qquad\qquad \boldsymbol{A} = \frac{1}{a}\left(1, \frac{1}{\sqrt{3}}, 0\right)$$

$$\boldsymbol{b} = a\left(-\frac{1}{2}, \frac{\sqrt{3}}{2}, 0\right) \qquad \boldsymbol{B} = \frac{1}{a}\left(0, \frac{2}{\sqrt{3}}, 0\right)$$

$$\boldsymbol{c} = c\,(0, 0, 1) \qquad\qquad \boldsymbol{C} = \frac{1}{c}\,(0, 0, 1)$$

**(b)** What is the shape of the reciprocal lattice?
*Answer:* Hexagonal.

**(c)** The element beryllium crystallizes in the hcp form with lattice constants $a = 2.27$ Å and $c = 3.59$ Å. If one had a powder of Be in a Debye-Scherrer camera and the illuminating wavelength were 1 Å, what would be the first two scattering angles that would be observed on the film?

*Answer:* 0.14 and 0.257 radians.

(*d*) What is the longest wavelength that would still give at least one line in the Debye pattern?

*Answer:* About 7.18 Å, which means that the electron target in the X-ray tube must have atoms at least as heavy as Si to have its characteristic wavelength give a line in this experiment. The 1 Å line would require a target at least as massive as Se. This dependence on atomic number reflects the fact that the most energetic line is at the binding energy of the innermost electron. Since the innermost electron sees the full nuclear charge, the binding energy of this electron goes approximately as the square of $Z$.

3. Suppose you had a hexagonal crystal such as the one described in problem 2. Let the minima in the conduction band lie on the axes of the reciprocal-lattice hexagon, with the major axes of the constant energy surfaces parallel to the hexagon's axes and with $m_\| = 3m_\perp = m_0$. Answer the following questions about the conduction due to electrons in the bottom of the conduction band.

(*a*) Will the conductivity be isotropic?
(No. The conductivity will clearly be different in the hexagonal plane from what it will be in the $c$-axis direction.)

(*b*) What is the average effective mass in the $c$-axis direction?

*Answer:* Each minimum contributes equally to the acceleration, so the effective mass for each minimum is the average effective mass: $m_c = = m_\perp = m_0/3$.

(*c*) Find the average effective mass as a function of the angle of applied force in the hexagonal plane.

*Answer:* By finding the average acceleration for forces applied in the $x$ and $y$ directions, one obtains the nonobvious result that the acceleration in the hexagonal plane is independent of the direction of the force in the plane. The effective mass in the plane averages to $(1/m)_{av} = \frac{1}{2}(1/m_\| + 1/m_\perp) = (2/m_0)$.

(*d*) Would the conductivity be greater in the $c$-direction or in the plane of the hexagon?

*Answer:* From the answers to (*b*) and (*c*) it is clear that the effective mass in the plane is 1.5 times the effective mass perpendicular to the plane. Thus, all other things being equal (which they may well not be), one would expect that the conductivity would be 1.5 times as large along the c-axis as perpendicular to it.

4. A sample of germanium is placed in a cyclotron resonance apparatus with $B$ in the [100] direction. The R-F generator and the cavity are tuned to $2 \times 10^{10}$ Hz. Draw a curve of relative absorption versus $B$, giving the $B$ values of the peaks.

*Answer:* In the [100] direction, the electrons contribute only 1 peak at

$$B = \frac{\omega}{q} \left[ \frac{3m_\parallel m_\perp}{m_\parallel + 2m_\perp} \right]^{1/2} = 0.105 \text{ webers/m}^2.$$

For the holes, the resonances come out to be—ignoring the nonsphericity —0.032 and 0.267 webers/m$^2$.

5. Stress applied to a crystal shifts the energy bands a small but significant amount. If a silicon crystal is stressed along a [100] axis, the stress raises the energy of the two conduction-band minima in that direction and slightly lowers the energy of the other four equivalent minima. The valence-band maximum is distorted so that the constant-energy surfaces become very slightly oblate, the short axis in the direction of stress. The electrons redistribute themselves in each case, filling the lower-energy states in preference to the higher ones. (Under the proper circumstances, this shift in the bands produces a very substantial change in the conductivity of a Si resistor. This large *piezoresistance* is the effect employed in silicon strain gauges.) Support your answers to the following questions:

(*a*) All other things being equal, is the effect bigger in an *n*- or *p*-type sample?

*Answer:* *n*-type. In the *p*-type sample, the small band distortion produces almost no change in effective mass. For an *n*-type sample, on the other hand, the shift from the minimum along the stress axis to the minima along the other {100} axes reduces the fraction of electrons with heavier effective mass ($0.97 \, m_0$ versus $0.19 \, m_0$).

(*b*) For an *n*-type sample, is the conductivity of a sample strained in the [100] direction a scalar quantity?

*Answer:* No. The resistance along the strain axis has been reduced, while along the other [100] directions it has been increased. In an arbitrary direction, the electric field and the current will not align and the conductivity must be represented by a tensor.

(*c*) Let the stress applied in the [100] direction of an *n*-type silicon crystal be sufficient to shift 1% of the electrons from the minima aligned with the stress to the other four minima. Find the percentage change of the conductance in each of the [100] directions.

*Answer:* With the effective mass ratio of 5.11, the 1% shift increases the conductance along the axis of stress by 4.5% and decreases the conductivity normal to the axis of stress 0.28%. Note that for the {100} directions normal to the stress, half of the displaced electrons contribute a heavy mass and the other half have the same light mass they would have had in the minimum they came from.

## CHAPTER 3

# DISTRIBUTION FUNCTIONS AND EQUILIBRIUM STATISTICS

## INTRODUCTION

The last two chapters have been devoted to obtaining and exploring the $E:k$ plot. That worthy object is really a catalog of all the solutions to the perfectly periodic crystal problem. We have been asking the question: Given a particular solution to the wave equation, what behavior will we observe? That is a necessary question to ask and answer, but it is certainly rather remote from the question: Given a piece of silicon lying on the table, what behavior will we observe? This question really asks which of the solutions from our extensive catalog will we observe, given the external stimuli such as heat, light, and electromotive forces. Eventually, the answer to this question must be that these parts will act like a diode and those like a transistor. Actually, the answer that we will obtain will be probabilistic; the most probable solution will be found as well as the range of deviations that would not be unlikely to occur.

An illustrative comparison can be made between what we are about to do and useful statistics in the game of craps. The "eigenstates" for the pair of dice are the 36 combinations of 1 to 6 for each of the dice. To get an analog of the "$E:k$" plot, note that as far as the game is concerned, all states with the same sum are equivalent. The sum is similar to $E$ in this respect. We can use the value on one of the dice as the "$k$" variable. Figure 3.1 is the result. It represents, in a rather trivial fashion, the analog of the information that we obtained with much effort in Chapters 1 and 2. The question that we now need to ask is: What value of "$E$" are we most likely to observe in a given throw? This is really a question that asks for the most frequent result in a series of equivalent experiments. The answer to the question can be obtained by a large number of rolls of a pair of dice, or equally well, by one roll of a

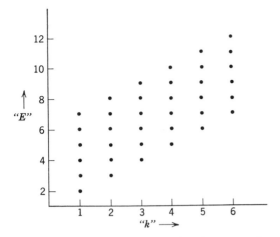

**Figure 3.1** A pseudo $E:k$ plot for the game of craps. "$E$" is sum of the dice values and "$k$" is the value of one of the dice. Each point represents a different "time-invariant solution" or "state."

large number of pairs. Although you would probably use one pair if you were actually performing this experiment, in crystals we will be asking for the probability that an electron has a particular energy, with $10^{22}$ or so electrons in one state or another at all times. Thus, for the crystal, an entire distribution (although not necessarily the most likely one) is always presented. Let us assume, for a more accurate analogy, that we have before us $10^9$ pairs of dice, fairly rolled and ready to be counted.

Which value of "$E$" appears most often? (Answer: 7. This answer is obtained either "from knowledge of the game" or by observing in Figure 3.1 that there are more states with value 7 than with any other particular value. Since all states have equal probability of coming up, the value of "$E$" representing the most states will be observed most often.) It is extremely important—in fact, this is the crux of the chapter—to recognize that the states are all equally probable but that of specifying "$E$" determines a *group* of states, the group size varying with the "$E$" chosen. You must give up attributing anthropomorphic qualities to either dice or electrons. Dice do not *prefer* to come up 7, nor do electrons *prefer* to be in their ground state. All possible individual arrangements are equally probable. If six arrangements happen to have a common property (such as the sum being 7) while another state is the only one with some other value of that property (say, 2), then we will observe 7 six times as many times as 2, *on the average*.

The phrase *on the average* is loaded. Saying that the average number of 7's is 6 in 36 throws does not imply that a given 36 rolls will contain exactly six 7's. Far from it. By some arithmetic that we are going into below, it is easy to determine that the number of times five 7's would come up in 36 throws is only slightly less than the number of times six 7's would come up. The ratio is 30/31. Even no 7's in 36 throws has reasonable frequency, about 1% of the frequency of six 7's.

You should now be prepared to look at the two questions we will want to answer about a crystal. We first want to know what group of allowed electron distributions is most likely to occur. The answer is the same as in the craps example: the most likely group is the one with the greatest number of equivalent distributions. The first part of this chapter is devoted to finding this most probable group. The second question to be answered is: How much deviation from the most probable distribution is reasonable to expect? This second question introduces the subject of noise. It is taken up briefly in the latter half of the chapter.

## THERMAL EQUILIBRIUM

The word *equilibrium* suggests balance and at least a local steady state. *Thermal equilibrium* carries these concepts to their limit. Between two bodies or systems in thermal equilibrium there can be *no* net transfer of any sort. This statement is frequently called the *law of detailed balancing*, and it can be shown to yield (or alternately, can be derived from) the Second Law of Thermodynamics. A few examples are in order. If I mix two chemical constituents that react to form a third, such that

$$A + B \Rightarrow C \tag{3.1}$$

thermal equilibrium will not be achieved until the proportions of $A$, $B$, and $C$ are such that the formation of $C$ is everywhere balanced by an equal rate of dissociation of $C$. What if $C$ does not dissociate at the given temperature? (Answer: Then the reaction must go to completion; that is, either $A$ or $B$ or both must be exhausted, depending on the proportions of each available and the demands of equation 3.1.)

As a second example, consider two hot bodies that come to the same temperature by transferring energy radiatively. When they come to equilibrium, the net energy flow must cease. More than that, the law of *detailed* balancing says that the net transfer stops for each point in both bodies and for each frequency. By detailed balancing we mean that absorption of energy in the red cannot be balanced by radiation in the green, nor can absorption at point $A$ be balanced by emission from point $B$. If detailed balancing did not hold, it would be easy to violate the Second Law of Thermodynamics by interfering with one of the flows but not the other. For example, if the emission at point $B$ could compensate for the absorption at point $A$, all I would have to do is place a mirror that reflected the emission at $B$ onto $A$. Then, the body that contained $A$ and $B$ would start absorbing net energy from the other body, and heat would flow from cold to hot with no work being done. An equivalent experiment using dichroic mirrors that reflect one color and pass another shows that the Second Law of Thermodynamics could be violated if the red absorption were compensated by the green emission.

An interesting result was derived from these rules for radiation by your old friend Kirchhoff, of current and voltage fame. (He did his bit for several sciences. Along with voltage, current, and radiation laws, you can also blame him for contributing a few pieces to that elegant object, the periodic table.) He noted that if two bodies of different absorptivity were to be in equilibrium according to the law of detailed balancing, the ratio of the energy absorbed to the energy emitted by each body must be unity, *independent of any property of the material*. The amount of energy absorbed is proportional to the radiation falling on the surface $I(\lambda, T)\,d\lambda$, the constant of proportionality being called the absorptivity, $a(\lambda, T)$. Thus, we have by definition:

$$\text{energy absorbed per second} = I(\lambda, T)\,a(\lambda, T)\,d\lambda \qquad (3.2)$$

The amount emitted is obviously a function of the material and the temperature and is described by a coefficient called the emissivity, $e(\lambda, T)$. Thus:

$$\text{energy emitted per second} = e(\lambda, T)\,d\lambda \qquad (3.3)$$

According to Kirchhoff's law the ratio of (3.3) to (3.2) is unity; thus

$$e(\lambda, T)/a(\lambda, T) = I(\lambda, T) \qquad (3.4)$$

A *blackbody* is one that absorbs all the radiation incident upon it (i.e., $a = 1$). Since $e/a$ is independent of the material, $I(\lambda, T)$ must be a *universal* function of $T$ and $\lambda$. Furthermore, the universal function is equal to the emissivity of a blackbody and must also be the radiative flux in a closed system at thermal equilibrium. If you peek into a carefully equilibrated oven, the radiation coming out of the tiny hole will be that of a blackbody at the oven temperature *regardless of the materials from which the oven is made.* (The "blackness" is a measure of absorptivity in the intuitive sense. If you shine light in the small hole in the oven, very little of that light will get back out of the hole. Hence, the hole absorbs the light incident upon it.) The furnace with a hole in it is a very direct way of making a blackbody source, and it is the primary standard for radiative measurements. Later on in this chapter, you will derive $I(\lambda, T)$ from impressively fundamental considerations. Meanwhile, you have Kirchhoff's law that good absorbers are good radiators and blackbodies are the best of the lot. These results depend on nothing but the Second Law of Thermodynamics for this derivation.

It is important to notice one very distinct limitation on the law of detailed balancing: it applies only when the system is in equilibrium. The process of coming to equilibrium can be so slow that for all practical purposes, thermal equilibrium is never achieved. The earth, for example, reradiates only a miniscule fraction of the energy it receives from the sun. As you have undoubtedly noticed the earth is not in equilibrium with the sun, although it is not entirely inappropriate to use the words *steady state* to describe the flow of energy from the sun to the earth. Since the earth has been whirling around the sun for billions of years without achieving thermal equilibrium, it should not surprise you to find that most systems are not in equilibrium. In Chapter 4, we will consider a variety of steady-state systems, systems in which the flow of particles or energy is either constant or perfectly periodic. In this chapter, we deal only with thermal equilibrium, defined as the limit where all net flows cease.

Thermal equilibrium is a very uninteresting state. It is static, endless, and useless. Our interest in it derives from the fact that systems near thermal equilibrium tend to come to thermal equilibrium in predictable ways. The predictable behavior of systems not quite in equilibrium allows us to construct useful devices.

## "THERMAL EQUILIBRIUM" IN A SHUFFLED DECK OF CARDS

Consider a deck of cards as they come from the manufacturer. They are stacked sequentially by rank and suit. Follow the "flow" of aces within the deck as it is reordered by quasi-random shuffling. The four aces are on top to begin with, so the only possible motion is down. If the shuffle is of the usual variety—cut the deck in half and interlace the two halves—the aces would still lie near the top of the deck. After a number of shuffles, there is a reasonably good chance that one or two of the aces have gone beyond the 1/4 point and stand an equal chance of ending in the upper or lower halves of the deck on the next shuffle. Finally, after many shuffles, the probability of finding an ace in any portion of the deck is independent of where that portion was drawn from. When this state has been achieved, the regular downward motion of the aces has ceased. Since the aces are no different from other cards, it is reasonable to assert that the deck has been completely randomized and that further shuffling will intoduce no change in the *probable* distribution of the ensuing deal. Since there is no net flow of any species of card *on the average*, according to the law of detailed balancing, equilibrium has been achieved.

Objections can be raised to this example, since it is possible to arrange a deck with two aces on top and two on the bottom so that no net *overall* flow ever exists. Yet no one would deny that shuffling such a deck would change the a priori probabilities in the distribution of the next deal. What does the law of detailed balancing say about this? (Answer: The law says that *all* flows must average to zero. If you can choose a volume in which there are no aces a priori, the net flow of aces must be into this volume. The important step in the above example is that sufficient shuffling has been done so that there is equal probability of finding an ace in *any* location within the deck.)

Another fact that should be clear in the card example is that there is some net motion of the aces in most shuffles even after thorough randomization. The probabilities are not changing but the aces are moving. This is nothing more than the usual sampling problem that arises in taking an average. The accuracy of the average increases slowly with the number of sample points. For example, if a million decks were shuffled together, the expected r.m.s. error in the average would be $10^{-3}$ times the expected r.m.s. error for only one deck. The important conclusion to draw from this is that the equilibrium distribution gives only an expectation value for finding some object in some place. In general, the probability is neither 0 or 1. We have already run across the concept of expectation values for certain measurements in Chapter 1. The new element added in this chapter is the selection of the most random distribution, the one that gives equal probability to all equivalent states. In Chapter 1 or 2, we chose which states we wanted to fill.

## MAXIMUM PROBABILITY AND THERMAL EQUILIBRIUM

The probability that a given event will occur, such as an ace being located on top of a deck or an electron being found in a particular state, depends on the events that preceded the measurement. For example, if you put the ace on top of the pile, the probability of there being an ace on top is 1. The discussion above shows that shuffling would decrease the expectation value. (Eventually it would decrease to 1/13.) It is important to note that much shuffling is needed to achieve this change, and as long as this mixing does not go on, the probability would remain 1.

Once we admit the possibility of adequate shuffling, it is possible to consider the achievement of equilibrium from what appears at first to be a completely different approach. From the statistical point of view, thermal equilibrium represents *the distribution of maximum probability*. The connection between probability and detailed balancing is obtained by relating the effect of shuffling on the variables to be measured. In studying detailed balancing, the variable observed is the net flow of aces into or out of a given volume —say, the first thirteen cards—during successive shuffles. For probability measurements, the variable is the number of aces in the volume at the end of a shuffle. If there is no net flow on the average, the average number of aces in the volume is constant. Now if the average number of aces were not the most probable number of aces, you could conclude that the aces had some special preference for occupying or not occupying your particular sample. This is more than contrary to hypothesis; I suggest that if it occurs, your cards have spent too much time in the cafeteria or your dealer is getting too clever.

## THE ARITHMETIC OF FINDING THE MAXIMUM PROBABILITY

Our procedure at this point is to derive a general technique for finding the distribution of maximum probability. Applying this technique to a few classic problems will fill out the remainder of this chapter.

The problem that lies before us is best thought of in two parts. First, we specify a system that represents all possible solutions to a wave equation and a set of appropriate boundary conditions. Included in the description of the system must be the following.

**1.** The possible eigenstates of the system; in other words, the $E:k$ plot.

**2.** The total internal energy of the system.

**3.** Any rules that describe how the states may be filled, such as the Pauli exclusion principle.

**4.** Any rules concerning conservation of particles.

Second, we must have a procedure to find the most likely distribution of particles among the states that does not violate any of the rules of the system.

Consider, for example, the simple system shown in Figure 3.2. There are 9 electrons to be distributed among some 16 states on 3 energy levels. The exclusion principle holds, so only one electron may be in any given state. For the purpose of illustration, let $E_1$ be four units of energy, $E_2 = 2$, and $E_3 = 1$. Let the internal energy of the system be 21 units. The $x$'s in the states show one distribution that will satisfy all the system requirements. Since all the states with the same energy are equivalent, we can get a number of equivalent distributions by rearranging the $x$'s on any given level. A pair of distributions is *distinct* if you can discriminate between them. Thus, interchanging two $x$'s does not produce a distinct rearrangement. The question is, how many distinct equivalent rearrangements of Figure 3.1 are possible? (Answer: The surprisingly large number of 700.) The answer is obtained as follows. Consider the $E_1$ band of levels with seven states and three electrons. The first electron can go in any of seven states, the second in any of the six remaining, and the third in any of the remaining five. Thus the total number of ways to put the three electrons in seven states is $7 \times 6 \times 5 = (7!/4!)$. However, not all of these arrangements are distinct. Specifically, there are

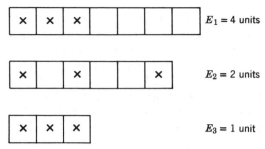

$E_1 = 4$ units

$E_2 = 2$ units

$E_3 = 1$ unit

**Figure 3.2** A distribution of electrons in the system states that satisfies the exclusion principle and $\sum E$'s $= 21$ units.

3! indistinct rearrangements of each choice. Thus, the total number of distinct rearrangements, $P(7, 3) = 7!/(4!3!) = 35$. In general, the number of distinct rearrangements of $n$ objects in $m$ equivalent states is

$$P(m, n) = \binom{m}{n} \equiv m!/(m - n)!n! \qquad (3.5)$$

(N.B. $0! = 1$) Since the rearrangements in the second row are independent of what goes on in the third row, the number of distinct arrangements of the second and third rows is the product of the number of arrangements of each. Thus, the total number of distinct rearrangements, $P_t$, is given by

$$P_t = P(m_1, n_1)\, P(m_2, n_2)\, P(m_3, n_3) \equiv \prod_i P(m_i, n_i) \qquad (3.6)$$

For the case shown, insertion of the numbers gives

$$P_t(3, 3, 3) = \left(\frac{6!}{3!3!}\right)\left(\frac{7!}{4!3!}\right)\left(\frac{3!}{0!3!}\right) = 700$$

The next question is: Are there any other arrangements that satisfy all the system rules? (Answer: Yes. The internal energy is correct for any solution of $4n_1 + 2n_2 + n_3 = 21$. Conservation of electrons requires $n_1 + n_2 + n_3 = 9$. And finally, $n_1 \leq m_i$ ensures an open state for each electron. There are only two solutions: the one given and $n_1 = 2$, $n_2 = 6$, and $n_3 = 1$.) Which of the two distributions is more probable? (Answer: The number of distinct rearrangements of the second solution is only 63. Since the total number of satisfactory rearrangements is the sum of the two types of rearrangements, the odds of finding the system in the 3, 3, 3 configuration is $700/63 = 11.1$ times as great as the odds for finding it in the 2, 6, 1 configuration. Accordingly, the equilibrium distribution is the 3, 3, 3 distribution.)

In principle, it is always possible to denumerate all possible solutions, compute their frequency of occurrence, and select the most probable from the table so generated. But it is obvious that a more efficient technique is called for, especially when you consider that the typical problem we want to investigate is the distribution of something like $10^{15}$ electrons in $10^{23}$ states. Since the object of this exercise is to find the distribution of *maximum* probability, we can take the derivatives of $P_t$ with respect to the $n_i$ and set each derivative equal to zero. We then have an extremum that will prove to be a maximum. Specifically, we wish to maximize

$$P_t = \prod_i P\,(n_i, m_i) \tag{3.6}$$

But to conserve the internal energy and the number of electrons, we must also require

$$\sum_i n_i = N \tag{3.7}$$

$$\sum_i (n_i E_i) = E_t \tag{3.8}$$

Note that (3.5) already includes the exclusion principle, since it is the expression for the number of distinct rearrangements of $n$ things in $m$ boxes with only one in a box. If the exclusion principle did not hold, $P(m, n)$ would have a different form.

The usual procedure for finding a maximum, subject to constraints such as (3.7) and (3.8), is Lagrange's *method of undetermined multipliers*. The method takes advantage of the fact that adding constants to the function being maximized does not change the derivatives of the function. Thus, I am permitted to add $[-\alpha N - \beta E_t]$ to the function I am maximizing, with $\alpha$ and $\beta$ being as yet undetermined multipliers. As I show below, when I take the derivatives, the $\alpha$ and $\beta$ are retained in the equation I set equal to zero. I thus will find a distribution with two unknown constants in it. Returning with this distribution to (3.7) and (3.8) will permit me to solve for the two unknown coefficients. By this bootstrap procedure I have then found the maximum of (3.5) subject to the constraints of (3.7) and (3.8).

As it turns out, it is mildly convenient arithmetically to consider not (3.5) but rather the logarithm of (3.5). The log of $P_t$ also turns out to be the entropy of the system, so there is more than just arithmetic convenience involved. We also will want to take advantage of Sterling's approximation, which in its logarithmic form states that

$$\log n! \cong n \log (n) - n \tag{3.9}$$

Thus, we seek the maximum of

$$\tag{3.10}$$

$$R = \log (P_t) - \alpha N - \beta E_t = \sum_i \left\{ \log \left[ \frac{m_i!}{(m_i - n_i)!n_i!} \right] - \alpha n_i - \beta n_i E_i \right\}$$

Using Sterling's approximation:

$$R = \sum_i \left[ m_i \log m_i - n_i \log n_i - (m_i - n_i) \log (m_i - n_i) - \alpha n_i - \beta n_i E_i \right]$$

To maximize $P_T$, I differentiate $R$ w.r.t. $n_i$ and set each derivative equal to zero.

$$dR/dn_i = 0 = -\log n_i + \log (m_i - n_i) - \alpha - \beta E_i$$

Combining the logs yields

$$\log \left( \frac{m_i}{n_i} - 1 \right) = \alpha + \beta E_i$$

or alternately:

$$n_i/m_i = \frac{1}{1 + e^{\alpha + \beta E_i}} \equiv f(E_i) \tag{3.11}$$

$f(E_i)$ is called the *Fermi function*.

To find the constants $\alpha$ and $\beta$, I advertised above that I was going back to (3.7) and (3.8). That is:

$$\sum_i n_i = \sum_i m_i f(E_i) = N \tag{3.7'}$$

$$\sum_i n_i E_i = \sum_i m_i f(E_i) E_i = E_t \tag{3.8'}$$

However, to avoid becoming embroiled in a long development of thermo-dynamics, I am going to cheat on the solution of (3.8'), and simply announce, deus ex machina, that $\beta = 1/kT$. The real justification for setting $\beta = 1/kT$ is that it is the only function that will yield the proper thermodynamic relationship between internal energy and entropy. The reason why the derivation is so lengthy is the effort necessary to relate $E_t$ to $T$, the absolute temperature. $k$, by the way, is Boltzmann's constant equal to $0.863 \times 10^{-4}$ eV/°K.

The solution to (3.7') is done quite directly by counting. For semiconductor problems, this counting entails setting the number of electrons that have been thermally excited equal to the number of empty spaces left behind. Such solutions are intuitively satisfying and rather simple to perform once the meaning of $f(E)$ is clear. Accordingly, let us first have a look at $f(E)$.

## THE USE AND MEANING OF THE FERMI FUNCTION

Equation 3.11 gives the definition and meaning of the Fermi function in the most direct terms. It states that $f(E_i)$ is the fraction of the states having energy $E_i$ that are filled if the system is in thermal equilibrium. $f$ is a continuous function of $E_i$ even though $m_i$ is not. As a matter of fact, the only difference between the Fermi functions for different materials (assuming the same temperature in all cases) is the value of α. This means that you have a universal function which describes all systems that obey the usual Pauli exclusion principle, conserve the total number of particles, and conserve internal energy. Such systems are said to obey *Fermi statistics*.

It is customary to replace the constant α by another constant divided by $kT$. That is, $\alpha \equiv - E_f/kT$. $E_f$ is called the *Fermi level*, and, like α, it is a constant for a system that assures you that (3.7') holds. It is rather important that you see the Fermi level as nothing more than an accounting trick, something that is used to assure that the assets (in electrons in the upper states) equal the liabilities (in empty states left behind). Sometimes in nonequilibrium situations—a place where equilibrium statistics obviously do not apply —it becomes convenient to invent similar accounting devices called quasi-Fermi levels. As long as you realize that these accounting devices are conveniences to the scientist and not physical properties of the material, you should find it reasonably easy to see what is wanted and how the Fermi level helps in its calculation.

Three tasks remain in the development of some facility with the use of equilibrium statistics. These include the calculation of the equilibrium number of electrons and holes in a pure and an impure semiconductor, a detailed look at the *p-n* junction in equilibrium, and finally, a quick look at one other important type of statistical system.

## EQUILIBRIUM STATISTICS IN PURE SEMICONDUCTORS

If I wanted to find the equilibrium number of electrons in the conduction band of a semiconductor, all I should have to do is sum $n_i = f(E_i)m_i$ over all the states in the conduction band. One important question must be answered before I proceed. From the information in Chapter 2 about the band structure and the number of states per band, where did the electrons come from that are in the conduction band? (Answer: The only source of electrons is the valence band. Accordingly, I can state that for every electron in the conduction band, there is a hole in the valence band.) Now for an easy question: If $f(E)$ is the probability that a state with energy $E$ is filled, what is the probability that it is empty? [Answer: $1 - f(E)$.] Since the sum of the holes in the valence band must be equal to the sum of the electrons in the conduction band, we have

$$\sum_i m_i f(E_i) = \sum_j m_j [1 - f(E_j)] \tag{3.12}$$

where $i$ runs over all the values in the conduction band and $j$ runs over all values in the valence band. Note that this equation has only one unknown, $E_f$. This we now proceed to find.

Doing the summations in (3.12) would be a terrible chore. It is considerably easier to take advantage of the very small spacing between levels and go from the summation form to an integral form. All that is required is to replace $m_i$ by the differential $dm$, to wit:

$$dm = g(E)\,dE = \text{the number of states between } E \text{ and } E + dE$$

The number of electrons with energy between $E$ and $E + dE$ is then

$$dn = f(E)\,dm = f(E)g(E)\,dE \tag{3.13}$$

$g(E)$ is the *density of states* function that was introduced in the last chapter. (See equation 2.32.) Summing (3.12) has now become

$$\int f(E)g(E)\,dE = \int [1 - f(E)]g(E)\,dE \tag{3.14}$$

$$\underbrace{\qquad\qquad}_{\text{conduction band}} \qquad \underbrace{\qquad\qquad}_{\text{valence band}}$$

Two things are needed for performing the integrations: $g(E)$ for each of the bands, and a simplified form of $f(E)$.

The density of states function was found in (2.32') to be

$$g(E) = (1/4\pi^3)\,dV(k)/dE \tag{2.32'}$$

The volume in $k$-space enclosed by an ellipsoidal surface of constant energy is given by the standard expression for the volume of an ellipsoid. (Note that in using ellipsoids I have already assumed parabolic bands. If I did not, I would have to foreswear any hope of completing the integral in closed form.)

$$V(k) = \frac{4\pi}{3} k_{1\max}k_{2\max}k_{3\max} \tag{3.15}$$

where the $k_{i\max}$'s are the semiaxes of the ellipsoid. For parabolic bands, we have at once from the discussion in Chapter 2 (equation 2.30):

$$E - E_c = \sum_i \frac{1}{2}\left(\frac{\partial^2 E}{\partial k_i^2}\right)k_i^2 = \sum_i \frac{1}{2}\frac{\hbar^2}{m_i^*}k_i^2 \tag{3.16}$$

where $E_c$ is the energy at the bottom of the conduction band. Solving (3.16) for the $k_{i\max}$ yields

$$k_{i\max} = \left[\frac{2m_i^*}{\hbar^2}(E - E_c)\right]^{1/2} \tag{3.17}$$

Putting (3.17) into (3.15) and then differentiating gives

$$\frac{dV}{dE} = 2\pi(E - E_c)^{1/2}\left(\frac{2}{\hbar^2}\right)^{3/2}(m_i^*m_j^*m_k^*)^{1/2} \tag{3.18}$$

The $dV$ we have in (3.18) is for one minimum. If there are $M$ equivalent minima ($M = 4$ in Ge and 6 in Si) in the conduction band, then (3.18) must be multiplied by $M$. This finally leads to the evaluation of (2.32) as

$$g(E)dE = \left(\frac{1}{2\pi^2}\right)\left(\frac{2}{\hbar^2}\right)^{3/2} M(m_i^* m_j^* m_k^*)^{1/2}(E - E_c)^{1/2}dE \qquad (3.19)$$

You will frequently see the quantity $\{M^{2/3}(m_i^* m_j^* m_k^*)^{1/3}\}$ listed as the effective mass of the electron. The proper name for this composite quantity is the *density of states effective mass*, for which I use the symbol $m^{**}$. The name derives from the fact that if $k_1 = k_2 = k_3$ and $M = 1$, $m^{**} = m^*$. Evaluation of $m^{**}$ for Ge and Si leads to the values in Table 3.1. Note that the analysis

**Table 3.1.** Density-of-States Effective Masses

|  | Ge | Si |
|---|---|---|
| Electrons | $0.55\,m_0$ | $1.1\,m_0$ |
| Holes | $0.37\,m_0$ | $0.57\,m_0$ |

in the valence band would be identical except that $(E - E_c)$ would be replaced by $(E_v - E)$, where $E_v$ is the top of the valence band. We now have $g(E)dE$ in a form that is reasonably integrable. All that remains is to beat $f(E)$ into the proper shape, put $f$ and $g$ into (3.14), and do the integration.

## THE BOLTZMANN APPROXIMATION TO THE FERMI FUNCTION

The exact form of the Fermi function (3.11) is easy enough to evaluate, but quite unreasonable to integrate. However, if $E - E_f$ happens to be several times $kT$ (which is 0.0259 eV at 300°K $\cong$ 27°C $\cong$ 86°F), then

$$1 + \exp(E - E_F)/kT \cong \exp(E - E_F)/kT \quad \text{and} \quad f(E) \simeq \exp-(E - E_f)/kT$$

$$(3.20)$$

Equation 3.20 is the Boltzmann approximation. The necessary and sufficient condition for its being valid is that the exponential term in (3.11) be much larger than 1.

It is important to see exactly what the Boltzmann approximation describes. Figure 3.3 shows the Fermi function. The portion in which the Boltzmann approximation is accurate to better than 1% is drawn in a heavy line. Note that it is the tail end of the curve—the region where the probability of a state being full is very small. Before the era of quantum mechanics, when exclusion principles and discrete energy states were unrecognized, the only statistical law that could be derived "from logical considerations"

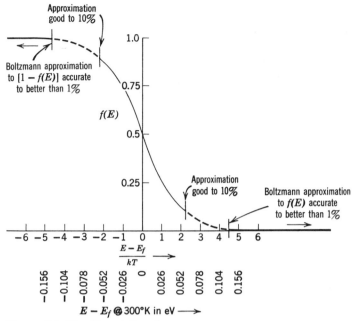

**Figure 3.3** The Fermi function.

was Boltzmann's. As you have just seen, it is the proper law for *dilute* systems, systems in which the probability of finding a state occupied is very small.

Since we will also want to compute the probability that a state in the valence band is empty, it is worthwhile noting here that $1 - f(E)$ is also approximated by a Boltzmann distribution if $E_f - E \gg kT$, to wit:

$$1 - f(E) = 1 - \frac{1}{1 + e^{(E - E_f)/kT}} = \frac{1}{1 + e^{(E_f - E)/kT}} \cong e^{-(E_f - E)/kT} \quad (3.21)$$

Once again, the Boltzmann approximation is good to better than 1% for $(E_f - E) \geq 4.6kT$ and to better than 10% for $(E_f - E) \geq 2.2kT$. These regions are also shown on Figure 3.3. Note carefully from the bottom scale on the abscissa that at 300°K, having $E - E_f = 0.1$ eV is sufficient for 2% accuracy. As you will see below, this condition is met for most practical problems.

## CALCULATION OF THE NUMBER OF CARRIERS USING THE BOLTZMANN APPROXIMATION

There is very little left to do but combine (3.19) with (3.20) or (3.21) in the manner indicated by (3.14). It is important to note that the exponential

character of $f(E)$ guarantees that the integrand will become negligible
BEFORE $g(E)$ deviates significantly from its parabolic form. Actually, the
integrand vanishes so rapidly in the Boltzmann approximation that we need
not be concerned with the exact width of the bands; integrating from the
band edge to infinity is the same as integrating over the band. Note that if
this were not true, we would have to take into account the fact that $g(E)$
must deviate substantially from parabolic form. Such difficulties raise their
ugly heads when you consider calculations in typical metals (for example
Figure 2.26). However, for typical semiconductors, the Boltzmann approximation holds quite well, the bands are sufficiently parabolic, and (3.14)
is readily evaluated. Combining (3.19) with (3.20) yields

$$\int_{E_c}^{\infty} g(E) f(E) \, dE = \frac{\sqrt{2} (m^{**})^{3/2}}{\pi^2 \hbar^3} \int_{E_c}^{\infty} (E - E_c)^{1/2} e^{-(E - E_f)/kT} dE = n \qquad (3.22)$$

The easiest way to evaluate (3.22) is to substitute $u^2 = (E - E_c)/kT$.
Putting this into (3.22) yields

$$n = \frac{\sqrt{2} (m_e^{**})^{3/2}}{\pi^2 \hbar^3} (kT)^{3/2} e^{-(E_c - E_f)/kT} \int_0^{\infty} 2u^2 e^{-u^2} du \qquad (3.22')$$

The definite integral in (3.22') can be found readily in integral tables to be
$\sqrt{\pi}/4$. Accordingly, we have found the number of electrons in the conduction
band in thermal equilibrium to be

$$n = \left( \frac{m_e^{**}}{\pi} \right)^{3/2} \frac{1}{\sqrt{2}\hbar^3} (kT)^{3/2} e^{-(E_c - E_f)/kT} \equiv N_c e^{-(E_c - E_f)/kT} \qquad (3.23)$$

$N_c$ is called the *effective density of states at the band edge*. The name derives
from the fact that if there were $N_c$ states located at $E_c$, $n$ in (3.23) would be
the number of such states filled.

An equivalent derivation for holes in the valence band leads to an expression
similar to (3.23). The only differences are in $m^{**}$ and the fact that we would
now integrate from $-\infty$ to $E_v$ using $(1 - f)$ rather than $f$. The result is

$$p = \left( \frac{m_h^{**}}{\pi} \right)^{3/2} \frac{1}{\sqrt{2}\hbar^3} (kT)^{3/2} e^{-(E_f - E_v)/kT} \equiv N_v e^{-(E_c - E_f)/kT} \qquad (3.24)$$

$N_v$ has the same meaning for the valence band that $N_c$ does for the conduction band. They are both functions or temperature and the curvatures of their respective bands.

## FINDING THE FERMI LEVEL IN A PURE SEMICONDUCTOR

To find the Fermi level, we are really obliged to satisfy equation 3.7'. However, since all the electrons in the conduction band came from the valence band, we will have accounted for all the electrons in the crystal if we simply set the number of electrons in the conduction band equal to the number of holes in the valence band, to wit:

$$n = p = n_i \qquad (3.25)$$

$n_i$ is the number of electrons or holes in equilibrium in an *intrinsic* (i.e., pure) semiconductor. Putting (3.23) and (3.24) into (3.25) yields

$$N_c\, e^{-(E_c - E_f)/kT} = N_v\, e^{-(E_f - E_v)/kT} = n_i \qquad (3.26)$$

Solving for $E_f$ yields:

$$E_f = \frac{E_c + E_v}{2} - \frac{kT}{2} \log\left(N_c/N_v\right) \equiv E_i \qquad (3.27)$$

Equation 3.26 gives the intuitively sensible result that the Fermi level in an intrinsic semiconductor lies close to the middle of the forbidden gap. The "intuitive" part comes from a reexamination of Figure 3.3. In essence, *if everything is symmetric, the balance point is the middle*. The rather small log term in (3.27) is the correction for asymmetry in the band shape. Note once again that (3.27) is the choice to make to satisfy $\sum_j n_j = N$ and that we obtained it by setting the number of holes equal to the number of electrons.

$E_i$ in (3.27) is a characteristic of a pure material and is named the *intrinsic Fermi level*. Associated with it, through (3.26), is the *intrinsic carrier density*, $n_i$. From (3.26), (3.23), and (3.24) we can obtain a result that is an extremely important tool in solving device problems:

$$pn = n_i^2 = N_c N_v\, e^{-(E_c - E_v)/kT} \qquad (3.28)$$

Note that in (3.28) there is no mention of $E_f$. Thus, (3.28) is characteristic of the material and independent of the Fermi level. The only requirement to have (3.28) hold is that the Boltzmann approximation be valid.

Since I have just introduced four new parameters—$N_c$, $N_v$, $E_i$, and $n_i$—it is worthwhile to get some idea of how big each of these might be. From

(3.23) and (3.24) we have

$$N_{c,v} = \left(\frac{kT}{\pi}\right)^{3/2} \frac{(m^{**}_{e,h})^{3/2}}{\sqrt{2}\hbar^3} = 2.5 \times 10^{19} \left(\frac{m^{**}_{e,h}}{m_0}\right)^{3/2} \quad \text{at} \quad T = 300°K \quad (3.29)$$

Since $m^{**}$ is usually not very different from $m_0$, there are effectively about $10^{19}$ states/cm³ near the band edges of most materials. There are a few extreme exceptions such as InSb (one of the so-called III-V compound semiconductors comprised of elements from periods 3a and 5a). $m^{**}_e$ for InSb is only $0.013\, m_0$, giving an $N_c = 3.71 \times 10^{16}$ cm⁻³. For holes, there do not appear to be such large deviations in any of the materials so far investigated. It is worthwhile to recall that $10^{16}$ states/cm³ is an extremely small number, since there are about $10^{22}$ atoms/cm³.

Usually $E_i$ is approximately halfway from $E_v$ to $E_c$. Since $kT/2 \cong .013$ eV at 300°K, the logarithmic term is a very small correction for the usual range of $E_g$'s. An interesting problem arises with InSb, $N_v = 1.16 \times 10^{19}$ cm⁻³ and $E_g = E_c - E_v = 0.16$ eV. Find $E_i$ by (3.27) and comment on why the result is worthless. [Answer: $E_i = E_v + 0.08 - (.013)\ln(3.2 \times 10^{-3}) = = E_v + 0.15$. The result is worthless since it places the Fermi level essentially at the bottom of the conduction band.] Although the Fermi level is undoubtedly very close to $E_c$, (3.27) was derived assuming that the Boltzmann approximation was valid and, unless $E_c - E_f \gg kT$, that assumption is false. A material in which the Boltzmann approximation does not hold is called *degenerate*. This use of the word is only remotely related to our use of it in Chapter 2 to describe states with the same energy. In statistical mechanics, "degenerate" means that the electrons are crowding into the lowest possible states and the effects of the exclusion principal are quite apparent. Thus, a large number of electrons have almost the same energy, and in this sense they are degenerate.

To see that we are normally on safe ground using the Boltzmann approximation, find $E_i$ for Ge. ($E_g = 0.66$ eV, the rest of the data from Table 3.1.) (Answer: $E_i = E_v + 0.342$ eV.) Since $E_i$ is more than $10\, kT$ from the nearest band edge, the Boltzmann approximation is extremely accurate.

Now compute $n_i^2$ for Ge. (Answer: $6.25 \times 10^{26}$ cm⁻⁶, making $n_i = = 2.5 \times 10^{13}$ cm⁻³.) Thus, only about one atom in $10^9$ contributes an electron to the conduction band. For Si, $n_i = 1.4 \times 10^{10}$ cm⁻³. For diamond, with $E_g \cong 5.47$ eV, $n_i < 10^{-27}$ cm⁻³! Since the mass of the earth is only $6 \times 10^{27}$ g, it is evident that a pure diamond the size of the earth could be expected to have barely a single free electron it! Small wonder then that pure diamond is an excellent electrical insulator. However, if we do get some electrons into the conduction band or some holes in the valence band, the very same material, diamond, makes a pretty good conductor. As you will see immediately below, getting more carriers than $n_i$ is the rule rather than the exception.

## THE BEHAVIOR OF IMPURITIES IN A SEMICONDUCTING CRYSTAL

Each Ge atom in a Ge lattice has four valence electrons that it shares with its four nearest neighbors. To free a Ge atom from the crystal takes a substantial amount of energy, more than from the free atom, indicating unambiguously that the shared electrons are at a lower energy than they would be in the free atom. A useful way to look at the potential energy of the electron is to observe that there is a certain amount of binding energy due to the charge on the nucleus and then some more binding energy associated with the electron's interaction with the rest of the crystal. Let us say that we put an aluminum atom in in place of a Ge atom. Al brings only three valence electrons with it. These three form bonds with three of the nearest neighbors, leaving the fourth with a dangling bond. There is no central charge (from the nucleus) available for binding another electron, but there would still be the binding energy associated with the rest of the crystal. Thus, we might expect that if an electron happened by, it might become attached to the site occupied by the Al atom. If this does happen (if it didn't, I wouldn't bring the subject up) the foreign atom is said to *accept* an extra electron and it is termed an *acceptor* impurity. The requirement for being an acceptor impurity is to have less than the number of electrons required to saturate the crystal bonds. For period IV crystals, impurities from periods I, II, and III fit this requirement. Since the accepted electron is bound to the impurity site but not to the impurity, it is evident that the energy of the extra electron will be greater than $E_v$. To say more than that requires more than the hand waving I've just done. But first, let us look at periods V through VIII.

For example, what would happen if I put an As (period V) atom in where a Ge atom belonged? The four bonds to the nearest neighbors are filled, leaving one electron left over. This electron has only the charge of the nucleus to attract it; there is no crystal binding energy. Accordingly, the binding energy of the extra electron will be less than the other four and it, too, will have an energy greater than $E_v$. (To get the sense of this discussion it is necessary to recognize that binding energy is represented by how deep a potential well the electron is in. To have *less* binding energy is to be in a *shallower* hole. Thus, the electron with the least binding energy has the highest potential energy.) If the extra electron happens to have an energy that is not too far below $E_c$, then thermal "reshuffling" may shake the electron loose, putting it into the conduction band. An impurity with extra electrons thus donates some electrons to the crystal. It is called, accordingly, a *donor* atom.

It would be very surprising indeed if all donor atoms behaved the same and all acceptors were alike. Some differences ought to be there, a priori. For example, Au goes into a Ge site with only one valence electron, Cd with two, and Ga with three. Gold could accept three electrons, but with

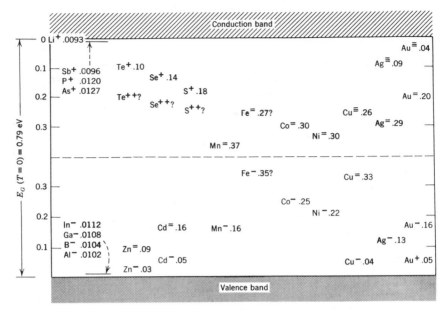

**Figure 3.4a** Energy levels of donors and acceptors in Ge. The ionization energy indicated is measured from the nearest band edge (in eV). (After T. H. Geballe, Figure 8.11 in N. B. Hannay, Ed., *Semiconductors*, Reinhold Publishing Corp., New York, 1959.)

the addition of each extra electron, there is some uncompensated charge that repels other electrons. Thus, the second electron should be less strongly bound, the third even less so. You could correctly guess from this sort of argument that the energy levels for the various acceptors would cover the entire forbidden gap.

Figure 3.4 presents measured values for the impurity levels found in Ge and Si. (These graphs, by the way, represent a prodigious amount of careful experimental work by a number of scientists.) By and large, nothing very much more than what we have already said can be added as a generalization. A few interesting groupings show up when we look for elements that act alike. Chemically Au and Cu are very much alike. Note that both of them act rather much alike as dopants also. For example, in Si both of them can act as either donors or acceptors. (The elements that can act in both capacities are said to be *amphoteric*.) A surprising and interesting pair of groups is the period IIIa and period Va elements, especially in Ge. Note that they have almost exactly the same energy in spite of some rather wide differences in chemical activity. Also note how small a separation there is between these two groups of elements and the nearest band edge. These elements could

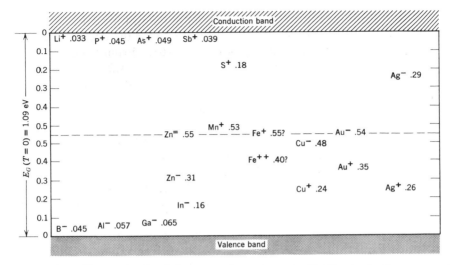

**Figure 3.4***b* Energy levels of donors and acceptors in Si. The ionization energy is measured from the nearest band edge (in eV). The figure is Figure 8.12 in Geballe (loc. cit. in Figure 3.4*a*). The data for Ag are from F. L. Thiel and Sorab K. Ghandi, *J. Appl. Phys.* **41**, 254, 1970.

act so similarly only if the property we were measuring had much more to do with Ge than with the impurity. This turns out to be the central clue to their behavior.

Briefly, the argument proceeds this way. Consider first a donor atom from the Va group. If the extra electron that is bound to the donor atom is very loosely bound (that is, if the extra electron is almost in the conduction band), the electron spends most of its time far from the impurity atom. Accordingly, any local structure that characterizes the impurity atom is averaged out by the intervening Ge. The impurity atom is thus nothing more than a unit electron charge located at the center of the extra electron's orbit. This view immediately shows why the different impurities would behave similarly. It even does a pretty good job of predicting the binding energy and showing why the energy is less in Ge than Si.

One of the central results in the usual course in atomic physics is the energy of the ground state of the hydrogen atom. We have not obtained that solution, but it is similar in some respects to the particle-in-a-box problem that we did in Chapter 1. The solution for the ground state gives

$$E_{\text{gnd}} = [- m_0 q^4 / 8h^2 \varepsilon_0^2] = - 13.6 \text{ eV} \tag{3.30}$$

where $q$ is the electric charge and $\varepsilon_0$ is the permitivity of free space. The hydrogen atom is characterized by a heavy charge of $+1$ at the origin and a light charge of $-1$ in orbit about it. This is just the situation I have described for the donor electron above. To find the ground state for the donor, however, I should use $m^*$ and the permitivity of Ge ($\varepsilon = 16\varepsilon_0$). From Chapter 2 you can obtain the average effective mass: $\overline{(1/m^*)} = 1/3(2/m_\perp + 1/m_\parallel) = (1/0.12\ m_0)$. Substituting these numbers into (3.30) gives a binding energy of

$$E_{\text{donor}} = E_c - 0.0064\ \text{eV}$$

This is a factor of $\sim 2$ too little, as you can see from Figure 3.4. The correction comes about because the electron does spend some of its time in the vicinity of the donor ion. Nonetheless, though it is an oversimplification, the hydrogenic model gives a remarkably good answer. As you might expect, a more detailed and accurate solution has been worked out. For our purposes, however, the data in Figure 3.4 is sufficient.

With Si, $\varepsilon = 12\varepsilon_0$ and $m^* = 0.259 m_0$. Putting these values into (3.30) yields $E_{\text{donor}} = E_c - 0.024$ eV. Once again, the hydrogenic prediction gives a binding energy that is low by a factor of $\sim 2$.

The impurities we have just discussed are called "shallow" impurities because they are only a small distance from the band edge. Those further away are called "deep" impurities. Obviously, these are inexact designations, but they prove useful in discussing the effects of doping on the characteristics of semiconductors.

## THE EFFECTS OF IMPURITIES ON THE NUMBER OF EQUILIBRIUM CARRIERS

When we found $E_i$ we took advantage of the fact that the only source of electrons for the conduction band was the valence band. This led directly to (3.25). Even if we add other sources (donors), we still must find $E_f$ by accounting for all the electrons. To accomplish this we must add terms for another source (donors) and another sink (acceptors). The new version of (3.25) is

$$p + N_D^+ = n + N_A^- \tag{3.31}$$

$N_D^+$ is the density of ionized donors (donors that have given up an electron), and $N_A^-$ is the density of ionized acceptors (acceptors that have acquired an electron). Just as in (3.25), the only unknown in (3.31) is the Fermi level, $E_f$. To show this, rewrite (3.31) in the same fashion as (3.26):

$$N_v e^{-(E_f - E_v)/kT} + N_D(1 - f'(E_D)) = N_c e^{-(E_c - E_f)/kT} + N_A f'(E_A) \tag{3.32}$$

Three new symbols have just been added to your already copious store. $E_D$ and $E_A$ are the energies of the donor and acceptor levels, respectively.

$f'$ is the Fermi function for impurities. Until we either find $f'$ or convince ourselves that impurities obey the same statistical rules as the conduction and valence bands (i.e., $f = f'$), we must allow for a new distribution function. Once we have $f'$, we can solve for $E_f$ in much the same way as we did in (3.27).

## STATISTICAL RULES FOR IMPURITIES

In the usual situation found in semiconductor technology, the impurity density is *EXTREMELY* small. As a matter of fact, the pure semiconductors probably represent the purest crystals on earth. Impurity contents as low as one part in a billion are not entirely out of the ordinary, and typical devices have impurity concentrations of one part in $10^4$ or less. It is thus not unreasonable to treat each impurity as if it were entirely isolated in a sea of pure semiconductor. Consider a period V donor in Ge. Each of these isolated donors have one spare electron, but that electron could be in one of two states—spin up or spin down. Therefore, for each donor in the crystal we should have two states. So far so good; it looks pretty much like the states in the bands. However, our central assumption in finding the band structure was that the electrons were so spread out that they did not interact. Thus, putting an electron into state $A$ did not affect the electron in state $B$. Such an assumption is obviously untrue for our isolated impurity atom. (Note once again that the Bloch solutions are not localized, whereas the impurities are localized. Thus, electrons on an impurity cannot contribute to conductivity; electrons in a band can.) If a period III impurity atom acquires one electron, it cannot acquire another. There is no binding energy for the second electron. Thus, every time we fill one state, we must take two off the available list. Consider what happens to (3.5) with this new "exclusion" principle. Take a case of putting three electrons into eight equivalent states. The first could go in any of eight places, the second in six, and the third into any of the remaining four. Thus,

$$P'(8, 3) = (8 \times 6 \times 4)/3! \tag{3.33}$$

Remember that the 3! removes the indistinct permutations. Generalizing (3.33) to the case of $n$ electrons to be put on $L$ impurity sites (note that the number of states is then $2L$), we get the expression:

$$P'(2L, n) = (2^L L!)/[2^{(L-n)}(L - n)!n!] = (2^n L!)/(L - n)!n! \tag{3.34}$$

To convince yourself of (3.34), check it against (3.33) and another example of your own choosing. Once again we want $P_t$ as in (3.6):

$$P_t = \prod_j P'(2L_j, n_j) \tag{3.35}$$

To practice the techniques presented in the beginning of this chapter, derive $f'$ from (3.35) above. Result:

$$f' = \frac{1}{1 + \frac{1}{2}e^{(E_0 - E_f)/kT}} \tag{3.36}$$

Although $f' \neq f$, you can see that (3.32) will contain only one unknown — $E_f$. Thus, in principal, we can solve for the equilibrium carrier density in the presence of donors and acceptors. Once again, actually solving for $E_f$ is simple only when the Boltzmann approximation holds. Since that is the case we will be most interested in and it's also the easiest, let us attend to it first.

## FINDING THE EQUILIBRIUM CARRIER CONCENTRATION FOR A CASE WHERE ALL LEVELS SATISFY THE BOLTZMANN APPROXIMATION

The case we are about to consider is a semiconductor containing only shallow impurities. Actually, no crystal is ever so pure as to have no deep impurities, but if the number of deep impurities is quite small (a relative term, please note), they can be ignored. Since the shallow impurities lie close to the band edges, if the Boltzmann approximation holds for the band edge, it will probably hold equally well for the impurities too. The first question is: What is the form of $f'(E)$ in the Boltzmann approximation? [Answer: $f'(E) = 2e^{-(E - E_f)/kT}$, from equation 3.36.]

We can now write (3.32) in the Boltzmann approximation; however, before cranking through a lot of algebra, take a look at the terms describing $N_D^+$ and $N_A^-$. When the Boltzmann approximation holds, $f(E_c)$ and $[1 - f(E_v)]$ are both much less than 1. Thus, the same thing must apply to $f'(E_D)$ and $[1 - f'(E_A)]$. Noting that $f'(E_D)$ is the fraction of unionized (i.e., filled) donors, the Boltzmann approximation is the same thing as saying $N_D^+ \cong N_D$. Similarly, with $[1 - f'(E_A)]$ being the fraction of acceptors still unionized, in the Boltzmann approximation, $N_A^- \cong N_A$. Thus, the Boltzmann approximation reduces (3.32) to

$$p + N_D = n + N_A \tag{3.37}$$

or

$$N_v e^{-(E_f - E_v)/kT} + N_D = N_c e^{-(E_c - E_f)/kT} + N_A \tag{3.38}$$

Plugging the all important $pn = n_i^2$ (equation 3.28) into (3.37) yields a form that is particularly easy to solve for $n$ or $p$:

$$p + N_D = (n_i^2/p) + N_A$$

Clearing fractions and solving the resulting quadratic yields

$$p = \frac{(N_A - N_D) \pm \sqrt{(N_D - N_A)^2 + 4n_i^2}}{2} \qquad (3.39)$$

The negative sign in (3.39) gives a positive value to $p$ when $N_D$ is greater than $N_A$. Note that for $N_A$ much larger than $N_D$ or $n_i$, $p \cong N_A$ and thus $n \cong n_i^2/N_A$. Such a material is called p-type. On the other hand, if $N_D \gg N_A$ and $n_i$, equation 3.39 (or its obvious counterpart for $n$) yields $p \cong n_i^2/N_D$ and $n \cong N_D$. A material with $n > p$ is called n-type. A material with $p \gg n$ or $n \gg p$ is called *extrinsic* to contrast it to the pure (intrinsic) material where $n = p = n_i$.

Finding $E_f$ directly from (3.38) is easy only if you can assume that either $p$ or $n$ is negligible. If $n$, for example, is much much larger than $p$, (3.38) says quite directly that

$$-\log\left[(N_D - N_A)/N_c\right] = (E_c - E_f)/kT \qquad (3.40)$$

There is another way of writing (3.38) that can be conveniently solved for all impurity concentrations. It will also serve us well in the discussion of contact potentials and p-n junctions. Using (3.26) and (3.27) in (3.23) and (3.24) yields

$$\frac{n}{n_i} = \frac{N_c e^{(E_c - E_f)/kT}}{N_c e^{(E_c - E_i)/kT}} = e^{-(E_i - E_f)/kT} \qquad (3.41)$$

$$\frac{p}{n_i} = e^{+(E_i - E_f)/kT} \qquad (3.42)$$

Putting these into (3.37) yields the new form:

$$N_D - N_A = n_i\left[e^{(E_f - E_i)/kT} - e^{-(E_f - E_i)/kT}\right] = 2n_i \sinh\left(\frac{E_f - E_i}{kT}\right) \qquad (3.43)$$

Note that both (3.43) and (3.39) depend only on the difference between the donor and acceptor densities, not on the densities themselves. Although there is no way to label an electron so that you know where it came from, a not unreasonable way to think of the effect of putting both donors and acceptors in the same crystal is to say that the acceptors absorb electrons from the donors until the smaller of the two is completely ionized. (Note further that an electron going from a donor to an acceptor ionizes BOTH impurities.) Therefore, the assumption of total ionization for shallow impurities is still valid even with the two species present. This result should be rather satisfying, because there is no such thing as a perfectly pure crystal. All crystals are doped to at least some extent with *both* donors and acceptors.

The only real difference between (3.43) and (3.23) or (3.24) is that the Fermi level is measured from the intrinsic level rather than from one or another band edge. Knowing one is knowing the other, but (3.43) is frequently more convenient.

## THE LAW OF MASS ACTION

Take a look at what (3.39) says about Ge with only about one part per million of As. This miniscule amount of impurity is approximately $10^{16}$ donors/cm$^3$. Accordingly, $n \cong 10^{16}$/cm$^3$ and $p = n_i^2/N_D = 6.25 \times 10^{10}$/cm$^3$. Thus, the addition of a second source of electrons (the donors) has drastically reduced the density of holes. Equation 3.39 says this quite directly, but it is also helpful to consider the dynamics that lead to such a "law of mass action." [The phrase "law of mass action" should take you back to your chemistry courses, and with good reason. What we are looking at here is the "chemistry" of holes and electrons with the reacion being $e^- + h^+ \Leftrightarrow$ (filled valence state).] Two separate and reversible processes are going on. One is the generation of hole-electron pairs, which comes to equilibrium with the recombination of the hole-electron pairs. The other process is the generation of electron ionized-donor pairs, which comes to equilibrium with the recombination of electrons and ionized donors. The two equations describing the equilibrium between the generation and recombination rates are coupled, since it is the same electron population in both equations.

Consider the generation rate of hole-electron pairs. At a given temperature, the amount of thermal energy available anywhere in the crystal is pretty independent of the doping level (i.e., the concentration of impurities), since almost all the energy is stored in the vibrations of the massive crystal lattice. What other factors that determine the generation rate could depend strongly on the carrier concentration? (Answer: In order to generate a hole-electron pair, it is necessary to have both a filled valence band state and an empty conduction band state.) Here is something that could depend strongly on the doping. The percentage of states that are filled at any value of energy, $E$, is simply $f(E)$. If the Boltzmann approximation holds for both holes and electrons, the valence band is *essentially* filled and the conduction band is *essentially* empty. Accordingly, over the range of impurity concentrations that correspond to the validity of the Boltzmann approximation, we should expect that the generation rate would remain rather constant. Of course, if the sample is impure enough so that the Fermi level is close to the conduction or valence bands, the generation rate would be expected to fall off. Also, if an electron passes through some intermediate states on the way to the conduction band, it is possible to develop some "traffic jams" that make the equilibrium generation rate dependent on the impurity concentration. More of this appears in Chapter 4.

The same sort of reasoning applies to the ionized-donor-electron generation rate, with the added feature that for shallow donors, the "reaction" goes to completion. Once again, this added feature applies only if the Fermi level is far from the donor energy level.

If you think of the Boltzmann approximation as the description of an almost-empty system, you can get some feeling for how much of a given impurity you can put into a crystal before it begins to "fill up." Consider the case of reasonably shallow donors. The total ionization of these donors can only occur if the density of states in the conduction band vastly exceeds the density of electrons being put into the states. In other words, if the Boltzmann approximation is to hold, the volume (in $k$-space) of the conduction band must be much larger than the volume occupied by the conduction electrons. The "volume" above the donor states is $N_c$. Thus, if $N_D \ll N_c$ you should expect the donors to be totally ionized. For Ge or Si, total ionization of shallow impurities is a reasonable assumption for impurity densities of the order of $10^{18}/\text{cm}^3$ or less. To get some idea of how shallow "shallow" is, compute $E_c - E_f$ in Ge with $N_D = 10^{18}/\text{cm}^3$. Compare this with $kT$ at $300°$K. [Answer: Using $N_c = 1.09 \times 10^{19}/\text{cm}^3$ and $kT = = 0.026$ eV, $E_c - E_f = 0.062$ eV. This is only $2.4\,kT$, making the Boltzmann approximation a bit weak at best. Accordingly, for moderate accuracy, the shallow impurity would have to be essentially at the band edge. For the typical donor from period V, the donor level is 0.01 eV below $E_c$, which is a bit too much for the Boltzmann approximation to be acceptably accurate.]

Having accepted the idea that the generation rate should be independent of doping over the range corresponding to the Boltzmann approximation, the question to ask is: What happens to the recombination rate as the doping is increased? The answer comes directly from the law of detailed balancing: The recombination rate at equilibrium must be the same as the generation rate. (Remember that the *rate of generation* is the number generated per unit volume per second.) Since we have just decided that the generation rate is independent of the doping level, the recombination rate must be doping-independent too. What then has changed? [Answer: If the equilibrium rate (either up or down) is $R_E$, then the ratio of $R_E$ to $n$ or $p$ is very much a function of the impurity concentration. This ratio has the units (seconds)$^{-1}$. It is the inverse of the time that the average electron or hole exists before it recombines. Note that if $n \neq p$, $\tau_e \equiv n_0/R_E \neq p/R_E \equiv \tau_h$.] It is evident that the carrier in the majority is much longer lived than the carrier in the minority. This makes most exceedingly good sense. In order for an electron to recombine, it must find itself a hole. If electrons vastly outnumber holes, very few electrons can find a hole to recombine with. On the other hand, all the holes have ample opportunity to recombine with electrons, since there are so many of them around. Here lies the secret behind the establishment of equilibrium.

If we generate $R_E$ hole-electron pairs per second, and the average lifetime of a hole is $\tau_h$, the average density of holes is $R_E\tau_h = p$. Similarly, the average number of electrons is $n = R_E\tau_e$. The addition of a large number of donors makes a great many electrons available without providing an equivalent number of holes. Thus, the holes have many more opportunities to recombine than they had before, while the electrons have proportionately fewer. The equilibrium number of holes goes down as the number of electrons goes up. As we have seen, the conditions required by the law of detailed balancing lead to the result that the two lifetimes change in just the right proportions to yield $pn = n_i^2$.

It is extremely important that you realize that the argument given above gives little information about the actual lifetime of a hole or electron. It was merely a discussion of the dynamics that lead to a law of mass action. All the conclusions are based on what the equilibrium state is for a given impurity distribution; in a sense, it was making up a problem to fit an answer. The discussion of the lifetime of excess holes or electrons is properly left to Chapter 5, in which several nonequilibrium problems are examined.

## SOLVING FOR $E_f$ IN THE CASE OF DEEP-LYING IMPURITIES

For the case in which all the impurity levels satisfied the Boltzmann approximation (the shallow-impurity case), we have a particularly simple method of finding the Fermi level and the carrier concentration. In brief, the technique for impurity concentrations $\gg n_i$ comprises the following.

**1.** Set the majority carrier density equal to the net impurity concentration (e.g., $n = N_D - N_A$).

**2.** Set the minority carrier density equal to the intrinsic carrier density squared divided by the majority carrier density (e.g., $p = n_i^2/n$).

**3.** Find $E_f$ using either (3.41), (3.42) or their equivalents, (3.23) and (3.24) [e.g., $E_f - E_i = kT \log(n/n_i)$].

For the case of the intrinsic or nearly intrinsic material, this routine has to be modified slightly to include the intrinsic carrier concentration. In that case, (3.39) replaces step 1 and (3.43) is used in step 3. (They can, of course, be used in all "shallow-impurity problems," but the method outlined is even easier.)

Since most impurities are not "shallow," it is necessary to have a technique or two for solving for the carrier concentration in the presence of "deep" impurities. Actually, the problem is precisely the same, but the Boltzmann approximation may not be used for the impurities. We thus end up with a transcendental equation, one that must be solved by either a graphical or successive approximation technique.

You might ask how often such problems are actually met in practice, and the answer is rather frequently. The most common problem is the heavily doped semiconductor rather than the intrinsically deep impurity such as Au or Cd. For reasons that show up in Chapter 7, the emitter region of most junction transistors is quite heavily doped. This requires a Fermi level relatively close to one of the band edges and results in only partial ionization of the impurities. In other words, the Boltzmann approximation for the impurities is no longer valid. At sufficiently high concentrations of impurities, the Fermi level must lie within the band itself, leading to a complete breakdown of the approximations we have made, but this situation is too complicated for any simple analytical method. Accordingly, let us take a quick look at the solution of (3.32) when the Boltzmann approximation does not hold for the impurity levels. In particular, let us consider the solution of the practical problem of finding the carrier concentration for Si with $10^{18}$ phosphorus atoms per $cm^3$ and no more than $10^{16}/cm^3$ of other donors or acceptors. The first question to answer is: With this doping, *approximately* where does the Fermi level lie and what is the "type" of the Si? [Answer: Since phosphorus is the dominant dopant, the material is *n*-type. Furthermore, with such a heavy concentration of P, the Fermi level must lie close to the conduction-band edge. If the donors were totally ionized, $n = N_D = 10^{18} = N_c e^{-(E_c - E_f)/kT}$. From Table 3.1 and equation 3.19, we get $N_c = 2.88 \times 10^{19}/cm^3$. Thus, $(E_c - E_f) = -kT \log(n/N_c) = 0.026$ $\log(28.8) = 0.0875$ eV.] Since P gives a level of 0.045 eV from the band edge, it is pretty evident that $E_D - E_f$ is not $\gg kT$. Therefore, the donors are not completely ionized, and the Fermi level must be somewhat further from the band edge than the calculation above indicates.

Putting in the exact Fermi function in (3.32) (which still retains the Boltzmann approximation for the bands themselves) yields

$$p + N_D \left( \frac{1}{1 + \frac{1}{2} e^{(E_f - E_D)/kT}} \right) = N_c e^{-(E_c - E_f)/kT} + N_A^- \qquad (3.44)$$

Since $p$ is the order of $100/cm^3$, that term can be dropped. Similarly, if $N_A$ is of the order of $10^{16}$, that term also can be ignored. (Note that the Boltzmann approximation WOULD hold quite well for the acceptors. The Fermi level is certainly quite far away from the acceptor levels.) That leaves us with

$$N_D \left[ \frac{1}{1 + \frac{1}{2} e^{(E_f - E_D)/kT}} \right] = N_c e^{-(E_c - E_f)/kT} \qquad (3.44')$$

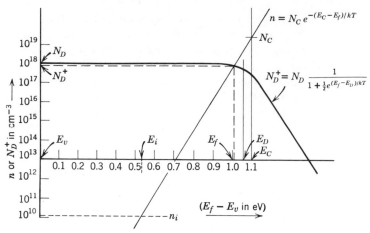

**Figure 3.5** Graphical solution of equation 3.44.

If you plot the log of both sides of (3.44′), you get the solution quite directly. Figure 3.5 is the required graph. The straight line is the plot of the log of the right side of the equation. The log of the left side of the equation is the other curve. Their intersection marks the solution. The numbers on the abscissa are the values of $(E_f - E_v)$.

Unfortunately, for impurity energy levels that are neither shallow nor really deep, you do not obtain much accuracy on a log plot. This proves to be the case for the problem we are considering. About the best you can do in determining $n$ from the graph is $n = N_D^+ \simeq 0.9 N_D$. If you really want an answer to several decimal places, you must obtain the answer by successive approximation. By that technique, the answer to two decimal places is $(E_c - E_f) = 0.089$ eV and $n = 0.93 \times 10^{18}$ cm$^{-3}$. From this you can see that you would have introduced a 7% error by using the Boltzmann approximation, an error that would usually not be significant. However, do not conclude from this result that you can use the Boltzmann approximation at all times. For example, compute $n$ for the case of $10^{18}$ sulfur atoms/cm$^3$. (Answer: Using the graphical technique, $n \simeq 0.33 \times 10^{18}$ cm$^{-3}$. If you used the Boltzmann approximation here, you would get an answer that was approximately 300% too large. That IS a significant error!)

## THE IMPORTANT CASE OF NONUNIFORM DOPING

All semiconductor devices have at least some purposeful nonuniformities built in. The most obvious of these are junctions between one type of material and another. For example, a metal-semiconductor contact represents an abrupt change between the distribution of states characteristic of the metal to the distribution of states characteristic of a doped or pure semiconductor. If the two materials are brought together and left to come into equilibrium with each other, what does our equilibrium statistics tell us about the distribution of electrons within the whole system?

To answer this question we must return to our derivation of the Fermi distribution function to see what differences, if any, this composite system would produce. The first question and the most important is: Did any step in the derivation of the distribution function depend in any way on whether the allowed states were localized or distributed throughout the crystal? (Answer: No. In considering the distribution of electrons among the states, it was assumed that electrons could get from one state to another and that all states of a given energy, no matter how they are distributed spatially within the system, have an equal probability of being filled.) We can draw two fundamental conclusions from the answer above: (1) The same distribution function applies to the heterogeneous system as applied to the simple, single-species system we considered at first, and (2) the Fermi level is a constant for the whole system, although if you change the system in any way, the Fermi level may change to a different value. This second point is actually covered by (1), but it is important that you recognize explicitly that the Fermi level is a single number for the whole system.

The techniques we have developed so far are still useful, but, with the added complexity of the heterogeneous system, we will need some additional tools to obtain the carrier concentration as a function of position throughout the system. The development is clearer with an explicit example, so let us consider a system comprising a piece of $p$-type Ge ($10^{16}$ Ga atoms/cm$^3$) being brought into intimate contact with a piece of $n$-type Ge ($10^{17}$ As atoms/cm$^3$). First, letting $T = 300°$K, find the Fermi levels in the two materials before they come in contact. [Answer: For the $p$-type material, $(E_i - E_f) = = 0.155$ eV. For the $n$-type material, $(E_f - E_i) = 0.225$ eV.]

Now we must consider the positions of the band edges with respect to each other. Remember, of course, that until we bring the two pieces together, they are not in equilibrium with each other. A convenient energy from which to measure the various band edges is the famous "point at infinity." That is, we wish to measure the energy it takes to remove an electron from the band edge completely out of the crystal. If we made this energy measurement on the two isolated crystals above, we should expect to get the same result. That is, reasonably pure homogeneous Ge has a particular band structure and none other. Of course, the $n$-type crystal would have a much larger number of occupied conduction-band states, but they would be the same states that would be found empty in the $p$-type crystal. If we make a plot of the energy levels that we find as a function of position throughout the crystal, we get a plot looking something like Figure 3.6. The solid areas represent

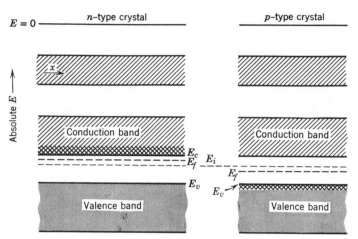

**Figure 3.6** Energy eigenstates as a function of position in two isolated Ge crystals.

filled bands; the cross-hatched areas are partially filled states; and the empty bands are shown as lightly hatched. Note that the only differences between the two diagrams are (1) the position of the Fermi level is not the same in the two systems and (2) the partially filled regions are not the same in the two regions. Of course, (1) implies (2), but it is important to be aware of both differences.

Now what happens if I bring the two systems together? (Answer: You might be tempted to say that nothing much would happen, this answer being based on the lack of activity usually observed when two dissimilar but nonreactive materials are brought into contact. Actually, however, if two materials are brought into intimate contact—and this involves getting rid of all the usual layers of surface oxides and adsorbed gases—a transitory reaction does take place.) After all, as soon as the $n$-type material is in contact with the $p$-type material, the electrons in the conduction band of the $n$-type material can wander freely into the conduction band of the $p$-type material. Similarly, the holes in the valence band of the $p$-type sample can move into the $n$-type material's valence band. As a matter of fact, the principle of detailed balancing suggests that they would have to distribute themselves rather equally between the two samples were it not for one factor: The electrons moving into the $p$-material constitute a net flow of charge. Associated with this net flow of charge must be an electric field and an electrostatic potential. Thus, the first charge goes over without difficulty, but the next one to follow it must do work to overcome the potential that the first has created. The third must overcome the potential of the first two and so forth. Note how different this is from the case of homogeneous doping. There, for every electron that went into the conduction band, there was a compensating positive charge left behind—either a hole or an ionized donor. The electrons that move from the $n$ to the $p$ material are completely uncompensated. Thus, it becomes increasingly difficult to move electrons into the $p$ material and the process brings itself to a halt long before a uniform distribution of electrons is achieved. Of course, if there is no net flow, the crystal has reached thermal equilibrium, so the self-generated electrostatic field must be part of the equilibrium state. Let us now take a look at how that fits into the Fermi distribution.

The static potential is very real; it is not an accounting convenience like the Fermi level. Thus, if I move an electron from the bottom of the conduction band in the $n$-type material to the bottom of the conduction band in the $p$-type material, I *must do work on it*. Therefore, the bottom of the conduction band in the $p$-type material is now at a higher potential than the bottom of the conduction band of the $n$-type material. In other words, there has been a shift in the distribution of energy levels. Note that the shift described is in the correct direction for bringing the Fermi levels of the two samples into proper alignment. We must now take advantage of our knowledge that there is but one Fermi level for a system to obtain the equilibrium distribution and the electric field throughout the two-crystal sample.

## POISSON'S EQUATION AND THE DISTRIBUTION FUNCTION

Figure 3.7 shows the sort of thing we should expect from the discussion above. On the left side is the $n$-type material, on the right the $p$-type. Because of the potential between the two materials, an electron must be well above the conduction-band edge on the $n$-side to be able to travel into the $p$-side. Because electric fields are continuous and have continuous derivatives, we

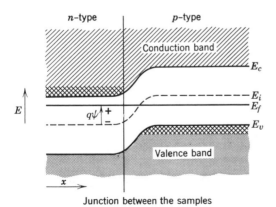

Junction between the samples

**Figure 3.7** Energy of the bands as a function of position for a function between a $p$-type semiconductor and an $n$-type semiconductor. Compare with Figure 3.6.

should expect a smooth shift in the potential energy as the position changes. Furthermore, we should expect the change to take place reasonably close to the junction, since the shift should be moderately independent of sample size. In a sense, the electrons should not know that there is a junction there until they get reasonably close to it. So much for what to expect; what can we do to get an analytic description?

In principle, we can solve the problem exactly. We now have a small charge imbalance, so equation 3.37 no longer holds. Instead, we must write Poisson's equation relating the charge imbalance to the second derivative of the electrostatic potential:

$$\frac{d^2\psi}{dx^2} = \frac{-q}{\varepsilon}(p + N_D - n - N_A) \tag{3.45}$$

$\psi$ is the electrostatic potential and $-q\psi$ is thus the electrostatic energy per electron. This electrostatic energy is just the shift in energy levels that we have been discussing above. Singe $E_f$ is constant for the system, it makes sense to measure the energy differences with respect to $E_f$. Furthermore, it is only the *change* in $E$ with respect to the Fermi level that is important, so we could choose the shift of $E_c$, $E_v$, or $E_i$ as the variable to observe. The most convenient turns out to be $E_i$, so let us define $q\psi = (E_f - E_i)$. Putting $q\psi$ into (3.41) and (3.42) and then substituting them into (3.45) gives the set of equations that describe the distribution of electrons completely:

$$p = n_i e^{-q\psi/kT} \tag{3.42'}$$

$$n = n_i e^{+q\psi/kT} \tag{3.41'}$$

$$\frac{d^2\psi}{dx^2} = -\frac{q}{\varepsilon}\left[n_i e^{-q\psi/kT} + N_D(x) - n_i e^{q\psi/kT} - N_A(x)\right] \tag{3.45'}$$

These equation are completely general. To discuss a particular problem, you have to specify the doping profile [i.e., $N_D(x)$ and $N_A(x)$] and boundary conditions. However, before launching into specifying the problem, it is worthwhile noting what an awful equation (3.45') is. It is a second-order transcendental differential equation. No general solution to it has been found. Fortunately, if one tromps hard enough on (3.45'), it is possible to beat it into tractable form by some drastic approximations. These approximations lead to what is generally called the *depletion model*.

## SOLUTIONS IN THE DEPLETION APPROXIMATION

What we would like to do with (3.45') is turn it into a linear differential equation. Then we could undoubtedly solve it. There are two ways of accomplishing this purpose. One would be to assume that $p = n = 0$. That is, assume there is no free charge at all; only the donor and acceptor ions remain and they are locked in place in the crystal lattice. The volume in which this approximation is valid is "depleted" of free charge; hence it is called the *depletion region*. (This depletion approximation should seem like a rather drastic assumption; it is. However, it is moderately defensible for very abrupt junctions, and it does lead to a solvable equation.) The depletion assumption leads to the equation

$$\frac{d^2\psi}{dx^2} = \frac{-q}{\varepsilon}\left[N_D(x) - N_A(x)\right] \tag{3.46}$$

The other possibility for linearizing (3.45′) is to set the right side of the equation equal to zero. That is what we used before to write (3.37). If we set the right side of (3.45′) equal to zero, we are assuming that there is no *net* charge in the region—that the crystal is neutral. Such is certainly the case in the typical conductive crystal, so this assumption should not be too bad at the right time and place. However, it should be pretty obvious that you cannot go from one assumption to the other without passing through a region where neither assumption is valid. Our defense for the use of these equations must really rest on the fact that for certain problems, the transition region does not significantly affect the answer. Unfortunately, such an argument must be considered separately for each specific case. In cases in which we are not trying to find the spatial distribution of the charge in detail, the two extreme assumptions will stand us in good stead. On the other hand, for the calculation of certain parameters that depend critically on the spatial distribution of the uncompensated charge (notably the junction capacity), these assumptions could be expected to yield rather large errors.

Take a look at what these approximations mean for our present junction as illustrated in Figure 3.7. Far to the right, the density of holes is $10^{16}$ cm$^{-3}$, while far to the left the hole density is $6.25 \times 10^9$ cm$^{-3}$. The figure suggests that this change in density occurs over a very small region—less than $1/3$ of a micrometer. If instead of "very small" we were to use "infinitesimal," we would get the essentially-no-free-charge approximation in the interval between where the high density of $p$ ends and the high density of $n$ begins.

Once we have made this pair of assumptions, solutions to (3.45′) become rather easy. All of the volume containing a net charge density lies in a narrow region surrounding the metallurgical junction between the $n$- and $p$-sides. The rest of the crystal has the "bulk" property of being neutral. Since the whole crystal was electrically neutral before the formation of the depletion region, it must be neutral afterward. Thus, the charge due to the ionized donors on the $n$ side is equal to minus the charge due to the ionized acceptors on the $p$ side. (The sum of the charges adds up to zero without the depletion assumption. What the assumption does for us is to allow us to attribute the charge distribution to the ionized impurities whose distribution we presumably know a priori.) The solution for the electric field is accomplished simply by integrating equation 3.46. Figure 3.9*a* shows the charge distribution from Figure 3.8 in the depletion approximation. The only unknowns, according to the depletion approximation, are the *depletion lengths*, $l_n$ and $l_p$. However, only one is unknown, since neutrality requires $N_A l_p = N_D l_n$. Integrating (3.46) once gives the electric field as

$$E = -\frac{d\psi}{dx} = \begin{cases} 0 & x < -l_n \\ \frac{q}{\varepsilon} N_D(x + l_n) & -l_n \leq x \leq 0 \\ \frac{q}{\varepsilon} N_A(l_p - x) & 0 \leq x \leq l_p \\ 0 & l_p < x \end{cases} \tag{3.47}$$

This function is plotted in Figure 3.9*b*. A second integration yields the electric potential as

$$\psi = \begin{cases} \psi_n = \frac{kT}{q} \log\left(\frac{N_D}{n_i}\right) & x < -l_n \\ & \text{from equation 3.41} \\ \psi_n - \frac{q}{2\varepsilon} N_D(x + l_n)^2 & -l_n \leq x \leq 0 \\ \psi_p + \frac{q}{2\varepsilon} N_A(l_p - x)^2 & 0 \leq x \leq l_p \\ & \text{from equation 3.42} \\ \psi_p = -\frac{kT}{q} \log \frac{N_A}{n_i} & l_p \leq x \end{cases} \tag{3.48}$$

Note in (3.48) that I have used the fact that the difference in potential energy between the two bulk regions must be the difference in the Fermi levels that the two bulk regions would have if they were separated. In other words, independent of the depletion approximation, the total change in the electric potential in going from one bulk region to the other is just sufficient to align the Fermi levels. This shift is easily written down from (3.48). It is called the diffusion potential, $V_D$.

$$V_D \equiv \psi_n - \psi_p = \frac{kT}{q} \log \frac{N_D N_A}{n_i^2} \tag{3.49}$$

Evaluating $V_D$ for our example yields $V_D = 0.38$ V. Thus, in (3.48) only the diffusion lengths are unknown. By setting the two expressions for $\psi$ equal at $x = 0$, we obtain

$$\psi_n - \psi_p = V_D = \frac{q}{2\varepsilon}(N_D l_n^2 + N_A l_p^2) \tag{3.50}$$

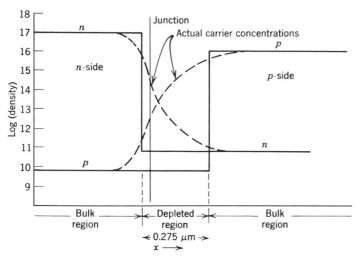

**Figure 3.8** Hole and electron concentrations in the depletion approximation (solid lines) and the real distribution by the solution of (3.45′) (dashed lines). Notice that the plot is semilogarithmic. Values are typical for Ge.

$l_n$ and $l_p$ are related by charge neutrality, so equation 3.50 can be solved for the depletion lengths to yield

$$l_n = \left[ \frac{2\varepsilon V_D}{q} \frac{N_A}{N_D(N_A + N_D)} \right]^{1/2} \tag{3.51}$$

$$l_p = \left[ \frac{2\varepsilon V_D}{q} \frac{N_D}{N_A(N_D + N_A)} \right]^{1/2} \tag{3.52}$$

Evaluating the diffusion lengths for our present example yields $l_n = .025 \ \mu m$ and $l_p = 0.25 \ \mu m$. Going back to Figure 3.9*b* we obtain $E_{max} = 28.2$ kilovolts/cm! This should strike you as a very large field to find in a reasonably conductive material. It is. The secret is that in the depletion region there is *essentially no mobile charge.* Thus, one way of looking at the depletion region is to describe it as a thin insulating layer separating two conductive regions. This view is useful but not entirely accurate.

Remember that the dynamics that led to the diffusion potential involved the constant thermal "reshuffling" of the electrons among the states. Thermal equilibrium was attained when this reshuffling resulted in no net flow. However, no *net* flow does not imply that there is no flow. In point of fact, there is a substantial flow of electrons and holes through the junction region. What the law of detailed balancing requires is that the flow to the right

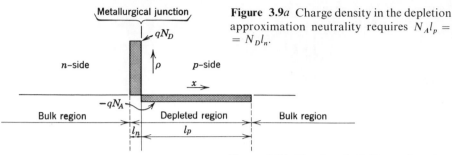

**Figure 3.9a** Charge density in the depletion approximation neutrality requires $N_A l_p = N_D l_n$.

**Figure 3.9b** The electric field as a function of position. This function is minus the integral of the right side of (3.46) using Figure 3.9a as the charge distribution.

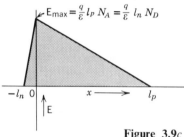

**Figure 3.9c** The electric potential as a function of position. This function is the second integral of equation 3.46. The values in the bulk regions are determined from equations 3.41 and 3.42.

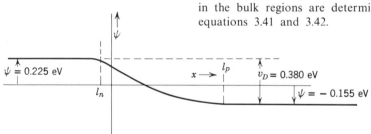

equal the flow to the left. If you wish to think in terms of the flows rather than in terms of the equilibrium statistics, what you should consider is this: Since most of the conduction electrons are on the left, a reshuffling tends to move electrons to the right. However, the electric field is directed to move electrons toward the left. Thus, the flow due to the electric field (called *drift*) is exactly balanced by the flow due to thermal reshuffling (called *diffusion*). Chapter 4 is devoted to the interesting consequences of meddling with one or another of these flows to get a net flow in one direction. It is only

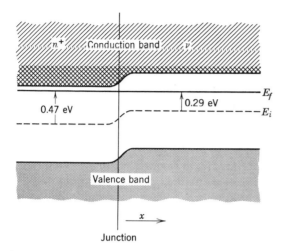

**Figure 3.10**   Energy of the bands as a function of position for an $n^+v$ junction. Compare with Figure 3.7.

by departing from equilibrium that we can make devices useful. What we have just derived is the point of departure. Chapters 4, 5 and 6 will show that the equilibrium distribution enters very strongly into the description of the behavior of devices operating in nonequilibrium situations.

To get some feeling for the limitations on the depletion approximation, consider the case of a junction between a heavily doped, $n$-type region (labeled $n^+$) and a lightly doped $n$-type region (labeled $v$). Let the $n^+$ region contain $10^{18}$ donors/cm$^3$ and the $v$ region $10^{15}$ donors/cm$^3$. If the material is Si, what would the energy bands look like as a function of position (i.e., draw the equivalent of Figure 3.7)? (Answer: If the $n^+$ region is on the left, then $\psi = (kT/q)\log(N_D/n_i) = 0.47$ V on the left and 0.29 V on the right. The transition should still be rather abrupt. Figure 3.10 illustrates these conclusions.)

Does a suitable depletion approximation exist for this $n$ junction? [Answer: In general, no. Although the $n^+$ region has a depleted volume on the left side of the junction, the net charge on the right side of the junction must be *conduction-band electrons*. They are the only source of negative charge in the $n$-type material. This means that on the right-hand side—the $v$-side—of the junction, there is the opposite of a depletion region; there is an *enhancement* region with *more* electrons than donors. Since this charge is free to move about, the charge distribution is not known a priori, and we cannot simply integrate a known function to solve (3.45′).]

This illustration should demonstrate how limited a tool the depletion approximation really is. It solves one problem, and not very exactly at that. However, that one problem—the relatively abrupt $p$-$n$ junction—is an extremely important one in device technology. Since $p$-$n$ junction characteristics are central to this text, take a moment to catalog the facts we have just treated:

**1.** A contact or diffusion potential builds up between dissimilar materials. This potential is just sufficient to align the Fermi levels in the two materials. For a $p$-$n$ junction, the diffusion potential is given by (3.49).

**2.** The diffusion potential is generated by an actual charge transfer from one material to the other. Since the two materials are presumably neutral to begin with, the interchange of charge still leaves the overall crystal neutral. In the depletion approximation this leads to $N_A l_h = N_D l_n$. Thus, the depletion zone is widest in the most lightly doped crystals and narrowest in the most heavily doped crystals.

**3.** By combining (3.49), (3.51), and (3.47), you can obtain an expression for $E_{max}$ in the depletion approximation. The result is

$$E_{max} = \left[ \frac{2kT}{\varepsilon} \frac{N_A N_D}{N_A + N_D} \log \left( \frac{N_D N_A}{n_i^2} \right) \right]^{1/2} \qquad (3.53)*$$

Equation 3.53 points out the fact that the maximum fields occur in junctions between heavily doped samples. This makes eminently good sense when you consider that junctions between heavily doped samples have large $V_D$'s and very short depletion widths. For example, a $p$-$n$ junction between two samples of Ge with $10^{18}$ cm$^{-3}$ impurity concentration would give an $E_{max} = 2.49 \times 10^5$ V/cm.

## ON THE EXPERIMENTAL OBSERVATION OF THE EQUILIBRIUM FLOWS AND THE DIFFUSION POTENTIAL

It might seem interesting to you to try to measure $V_D$ by putting a voltmeter across a junction diode (i.e., a $p$-$n$ junction). Think very carefully before answering the following question. Given any kind of voltmeter that suits your fancy, is it possible to observe a potential across the diode if the diode is in thermal equilibrium? (Answer: Yes, but not $V_D$. You may have been properly brainwashed to answer no, and that answer is correct if your meter measures the *average* potential. After all, the law of detailed balancing as-

---

* N.B. To use (3.53) you have to keep your units honest. If $\varepsilon$ is in farads/cm, $kT$ has to be in joules. If you put $kT$ in electron volts, keeping $\varepsilon$ in farads/cm, you can get some truly remarkable numbers. If you get an answer that isn't between $10^4$ and $10^6$ V/cm, you have probably made an error.

sures us that the current flowing out of one lead of the diode is exactly cancelled *on the average* by the current flowing into that lead. But every thermal shuffling cycle results in some net flow one way or the other. Accordingly, if the meter is a very fast, wide-bandwidth oscilloscope, we should expect to see a very wiggly trace best described as "hash" or noise. Although the average value of this signal is zero, its mean square value is not. Thus, a sensitive, wide-bandwidth, true R.M.S. meter will give a reasonably stable reading. This R.M.S. signal is called *thermal noise*, and its spectrum is that of a blackbody. We will return to this shortly, but first let us settle the question of measuring the diffusion potential.)

The problem with measuring the diffusion potential in an equilibrium system is a consequence of the Second Law of Thermodynamics. Essentially, $V_D$ is just the right potential to result in no flow and thus no measurement. To measure the "built-in" field in a junction, you have to disturb equilibrium a little. "A little" is here defined as a small enough perturbation to keep the diffusion potential approximately equal to its equilibrium value. Let us say, for example, that I had a bar of Ge in which the doping was linearly graded from one side to the other. Into the center of this bar, I inject a number of holes and electrons with a bright pulse of light. These holes and electrons are *excess* carriers, so there is no flow to counteract their motion. As Chapter 5 will show, the pulse of excess carriers will drift together in one direction or the other depending on which carrier is in the majority in equilibrium. This drift can be detected, so it is possible to measure the presence of the field due to the inhomogeneous doping. This acceleration of excess charges is actually quite useful in device technology. It is what pumps the charge around in a solar cell and it is also very important in high-frequency transistors. One of the limitations on high-frequency performance of a junction transistor is the time it takes an electron or hole to traverse the base region of the transistor. By establishing a substantial impurity concentration gradient across the base region, the transistor designer can considerably shorten the transit time for the excess carriers. As it turns out, the modern technique for making transistors leads to just such a doping gradient, so the shorter transit time is more or less free today.

## THERMAL NOISE AND BOSE-EINSTEIN STATISTICS

Let us now return to those random signals that appeared across the semiconductor bar. The analysis of these signals was first done by Nyquist in 1928, following the measurement of the noise voltage by J. B. Johnson in the same year. (Another common name for thermal noise is Johnson noise.) Nyquist's analysis was so pretty that I follow it quite directly here. [For the original paper, see *Phys. Rev.* **32**, 110 (1928).]

Let us say that we wish to predict the noise voltage would appear across a resistor at temperature $T$. The resistor might be our piece of germanium, but that really does not matter. All that we demand is that the sample have finite conductivity. Nyquist's clever scheme was to connect two equal resistors on opposite sides of a matched transmission line (see Figure 3.11). After the transmission line and the two resistors have come to equilibrium with each other, short circuits are applied at both ends. We now ask two questions: (1) What is the equilibrium distribution of energy in the transmission line? (2) If the shorts are removed, how rapidly will the energy flow into the resistors? (Note that no net energy flows into the resistors; the flow from the line to the resistors is balanced by the flow from the resistors to the line.) When these two questions are answered, we will know the power flow either into or out of the resistors. If I replace the noise-generating resistors by generators in series with noiseless resistors, I can calculate the average power flow from the generator into the line as $V_n^2/4R$ (the 4 comes from the fact that in the second circuit in Figure 3.11, the noise voltage from the generator is divided equally between the series resistor and the line.) Thus, if the answer to (2) gives the power as $P_n$, we have:

$$V_n^2 = 4RP_n \qquad (3.54)$$

Equation 3.54 is the result we are seeking. What we must do is to find $P_n$ as a function of $T$. Before launching into that endeavor, take a look at a few of the assumptions inherent in the reasoning above. First and certainly foremost, if we are to have a voltage of any reasonable sort across the resistor, $R$, the wavelength of the oscillations we are discussing must be quite large compared to the dimensions of our resistor. This assures us not only that we have a sensible voltage to work with but also that the resistor will not radiate much energy at these wavelengths. The resistor is not an antenna (or at least, it is a very, very poor one); from the blackbody point of view, its emissivity is negligible. Second, we want to treat the transmission line in its simplest form; in other words, we want to deal solely with the $TEM_{00}$ mode. Thus, the lateral dimensions of the line must be small compared to the wavelength. It is not that the problem could not be solved without these small-compared-to-the-wavelength assumptions; what we save ourselves is a very hairy field-theory problem — one that really serves little function, since the circuit and voltage aspects of the solution are lost in any case.

Let us start by answering question (1). First, what are the Eigenfunctions of the shorted line? [From elementary transmission line theory, $v(x) =$ $= v_n \sin(k_n x) e^{j\omega_n t}$ where $k_n = n\pi/L$, and $\omega_n = k_n v_p$, with $v_p$ being the phase velocity in the line.] What is the energy for each photon in the $m$'th eigenstate? [Answer: $E = \hbar\omega_n = \hbar m\pi v_p/L$.] Does equation 3.8 hold for this system? [Answer: Yes. Equation 3.8 is a statement of conservation of energy

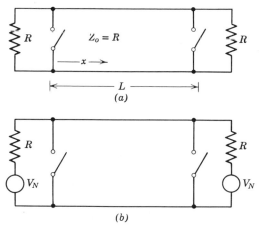

**Figure 3.11** Circuit used for Johnson noise derivation. (*a*) Nyquist's circuit; (*b*) Equivalent circuit using noise generators.

that certainly holds for this system.] Now for a somewhat harder question: Does the exclusion principal apply to this system? [Answer: No. You may excite any mode on a transmission line to any integer number of photons that suits your fancy. In other words, the voltage on a shorted transmission line can be arbitrarily large. This is certainly an intuitively agreeable result. It can be obtained directly from the discussion of spin in the previous chapter (shortly after equation 2.16′), in which it was stated that the Pauli exclusion principle applied only to particles with half-integer spin. The measured spin of the photon is one. Thus, an exclusion principle does not apply.] If you have $m$ states with $n$ particles in them, what is the form of (3.5) without the exclusion principle? [*Hint:* The easiest way to develop the expression for $P(m, n)$ with no exclusion principle is to consider in how many distinguishable ways you could draw the $n$ particles and $(m - 1)$ interior separators from a scrambled set. Consider $m = 2$ and $n = 2$, for example. There are three distinguishable arrangements: (2, 0), (1, 1), and (0, 2). You could consider these as the arrangements of (1), (1) and ( , ) (i.e., the three objects: a one, a one, and a comma).] Now what is the general form of $P(m, n)$? [Answer: The number of arrangements of $[n + m - 1]$ objects is $(n + m - 1)!$. $n!$ of these represent rearrangements of the particles and are therefore indistinguishable. Similarly, $(m - 1)!$ of these represent indistinguishable rearrangements of the ( , )'s. Thus, the number of distinguishable rearrangements of $n$ particles in $m$ states with no exclusion principle is

$$P(m, n) = \frac{(n + m - 1)!}{n!(m - 1)!}$$
(3.55)

The next step is to find the most probable distribution in the same fashion that we employed to find the Fermi distribution. This involved maximizing (3.6) subject to conditions (3.7) and (3.8). We have already decided that (3.8) still holds. How about (3.7)? [Answer: For photons the answer is no; the number of particles is not conserved. However, this is not necessarily true for all particles of integer spin (so-called Bosons). Light comes and goes with great regularity. For example, if a photon is absorbed by a black object, the photon is gone. Its energy is still there, but not in the form of a moving photon. Similarly, a hot object can emit a photon without the photon having been there before. Thus, the number of photons in a system is not necessarily constant. If we leave out the entire class of nuclear reactions, the world of particles can be neatly divided into those that are and those that are not conserved. The first class inevitably has rest mass (i.e., an intrinsic mass), while the second has mass only by virtue of its momentum. The first class of objects includes particles with half-integer (Fermions) and integer spins (Bosons)— for example, the electron (spin 1/2) and the helium atom of atomic weight 4 (spin 1 or 3). Similarly, among the nonconserved particles, you find both Fermions and Bosons—for example, the photon (spin 1) and the neutrino (spin 1/2). Thus, the answer on whether (3.7) holds depends on the particle.]

Let us consider the case of the photon, since that is the particle we would find traveling up and down the transmission line. To complete the task of finding the distribution function, rewrite equation 3.10 using the appropriate $P(m, n)$ and dropping (3.7). This yields

$$R = \log(P_t) - \beta E_t = \sum_i \left[ \log \frac{\{n_i + m_i - 1\}!}{n_i!(m_i - 1)!} - \beta n_i E_i \right] \quad (3.56)$$

Applying Sterling's approximation to (3.56) gives:

$$R = \sum_i [(n_i + m_i - 1)\log(n_i + m_i - 1) - n_i \log n_i - (m_i - 1)\log(m_i - 1) -$$
$$- \beta n_i E_i]$$

By differentiating w.r.t. $n_i$ and setting the result equal to zero, you obtain the most probable distribution as

$$\frac{\partial R}{\partial n_i} = \log(n_i + m_i - 1) - \log n_i - \beta E_i = 0$$

or, noting that $m_i \gg 1$,

$$\frac{n_i}{m_i} = 1/[e^{\beta E_i} - 1] \equiv f_B \quad (3.57)$$

Equation 3.57 is referred to as the Bose-Einstein distribution function. Once again, thermodynamics leads to replacing $\beta$ by $1/kT$. Note that (3.57) is quite different from the Fermi function (3.11). Most important to note is that the Bose function diverges at $E = 0$. Thus $f_B$ can be $\gg 1$, which is certainly different from the Fermi distribution. There is one striking similarity to the Fermi distribution: for $E \gg kT$, the Bose function can be approximated by the Boltzmann distribution. On the other hand, for energies much less than $kT$, $f_B \simeq kT/E = kT/h\omega$. At room temperatures, $kT$ represents a photon from the middle infrared spectrum. Thus, the radio frequencies that we are considering in the noise-spectrum calculations lie in the region where $E \ll kT$. We are now in a position to complete the Nyquist noise power derivation.

We know that the modes (eigenstates) in the transmission line are spaced at frequency intervals: $\Delta v = v_p/2L$. Thus, in any band of frequencies $dv$, the number of modes is given by

$$dm = (2L/v_p)dv. \tag{3.58}$$

If we multiply (3.58) by the Bose distribution function and the energy per photon, we will obtain the energy stored in the line between $v$ and $v + dv$:

$$dE(v) = hvf_B dm \simeq hv \left( \frac{kT}{hv} \right) \left( \frac{2L}{v_p} \right) dv \tag{3.59}$$

Now consider what happens if I open the switch on the left-hand side of the transmission line. Since the resistor matches the line, the line dumps all of the energy stored in it into the resistor in a time $\tau = 2L/v_p$. (The 2 comes in since half of the photons are going to the right as the switch is opened. Some of them will have to travel all the way down the line and back to get to the resistor. Hence the $2L$.) Thus, the noise power that flows into the resistor in the frequency range from $v$ to $v + dv$ is given by

$$dP_n = \frac{dE}{\tau} = kTdv \tag{3.60}$$

Putting (3.60) into (3.54) completes the derivation, giving the expression for Johnson noise as

$$\bar{V}_n^2 = 4kTR\Delta v \tag{3.61}$$

Note that (3.59) takes explicit advantage of the fact that, for the problem we are considering, $hv \ll kT$. This makes (3.61) the correct expression only for the low-frequency case in which low is defined as satisfying both the $hv \ll kT$ requirement as well as the constraint that $\lambda$ must be quite large compared to the dimensions of the resistor and the lateral dimensions of the transmission line.

## PLANCK'S LAW

If all we want is the power radiated by a body, we need not restrict ourselves to the low-frequency case. Rather than getting involved in an elaborate field-theory exercise, let us take the simple problem of finding the equilibrium radiation distribution in a cubical cavity one meter on a side. By considering only the equilibrium situation within the cavity, we avoid the problem of getting the radiation out of the cavity and coupled to some measurement apparatus in the outside world.

The distribution of modes within the cavity is given by

$$dm = 8\pi v^2 dv \ (V/c^3) = g(v)dv \tag{3.62}$$

For our special case of the 1-meter cavity, $V = 1 \text{ m}^3$. Note that (3.62) is really just a three-dimensional form of (3.58). Find the energy stored in the interval between $v$ and $v + dv$. Answer:

$$dE(v) = f_B(v) \ hvdm = \frac{8\pi h v^3 dv}{c^3(e^{hv/kT} - 1)} \tag{3.63}$$

Equation 3.63 is the famous Planck radiation law. It was the first great triumph of the quantum concept in physics, describing exactly the radiation that had been observed from hot bodies.

## A BRIEF EXERCISE

To make sure you understand the steps involved, find the distribution law for particles like He, which are Bosons that obey a particle conservation law. (Answer: The introduction of particle conservation merely reinstates the $\alpha$ term that we had in the Fermi function. Thus, for conserved Bosons,

$$f_B = 1/[e^{\alpha + \beta E} - 1] \Rightarrow 1/[e^{[(E - E_B)/kT]} - 1] \tag{3.64}$$

$E_B$ is an accounting device quite similar to $E_f$. You must select the Bose level so that $\sum n_i = N$.)

## CHAPTER SUMMARY

This chapter concerned itself with that peculiar and infrequently observed state called *thermal equilibrium*. Several equivalent definitions of thermal equilibrium were introduced, all of them derived from the law of detailed balancing. The two most important viewpoints are that thermal equilibrium represents the distribution with the highest probability of occurrence and that thermal equilibrium represents the distribution in which no net flows of any sort occur.

Several specific systems of particular interest to semiconductor technology were considered. The first and most important was the case appropriate to fermions—particles with $(n + 1/2)$ units of spin. In particular, we first considered the specific example of electrons. The rules appropriate to the distribution of electrons among a set of independent states were:

**1.** The exclusion principle allowed only one electron per state (counting spin).

**2.** The number of electrons was a constant $(\sum_i n_i = N)$.

**3.** The total energy of the system was conserved $(\sum_i n_i E_i = E_t)$.

We then found the probability that a particular set of equivalent states would occur and found the set with the maximum probability subject to the rules above. Our result was the Fermi distribution function: $f(E) = 1/(1 + e^{(E - E_f)/kT})$.

Using the Fermi distribution law and assuming parabolic bands and the applicability of the Boltzmann approximation to $f(E)$, several useful relationships were found between the temperature, the band shape, and the number of carriers in an intrinsic (pure) semiconductor. These included

**1.** $$N_{c,v} = \left( \frac{m_{e,h}^{**}}{\pi} \right)^{3/2} \frac{1}{\sqrt{2\hbar^3}} (kT)^{3/2}$$

**2.** $$E_i = \frac{E_c + E_v}{2} - \frac{kT}{2} \log\left( \frac{N_c}{N_v} \right)$$

**3.** $$n_i = N_c e^{-(E_c - E_i)/kT} = N_v e^{-(E_i - E_v)/kT}$$

In the process of finding the above relationship, the general expression for $n$ and $p$ were derived. These were

$$p = N_v e^{-(E_f - E_v)/kT} = n_i e^{(E_f - E_i)/kT}$$
$$n = N_c e^{-(E_c - E_f)/kT} = n_i e^{(E_i - E_f)/kT}$$

From the expressions for $n$ and $p$ it is evident that a law of mass action applies to their product, to wit:

$$pn = n_i^2$$

All of these relationships depend on the validity of the Boltzmann approximation. If the material is degenerate, they are worthless.

We then considered a somewhat different Fermi system in which pairs of levels were not independent. These levels were attributed to impurities in the semiconducting crystal and were found to lie in the forbidden band.

The pairing of these levels occurred because they were "localized," and the filling of one level changed the potential to a degree that obliterated the other level. Thus, the first rule listed above became:

**1′.** The exclusion principle allowed only one electron per *pair* of states (counting spin).

For this system we found that the Fermi function was given by

$$f' = \frac{1}{1 + \frac{1}{2}e^{(E - E_f)/kT}}$$

We then discussed the method for finding $E_f$, $n$, and $p$ for the very important case of uniform doping with "shallow" impurities. Here the Boltzmann approximation applies to all levels. We concluded that all the impurities would be ionized under these conditions, and the following rules could be used to find $n$, $p$, and $E_f$, if, for example, the material was $p$-type:

**1.** $p = N_A - N_D$

**2.** $n = n_i^2/p$

**3.** $E_f - E_i = kT \log (N_A/n_i)$

So-called deep-lying impurities (i.e., impurity levels for which the Boltzmann approximation does not happen to hold because of the temperature or the impurity distribution) were discussed. A graphical or cut-and-try technique was required to solve those problems, since the degree of impurity ionization was not known a priori.

Nonuniform doping was the next important subject. The first step was to establish the fact that an internal electric field would exist within the non-uniformly doped sample. Thus, Poisson's equation was coupled with the distribution law. The result was horrendous, but some drastic and only partially justified approximations allowed us to solve at least one important problem — the abrupt junction between an $n$-type and a $p$-type sample. By the use of the *depletion approximation*, we obtained the depletion lengths [ (3.51) and (3.52) ] and the maximum field within the junction (3.53). Without resorting to any assumptions, we were able to obtain the contact potential, $V_D$, from the requirement that in equilibrium, a system can have only one Fermi level. This led to (3.49). Typical numbers for $V_D$ are a few tenths to one volt, for $E_{max}$ are of the order of $10^5$ volts/cm, and for the depletion width are of the order of a few tenths of a micrometer.

The final topic considered in this chapter was the Nyquist derivation of the power spectrum for Johnson noise. To establish that relationship, we had to find the distribution function for Bosons (particles with integer amounts of spin). We obtained the Bose-Einstein distribution functions for particles that were and were not conserved. The central feature for Bosons is that no exclusion principle applies. The Nyquist formula applied to the low-frequency portion of the spectrum where the power per unit spectral width in a one-dimensional structure was proportional to the temperature. The more general case, which included all frequencies and was generalized to three dimensions, gave the Planck radiation law for blackbodies.

## Problems

1. The depletion approximation is based on the assumption that the density of mobile carriers in the high field region in a $p$-$n$ junction is negligible compared to the fixed charge density. As a test of the self-consistency of this approximation, use the approximation and (3.41') to calculate the fraction of the depletion region in the diode described below in which the free charge is less than $10\%$ of the fixed charge. The diode is an alloy junction that gives a very abrupt transition from $n$ to $p$ type. Let the $n$-type layer be Sb-doped germanium with $10^{16}$ Sb atoms/cm$^3$. The $p$-type layer is gallium-doped germanium with $10^{18}$ Ga atoms/cm$^3$. Let $q/kT = 40$ V$^{-1}$.

   *Answer:* Only $74\%$.

   Would the result differ substantially for a similarly doped silicon diode?

   *Answer:* No.

   **Comment.** The depletion approximation looks much better when it is used on reverse-biased diodes, and it is there that it finds its principal application.

2. In the classical theory of specific heat, each electron should contribute $3kT/2$, on the average, to the energy of the solid. This result comes from the general principal (based on Boltzmann statistics) that each dynamical variable has an average value of $(1/2) kT$ and each electron has three degrees of freedom. (Note well: a degree of freedom means that the electron may have any value for a given dynamical variable, such as velocity in the $x$-direction.) Since there are roughly $N_0 = 10^{22}$ electrons/cm$^3$ of most solids, and at room temperature $kT = 1/40$ eV $\cong 4.0 \times 10^{-21}$ joules, one would expect that at room temperature the electronic energy/cm$^3$ would be of the order of 400 joules. Furthermore, with the specific heat defined as $dE/dT = C$, one would expect $C = (3/2) kN_0 = (3/2)(1.33)$ joules/$^\circ$C cm$^3$. In practice, the electronic specific heat at room temperature is so small that it can hardly be observed (compared to the crystal lattice specific heats). Using Ge as an example, discuss why the electronic specific heat should be much less than the classical result. Also, state what the ratio of the electronic specific heats would be for $10^{17}$ $n$-type Ge and intrinsic Ge.

3. What you call a "shallow" and "deep" impurity level depends to some extent on what effect you are measuring. For example, consider the conductivity a sample of Si ($E_g = 1.1$ eV) with $10^{16}$ donors lying 0.005 eV below the conduction band and $2 \times 10^{16}$ so-called deep donor levels. Since the conductivity is simply proportioned to the number of carriers, how far below the conduction band must the deep impurities lie so that they contribute 50% of the total conductivity at 300°K? Let:

$$N_c = 2.88 \times 10^{19}/\text{cm}^3$$

$$N_v = 1.08 \times 10^{19}/\text{cm}^3$$

$$n_i = 1.4 \times 10^{10}/\text{cm}^3$$

4. (a) Under what circumstances is the Fermi distribution the appropriate description of the occupation number versus the state energy?
(b) Given the conditions you specified in (a), under what circumstances may one substitute the Boltzmann distribution for the Fermi distribution?
(c) What purpose did the transmission line serve in the Nyquist derivation of the Johnson noise formula?

5. Ohmic contacts are made to a p-type resistor (shown below) by diffusing boron in at both ends to give the impurity profile shown. Find the diffusion potential between the ohmic contacts.

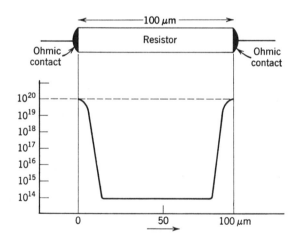

6. Silicon carbide is a compound semiconductor of great hardness, refractoriness, and chemical stability. It is exceedingly hard to prepare properly, but some of its properties inspire one to try. (No one has yet succeeded in creating large single crystals.) Some of its properties are listed below.

You are asked to comment on the implications of these properties to two specific devices.

*Properties of Cubic SiC* (there are noncubic forms with other properties):
Type of lattice—zincblende (ZnS) with lattice constant 4.358 Å
Band structure—indirect with a band gap of 2.6 eV
$n_i^2$ (300°K) $= 0.2 \times 10^{-10}/cm^{-6}$
$\mu_e$ (best value to date) $= 1100 \ cm^2/V \ sec$
$\mu_h = 0.25 \ \mu_e$
Typical minority lifetime $= 10^{-7}$ seconds
Impurity activation energy: N 85 meV
                            Al 275 meV
Best purity available—about $10^{16}$ active impurities/cm³
$kT$ (300°K) $= 0.026$ eV

**Problem:** If I put a net concentration of $10^{17}$ N atoms/cm³ into the crystal, will they be "thorougly" ionized (thoroughly $= 99\%$)? Would the crystal be *n*- or *p*-type?

7. A simple sensitive differential thermometer can be made by connecting a pair of diodes as shown below. Assuming that both diodes are made on silicon substrates with $10^{16}$ As atoms/cm³ and that the *p* side has $10^{18}$ boron atoms/cm³, compute the contact potential difference between the two diode leads for a temperature difference of 1°C at 300°K. (If you assume that any of the variables do not change significantly over the 1° change, show that you are correct.) Would a more heavily doped substrate give a larger or smaller voltage?

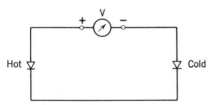

*Answer:* $dV/dT = 0.61$ mV/°C; larger.

# 4

# THE MOTION OF ELECTRONS
# IN REAL CRYSTALS

Chapters 1 and 2 were devoted to developing some solutions to the time-independent wave equation for the case of the perfectly periodic crystal. Entertaining as these solutions might be, they belong to that never-never world of weightless pulleys and frictionless inclined planes that came and went in freshman physics. Before we can solve real problems, we must account for the fact that there is no such thing as a perfectly periodic structure nor even a truly time-independent potential. Even the best of crystals is loaded with impurities, laced with imperfections, and in a state of constant vibration as a consequence of its internal energy (heat). What good are our idealized solutions? Well, they cannot be too bad for some uses; after all, they have already predicted insulating and metallic crystals, the effective mass behavior of electrons and holes, the hole itself, and the wavelength dependence of the internal photoeffect. All of these effects are observed by very real and physically meaningful experiments. On the other hand, the ideal crystal would have some properties that are not observed. Most conspicuous of these is the weird motion of electrons under the influence of an electric field. With reference to Figure 2.18, note that the application of a constant force, $q\mathsf{E}$, causes $k$ to increase continually. Figure 2.18 shows that this would cause the electron's energy to oscillate. In the reduced diagram (Figure 2.19), $k$ also would oscillate, appearing at $-\pi/d$ as it left $\pi/d$. Thus, the electron would alternate between doing work on the field and having work done on it by the field. Since such oscillations (called Zener oscillations) are not observed, the perfect-crystal idealization must break down. You should also note that I have comfortably introduced and used transitions between states even though such a time-dependent change demands the time-dependent wave equation. How is it that we can use the time-independent equation where the time-dependent one belongs or the perfect lattice as a model of the imperfect? Or even more to the point, when will such approximations work?

The answer to the question above requires that we separate our nonconstant parameters into those that change very slowly (either with position or time) and those that vary rather abruptly. The abrupt change has already entered in the form of boundary conditions on finite-sized crystals. Our argument for the use of Bloch waves within a finite but otherwise perfect crystal should be as follows.

**1**. Within the crystal, only allowed solutions can propagate very far, so at any reasonable distance from the surface all the observed waves are allowed Bloch waves.

**2**. The surface reflects a certain fraction of the incident waves (100% for the isolated crystal) which implies that time-independent solutions exist only for the set of discrete wavelengths that "fit" the crystal as a whole.

The use of the word "abrupt" here means that the change from being "inside" to being "on the surface" takes place in a distance that is very small compared to the smallest crystal dimensions. We can then ignore the exact nature of the surface as long as we do not get too close to it. (We WILL get too close to ignore the surface in Chapters 6–10.)

For a time-dependent potential, "abrupt" means a change that takes place in a time that is short compared to the period of the wave, namely $h/E$. Although we can solve such problems, they do not assume any importance in most solid-state physics, so there is little point in our pursuing them here.

The slowly varying potential is of central importance in semiconductor theory, since it includes all of the applied and thermally generated potentials that we normally deal with. For example, when you apply a voltage across a bar of Ge, you are introducing a distinctly nonperiodic term to the periodic crystal potential. What permitted us to use periodic-potential solutions when the potential was not periodic? The usual argument here is that the applied fields are so small compared to the interatomic fields that the electron always sees itself in a locally periodic field. To get some idea of how big *local* is, note that the distance between atoms in Ge is of the order of a few tenths of a nanometer. The field at a distance of $2 \times 10^{-10}$ m from a unit nuclear charge is (from Gauss's law) $4 \times 10^8$ V/cm. Even in the depletion region of a heavily doped aburpt junction, fields seldom rise much above $10^6$ V/cm. Thus, even at these enormous field strengths, you have to go 100 atomic diameters before the change in nonperiodic potential is of the order of a single atomic step in the periodic term.

This argument that the applied potential is trivial compared to the periodic lattice potential is not entirely sufficient to assure us that Bloch waves are the proper solutions. A further requirement is that the wave solutions are not highly localized. If the wave functions are sufficiently localized, they can interact so strongly with the nearby sites that they themselves induce very important local distortions in the periodicity. If the induced polarization is not too large, the electron may still be distributed and potentially mobile. However, the electron is now obliged to carry the lattice polarization along with it. The combined entity, with its lower energy and higher mass than the electron, is called a *polaron*. Polarons are of particular importance in the very-wide-band-gap, ionic materials such as the alkali halides.

The existence of the polaron is indicative of a basic flaw in the Bloch formalism that is a subject of some current excitement. In the derivation of Bloch's theorem, the periodicity of the potential is fundamental to the argument. If the electron under consideration induces a reaction from neighboring electrons, this periodicity is destroyed. If the electrons are spread out in both *k*-space and *r*-space, it is an excellent approximation to assume that the electron under consideration sees only the average time-independent periodic potential of all of the other electrons. In such circumstances, the Bloch formalism should provide an excellent model. For metals this is generally the case. As the electrons become more tightly bound in the sense that their expectation values become large at certain points in space, nearby electrons and ions will react to the presence of an electron, destroying the periodicity. At first, one can patch up the Bloch solutions by inventing some new particles like the polaron or the *exciton* (a hole and an electron bound together like a hydrogen atom by their electrostatic interaction). These particles (or interactions, if you prefer) are quite readily observed and occasionally have important consequences in devices. As you might expect, the patching works only up to a point. When the localization of the electron is too great, the whole house of cards collapses and a totally different type of solution must be invoked. The new solution is effectively a collection of adjacent individual atoms with the electrons bound to their own atoms and forbidden from making transitions to those nearby. Such materials are well known. Their most prominent members are the very weakly connected Van der Waals solids of the noble gases. The current interest in the failure of the Bloch formalism centers upon a prediction made by N. F. Mott [see N. F. Mott and R. S. Allgaier, *Phys. Status Solidi* **21**, 343 (1967)] that certain materials might show the transition from metallic conductivity to insulator behavior with a very small variation in composition. This *Mott transition* is now being eagerly pursued. Note, however, that moving atoms together or apart makes gross changes in the local field. The largest fields that we are able to impose from the outside are trivial in comparison.

Although the essentially periodic potential of the lattice plus the applied field may have solutions that are almost indistinguishable from the perfectly periodic potential, this does not answer all of the more obvious objections.

For example, we have used only solutions to the *time-independent* equation. How can I talk about accelerations or changes of state in a time-independent system? In a strict sense, I cannot, but in a strict sense there are no truly time-independent systems. A very accurate analogy to our present question occurs when you solve for the modes of vibration of a violin string. Given that problem, you would solve first assuming that there was no damping. That would lead to a time-invariant equation with the modes as eigenfunctions. Having obtained that equation by ignoring the lossy part of the system, you would not be surprised to see the oscillations disappear with time. On the other hand, if the damping terms are reasonably small, you would expect your solution for the modes to be reasonably accurate. On the basis of this expectation, the standard way to solve for the loss rate is to see how rapidly the damping terms extract energy from the string, assuming that the string is in one of the idealized eigenstates.

The same technique is employed in more conspicuously quantized systems. In our derivation of the wavelength dependence of the internal photoelectric effect (Figure 2.30, etc.), we eschewed derivation of the transition rate from first principles. However, if you will remember what we did there, you will see that we assumed that the electron passed from one eigenstate to another. We assumed that we knew exactly what those eigenstates were in spite of the perturbation of the electromagnetic field. That is perfectly reasonable if (and only if) the electromagnetic field is quite small compared to the crystal potential. Had we wanted to derive the transition rate, we would then have had to examine how the known eigenstates interacted with the electromagnetic field.

The last paragraph should make you wonder about the derivation of the effective mass in Chapter 2 (equations 2.18 through 2.21). The states in the band are discrete, yet we never examined the transition rate between states. In point of fact, I very explicitly treated the electron as if it were a free particle in a *continuum* of states. There are many ways of defending this "essentially free" approximation. They all depend on the fact that the energy difference between states is very, very small compared to thermal free energies ($kT$), even for very low temperatures. The electrons are continually being redistributed among the states in the band, so no state is occupied for a very long time. Since frequency (energy) and time are Fourier transform pairs (in physics as well as signals and system theory), decreasing the time spread increases the frequency spread. Thus, the thermal reshuffling broadens the individual energy levels sufficiently to make them continuous. The continuity results only because the energy levels were so close to begin with. To see how close, note that a typical band may have a spread of 2 eV with something like $10^{23}$ states per $cm^3$ per band. Thus, the average spread between states is of the order of $10^{-23}$ eV. For such a small energy difference, a little spreading goes a long way. It should be equally obvious that the 20 to 23 orders of magnitude difference between this tiny energy and all the other energies we have dealt with is enough to justify treating band gaps and impurity levels as quite discrete, even with the thermal spreading.

Once you have accepted the concept that the interaction between the electron wave and the lattice gives a quasi-continuous set of states, the effective mass concept is established. Within the band, we can treat all of the other man-made, thermal, or crystal perturbations as if they were acting on a free particle with this peculiar mass, $m^*$. What we now must do is examine how the various perturbing fields interact with each other to produce the observed motion of electrons in crystals. We will find that the thermal motion of the crystal and the crystalline imperfections tend to scramble regular or coherent electron motion into random motion. When we accelerate electrons in a particular direction with an electric field, the energy gained is quickly converted into purely random, thermal motion. Thus, our effort to accelerate the electrons is constantly frustrated, leading to a uniform drift of the electrons in the direction of the field and an associated heating of the electrons (and through them, the crystal). We will proceed in the following way.

First, we will consider individually the effects of the two most important sources of electron scattering in semiconductors: phonon scattering and ionized-impurity scattering. Next, another equation bearing the imposing name of Boltzmann will be introduced, which will allow us to calculate the average motion of electrons subject to a variety of fields and concentration gradients. From the average motion of the electrons we will obtain the conductivity and diffusion coefficients and equations so necessary for semiconductor device analysis. *Allons!*

## SCATTERING FROM A CHARGED IMPURITY

If we can assume that the electrons within a band are very much like free particles—that is, that the states are not only essentially continuous but also mostly empty (otherwise the electron would not be free to move from state to state)—then the scattering of a moving electron by a fixed charge is essentially the classical Rutherford scattering problem. (Rutherford was throwing $\alpha$-particles at charged nuclei, but his analysis is equally applicable to throwing electrons at donors or television cameras past the moon.) Figure 4.1 illustrates the simplest model. A free electron is approaching an ionized donor along a path that would bring it within a distance, $b$, of the donor if there were no attraction between them. The electron experiences an acceleration toward the donor that is inversely proportional to the square of the distance from the donor. This force curves the trajectory of the electron,

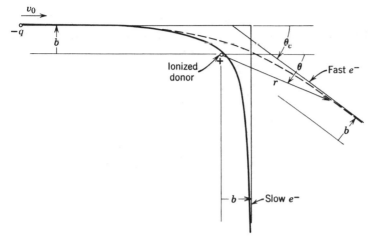

**Figure 4.1** Electron scattering by an ionized donor. Trajectories for fast and slow electrons are shown.

the angular deviation being greatest for the electron that spends the longest time being accelerated in the "reasonably close" region. Thus, the slowest electrons have their trajectories bent the most.

The Rutherford solution is readily obtained from $F = ma$ and the Lorentz force equation. In the plane of the trajectory, the latter gives

$$F = \frac{-q^2}{4\pi\varepsilon r^2} \, u_r = \frac{-q^2}{4\pi\varepsilon(x^2 + y^2)^{3/2}} \, (xu_x + yu_y) \tag{4.1}$$

The $u$'s are unit vectors in the indicated directions. The origin is the donor ion, and the $x$-$y$ plane is the plane containing the origin, the electron, and the electron's initial velocity. Since the force vector lies in the $x$-$y$ plane, the electron's trajectory will also be in the $x$-$y$ plane. Writing the $x$-component of $F = ma$ gives

$$\frac{F_x}{m^*} = \frac{d^2x}{dt^2} = \frac{-q^2}{4\pi\varepsilon m^*} \frac{1}{(x^2 + y^2)} \frac{x}{(x^2 + y^2)^{1/2}} \tag{4.2}$$

Converting (4.2) to polar coordinates yields

$$\frac{-q^2}{4\pi\varepsilon m^*} \frac{\cos\theta}{r^2} = \frac{d^2 (r\cos\theta)}{dt^2}$$

$$= \cos\theta \left[ \frac{d^2r}{dt^2} - r\left(\frac{d\theta}{dt}\right)^2 \right] - \sin\theta \left[ r\frac{d^2\theta}{dt^2} + 2\frac{dr}{dt}\frac{d\theta}{dt} \right] \tag{4.3}$$

Since the choice of the $x$ and $y$ axes was really quite arbitrary, equation 4.3 must be completely independent of this choice. The only way this independence could be achieved is to have the terms multiplying the cosine sum to zero and the terms multiplying the sine also do so. Thus, (4.3) is really two equations, to wit:

$$r \frac{d^2\theta}{dt^2} + 2 \frac{dr}{dt} \frac{d\theta}{dt} = 0 \tag{4.4}$$

$$\frac{d^2r}{dt^2} - r \left( \frac{d\theta}{dt} \right)^2 = \frac{-q^2}{4\pi\varepsilon m^*} \left( \frac{1}{r^2} \right) \tag{4.5}$$

Note that (4.4) can be written as

$$\frac{d}{dt} \left( r^2 \frac{d\theta}{dt} \right) = 0 \tag{4.4'}$$

Since $r^2(d\theta/dt)$ is the angular momentum, equation 4.4' states the very satisfying result that angular momentum is conserved. Figure 4.1 shows that the initial angular momentum is

$$r^2 \frac{d\theta}{dt} = - v_0 b^2 \tag{4.6}$$

Equation 4.6 can be used in (4.5) to yield

$$\frac{d^2r}{dt^2} - \frac{v_0^2 b^2}{r^3} = \left( - \frac{q^2}{4\pi\varepsilon m^*} \right) \frac{1}{r^2} \tag{4.7}$$

Since the variable, $r$, appears in the denominator of (4.7), it is convenient to substitute $w = 1/r$. Then, by differentiating $r$ w.r.t. time and substituting into (4.6), a much simpler differential equation is obtained with $\theta$ as the independent variable. Thus:

$$\frac{dr}{dt} = - \frac{1}{w^2} \frac{dw}{d\theta} \frac{d\theta}{dt} = v_0 b \frac{dw}{d\theta} \tag{4.8}$$

$$\frac{d^2r}{dt^2} = v_0 b \frac{d^2w}{d\theta^2} \frac{d\theta}{dt} = - \frac{1}{r^2} (v_0 b)^2 \frac{d^2w}{d\theta^2} = - w^2 (v_0 b)^2 \frac{d^2w}{d\theta^2}$$

Substitution into (4.7) yields a simple second-order differential equation with constant coefficients:

$$\frac{d^2w}{d\theta^2} + w = \left( \frac{q^2}{4\pi\varepsilon m^*} \right) \frac{1}{v_0^2 b^2} \tag{4.9}$$

The most convenient form for the solution to (4.9) is

$$w = \frac{1}{v_0^2 b^2}\left(\frac{q^2}{4\pi\varepsilon m^*}\right) + A\cos(\theta + \phi) \qquad (4.10)$$

where $A$ and $\phi$ are to be determined by the boundary conditions. What are the initial conditions? (Note that initially the electron in Figure 4.1 is at $r = -\infty$ and $\phi = \pi$.) [Answer: Since $w = 1/r$, $w(\pi) = 0$. The initial electron velocity is $v_0$. From (4.8) we have $(1/v_0 b)(dr/dt)_{t=0} = 1/b = dw/d\theta|_{\theta=\pi}$.] Applying these boundary conditions to (4.10) yields

$$A = \frac{1}{v_0^2 b^2 \cos\phi}\left(\frac{q^2}{4\pi\varepsilon m^*}\right); \quad \tan\phi = v_0^2 b\,\frac{4\pi\varepsilon m^*}{q^2}$$

Putting these into (4.10) gives the trajectory of the electron in terms of $w$ and $\theta$:

$$w = \frac{1}{v_0 b^2}\left(\frac{q^2}{4\pi\varepsilon m^*}\right)\left[1 + \frac{\cos(\theta + \phi)}{\cos\phi}\right] \qquad (4.10')$$

Considering the small dimensions of the volume over which the scattering event takes place, we are really interested only in the total angle through which the electron is turned, $\theta_c$. Accordingly, we want to evaluate (4.10') at $r = \infty$ (i.e., $w = 0$). From our boundary conditions above, we know that $\theta = \pi$ satisfies $w = 0$. By inspection of (4.10'), it is evident that an argument of $(\pi - \phi)$ will also make $w = 0$. Thus, $(\theta_c + \phi) = (\pi - \phi)$ yields the other root to (4.10'). From the second root and the value for $\tan\phi$:

$$\tan\left(\frac{\theta_c}{2}\right) = \tan\left(\frac{\pi}{2} - \phi\right) = \cot\phi = \left(\frac{q^2}{4\pi\varepsilon m^*}\right)\frac{1}{v_0^2 b} \qquad (4.11)$$

Equation 4.11 is the Rutherford Scattering Formula. From our point of view, the most important aspect of (4.11) is that the angle of deflection is inversely proportional to the *square* of the incident particle's initial speed. This means that on the average, slow particles will be much more strongly scattered per collision than fast particles. From this fact, we will be able to obtain the thermal dependence of impurity scattering and an estimate on the relaxation or "reshuffling" time.


## THERMALIZATION BY IMPURITY SCATTERING

Since the impurities are distributed in a reasonably random fashion through-out the crystal, you cannot use (4.11) to predict where the next scattering event will send a particular electron. The randomness of the encounters between electrons and ions means that the scattering tends to scramble the velocities just as shuffling scrambles cards. What (4.11) will do for us is enable us to calculate the r.m.s deflection just as a study of the dynamics of shuffling would permit you to predict the r.m.s. motion of a card in one shuffle. As you are undoubtedly aware, a single shuffle does a very inadequate job of randomizing a deck of cards. A study of shuffling dynamics would enable you to establish some criterion for a proper number of shuffles — say, enough so that the r.m.s. travel of a card was of the order of 13 spaces from its original location in the deck. In a similar way, we will find the time it takes for the average particle to be deflected through some rather arbitrary angle, if its initial speed was $v_0$. This time will be referred to as the thermal relaxation time and it will be a function only of $v_0$ and the density of scattering centers.

The first step is to obtain the average time between collisions for our particle with speed $v_0$. If there are $N_D$ donors and $N_A$ acceptors per cm$^3$, what is the mean distance between scattering centers? [Answer: $(N_D + N_A)^{-1/3}$.] Since the speed is $v_0$, the time between collisions must average to $\tau_c = (N_A + N_D)^{-1/3}/v_0$.

The next step is to find how the collisions just described scramble the velocity. Since the distribution of impurities is both random and locally isotropic, it is reasonable to conclude that there is no favored direction of scattering. For example, as many electrons with an initial velocity in the horizontal plane will get scattered upward as downward. Thus, the collisions do not cause the average number of electrons going in any *particular* direction to increase. It is equally obvious that after being deflected, an electron has less velocity in its original direction than it did before (see Figure 4.2). Note that the electron's speed has not changed, only the direction of travel. Why? (Answer: The electron's mass is four to five orders of magnitude less than the ion's mass, so very little energy can be exchanged in a collision. A perfectly appropriate way to describe this fact is to say that the rate of heat transfer between the impurities and the electrons is relatively slow.)

Since the initial velocities vanish after a number of collisions, the collisions must be shuffling the a priori distribution of electron velocities. We can obtain the average rate at which the a priori distribution disappears by averaging the loss of initial velocity in a particular collision times the relative frequency of that collision.

The loss of velocity in a given collision is simply

$$\Delta v_0 = - [v_0 - v_0 \cos \theta_c] \qquad (4.12)$$

where the collision is characterized according to the particular $b$ (and hence the $\theta_c$) that happened to have occurred. Since the collisions occur, on the

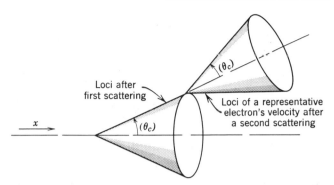

**Figure 4.2** Cones representing the foci of an average $x$-directed electron after one and two successive scattering events.

average, with frequency $\tau_c$, the average rate of change of the initial velocity is given by

$$\left\langle \frac{dv_0}{dt} \right\rangle = - \frac{v_0}{\tau_c} \frac{\displaystyle\int_0^{b_m} \{ 1 - \cos\theta_c(b) \} \, dn(b)}{\displaystyle\int_0^{b_m} dn(b)} \tag{4.13}$$

where $dn(b)$ is the number of electrons that would pass between $b$ and $b + db$ if they were not deflected, and $b_m$ is the maximum $b$, a somewhat fuzzy constant that is roughly half the mean distance between scattering centers. Since the electrons are assumed to be randomly distributed, the number passing through any volume must be proportional to the volume. Accordingly, $dn(b) =$ $= (\text{constant}) \, b^2 db$. Let

$$a = \frac{q^2}{4\pi\varepsilon m^*}$$

Putting this and $b = (a/v_0^2) \cot(\theta_c/2)$ [from (4.11)] into (4.13) yields

$$\left\langle \frac{dv_0}{dt} \right\rangle = - \frac{v_0}{\tau_c} \frac{\displaystyle\int_{\theta_m}^{\pi} (1 - \cos\theta) \frac{\cos\dfrac{\theta_c}{2}}{\sin^3\dfrac{\theta_c}{2}} \, d\theta_c}{\displaystyle\int_{\theta_m}^{\pi} \frac{\cos\dfrac{\theta_c}{2}}{\sin^3\dfrac{\theta_c}{2}} \, d\theta_c} \tag{4.14}$$

After some diligent application of the integral tables and some integration by parts, (4.14) becomes

$$\frac{1}{v_0} \left\langle \frac{dv_0}{dt} \right\rangle = - \frac{1}{\tau_c} \frac{2a^2}{v_0^4 b_m^2} \log\left( 1 + \frac{v_0^4 b_m^2}{a^2} \right) \tag{4.15}$$

Note that *if* the second term in the log argument is much less than 1 (it usually is not), equation 4.15 reduces quite nicely to

$$\frac{dv_x}{dt} = - 2 \left( \frac{v_0}{\tau_c} \right)$$

which seems like a rather nonsensical result. It would imply that the average velocity change was twice the initial velocity per collision. If you will recall that slow particles are very strongly deflected ($\theta_c \simeq \pi$), this result makes some sense. It is really a statement that the velocity is reversed in a collision in the limit of zero initial velocity (or zero $b$, for that matter).

Since equation 4.15 is of the form

$$\frac{dv}{dt} = - \frac{v}{\tau_R} \Rightarrow v = v_0 e^{-t/\tau_R}$$

it is plausible to interpret

$$\tau_c \left\{ \frac{v_0^4 b_m^2}{2a^2 \log (1 + v_0^4 b_m^2/a^2)} \right\}$$

as the randomization or relaxation time, $\tau_R$. Putting in the value of $\tau_c$ (4.15) then says that

$$\tau_R \propto \frac{v_0^3}{\log (1 + v_0^4 b_m^2/a^2)} \tag{4.15'}$$

Thus, very fast electrons are redistributed rather slowly and vice versa. Note that this result applies ONLY to scattering by impurities.

## LATTICE VIBRATIONS

Except at low temperatures or in very impure crystals (e.g., typical metal alloys), the dominant electron scattering mechanism is phonon scattering. A phonon is a quantum of mechanical vibration, quite equivalent in most respects to the quantum of electromagnetic vibration, the photon. The principal differences are that the phonon has no existence outside the crystal and, as shown below, the spectrum of phonons in any lattice is quite finite. The important similarities to keep in mind are that both are bosons and both share the universal wave properties including the Planck and de Broglie rules.

It is fairly easy to see that a photon can interact with an electron through the electric field, but putting the oscillating (photon) field into the wave equation introduced problems that were beyond the techniques in this book. We obtained the needed information about the photon absorption by invoking the classical concepts of conservation of energy and momentum. Wave mechanics were introduced only in counting discrete states and in the relationship between $k$ and momentum.

The situation with phonons is similar but not quite so simple. The interaction of phonons with electrons is through the potential term in the wave equation. Our solutions presumed that the potential term was perfectly periodic and time independent. With a phonon (mechanical vibration) traveling through the lattice, a small time-dependent term must be added to the potential. The addition of this term results in transitions between states just as it did with the internal photoelectric effect. Once again, no attempt will be made to derive the transition rate. Instead, let us assume that the transitions will take place and attempt to find how the transition rate depends on the electron velocity and the crystal temperature. The steps to be followed are (1) derivation of the $E:k$ plot for phonons; (2) derivation of the number of phonons as a function of $T$ [essentially the application of Bose statistics to the $E:k$ plot in (1)]; (3) denumeration of the rules that must be satisfied so that scattering may take place; and (4) counting the number of ways each scattering event could take place according to these rules. We begin with the $E:k$ plot.

The derivation of the $E:k$ plot for phonons follows the same steps as the $E:k$ plot for electrons in crystals (only it seems easier because it is much simpler to visualize the wave motion involved). First the relationship between energy ($hv$) and the wavelength must be obtained. Then the system is quantized by requiring the waves to fit within the crystal. The least complicated system is a one-dimensional crystal of equally spaced atoms of equal mass. If an atom is forced from its equilibrium position, the attraction between the atoms in the chain pulls the displaced atom back toward its rest position. The simplest model with this behavior is a line of masses bound to the nearest neighbors by a linear spring. In other words, we will ignore all forces except those between nearest neighbors. Let $x_p$ be the displacement of the $p$'th atom. The force acting on the $p$'th atom depends on the relative displacements of the $(p-1)$'th, the $p$'th, and the $(p+1)$'th as

$$F_p = M \frac{d^2 x_p}{dt^2} = -K\left[(x_p - x_{p+1}) - (x_{p-1} - x_p)\right] \qquad (4.16)$$

If the mean separation of the atoms is $d$, then the rest position of the $p$'th atom is $pd$. Equations of the form (4.16) have sinusoidal solutions:

$$x_p(t) = A e^{j(\omega t - kpd)} \qquad (4.17)$$

Putting (4.17) into both sides of (4.16) yields the desired relationship between $\omega$ and $k$:

$$- M\omega^2 x_p = KAe^{j(\omega t - kpd)}\left[e^{jkd} + e^{-jkd} - 2\right] = 2Kx_p\left[\cos kd - 1\right]$$

or, in other words:

$$\omega^2 = 2\frac{K}{M}(1 - \cos kd) \tag{4.18}$$

As usual, the $E:k$ plot inherent in (4.18) is unchanged by adding $2n\pi/d$ to $k$. For the case of mechanical vibrations, it is very easy to see what this indeterminacy means. Figure 4.3 shows one particular set of atom displacements with two waves that would yield those displacements. In reality, there are no "lines" between atoms, so there is no way to discriminate between the two waves shown. Therefore, it is eminent good sense to state that the two waves are physically equivalent, since there is no physical way to tell

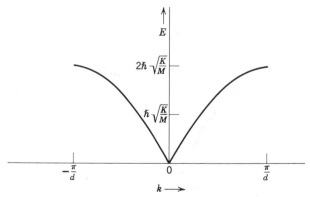

**Figure 4.3** Two waves that correspond to the same sinusoidal displacement of the atoms. Since there is no way to observe the wavelength directly, and since both yield the same physically observable effect, the two waves are equivalent.

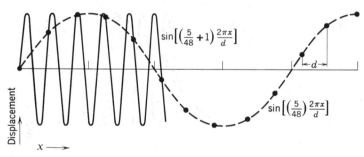

**Figure 4.4** The $E:k$ plot (also called a dispersion diagram) for one mode of a monoatomic lattice.

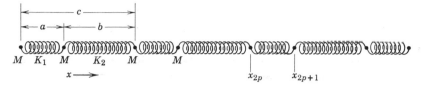

**Figure 4.5** A one-dimensional monoatomic lattice with two inequivalent sites. This is a one-dimensional analog of germanium and silicon.

which you have. It is customary to choose the first (lowest-order) set. What, then, is the range of $k$? (Answer: $-\pi/d \leq k \leq \pi/d$. Note that this is the first Brillouin zone.) The $E:k$ plot corresponding to (4.18) is shown in Figure 4.4.

Figure 4.4 is not quite the whole story. It is possible to excite three different types of waves on the string of atoms shown in Figure 4.3: There are two types of transverse (shear) waves—one that goes up and down on the paper shown in the figure and one that goes in and out of the paper. Then there is the longitudinal (compressional) wave in which the atoms move along the same axis as the wave. The first two are usually described by the same equation (i.e., the same $K$ and $M$), while the third will usually have a substantially different spring constant. Thus, a complete $E:k$ plot would usually have two branches, one of which is actually doubly degenerate.

The case we have just treated (when generalized to three dimensions) describes monoatomic crystals in which each atom along any axis is in an exactly equivalent position. Unfortunately, although there are many such lattices, it is much more common to have several types of atoms (different $M$'s) or several inequivalent sites (different $K$'s) or both. The two most commonly used semiconductors (Si and Ge) are monoatomic, but they have two inequivalent sites. The compound semiconductors (e.g., GaAs or PbS) suffer from both complications. The introduction of different $K$'s or $M$'s introduces a second set of branches that are of some importance in electron scattering.

To obtain the characteristic double-branched solution, consider the one-dimensional lattice shown in Figure 4.5. The lattice spacing is $c$, but each lattice cell contains *two* atoms. The spacing is as shown in the figure. Since $a \neq b$, it follows that $K_1 \neq K_2$. The two $K$'s require us to write two equations equivalent to (4.16). Otherwise, the solution is almost identical to the simpler case we have just considered. Consider the forces acting on the $2p$'th particle and the $(2p + 1)$'th:

$$F_{2p} = M \frac{d^2 x_{2p}}{dt^2} = K_1 (x_{2p-1} - x_{2p}) + K_2 (x_{2p+1} - x_{2p}) \tag{4.19}$$

$$F_{2p+1} = M \frac{d^2 x_{2p+1}}{dt^2} = K_2 (x_{2p} - x_{2p+1}) + K_1 (x_{2p+2} - x_{2p+1}) \tag{4.20}$$

Since we are seeking allowed wave solutions and since the equations above are linear, constant-coefficient equations, we should guess solutions like (4.17). However, with two different spring constants, we should allow the two different sites to have different complex amplitudes (i.e., amplitude and phase). Accordingly, our solutions should be

$$x_{2p} = A_e e^{j[\omega t - k(2pc)]} \tag{4.21}$$

$$x_{2p+1} = A_0 e^{j[\omega t - k(2pc + a)]} \tag{4.22}$$

Putting these into (4.19) and (4.20) and taking advantage of the fact that $a + b = c$, the lattice spacing leads to two equations equivalent to (4.18):

$$A_e[\omega^2 M + K_1 + K_2] = A_0[K_1 e^{jkb} + K_2 e^{-jka}] \tag{4.23}$$

$$A_e[K_1 e^{-jkb} + K_2 e^{jka}] = A_0[\omega^2 M + K_1 + K_2] \tag{4.24}$$

Taking the ratio of the two equations (i.e., setting the determinant of the coefficients equal to zero) yields the desired relationship between $E$ and $k$:

$$[\omega^2 M + K_1 + K_2]^2 = K_1^2 + K_1 K_2 \{e^{-jkc} + e^{+jkc}\} + K_2^2$$

or

$$\omega^2 = \frac{K_1 + K_2}{M} \pm \sqrt{\frac{K_1^2 + K_2^2}{M^2} + \frac{2K_1 K_2}{M^2} \cos kc} \tag{4.25}$$

Once again, we must consider (4.25) three times—twice as the characteristic equation for two transverse modes (one complete set of $K$'s) and once for the longitudinal mode (another complete set of $K$'s). The more energetic branch in each case [$+$ sign in (4.25)] is called the optical branch, the less energetic one ($-$ sign) the acoustical branch. It is very important to note that for $k = 0$, the acoustical phonon has zero energy but the optical phonon has a substantial energy. Figure 4.6 presents the measured data representing equation 4.25 for Ge and Si. Note that the highest frequencies ($10^{13}$ Hz) correspond to photons from the middle infrared (30 microns $\approx \lambda$) and energies of the order of $2kT$ at room temperature.

Since the optical phonons near $k = 0$ are like photons in that they have a lot of energy and very little momentum, you might guess that they would strongly interact with photons. They can. If you were to observe the light reflected from a piece of rock salt (NaCl) or any other ionic crystal, as the frequency of the light was tuned (downward) through the $k = 0$ frequency of the phonons, you would see the rock salt change from a remarkably transparent substance to a highly reflecting (and absorbing) substance and then back to a moderately absorbing medium. The high absorptivity is directly related to the generation of optical phonons. The coupling between the electromagnetic field and the lattice vibration (phonon) is through the dipole

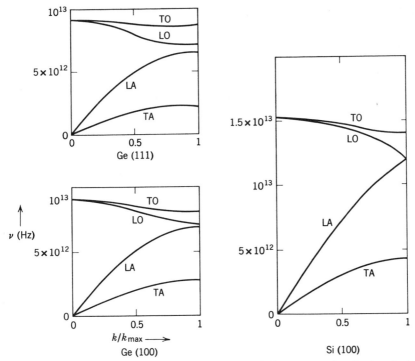

**Figure 4.6** $E:k$ plot for Si and Ge. The ordinate is frequency; the abscissa is the fraction of the maximum values of $k$ [different in the (100) and (111) directions]. TO = = transverse optical branch; TA = transverse acoustical branch; LO = longitudinal optical branch; and LA = longitudinal acoustical branch. [From B. N. Brockhouse and P. K. Iyengar, *Phys. Rev.* **111**, 747 (1958).]

moment of the ions. The oscillating electric field tries to push the $Na^+$ and the $Cl^-$ back and forth, but the ions can respond vigorously only when the field matches one of the natural modes of the lattice. Since the photons carry so little momentum, only those phonons near $k = 0$ can be excited. Thus, the peak in absorption and reflection is quite narrow. The German word *Reststrahl* (residual ray) is usually used to describe the reflected (and hence rather monochromatic) beam. If you happen to want a photon with the Reststrahl frequency, an ionic crystal makes a very good filter.

A short problem: Since equation 4.25 must become equivalent to equation 4.18 when $K_1 = K_2$, how is Figure 4.4 reconciled with Figure 4.6? (Hint: The lattice spacing would be $2d$ for Figure 4.6 but only $d$ for Figure 4.4. Yet both figures would describe the same lattice.) [Answer: For the case of equal masses and spring constants, Figure 4.6 is simply Figure 4.4 folded at the $\pm \pi/2d$ points. Note that the longitudinal branches in Si along the

(100) directions show this sort of degeneracy. All of the other branches shown for Si and Ge do not. They have an abrupt discontinuity in $E$ at the edge of the zone. Physically, the difference between the acoustic and optical modes is: In an acoustic vibration, adjacent atoms are moving in the same direction; in an optical mode, adjacent particles are moving in opposite directions. At the edges of the zone, one branch represents a mode where all the *odd* particles are stationary, the other a mode where all the *even* particles are stationary. Of course, if there is no way to tell an even from an odd particle, the optical branch must be degenerate with the acoustic branch at the band edge. In general, the frequency (energy) for only one type of particle moving will be different from that for only the other type moving. Therefore, you should expect to find a discontinuity in the $E:k$ plot at the band edge.]

## ALLOWED VALUES OF $k$

The number of modes, $g(k)$, between $k$ and $k + dk$ is obtained in much the same way as it was for electrons in Chapter 2. Since the phonons must be totally reflected at the surface of the lattice, only those phonons that fit the crystal can be solutions to the time-independent equation. If the crystal is of length $L$, what does this say about $k$ and $g(k)$? [Answer: The boundary conditions require that we combine solutions of the form of (4.21) to produce

$$x_{2p} = A'_e \sin \left[ \omega t - k(2pc) \right] \tag{4.21'}$$

with the boundary conditions satisfied if and only if

$$k_n = \frac{n\pi}{L} \tag{4.26}$$

Finally, since $g(k)$ is the density of states per cm of (one-dimensional) crystal, $g(k) = 1/\pi.$]

The three-dimensional crystal requires the specification of three $k$'s, each satisfying an equation like (4.26). This leads immediately to $g(k) = 1/\pi^3$. Using (4.26), find the number of modes (states) per branch. [Answer: The maximum $k$ is $\pi/c$. Equation 4.26 states that there is a mode every $\pi/L$ along the $k$ axis. Thus: $N = (\pi/c)/(\pi/L) = L/c =$ number of lattice cells in the one-dimensional crystal. For three-dimensional crystals the result is simply $N = L_1 L_2 L_3 / c_1 c_2 c_3 = V_{crystal}/V_{unit\ cell} =$ number of lattice cells in the crystal.]

Summarizing our results: The number of modes per branch is equal to the number of lattice cells in the crystal; branches come in groups of three (although not always "transverse" or "longitudinal"); and the number of groups of branches equals the number of atoms in the cell. Note that the last two statements taken together give the very satisfying result that each atom in the crystal can vibrate in any direction—the *total* number of modes is precisely three times the number of atoms in the crystal.

## THE NUMBER OF PHONONS VERSUS THE LATTICE TEMPERATURE

Our purpose in pursuing phonons is to determine how they limit the flow of electrons. If an electron is to be scattered from one state to another, both energy and momentum must be conserved. Although we normally treat conservation of momentum as a particle property, within the lattice both the electron and the phonon are so wavelike in character that it is difficult conceptually to treat their interaction on a strictly particulate basis. As you will see below, the wave description is both neat and simple, with nothing new but some insight added. For example, conservation of momentum and energy turns out to be a generalized Bragg's law. The one subject that really calls for a particle viewpoint is the denumeration of the number of phonons. Since we will need that number, let us proceed with that first and then turn to the interaction between the electron and phonon waves.

We want to know both the number of phonons per $cm^3$ and the average energy per phonon. The number of phonons per unit volume is

$$N_p = \int_0^{E_{max}} g(E)f_B(E)dE \tag{4.27}$$

Unfortunately, (4.27) is really a complicated mess; $g(E)$ is obtained from (4.18) and it certainly is not a very friendly-looking function. Equation 4.27 is further complicated by the fact that $f_B(E)$ diverges at the origin [see (3.57)]. [Actually, since there is a *minimum* phonon energy corresponding to the largest wavelength that can fit within a given crystal, the integration is from $E_{min}$ to $E_{max}$ and no divergence occurs. Notice that even without invoking the finite size of the lattice, the total energy of the phonons does not diverge. This is because the product, $Ef_B(E) \rightarrow kT$ as $E \rightarrow 0$, as you saw just after (3.57) was derived.] To avoid dealing directly with (4.27) we can consider the case in which $E_{max} < kT$. This region of relatively high temperatures is easier to work with because $f_B(E) \simeq kT/E$. That, in turn, says that the energy in each mode is $kT$ and the number of phonons in a particular mode is $kT/hv$. Since the number of modes is $3N$ (where $N$ is the number of atoms in the lattice), the total energy stored in the lattice is $3NkT$. This is nothing more nor less than the "freshman chemistry" law of Dulong and Petit for the specific heat of a solid. As you can see, it is applicable only for the condition that $E_{max} < kT$. How good an approximation is this at room temperature (300°K)? The data in Figure 4.6 should give you a reasonable idea. For the acoustic branch, numbers of the order of $5 \times 10^{12}$ seem appropriate for $v$, giving an energy $hv$ of .021 eV. Thus, $E_{max} \simeq kT$ and the approximation is barely adequate.

To obtain $N_p$ we need a formula for $g(E)$ that is at least as good as our approximation for $f_B(E)$. The route that we took in Chapter 2 was to find the density of states as a function of $k$ and then convert to $E$ with the $E:k$ plot. Since each state in our $1 \times 1 \times 1$ cm$^3$ crystal occupies a little cube in $k$-space, it is easy to write down $g(k)$ as long as we do not have to worry about fitting the cubes within the limits of the Brillouin zone. In other words, we are filling a funny-shaped box (Figure 2.21) with little cubes *starting from the center* of the box. As long as the cubes do not meet the box, they stack equally well in all directions. Within a radius of $k$ from the origin, there are $M(k) = 4\pi k^3/3/(\text{volume of a cube}) = 4k^3/3\pi^2$ states. As soon as the cubes meet the box, $g(k)$ must start to decrease. When will $g(k)$ get to zero? (Answer: when the box is filled.)

To get a slightly oversimplified view of what is going on, consider the simplest three-dimensional lattice—a simple-cubic monoatomic lattice. Furthermore, let us replace the acoustic dispersion curves (e.g., Figure 4.6 with only an acoustic branch) with straight lines. This last step is the same as setting the phase and group velocities equal to the same constant. ($v_p = \omega/k$, $v_g = \partial\omega/\partial k$, and for the straight line going through the origin $\omega/k = \partial\omega/\partial k$.) For the purposes of this illustration, let the phase velocity of the transverse waves be half that of the longitudinal waves. Then the $E:k$ plot is the simple affair shown in Figure 4.7. This figure also shows that first Brillouin zone, which happens to be a cube. Using (4.28) and some solid geometry to determine how much of the sphere is inside the cube, we can construct a graph for $g(E)$ (Figure 4.8). The separate curves for the transverse and longitudinal branches are shown as dashed lines. The first break in each dashed curve is where the appropriate sphere just touches the cube; the second is where the sphere first touches the edge of the cube. (If you were hollowing out the cube, the first break would represent the point where you would first puncture the surface; the second would represent the radius at which the structure would fall apart into eight pieces. Figure 4.7b shows the Brillouin zone at the second break.) To get the total density of states, we need the sum of the states from the one longitudinal and two (degenerate) transverse modes. This is shown as the solid curve in the figure. The total number of modes is the area under the solid curve: $3M$.

Before tying all of this together to get $N_p$, note that the face-centered cubic's Brillouin zone (Figure 2.21) is much more like a sphere than the simple cubic's zone. This means that for the f.c.c. lattice, the spherical filling ($g \propto E^2$) continues for larger radii and then turns off much more abruptly. In a truly spherical zone, each branch has a parabola for its entire $g(E)$ versus $E$ plot. Accordingly, it is a reasonable approximation for the f.c.c. case to use only the parabolic portion of the curve. Such an approximation certainly makes integration much easier.

For each of the parabolas, we must substitute into (4.28) the appropriate

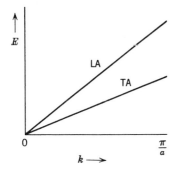

**Figure 4.7a** The $E:k$ plot, on the assumption that $V_p = V_g =$ constant for each branch. TA = = transverse acoustical branch and LA = longitudinal acoustical branch.

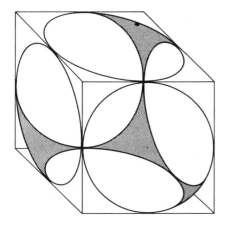

**Figure 4.7b** The first Brillouin zone for the simple-cubic lattice. The figure is shown hollowed out to the point where the second abrupt change in slope would occur.

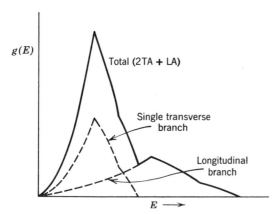

**Figure 4.8** The density-of-states function for the monoatomic simple-cubic lattice. The dashed lines are the curves for the individual modes. TA = transverse acoustical branch; LA = longitudinal acoustical branch.

$E:k$ relationship Since we have assumed a constant phase velocity, $E = \hbar\omega = \hbar v_p k$. The maximum value of $E$ is $E_{max} = \hbar v_p k_{max}$. From (4.27) and (4.28) we have for the $i$'th branch

$$dN_{p_i} = f_B(E) \frac{dM_i}{dk} dk = \frac{kT}{E} \left( \frac{4}{\pi^2} \right) \left( \frac{1}{\hbar v_p} \right)^3 E^2 dE \qquad (4.29)$$

Applying (4.29) to each of the branches and integrating over the appropriate energy range gives

$$N_p = 2 \int_0^{E_{Tmax}} kT \left( \frac{4}{\pi^2} \right) \left( \frac{1}{\hbar v_{pT}} \right)^3 E\, dE + \int_0^{E_{Lmax}} kT \left( \frac{4}{\pi^2} \right) \left( \frac{1}{\hbar v_{pL}} \right)^3 E\, dE$$

$$= kT \left( \frac{2}{\pi^2 \hbar^3} \right) \left[ \frac{2}{v_{pT}^3} E_{Tmax}^2 + \frac{1}{v_{pL}^3} E_{Lmax}^2 \right]$$

$$= kT \left( \frac{2}{\pi^2 \hbar} \right) k_{max}^2 \left[ \frac{2}{v_{pT}} + \frac{1}{v_{pL}} \right] \qquad (4.30)$$

To find the energy of the average phonon is now quite simple; it's the total energy, $3NkT$, divided by $N_p$. Noting that $M = 1/d^3$ ($d$ is the lattice spacing) and $k_{max} = \pi/d$, we obtain

$$\langle E_p \rangle = \frac{h}{2d} \left[ \frac{v_{pT} v_{pL}}{v_{pT} + 2v_{pL}} \right] \qquad (4.31)$$

Why is the average phonon's energy independent of temperature? [Answer: We have assumed that $f_B = kT/E$, which means that each mode has an energy, $kT$. Since each mode is equally excited, and since the *energy per phonon* is almost completely independent of temperature (thermal expansion of the lattice implies that there is some temperature dependence, but that is certainly a small effect), the percentage of the phonons that are in a given mode is constant. Remember that this result is heavily dependent on the assumption that $f_B(E) = kT/E$, an assumption that was marginal at best.]

Since (4.31) depends only on the lattice spacing and the velocity of sound in the crystal—numbers that do not vary that much from one solid to another—we can obtain a pretty good idea of the magnitude of the average phonon's energy in any solid. If you look up the velocity of sound in a data book such as *The American Institute of Physics Handbook*, you will see that most solids lie within the range of 2 to $10 \times 10^5$ cm/sec. The lattice spacing for Ge and many other semiconductors is in a not-very-wide range about $5 \times 10^{-8}$ cm. Setting both $v_{pt}$ and $v_{pl}$ equal to $5 \times 10^5$ cm/sec and $d \simeq 5 \times 10^{-8}$ cm, the average phonon energy turns out to be $0.11 \times 10^{-3}$ eV. Since $kT$ at

300°K is $26 \times 10^{-3}$ eV, you can see that $kT \gg E$ is certainly a very good assumption for the *average* phonon.

If the phonon velocity is reasonably constant (straight-line approximation for Figure 4.6), we have the useful approximation for the $E:k$ relationship:

$$E = \hbar v_p k \tag{4.32}$$

Equation 4.32 comes directly from the Planck and de Broglie laws with $\lambda = v_p/v$. [A rather interesting result occurs if you set $\hbar k = p = m v_p$ in (4.32); you get $E = m v_p^2$. Working backward from $E = mc^2$ to (4.32), you get de Broglie's hypothesis!]

## ELECTRON-PHONON SCATTERING

A phonon is a periodic disturbance in a crystal lattice; it is periodic both in time and position. If an electron were traveling through a spatially periodic array, it would obey all of the scattering rules we derived in Chapter 2. The fact that the atom positions are not constant in time, however, says that we can use those rules only if the scattering occurs in a time that is short compared to the motion of the phonon through the crystal. Note that if the atom positions were not time dependent, there would be no net scattering. Why not? (Answer: If the phonon merely complicated the potential function in the wave equation but still permitted us to use the time-independent form of the wave equation, we would obtain merely a new set of eigenvalues. In other words, both solutions would be standing waves—just different standing waves. To have transitions, we must have a time-dependent wave equation.)

Our first task, then, is to show that the average electron's speed is much larger than $5 \times 10^5$ cm/sec. The easiest route to such a goal is to find the average electron's kinetic energy. Since $E_{kin} = m^* v_e^2/2$, the average energy above the band edge gives us the mean square velocity. The average kinetic energy is given by

$$\overline{E - E_c} = \frac{\displaystyle\int_{E_c}^{\infty} (E - E_c) f(E)\, g(E)\, dE}{\displaystyle\int_{E_c}^{\infty} f(E)\, g(E)\, dE} \tag{4.33}$$

This equation looks very much like (3.22) [the denominator *is* (3.22)]. If we make the same substitution that gave (3.22′) $[u^2 = (E - E_f)/kT]$ we obtain

$$\overline{E - E_c} = kT \frac{\displaystyle\int_0^{\infty} u^4 e^{-u^2}\, du}{\displaystyle\int_0^{\infty} u^u e^{-u^2}\, du} = \frac{3}{2} kT \tag{4.33′}$$

Thus, $v_e^2 = 3kT/m^*$. Using $m^* = m_0$ and $kT = .0259$ eV, we find an r.m.s. electron velocity of $1.17 \times 10^7$ cm/sec. That is certainly much larger than $5 \times 10^5$ cm/sec.

Since the average electron's velocity is much so greater than the phonon's, it is a reasonable assumption to say that "for all practical purposes" the electrons see a standing plane wave. If the waves were truly stationary and the crystal were illuminated by a beam of electrons from the outside, we would have the situation discussed at some length in Chapter 2. The only difference would be the use of electrons in place of X-ray photons. For that particularly simple case we would get Bragg's law in its usual form:

$$2\lambda_p \sin \theta = n\lambda_e \qquad (4.34)$$

The standing phonon wave is the periodic structure and the plane electron wave the illumination. If we convert from $\lambda$ to $k$ in (4.34), conservation of momentum is quite evident:

$$2k_e \sin \theta = nk_p \qquad (4.34')$$

Equation 4.34' states that the electron's momentum in the direction of the phonon's momentum is equal to half the phonon's momentum. Figure 4.9 illustrates the collision and (4.34') states that if the electron is going to reflect from the phonon, momentum must be conserved. The $n$ in (4.34') allows the electron to bounce off of several phonons simultaneously.

Unfortunately, (4.34') describes a situation that is a little too simple. We must conserve energy as well as momentum. To do so, we must generalize Bragg's law to cover two possible situations. First, it is possible for an electron (or photon) to transfer considerable energy as it bounces from the "mirror." (The game of baseball is predicated on this possibility.) In that case, the reflected wave has a different $\omega$ (and thus a different $k$) from the incident wave. Second, whether or not energy is transferred, there is no reason why the $E{:}k$ plot in the reflected direction should be the same as in the incident direction. As we have seen in Chapter 2, electrons can have highly anisotropic $E{:}k$ plots. Crystals can frequently exhibit similar anisotropy for photons. In either case, the law of reflection (in essence, 4.34') is no longer the familiar form of Snell's law: the angle of incidence is equal to the angle of reflection.

We will always get the proper form if we proceed directly from the conservation laws. Physically, we are allowing for the Doppler shift if the mirror is moving (either before or after the collision) and for the anisotropy of the medium. The conservation rules are

$$k_e \pm k_p = k_e' \qquad \text{(momentum)} \qquad (4.35)$$

$$E_e + E_p = E_e' \qquad \text{(energy)} \qquad (4.36)$$

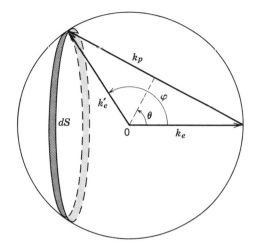

**Figure 4.9** Vector-scattering diagram for a collision that satisfies the simple form of Bragg's law. $k_e$ and $k'_e$ are the initial and final electron wave vectors. $k_p$ is the phonon-wave vector. In this example, the phonon is absorbed. The sphere is a constant-energy surface and the initial and final wave vectors lie in the same surface. $dS$ is the surface area of final states that correspond to the same deflection angle, $\phi$.

The ($\pm$) in (4.35) and (4.36) allow for absorption ($+$) and emission ($-$) of phonons. (If the ball hits the bat, it is the bat that gets energy; if the bat strikes the ball the reverse is true. Energy is always transferred in the direction of motion of the center of mass.)

If the phonon's energy is very small compared to the electron's, (4.36) reduces to $E_e \approx E'_e$. If the energy surfaces are spherical, $E_e = \hbar^2 k_e^2 / 2m^*$. Thus, $k_e^2 \simeq k'^2_e$; in other words, the initial and final wave vectors have equal lengths. From Figure 4.9 you can see that having the wave vectors of equal lengths lead at once to (4.34'). What we have just shown is that the conventional form of Bragg's law is appropriate to collisions in which negligible energy is transferred and for which the constant energy surfaces are spheres centered on the origin. All scatterings correspond to a change in location on the same constant energy surface, and the phonon's wave number is the cord connecting the initial and final states.

Now consider collisions that carry an electron over a spherical energy surface not centered on the origin. What is the form of Bragg's law in that case? Answer: With reference to Figure 4.10, conservation of energy is given by

$$E_e = \hbar^2 k_1^2 / 2m^* = \hbar^2 |k_e - k_c|^2 / 2m^* = \hbar^2 k'^2_1 / 2m^* = \hbar^2 |k'_e - k_c|^2 / 2m^* = E'_e$$

In other words, conservation of energy requires the magnitudes of the "kinetic" wave vectors, $k_1$ and $k'_1$, to be equal. By subtracting $k_c$ from both sides of (4.35), we get the form of Bragg's law appropriate to this case:

$$k_e - k_c \pm k_p = k'_e - k_c \qquad (4.37)$$

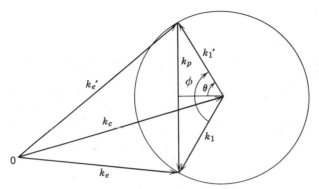

**Figure 4.10** Vector-scattering diagram for an electron on a spherical constant-energy surface *not centered* on the origin. The simple form of Bragg's law holds for the "kinetic" wave vectors, $k_1$ and $k_1'$. The phonon is emitted in this example.

or, in terms of the kinetic wave vectors:

$$k_1 \pm k_p = k_1' \tag{4.38}$$

Equation 4.38 leads immediately to a Bragg law such as (4.34′).

The ellipsoidal case and the case in which the electron goes from a surface near one minimum to a surface near another (called intervalley scattering) are really no different except that nothing so neat as (4.34′) results. The simplest presentation is one like Figure 4.10. The phonon required to connect the initial and final states is the one given by the wave vector that connects the two states.

When we examine cases where the energy transfer is significant, Bragg's law again becomes reasonably messy. The angle of reflection is either larger or smaller than the angle of incidence, depending on whether the phonon is absorbed or emitted, and the difference depends on the phonon's wave number. (See, for example, W. Shockley, *Electrons and Holes in Semiconductors*, p. 530, Van Nostrand, 1950).

## COLLISION DYNAMICS

In our analysis of the optical absorption of multivalleyed semiconductors, we were able to predict the variation of absorptivity with energy simply by counting the number of different ways a photon of a particular energy could induce a transition. We never had to examine the dynamics of the transition beyond the simple requirements that energy and momentum be conserved. Unfortunately, we cannot escape so easily from a direct confrontation with the electron-phonon collision dynamics. The problem is that there can be two very distinct kinds of phonons—transverse and longitudinal—that have roughly the same momentum and energy but different interactions with the electron. For example, longitudinal phonons are periodic variations in the crystal layer spacings. These density variations will scatter electron waves in a most classical way.

It is much less obvious that transverse phonons, which result in no first-order density fluctuations, could cause electron scattering. In fact, for the simplest form of Bragg's-law scattering, which is appropriate to the spherical energy surface, there is no first-order interaction between the transverse phonons and the electrons. Even though the transition is allowed and satisfies all the conservation laws, it just doesn't happen.

If life were that simple, we could add up all the collisions with longitudinal phonons and quit there. This, in fact, was just what Shockley did in his first analysis of the scattering problem. Unfortunately, although it was unknown at the time of Shockley's analysis, the spherical energy surface is somewhat akin to the weightless pulley and frictionless inclined plane. Once you introduce the more realistic ellipsoidal energy surface (or worse), the transverse phonons become effective as scattering agents. However, they are not equivalent to longitudinal phonons—they may be either more or less effective in scattering electrons—and an accurate analysis of the scattering rate demands a detailed determination of the scattering interaction. Efforts to be complete have not been outstandingly successful, although the progress in understanding the problem is substantial. The methods used for determining the scattering probabilities lie well outside those introduced in this text, so we must forgo a quantitative analysis of this part of the problem. A particularly clear and well-written introduction to both the methods and the results can be found in Frank Blatt's excellent book, *Physics of Electronic Conduction in Solids*, McGraw-Hill, 1968.

Since we cannot do a meaningful analysis of the several different competing modes of phonon scattering without taking an excursion far afield from our present course, I propose the following procedure. Let us first denumerate the several scattering modes, with the objective of determining the regime in which each might be important. One of our results will be that for low electric fields and room temperatures, the scattering should be dominated by low-energy acoustic phonons. Proceeding from that observation toward an analytical expression for the scattering rate, I will then make the completely unwarranted assumption that the scattering probability per transition is proportional only to the number of different ways in which that transition can occur. This was a good assumption for band-edge absorption, but it is demonstrably false for electron-phonon scattering. The purpose in making such an assumption is to permit us to exercise some of our other analytical tools to obtain what is at least a plausible scattering rate for the electrons. While I could declare such a rate deus ex machina, the derivation permits you to see how some of the other factors, quite apart from the collision cross-section, enter the analysis. This insight will be helpful in understanding the very nonohmic behavior one observes in semiconductors in moderate to high electric fields. Let us begin with the scattering modes.

Phonons come in two very distinct flavors. There are the acoustic phonons, which are characterized by their relatively small energies, and there are the optical phonons with their rather high energies. Both types have all possible values of crystal momentum. With optical phonons, the energy is large regardless of the momentum; for acoustic phonons, the energy is approximately linearly proportional to the momentum. Finally, at room temperature, there are relatively few optical phonons, but there are huge numbers of low-energy acoustic phonons.

Now consider an electron in the conduction band of germanium (Figures 2.25 and 2.29). It may readily scatter about in a shell of almost constant energy surrounding its own minimum along the (111) axis. Such scattering requires some momentum but very little energy, just the combination so readily available from the low-energy acoustic phonons. If more potent phonons enter the picture—phonons with energy of the order of $kT$ at room temperature—the electron may scatter into one of the other (111) minima. (Why does this require more energy? After all, all of these minima are energetically quite equivalent. [Answer: The obvious requirement for intervalley scattering is enough momentum to make it from here to there. These equivalent minima are quite far away from each other in $k$-space, so a substantial amount of $k$ is required for a transition. Figure 4.6 shows that substantial amounts of $k$ call for substantial amounts of $E$.]) Finally, with really energetic electrons or phonons, transitions become possible between inequivalent minima and between shells of very different energy around the same minima. Now let us consider the circumstances under which each of these scattering processes might be expected to take on important proportions.

If the electrons are in reasonably good thermal equilibrium with the lattice and $kT$ isn't very large, there will be very few energetic electrons or phonons. Under these rather loosely defined conditions, we should expect only the low-energy acoustic phonon scattering, scattering that carries electrons about in a shell of almost constant energy. As we raise the temperature of the system to the point where $kT$ is of the order of an energetic acoustic or an optical phonon—that is, about room temperature or above—we might expect the conditions required for intervalley scattering to be met in enough electron phonon collisions to make the intervalley scattering rate important. This new scattering process, superimposed over the already important low-energy intravalley scattering, would make itself evident by reducing the mobility below that predicted by intravalley scattering alone.

The next escalation in the scattering rate would seem to require that we heat the sample to temperatures well in excess of room temperature. However, if we relax our requirement that the electrons and the lattice be in thermal equilibrium, we can find ourselves in a very important regime where the electron "temperature" is startlingly high. You might well wonder how the electrons that seem to be in such intimate physical contact with the lattice could get so much hotter than the lattice. The answer is that physical proximity does not guarantee a strong interaction; you may shine light through a glass plate without inducing much net change in either the plate or the light. When you apply an electric field to a piece of $n$-type germanium, you are doing work on the electrons in the conduction band. These may transfer energy to the lattice only by the emission of phonons. Since the average phonon that is stimulating transitions has so little energy, it takes many, many collisions to carry the electron's new-found energy into the crystal lattice. A very fruitful way to look at this situation is to consider the electrons and the lattice as two separate ensembles with a weak thermal connection between them. Note carefully that we are discriminating between the few scatterings that it takes to thermalize (randomize) the velocity distribution itself and the many scattering events that it takes to bring the electrons and the lattice into equilibrium with each other.

It is worthwhile to note the enormous difference in the specific heats of a semiconductor lattice and the few conduction electrons that typically reside within it. At 300°K, a cubic centimeter of germanium takes 0.396 calories (1.65 joules) to heat it 1°C. That $cm^3$ contains roughly $4 \times 10^{22}$ atoms. If the cubic centimeter is $n$-type germanium, we might have something like $10^{16}$ conduction electrons. By Dulong and Petit's rule for specific heat (which should apply to the conduction electrons), the specific heat of the electrons alone is only $4.2 \times 10^{-7}$ joules/°C or about $10^{-7}$ times the specific heat of the lattice. To heat these electrons 1000°C above the lattice temperature takes little more than a third of a millijoule per $cm^3$! That's hardly any heat at all.

To get some idea of the power required to maintain a high electron temperature, let us make an educated guess that it requires about $10^{-9}$ seconds to transfer energy from the electron ensemble to the lattice. To maintain a 1°C temperature difference requires that $4.2 \times 10^{-7}$ joules be supplied to each $cm^3$ every $10^{-9}$ seconds, or 420 watts/$cm^3$. Noting that the number of cubic centimeters in a typical semiconductor device is usually exceedingly small, and that it is quite common to have high field regions with much surface and very little volume; it is not unusual to find some small but important regions of a device where the electrons have energies well in excess of $kT$ for the lattice. What might we expect from these regions?

Considering the number of different bulk effects that are observed, one might be tempted to answer, "almost anything." Actually, however, only three distinctly different events may transpire. The first new thing is that a substantial number of electrons obtain sufficient kinetic energy to be able to *spontaneously* emit optical or high-energy acoustic phonons. Why spontaneous? (Answer: Since we assume that the lattice is relatively cool, there are very few energetic phonons available to induce a transition. Although the dynamics of spontaneous emission are, at best, obscure, it is quite clear that hot bodies emit copious numbers of bosons quite spontaneously. Unless you are the kind of nut who reads by laser light, I may comfortably assert that the photons you are seeing were emitted quite spontaneously.) Since the hot electrons need not find a suitable phonon to rebound from—they more or less carry their collision around with them—the scattering rate for energetic electrons will be much larger than for the low-energy ones. What is more, the energy loss per collision will be much larger for these hot carriers. This means that it becomes harder and harder to accelerate the electron ensemble as the mean electron energy rises. Eventually, if nothing else happens first, this spontaneous emission rate gets so large that it is virtually impossible to increase the average drift velocity. The drift velocity is said to saturate.

Before this saturation occurs, a decidedly unexpected thing may be observed. If the carriers are presented with the possibility of scattering into inequivalent minima, as they are in Ge and GaAs among other materials, we have the distinct possibility that the inequivalent minima will have substantially different effective masses. If, as is usually the case, the more energetic minima have much higher densities of states, the hot electrons will tend to collect in the upper minima. Since these states will have a higher effective mass, one finds the rather peculiar result that the hot electrons suddenly become very heavy. If the applied field is sufficient to cause a substantial population transfer, the average effective mass may increase so much that the drift velocity will DECREASE with increasing field. This is the *Gunn effect* that is so copiously generating microwaves today [J. B. Gunn, *Solid State Comm.*, **1**, 88 (1963)]. The whole class of devices based on this effect are called *transferred-electron devices.*

Whether the Gunn effect occurs or not, if the field is raised high enough, a third effect will be observed. The electrons can gain enough energy between collisions to kick an electron from the valence to the conduction band, creating a hole-electron pair. Since this *impact ionization* takes place in the regime characterized by an almost constant, field-independent drift velocity, it does not concern us in this chapter. It will reappear in force when we consider generation-recombination mechanisms in the next chapter.

Summarizing what has been said, we expect that the phonon scattering rate at moderate temperatures and low electric fields will be dominated by low-energy acoustic phonon collisions. As the temperature and/or the field is increased, we should expect intervalley scattering to take on some importance. Since these transitions require phonons with energies of the order of $kT$ at room temperature, it is likely that intervalley scattering is already having some influence on the mobility at room temperature. If we apply substantial electric fields, we will observe the onset of drift saturation resulting from the spontaneous emission of optical phonons. Finally, if inequivalent minima are available, the transfer of the "hot" electrons to a higher minimum may lead to negative differential conductivity (the Gunn effect).

Let us now make a somewhat oversimplified analysis of the low-energy scattering rate to lead us into the very important Boltzmann transport equation.

## LOW-ENERGY PHONON SCATTERING

We are returning now to Figure 4.9 and equation 4.34'. Our approximation is that no energy is transferred in the collision; therefore, the collisions carry the electron about on a single spherical energy surface. We want the relaxation or randomization time, $\tau_R$. The calculation proceeds along lines similar to (4.15). The randomization time is given by

$$\frac{1}{\tau_R} = -\frac{1}{v_0}\left\langle \frac{dv_0}{dt}\right\rangle = \frac{1}{\tau_c}\left\langle 1 - \cos\phi\right\rangle \tag{4.39}$$

Equation 4.39 should be compared with (4.13), which was the expression developed for impurity scattering. Note that $\theta_c$ in (4.13) is the same as $\phi = 2\theta$ in Figure 4.9 and equation 4.39.

The derivation is in two parts. We will first obtain the expression for $\tau_c$, the mean time between electron-phonon collisions. We will then evaluate $\langle 1 - \cos\phi \rangle$, which will turn out to be 1. Thus, $\tau_c = \tau_R$.

Consider an electron whose initial wave vector points in the $x$ direction. The average rate, $dR(\phi)$, at which the electron is scattered into an angle between $\phi$ and $\phi + d\phi$ (i.e., into $dS$ in Figure 4.9) is proportional to the number of phonons, $dN'_p$, that can stimulate such transitions and to the velocity of the electron. The constant of proportionality, $\sigma$, is called the scattering cross-section. Thus:

$$dR(\phi) = \sigma v_e dN'_p(\phi) \tag{4.40}$$

The phonons that can stimulate the transition are those that are at an angle of approximately $(\pi/2) - \theta$ to the $x$-axis and have a wave vector of length approximately $2k_e \sin\theta$. Why *approximately*? [Answer: This is a wave scattering process, almost identical to the X-ray Bragg scattering that we discussed in Chapter 2. At that time—immediately following (2.4)—we examined how exactly the Bragg condition had to be met. We found that the scattering angle was not an exact number, that there was an *uncertainty* in its value proportional to the number of reflecting planes. A similar situation occurs in all wave-scattering processes; there is an uncertainty or "line width" associated with the finite interval, either spatial or temporal, over which the wave (or particle) is scattered.] Let us make the plausible, although not unassailable, assumption that the uncertainty in $k_p$ is proportional to $k_p$. This assumption is based on the requirement that the electrons scatter in a time that is short compared to the time it takes a phonon to change its position by $\lambda_p/4$. This time is $\tau_{p\varepsilon} = \lambda_p/4v_p = 2\pi/4v_pk_p$. Since the uncertainty in $E_p$ is inversely proportional to the scattering time, we have $\Delta E_p \propto k_p$. From (4.32) $E_p \propto k_p$, so the uncertainty in $k_p$ should also be proportional to $k_p$. Let us further assume that the scattering cross-section is independent of phonon energy. This completely arbitrary assumption yields the most commonly observed temperature dependence for the scattering time, but that only goes to prove that it is quite possible to get the right answer for the wrong reason.

The phonons contributing to scattering into the angle $\phi$ come from a toroidal volume element comprised of $dS$ in Figure 4.9 and a thickness proportional to the uncertainty in $k_p$. (Actually, there is also a mirror-image volume element corresponding to emission of phonons, but that will only add a constant numerical factor of 2 to the calculation. Since we are going to work with proportionalities, numerical factors are unimportant.) The number of phonons in that volume element is proportional to the product of the volume and the Bose factor. Using the low-energy form of the Bose factor $(kT/E)$ and (4.32), we have $dN'_p \propto (kT/E_p)k_pdS \propto kT\,ds$. Putting this into (4.40) yields

$$dR \propto v_ekT\,dS \qquad (4.41)$$

Integrating (4.41) gives the total scattering rate, which is inversely proportional to the mean time between collisions:

$$R = 1/\tau_c \propto v_ekT(4\pi k_e^2) \propto kTv_e^3 \qquad (4.42)$$

(In the conventional derivation (W. Shockley, *Electrons and Holes in Semiconductors*, Van Nostrand, 1950), it is assumed that the potential-energy term in the wave equation varies linearly with the lattice deformation. This is certainly a direct and correct approach, but it is appropriate only to longitudinal phonons and electrons on a spherical constant-energy surface. For this limited situation, the lattice-deformation potential analysis yields an expression for $R$ proportional to $kTv_e$ rather than as given in (4.42). A proper analysis must include the anisotropy of the effective mass that couples the transverse phonons into the scattering rate. We may restrain any overzealous urge to purity here by noting that no really successful analysis of the entire scattering process yet exists. Therefore, let us proceed with (4.42).

What we have from (4.42) is the collision rate, but what we want is the rate of change of the velocity as in equation 4.13. Putting $\Delta v = v[1 - \cos(\phi)]$ and computing the average value of $dv/dt$, we obtain

$$\frac{dv_e}{dt} = -Rv_e \qquad (4.43)$$

In other words, the average collision turns the electron through 90°. Accordingly, (4.42) gives us the randomization time as well as the mean time between collisions for electrons with an initial velocity, $v_e$. The same analysis would apply to the sphere not centered on the origin (Figure 4.10), except that we would have to use the "kinetic" wave vectors, $k_1$ and $k_1'$.

You should compare (4.42) to (4.15), which gives the randomization time for impurity scattering. For the impurity scattering, we found that the faster an electron went, the *less* likely it was to be deflected very much and the longer it would take to thermalize. In other words, as the sample is heated, impurity scattering disappears. Impurity scattering must therefore be dominant at "low" temperatures. For phonon scattering, (4.42) says that the faster an electron is going, the more likely it is to scatter. Thus, heating the sample increases the phonon scattering. We next wish to examine how this scattering will limit the conductivity of the sample and then see how this temperature dependence of $\tau_R$ will determine the temperature dependence of the conductivity.

## THE BOLTZMANN TRANSPORT EQUATION

The Boltzmann transport equation, if applied to a deck of cards, is a description of the conflict between shuffling and stacking. If a deck is being shuffled at a certain rate and stacked at a certain rate, the Boltzmann equation describes the rate of change of order within the deck. The distribution of cards throughout the deck is written in terms of the probability, $f$, of finding a card in a particular location. (In Chapter 3 we found the distribution function, $f_0$, that resulted when the system had been thoroughly randomized.) For a deck of cards, the Boltzmann equation is

$$\frac{df}{dt} = \frac{\partial f}{\partial t}\bigg|_{\text{shuffling}} + \frac{\partial f}{\partial t}\bigg|_{\text{stacking}} \qquad (4.44)$$

If $f = f_0$, $\partial f/\partial t|_{\text{shuffling}} = 0$. To maintain a steady-state distribution function other than $f_0$ requires continuous stacking so that $df/dt = 0$ even though $\partial f/\partial t|_{\text{shuffling}} \neq 0$.

Both shuffling and stacking can result in a net flow of a particular kind of card from one set of locations to another. For example, in the act of putting four aces on top of the deck, you generate a net flow upward. Subsequent shuffling will cause a net downward flow. It is equally possible to disturb the equilibrium distribution without causing a net upward or downward flow. Put two aces on top and two on the bottom. Shuffling then restores the distribution function to $f_0$, but no *net* vertical flow results.

To be sure that you understand exactly what is meant by $f$ and $f_0$, try the following questions. If we are interested only in the positions of the aces, and if we label the 52 positions 1 to 52 starting at the top, specify $f_0$. [Answer: For $f_0$ the probability of finding an ace in any location is the same: $1/13$. Thus, for the $n$'th position, $f_0(n) = 1/13$.] If the four aces are placed on top, what is $f$? [Answer: $f(n) = 1$ for $1 \leq n \leq 4$. $f(n) = 0$, $5 \leq n \leq 52$.]

Cards in a deck require only one coordinate for their specification. For electrons in a crystal, things are somewhat more complicated; we must give six coordinates to describe an electron completely—three to describe where it is and three more to specify where it is going (i.e., its momentum or velocity). The six-coordinate space is called *phase space*. With the added coordinates come a variety of shuffling possibilities. When you shuffle cards, an individual card may gain or lose "altitude." When the electrons scatter, they may gain or lose momentum, while between scattering events, they may change position. Somewhat different rules apply to the rate at which changes in position or momentum may take place. In order to write the Boltzmann equation for the electrons, we must be able to write each of the rate-of-change terms that should appear on the right side of (4.44). Our first task, then, is to see what changes in $f$ will take place.

Let us first begin with a one-dimensional crystal to simplify the pictorial representation of phase space. Phase space is now two-dimensional, the coordinates being $x$ and $v_x$. If we divide our two-dimensional space into little squares, the distribution function, $f(x, v_x)$, is the probability that the small square cell at $(x, v_x)$ will have an electron in it. If we include only one energy state per cell, $f \leq 1$ for electrons (or any other fermion); for bosons, $f$ may take on any value. In Chapter 3, we found that the energy that is characteristic of the cell completely determined $f_0$. However, it is quite possible and in some cases rather easy to achieve a distribution function that is completely unlike the equilibrium distribution function. For example, an electron gun, such as that in the display tube on an oscilloscope, emits electrons that all have the same velocity. Such a distribution function is the antithesis of a thermal-equilibrium distribution. Although it can be truly said that almost nothing useful operates in a state of thermal equilibrium, few get quite as far away as the electron gun. We will consider only those closely related to the equilibrium distribution.

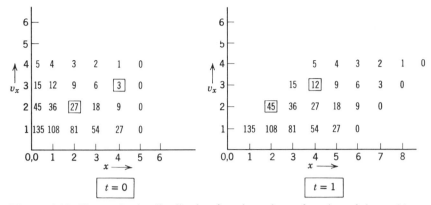

**Figure 4.11** Change in the distribution function, $f$, as a function of time with no applied force and no scattering, for a one-dimensional crystal. The numbers indicate the relative populations of each of the states, $(x, v_x)$. Two particular cells are "boxed" —one at $(2, 2)$ and the other at $(4, 3)$. The distribution is shown for $t = 0$ and again at $t = 1$.

If there were no scattering and no electric or magnetic fields, the electrons would continue indefinitely in uniform motion. If we wanted to know how the $f$ in some cell in phase space was going to change, we would look "upstream" to see what was coming into the cell. In other words, if the velocity is constant:

$$f(x, v_x, \delta t) = f(x - v\,\delta t, v_x, 0) \tag{4.45}$$

[The reason for using $\delta t$ in (4.45) rather than just plain $t$ is to make the relationship equally good for times, $\delta t$, that are small compared to the time it takes the velocity to change.] Equation 4.45 is illustrated in Figure 4.11. Integer numbers are used in the illustration even though typical occupation numbers are $\ll 1$. If you consider the numbers to be relative occupation numbers (e.g., all divided by, say, $10^5$), then they could very well represent a distribution function within a band.

If we were to apply some force to the electrons—say by an electric field, E—then the velocity changes with time according to Newton's second law: $dv_x/dt = F_x/m^* = -qE_x/m^*$. Picking a $\delta t$ small enough so that the velocity does not change much (i.e., ignoring the quadratic term in the relationship between $x$ and $t$), we can rewrite (4.45) to include forces:

$$f(x, v_x, \delta t) = f(x - v_x\delta t, v_x - F_x\,\delta t/m^*, 0) \tag{4.45'}$$

Notice that (4.45') includes the effects of macroscopic forces, but it does not include any of the effects of scattering. Figure 4.12 illustrates equation 4.45'. To write the Boltzmann equation, we need the time rate of change of $f$.

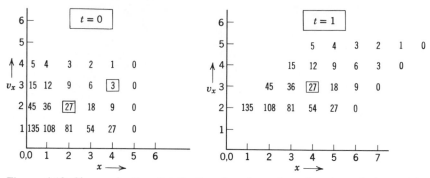

**Figure 4.12** Change in the distribution function, $f$, as function of time with an applied force, $F/m^* = 1$. The quadratic term ($\frac{1}{2}at^2$) has been ignored in computing the new positions. It would result in a shift to the right of $1/2$ for the right-hand graph's distribution.

It is fairly easy to see from (4.45') that to find out how $f$ is going to change, you look "upstream" to see what is coming. That leads to an equation whose signs may disturb your intuition about total derivatives, but when you note that we are talking about the divergence of the flux through the cell, it is clear that the signs are correct. The equation is:

$$\frac{df}{dt} = -\frac{\partial f}{\partial x} v_x - \frac{\partial f}{\partial v_x} \frac{F_x}{m^*} \tag{4.46}$$

The two terms on the right of (4.46) simply describe the difference in the rates at which particles flow in and flow out of the cell. Their velocity carries them in from the left and out to the right; the force carries them in from the bottom and out at the top.

We now must add the effects of scattering to (4.46). We do so formally by adding the term $(\partial f / \partial t)_{\text{coll}}$ to the right side of (4.46):

$$\frac{df}{dt} = -\frac{\partial f}{\partial x} v_x - \frac{\partial f}{\partial v_x} \frac{F_x}{m^*} + \left( \frac{\partial f}{\partial t} \right)_{\text{coll}} \tag{4.47}$$

Equation 4.47 is the Boltzmann transport equation in one dimension. To generalize to three spatial coordinates requires only that we write (4.45') using all six coordinates and then take all six derivatives. Using the symbol $\nabla f$ for the vector, $(\partial f / \partial x, \partial f / \partial y, \partial f / \partial z)$, and the symbol $\nabla_v f$ for the vector, $(\partial f / \partial v_x, \partial f / \partial v_y, \partial f / \partial v_z)$, the three-dimensional form of (4.47) is

$$\frac{df}{dt} = -\nabla f \cdot \boldsymbol{v} - \nabla_v f \cdot \frac{\boldsymbol{F}}{m^*} + \left( \frac{\partial f}{\partial t} \right)_{\text{coll}} \tag{4.48}$$

Now let us attempt to evaluate the collision term. How do collisions affect the spatial coordinates of the particles? (Answer: They don't. In a collision, particles exchange momentum and energy—functions of the velocity coordinates—but their spatial coordinates remain unchanged.) Thus, in a plot such as 4.12, collisions redistribute the particles in a vertical column but not in a horizontal row.

The various types of collisions have noticeably different results. For example, in an electron-electron collision, the sum of the momenta of the electrons is conserved. This means that such collisions preserve the average velocity of any ensemble of electrons. Such collisions are really very uninteresting in one dimension, since in one dimension all collisions are "head-on." The electrons exchange everything—energy and momentum—and since one electron is pretty much like another, you cannot tell whether a collision took place or whether the particles passed right through one another. In three dimensions, however, we have the possibility of a "glancing" collision. Energy and momentum are still conserved, but the initial energies for both particles are different from their final energies. On the average, energy is transferred from the faster to the slower electron. It is just such a process that leads to "thermalization" and an equilibrium distribution, a distribution in which the average electron is as likely to gain energy in a collision as lose it. However, even though the electron-electron collisions are just the sort that produce "a thoroughly shuffled deck," the average velocity of the whole ensemble—the velocity of the whole deck—is unchanged by these collisions.

To have the deck lose momentum, its constituents must run into something besides themselves. For electrons in a crystal, this "something else" is either a lattice defect or a phonon. These collisions, as we have seen in the previous sections, result in very little energy transfer, although large amounts of momentum can be involved. In the one-dimensional case, the only electron-energy conserving event that can occur is the replacement of $v_x$ by $-v_x$. Such collisions can very quickly reduce the ensemble's average velocity to zero, even though it does not do much randomizing. In the card example, this type of collision is equivalent to taking a distribution with four aces on the top to one with two aces on the top and two on the bottom. For example, if there were 10 electrons with velocity $+2$ and 5 electrons with velocity $-2$ in some small spatial volume, and if $20\%$ of them scattered in a given time, the average velocity of the 15 electrons would decrease from 2/3 to 2/5 in that time.

## FROM THE BOLTZMANN EQUATION TO THE
## ELECTRIC CURRENT

There is a fairly obvious relationship between the average velocity of an ensemble of electrons and the current due to that ensemble. First we need a definition of current. Current is the net flow of charge plus the rate of change of the electric displacement. (Since we are interested in the motion of the electrons, we wish to evaluate only the charge flow terms. There are relatively few places where we will need to consider displacement currents, and they have a way of taking care of themselves.) Thus, if the density of electrons is $n$, the current density at the point $r$ is given by

$$J(r) = -qn(r)\overline{v(r)} \tag{4.49}$$

where $\overline{v(r)}$ is the average electron velocity at the point $r$.

If you have the distribution function, $f(r, v)$, how do you obtain the average electron velocity? Answer: The average is obtained by summing all the electron velocities and dividing by the number of electrons:

$$\overline{v(r)} = \frac{\displaystyle\int_{\text{all } v} v \, dn(r, v)}{\displaystyle\int_{\text{all } v} dn(r, v)} \tag{4.50}$$

Equation 4.50 transfers our problem to finding $dn(r, v)$ given $f$, so we can resort to our familar $dn(r, v) = f(r, v) g(v) d^3v$ (where $d^3v$ is an appropriate volume element in velocity space such as $dv_x \, dv_y \, dv_z$). To have any useful way of going from our previous form, $g(k) d^3k$, to $g(v) d^3v$, we must assume a parabolic band. To make life as easy as possible, let us assume that the effective mass is isotropic (i.e., spherical constant energy surfaces). Then $E - E_c = = m^* v^2/2 = \hbar^2 (k - k_c)/2m^*$, and $g(v) = m^{*3} g(k)/\hbar^3 = (m^*/\pi\hbar)^3$.

We are now ready to determine the current once we have $f(r, v)$. This we will obtain from the Boltzmann transport equation with several reasonable but not necessarily true assumptions. Our first and most important assumption is that nothing that we do results in a velocity distribution very different in form from the traditional Boltzmann distribution. In other words, $f(r, v) = \alpha(r) e^{-m^* v^2/2kT}$, where $\alpha(r)$ is a variable that we will relate to $n(r)$ in a few moments. This assumption is equivalent to saying that the fields that are acting on the electrons in our small volume are sufficiently small that a very small fractional change in the Boltzmann distribution results in a balance between the rate at which the field carries particles out of the volume and the rate at which scattering carries them back in. Note, however, that some shift in the distribution function is necessary. Why? [Answer: The Boltzmann-like distribution of velocities (frequently called a Maxwellian distribution when used to describe the velocity distribution) is just the distribution that results in detailed balancing among the collision-induced changes. In other words, for a Maxwellian distribution, $(\partial f/\partial t)_{\text{coll}} = 0$. If we wish to have a steady state, the first two terms on the right of (4.48) must be balanced by $(\partial f/\partial t)_{coll}$. For example, in one dimension (Figure 4.12), the first term in (4.48) describes the horizontal trajectories of the particles in phase space and the second describes the vertical trajectories. Since these motions are orthogonal and independent, these terms cannot cancel. To have $df/dt = 0$, then, requires $(\partial f/\partial t)_{coll} \neq 0$.]

An extremely useful way to write the distribution function is in terms of a Maxwellian distribution plus a pair of terms to describe the non-Maxwellian character of the actual distribution, to wit:

$$f(r, v) = \alpha(r)\left[e^{-m^*v^2/2kT} + \gamma(r, v) + \delta(r, v)\right] \qquad (4.51)$$

The $\gamma$ term includes all of the excess electrons that are *not balanced* by excess electrons with oppositely directed velocity. $\delta$ includes all of the excess electrons that are balanced. The reasons for segregating them according to whether they are paired or not is that electron-phonon and electron-impurity collisions rapidly reduce $\gamma$ by pairing up the electrons (increasing $\delta$ in the process). These same collisions, as well as the electron-electron collisions, work at a different rate on the $\delta$ term, reducing it to zero if it is not constantly regenerated. Since we are pursuing the average velocity, only the unpaired electrons can contribute to our solution. We are thus relieved of the necessity of considering the details of thermalization; only the direction-randomizing collisions that we have discussed earlier in the chapter will enter our calculations.

Following our discussions of the randomization time for various scattering processes, we next assume that for each velocity, $v$, there is a characteristic relaxation time for $\gamma$. With that assumption we may write:

$$\left(\frac{\partial f}{\partial t}\right)_{coll} = \alpha(r)\left(\frac{\partial \gamma}{\partial t} + \frac{\partial \delta}{\partial t}\right) = \alpha(r)\left(\frac{-\gamma(r, v)}{\tau} + \frac{\partial \delta}{\partial t}\right) \qquad (4.52)$$

We have now almost completed our task. What remains to be done is to relate $\alpha(r)$ to $n(r)$ and then to solve the Boltzmann equation to obtain $v$ under our primary assumption that both $\gamma$ and $\delta$ are small compared to the Maxwellian term in (4.51).

To relate $\alpha$ to $n$, we use the fact that the sum of all the electrons in all the velocity states is the total number of electrons:

$$n(r) = \int_{all\ v} f(r, v)g(v)d^3v$$

Since our constant-energy surfaces are spherical, a suitable volume element is $d^3v = 4\pi v^2 dv$ and, as we have just seen, $g(v) = (m^*/\pi\hbar)^3$. Assuming that the dominant term in (4.54) is the Maxwellian term, we have

$$n(r) = \int_0^\infty \alpha(r)4\pi\left(\frac{m^*}{\pi\hbar}\right)^3 e^{-m^*v^2/2kT}v^2 dv = \left(\frac{2m^*kT}{\pi\hbar^2}\right)^{3/2}\alpha(r) \qquad (4.53)$$

Equation 4.53 is the relationship we are looking for. In typical problems, we will know $n(r)$ rather than $f(r, v)$, so (4.53) and (4.51) give us a way to find the distribution function, given $n$.

One important distinction should be made between the essentially Maxwellian distribution characterized by (4.51) and the equilibrium distribution that we have been calling $f_0$. The equilibrium distribution is Maxwellian; in point of fact, we can put it in the same form as (4.51), using $n_0(r)$ in (4.53): $f_0(r, v) = \alpha_0(r)e^{-m^*v^2/2kT}$. Our assumption is that the velocity distribution is approximately the same for the equilibrium and nonequilibrium distribution functions. We DO NOT assume that $f_0 \approx f$; in many cases of great practical interest, $n$ may differ from $n_0$ by six or eight orders of magnitude. This point is frequently obscured or missed in discussions of the Boltzmann equation.

We now proceed to the final step—finding $\overline{v(r)}$ for a steady-state, non-equilibrium situation. In the steady state, $df/dt$ is zero. From (4.48) we have $0 = -\nabla f \cdot v - \nabla_v f \cdot F/m^* + (\partial f/\partial t)_{\text{coll}}$. To simplify our analysis, let us examine only one component of $v$, say $v_z$. From (4.51), we have

$$\overline{v_z} = \frac{\int_{\text{all } v} v_z \, dn(v)}{\int dn(v)} = \frac{\int_v v_z g f \, d^3v}{\int g f \, d^3v} \tag{4.54}$$

Since by assumption $g$ is a constant, the two $g$'s cancel. We now take advantage of (4.52) to find $\alpha\gamma$, the only term of $f$ that will not average to zero in the numerator above:

$$\alpha(r)\gamma(r, v) = \tau(v)\left[ -\nabla f \cdot v - \nabla_v f \cdot \frac{F}{m^*} - \alpha(r) \frac{\partial \delta(r, v)}{\partial t} \right] \tag{4.55}$$

The last term in (4.55) will average to zero by assumption. Thus, for $\overline{v_z}$ we now have

$$\overline{v_z} = \frac{\int v_z \alpha\gamma d^3v}{\int f \, d^3v} = \frac{\int v_z \tau(v)\left[ -\nabla f \cdot v - \nabla_v f \cdot \frac{F}{m^*} \right] d^3v}{\int f \, d^3v} \tag{4.56}$$

We next assume that in taking the derivatives of $f$, the dominant term is the Maxwellian one. This need not be true, of course. Even if $\gamma$ and $\delta$ are everywhere small, were they to change very rapidly in phase space, their derivatives would be large. However, we would hardly expect such rapid variation for electrons in a crystal lattice. Accordingly, we may put in $\alpha(r) e^{-m^*v^2/2kT}$ for $f(r, v)$ in (4.56). That yields an integral that is readily evaluated:

$$\bar{v}_z = \frac{\displaystyle\int_{\text{all } v} v_z\tau(v)\left[-\nabla\alpha\cdot v - \alpha\nabla_v\left(-\frac{m^*v^2}{2kT}\right)\cdot\frac{F}{m^*}\right]e^{-m^*v^2/kT}d^3v}{\displaystyle\int_{\text{all } v}\alpha e^{-m^*v^2/2kT}d^3v}$$

(4.57)

Since we are averaging over all values of $v$, and since there are as many positive values of each component of $v$ as there are negative ones, only those terms with components of $v$ raised to an *even* power will not average to 0. Since each term in the bracket in (4.57) is multiplied by $v_z$, we need keep only those terms in the bracket that have $v_z$ to an odd power. For the first expression in the bracket, the choice is rather obviously the term containing

**Table 4.1.** Useful Integrals for Evaluating Boltzmann Transport Functions

---

1. $\displaystyle\int_0^\infty e^{-x^2}dx = \sqrt{\frac{\pi}{4}}$

2. $\displaystyle\int_0^\infty xe^{-x^2}dx = \frac{1}{2}$

3. $\displaystyle\int_0^\infty x^2e^{-x^2}dx = \frac{1}{4}\sqrt{\pi}$

4. $\displaystyle\int_0^\infty x^3e^{-x^2}dx = \frac{1}{2}$

5. $\displaystyle\int_0^\infty x^4e^{-x^2}dx = \frac{3}{8}\sqrt{\pi}$

6. $\displaystyle\int_0^\infty x^5e^{-x^2}dx = 1$

7. $\displaystyle\int_0^\infty x^6e^{-x^2}dx = \frac{15}{16}\sqrt{\pi}$

In general:

$$\int_0^\infty x^{2n+1}e^{-x^2}dx = \frac{n!}{2}$$

$$\int_0^\infty x^{2n}e^{-x^2}dx = \frac{(2n)!}{2^{2n+1}n!}\sqrt{\pi}$$

---

$(\partial\alpha/\partial z)v_z$. The expression containing $F$ presents more of a problem because some forces (notably the magnetic force) depend on the velocity. Since we will want to treat the magnetic-field case later, let us save the analysis with velocity-dependent forces for that time. Here let us consider only the electric force, $-q\mathbf{E}$, which is considerably simpler. The retained portion of the second term in the bracket is now $\alpha\,(m^*v_z/kT)\,(qE_z/m^*)$. Putting (4.57) into spherical coordinates $(v_z = v\cos\theta)$ and retaining only the terms that do not average to zero, we have

$$
\bar{v}_z = \frac{\displaystyle\int_0^\pi \int_0^\infty (v\cos\theta)\,\tau\,(v)\left[-\frac{\partial\alpha}{\partial z}v\cos\theta - \frac{\alpha qE_z}{kT}v\cos\theta\right]e^{-m^*v^2/2kT}v^2\sin\theta\,d\theta\,dv}{\displaystyle\int_0^\pi \int_0^\infty \alpha e^{-m^*v^2/2kT}\,v^2\sin\theta\,d\theta\,dv}
$$

$$(4.58)$$

Taking outside the integral sign all terms that do not depend on $v$ and $\theta$ and then performing the angular integration yields the final form:

$$
\bar{v}_z = \left[-\frac{1}{\alpha}\frac{\partial\alpha}{\partial z} - \frac{qE_z}{kT}\right]\frac{1}{3}\frac{\displaystyle\int_0^\infty \{v^2\tau(v)\}\,e^{-m^*v^2/2kT}v^2dv}{\displaystyle\int_0^\infty e^{-m^*v^2/2kT}\,v^2dv}
$$

$$(4.59)$$

Equation 4.59 can be made somewhat more useful by noting from (4.53) that $(1/\alpha)(\partial\alpha/\partial z) = (1/n)(\partial n/\partial z)$. The integrals represent the average value of $v^2\tau$ taken over a Maxwellian distribution. The conventional shorthand notation for this particular average is a pair of triangle brackets: $\langle v^2\tau\rangle$. Table 4.1 gives useful integrals for equation 4.59. With these modifications, (4.59) becomes

$$
\bar{v}_z = \left[-\frac{1}{n}\frac{\partial n}{\partial z} - \frac{qE_z}{kT}\right]\frac{1}{3}\langle v^2\tau\rangle
$$

$$(4.59')$$

It is customary to define two coefficients, $D_e$ and $\mu_e$, such that the average electron velocity is given by

$$
\bar{v}_z = \mu_e E_z - D_e(1/n)\left[(\partial n/\partial z)\right]
$$

$$(4.60)$$

From (4.59) it is easy to identify the meaning of these two coefficients. The one that relates the average velocity to the field is called the electron *mobility*; the other, which relates the velocity to the logarithmic derivative of the electron concentration, is called the *diffusion coefficient*.

Writing the current in terms of the mobility and diffusion constants (from equation 4.49) gives us one of the workhorse formulas of device analysis:

$$
J_e = qn\mu_e E + qD_e\nabla n
$$

$$(4.61)$$

A similar expression can be written for holes, but note the sign change that results from the hole's positive charge:

$$
J_h = qp\mu_h E - qD_h\nabla p
$$

$$(4.62)$$

These two formulas are vital to device analysis because currents and voltages are the terminal characteristics we will wish to relate. It is extremely important to note that current can arise from two distinct causes. If the charges flow as the result of the electric field, the current is said to be a *drift current*. If it is caused by a gradient in the concentration, it is said to be a *diffusion current*.

To see whether you can use (4.61) and (4.62), compute the conductivity of a piece of Ge with $10^{17}$ electrons/cm$^3$ and a mobility of 3000 cm$^2$/volt second. [Answer: Conductivity is defined as $\sigma = J/E$. From (4.61), $\sigma = = qn\mu_e = (1.6 \times 10^{-19})(10^{17})(3 \times 10^3) = 48$ mhos/cm.]

## THE EINSTEIN RELATIONSHIP

From (4.59′) you can see that the ratio of the mobility to the diffusion coefficient is always $q/kT$ [as long as the assumptions used to derive (4.59′) are valid]. This result can be derived from what looks like—but isn't—a completely different approach (originally developed by Einstein). In equilibrium both $J_e$ and $J_h$ must be zero. The equilibrium field can be obtained, as in Chapter 3, from (3.41′): $n = n_i e^{q\psi/kT}$. To get the field, the potential must be differentiated, giving

$$\mathbf{E} = -\nabla\psi = -\frac{kT}{q}\nabla[\log(n)] = \frac{-kT}{qn}\nabla n \qquad (4.63)$$

Putting (4.63) into (4.61) yields

$$0 = qn\mu_e\left[\frac{-kT}{qn}\nabla n\right] + qD_e\nabla n \qquad (4.64)$$

or in other words:

$$\mu_e = \frac{q}{kT}D_e$$

Why is the derivation of (4.64), using the equilibrium argument of Einstein, quite equivalent to obtaining it from (4.59′)? [Answer: The Einstein derivation depends on (3.41′), which is valid when the electrons are distributed in a Boltzmann distribution. Equation 4.59 was derived on the assumption that the electrons were distributed in a Maxwellian fashion. Since we are comparing coefficients that are averages over the velocity distribution, the Maxwellian distribution occurs in both cases and is central to both arguments. Accordingly, we should not expect the Einstein relationship (4.64) to hold when the equilibrium distribution is not a Boltzmann distribution. Of course, even in those cases, detailed balancing must hold, so if you have a replacement for (3.41'), you could readily derive a new form of (4.64) using the Einstein method.]

The utility of the Einstein relationship is that only one of the two co-efficients need be measured or remembered. The other is then immediately at hand.

## THE TEMPERATURE DEPENDENCE OF THE MOBILITY

If we insert the appropriate value of $\tau$ into (4.59) and perform the indicated averages, we should be able to determine the mobility and diffusion coefficients directly from fundamental considerations. Efforts to do this have not been spectacularly successful, because of all the difficulties in specifying $\tau$ versus $v$ for phonon scattering. When a whole loaf is not available, the standard ploy is to try for a half. Let us say that we measure the conductivity of an extrinsic semiconductor at 300°K (room temperature). Can we predict what it would be at 350°K? At these temperatures in Si and Ge, one is normally dealing with extrinsic samples with fully ionized impurities. Accordingly, the majority carrier density is almost independent of temperature, and the change in conductivity is given by $\partial\sigma/\partial T = qn(\partial\mu_e/\partial T)$ (for an $n$-type sample). From (4.59') we may write $\mu_e = (q/3kT)\langle v^2\tau_e\rangle$. We now come to the nub of the problem, which is what to put in for $\tau_e$. If we select a typical, reasonably pure sample, phonon scattering will predominate at 300°K, so, for lack of a better value, we should use (4.42). At lower temperatures or with very impure samples, (4.15) should be employed.

Putting (4.42) into (4.59') and performing the integration as specified in (4.59) gives

$$\mu_e \propto (1/kT)^2 \langle 1/v\rangle \propto (1/kT)^{2.5} \tag{4.65}$$

Considering the *ad hoc* nature of the constant cross-section assumption that led to (4.42), (4.65) turns out to be surprisingly good for several of the most common situations. Table 4.2 gives the generally accepted room-temperature values for the temperature dependence of the mobility for both electrons and holes in a number of semiconductors. Figure 4.13 shows $\mu$ versus $T$

Table **4.2.**[a]

|      | Electrons | Holes |
|------|-----------|-------|
| Ge   | 1.66      | 2.33  |
| Si   | 2.5       | 2.7   |
| GaAs | 1.0       | 2.1   |
| InSb | 1.6       | 2.1   |
| GaSb | 2.0       | 0.9   |

[a]The table lists $N$ in the relationship $\mu \propto T^{-N}$.

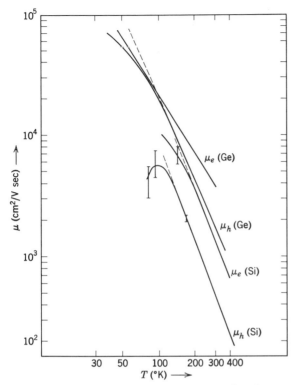

**Figure 4.13** Mobility versus temperature for electrons and holes in Ge and Si. The "best-fit" slopes of the straight-line portions are given in Table 4.1. The Ge data are from F. J. Morin, *Phys. Rev.* **93**, 62 (1954). The Si data are from G. W. Ludwig and R. L. Walters, *Phys. Rev.* **101**, 1699 (1956).

for Ge and Si. The table makes it perfectly clear that a successful universal theory of carrier mobility will have to take into account more complexities than we have been willing to undertake.

The most hopeful interpretation that we could place on (4.65)'s success in predicting several of the coefficients listed in Table 4.2 is that in the successful cases, our ad hoc assumption gives a good representation of the average scattering rate. Although this would give us no insight into the scattering dynamics themselves, it might well allow us to calculate the *field dependence* of the mobility. This is a very important relationship for device analysis, since most of the active devices in use today depend to some extent on carrier drift through regions of high electric field. As we shall see, this hopeful interpretation is not entirely unfounded. For the cases in which an approximate match to Table 4.2 is found, we will be able to predict the mobility for a reasonably large range of electric field. More of this will appear later in the chapter.

To see the importance of the thermal drift of the mobility to device performance, consider the problem of designing a monolithic microcircuit—that is, a circuit comprising both active and passive elements all made on the same small chip of semiconductor. For example, if you were to make a resistor out of an isolated strip of $n$-type silicon, its resistance would increase with temperature as $T^{2.5}$; a 10-ohm resistor at 300°K is a 15-ohm resistor at 350°K. Active devices suffer similar gross changes (although other factors may be much more important than $\mu$), making the design of temperature-stable circuits a most entertaining activity.

Figure 4.13 shows some signs of impurity scattering at the lower temperatures. For holes in Si, the mobility starts to decrease as $T$ falls below 100°K. The same effect would show up for the other three curves if they were continued to low enough temperatures. That the impurity-limited mobility should increase with increasing temperature is precisely what (4.15) predicts.

If we put the scattering time from (4.15) into (4.59), we obtain the rather awkward expression:

$$\mu \propto \frac{q}{kT} \frac{\displaystyle\int_0^\infty \frac{v^5 e^{-m^*v^2/2kT} v^2 dv}{\log(1 + v^4 b_m^2/a^2)}}{\displaystyle\int_0^\infty e^{-m^*v^2/2kT} v^2 dv} \tag{4.66}$$

If we assume that the argument of the logarithm in (4.66) is reasonably large compared to 1, the denominator of the upper integral will change very slowly over the narrow interval in which the integrand has substantial values. Thus, a reasonable approximation is to replace the variable denominator by its value at the velocity that maximizes the numerator. [This was the approximation used by Ester Conwell and Victor Weisskopf in their first derivation of the impurity-limited mobility, in *Phys. Rev.* **77**, 388 (1950).] With this simplification, (4.66) can be readily evaluated, leading to

$$\mu \propto (kT)^{3/2} / \log\left[1 + \left\{\frac{28\sqrt{\pi\varepsilon kT}}{q^2(N_A + N_D)^{1/3}}\right\}^2\right] \tag{4.67}$$

The constants in the denominator are the values that come from substituting for $a$ and $b$ and putting in the value of $v$ that maximizes the integrand in (4.66).

It is interesting to compare (4.67) with (4.65). Equation 4.67 says that resistance should decrease with increasing temperature when impurity scattering dominates; (4.65) says that the reverse is true when phonon scattering

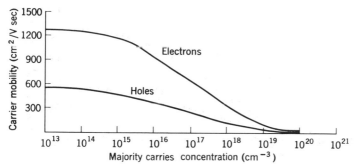

**Figure 4.14** Mobility versus majority-carrier concentration for Si. Data taken from E. Conwell, *Proc. IRE* **46**, 1281 (1958).

is most important. Be sure to remember that resistivity is also dependent on the number of carriers available, and in some cases that can be the most important term by several orders of magnitude. Since the resistivity is essentially the sum of the resistance caused by phonons plus the resistance caused by impurities, it follows that the addition of impurities always lowers the mobility. It should also be true that the conductivity of less pure samples will decrease less rapidly with the increasing temperature. Figures 4.14 and 4.15 bear this out.

## THE HALL EFFECT

Measuring $\sigma$ is easy. However, doing so does not give you either $\mu$ or the carrier concentration, only the product of the two. From what we have discussed so far, it should be clear that we need to know both terms individually. In the typical situation described by (4.61), $\mu_e$ and $D_e$ are characteristics of the material (presumably constant), while $n$ and its derivatives are variables that are adjustable from the outside. For example, shining a bright light on the material generates many hole-electron pairs but would not usually effect $D$ or $\mu$.

What we are looking for is a method that would measure either $\mu_e$ or $n$ individually. One technique for the latter measurement is to assay the impurity content of the sample and obtain $n$ directly from $N_d - N_a$. It should be pretty obvious that such a measurement on a sample with less than a part per million of total impurities would be distinctly nontrivial. Couple these inherent measurement difficulties with the fact that, at moderately high doping, the band gap and the impurity energy level become functions of the impurity concentration (so that the energy values to put into the Fermi or Boltzmann distribution function become uncertain), and you have an experimentalist's nightmare. Such experiments have been performed [e.g.,

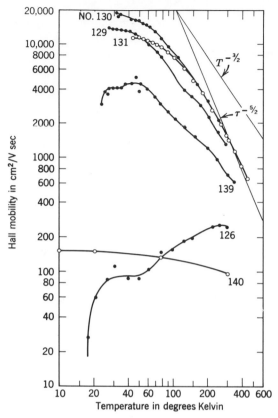

**Figure 4.15** Mobility versus temperature for some arsenic-doped silicon samples. The approximate total impurity concentrations are

*Sample Number    Impurity Concentration/cm³*

| Sample Number | Impurity Concentration/cm³ |
|:---:|:---:|
| 131 | $2 \times 10^{14}$ |
| 130 | $2 \times 10^{15}$ |
| 129 | $2 \times 10^{16}$ |
| 139 | $1 \times 10^{17}$ |
| 126 | $3 \times 10^{18}$ |
| 140 | $5 \times 10^{19}$ |

Data are from F. J. Morin and J. P. Maita, *Phys. Rev.* **96**, 28 (1954).

G. Backenstross, *Phys. Rev.* **108**, 1416 (1957); a somewhat improved interpretation of the Backenstross data by F. M. Smits appears in E. M. Conwell, *Proc. IRE* **46**, 1287 (1958)], but they hardly constitute either a desirable or an accurate method.

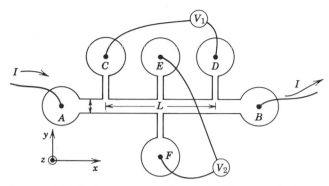

**Figure 4.16** Hall measurement sample. The sample is prepared from a thin, flat wafer by etching away all but the desired material (by a variation on the photoengraving technique). Wire leads are then attached to the six circular pads.

Several other techniques exist, but by far the most commonly employed is the Hall measurement. Let us make a very simplistic analysis of this experiment to show what we would like to achieve and how the experiment is performed. Then we will make a more realistic analysis to see what the Hall measurement really determines.

A sample of the material to be tested is prepared in the shape shown in Figure 4.16. Leads $A$ and $B$ are connected to a constant current source. Pairs $C$ and $D$ and $E$ and $F$ are connected to voltmeters that draw no current (typically, a sensitive potentiometer is employed). A D-C magnetic field is applied to the sample in the $z$-direction. (Note the coordinates in Figure 4.16; the $z$-axis comes out of the page.) The average electron is drifting from $B$ to $A$ with velocity $v_e$. The magnetic field, $\boldsymbol{B}$, exerts a force $\boldsymbol{F} = -qv_e \times \boldsymbol{B}$ on this average electron, trying to make it go in the $-y$ direction. Since there is no D-C path for current in the $y$-direction, charge must collect along the sides. (For electrons, the charge is $-$ on pad $F$ and $+$ on pad $E$. For holes, since both $v_h$ and the charge reverse sign, the positive charge must be on pad $F$.) This accumulation of charge continues only until the electric field due to the charge accumulation pushes the electrons in the $y$-direction as hard as the magnetic field pushes them in the $-y$ direction. Thus, in the steady state, which is reached with great speed, we have

$$- qv_x B_z = - qE_y \tag{4.68}$$

We may write $v_x$ in one of two ways: $v_x = -J_x/nq = -E_x\mu_e$. Putting either of these into (4.68) yields a useful relationship:

$$\frac{E_y}{J_x B_z} = \frac{-1}{qn} \equiv R_H \qquad \text{or} \qquad \mu = \frac{E_y}{E_x B_z} \tag{4.69}$$

$R_H$ is called the *Hall coefficient*. It's sign is the sign of the dominant carrier's charge and it is inversely proportional to the carrier density.

The shape of the Hall measurement sample is long and basically of constant cross-section, so the current will be very uniform throughout the central bar. Thus, the current density to be put in (4.69) is simply the total current divided by the area of the bar. The electric field in the $y$-direction will be constant across the bar if $J_x$ is constant. Thus, $E_y$ in (4.69) is $V_2/w$ (see Figure 4.16). Similarly, $E_x$ is $V_1/L$. These results, if realistic, would mean that the measurement of $\mu_e$ and $n$ would be very simple indeed. If life were that easy, it is unlikely that Messrs. Backenstross and Smits would have gone to all that trouble to try to measure the carrier concentration by a total impurity determination. Which step or assumption in the derivation above was improper? [Answer: Using the average velocity in (4.68) is improper. If you will recall, when we were computing the average velocity a few pages ago, we took advantage of the fact that the force due to the electric field was independent of the electron's velocity. Since that is not true for the magnetic force, we must return to equations 4.56 and 4.57 and proceed properly. Not only will we find that $R_H$ and the Hall-measured mobility, $\mu_H$, are nontrivial to interpret; we will also find that the sample's electrical conductivity is a function of the magnetic field!]

## AVERAGING THE VELOCITY IN A MAGNETIC FIELD

If we put the magnetic force into (4.57), we are in for a rude surprise. Try it and see what you get. [Answer: In (4.57) we assumed that the dominant term in $\nabla_v f$ was the Maxwellian one, giving $\nabla_v f \propto v$. Thus, $\nabla_v f \cdot F \propto v \cdot v \times B$. But, $v \cdot v \times$ (anything) $= 0$, so the magnetic field does not cause any change in a distribution function that is Maxwellian!] This result should seem pretty plausible. After all, the magnetic force is always perpendicular to the electron's velocity, so the field does no work on the electron. The magnetic field causes the electrons to travel about in closed orbits on a constant-energy surface, but since each state on such a surface has an equal chance of being occupied (for a Maxwellian distribution), as many electrons come in from the right as go out at the left.

Since the magnetic field does not disturb a Maxwellian distribution, we cannot use the same trick that got us from (4.56) to (4.57); we cannot ignore the non-Maxwellian terms. If there is an average velocity in the $x$-direction, a $z$-directed magnetic field will cause the electrons to turn, generating an average flow in the $-y$ direction. Thus, the magnetic field acts directly on non-Maxwellian terms in $f$ and these must be treated explicitly in evaluating (4.56). The traditional approach at this point is to expand $f$ in a Fourier series of spherical harmonics. The orthogonality and symmetry of these functions allows a reasonably easy evaluation of the integral, but unless you are really familiar with spherical harmonics, this procedure, however correct, looks like pure metaphysics. Instead, let us follow the procedure used by Harvey Brooks (*Advances in Electronics and Electron Physics*, Vol. VII, Academic Press, 1955). In this technique, we first find the trajectory in $v$-space of electrons subjected simultaneously to constant electric and magnetic fields. Interestingly enough, this will yield a constant velocity plus some oscillatory terms. The oscillatory terms represent the circular trajectory of the electron in $v$-space; the constant term is the displacement of the center of the circle from the origin. If we make our usual assumption about the electric field—namely, that it is small enough so that the disturbed velocity distribution is never very asymmetric or non-Maxwellian—we can assume that the magnitude of the velocity is almost constant. (We will thus be assuming that the circle is almost centered at the origin.) If $|v|$ is almost constant, the mean time between collisions will also be a constant (for a given $v$). We can find the average velocity as a function of the initial velocity and scattering time. Then we will average that result over all possible initial velocities, assuming as before a Maxwellian initial distribution.

Before performing the steps outlined above, take a moment to note how they will correspond to what we should do according to (4.56). The trajectory-in-$v$-space calculation followed by the determination of the average position along that trajectory will give us an effective (force $\times$ scattering time) product to put into (4.57) in place of $F\tau(v)$. The average over the Maxwellian distribution is then equation 4.57 again, with the new $F\tau$.

We consider first the simplest case: a single type of carrier (electrons) with spherical constant-energy surfaces (constant effective mass). Let the magnetic field, $B$, point in the $z$-direction. Then, from Newton's second law and the Lorentz force equation, we have

$$F = m^* dv/dt = -q(E + v \times B)$$

or, by components:

$$dv_x/dt = -\frac{q}{m^*}(E_x + v_y B_z) \tag{4.70}$$

$$\frac{dv_y}{dt} = -\frac{q}{m^*}(E_y - v_x B_z) \tag{4.71}$$

$$\frac{dv_z}{dt} = -\frac{q}{m^*}E_z \tag{4.72}$$

The simplest way to handle the arithmetic for the first two equations, since they are coupled by their last terms, is to form their complex sum. To this purpose, let $V' = v_x + jv_y$, $E' = E_x + jE_y$, and $\omega_0 = qB_z/m^*$. Notice that the units of $\omega_0$ are indeed $\sec^{-1}$. $B$ is in units of webers/cm$^2$ (or Gauss $\times\ 10^{-8}$ if you prefer) and a weber is the same as a volt second (from Faraday's law, $V = d\phi/dt$). Putting $q$ in as 1 electron and $m^*$ in units of eV/$c^2$, the units check. $\omega_0$ is, of course, nothing more than the cyclotron resonance frequency defined in (2.35).

Adding $j$ times (4.71) to (4.70) yields

$$dV'/dt = -\frac{q}{m^*}E' + j\omega_0 V' \tag{4.73}$$

This is easily solved to yield

$$V'(t) = \frac{qE'}{m^*j\omega_0} + Ae^{j\omega_0 t}$$

If the initial complex velocity is $V_0'$, substitution for $A$ yields

$$V'(t) = \frac{qE'}{m^*j\omega_0}(1 - e^{j\omega_0 t}) + V_0' e^{j\omega_0 t} \tag{4.74}$$

Note that (4.74) came out as advertised: there is a constant term plus a pair of oscillatory terms.

We now must assume that $|V'| \simeq |V_0'|$. Then, if the scattering time is a function only of the magnitude of the velocity, we may write the rate of change of an initial number of electrons with speed between $v$ and $v + dv$ as

$$dn(v)/dt = -n/\tau(v) \qquad \text{or} \qquad dn = \frac{n_0 e^{-t/\tau}}{\tau}dt \tag{4.75}$$

The second form of (4.75) should be read, "The number, $dn$, that travel for a time between $t$ and $t + dt$ before being scattered is given by...

To find the average complex velocity, we must take

$$\overline{V'} = \frac{\displaystyle\int_0^\infty V'(t)n_0 e^{-t/\tau}dt}{\displaystyle\int_0^\infty n_0 e^{-t/\tau}dt} \tag{4.76}$$

Putting (4.74) into (4.76) and performing the integration yields

$$\overline{V'} = \frac{1}{1 - j\omega_0\tau}\left[V_0 - \frac{qE'\tau}{m^*}\right] \tag{4.77}$$

We should also find the average value of $v_z$, since that is not included in $V'$. From (4.72), $v_z(t) = v_{0z} - qE_z t/m^*$. Inserting this in (4.76) yields

$$\overline{v_z} = v_{0z} - qE_z\tau/m^* \tag{4.78}$$

To get the $x$ and $y$ velocities we must take the real and imaginary parts of (4.77). Recall that all we are really seeking is the proper $F\tau$ product to put into (4.56). The initial Maxwellian distribution will contribute nothing. For example, in (4.78) we would use only $-q\mathsf{E}_z\tau$. [Note that that was exactly what we DID use in (4.57) when we put the force in directly.] In other words, the initial velocity does not enter the force time product at all. Thus, we are only interested in the second term in (4.77), which for convenience, let us call $\overline{V}''$. Then

$$Re(\overline{V''}) = \frac{1}{1 + \omega_0^2\tau^2}\left(\frac{q\tau}{m^*}\right)(-\mathsf{E}_x + \omega_0\tau\mathsf{E}_y)$$

$$= \frac{1}{m^*}\left[-q\mathsf{E}_x\tau + \frac{\omega_0^2\tau^3}{1 + \omega_0^2\tau^2}q\mathsf{E}_x + \frac{\omega_0\tau^2}{1 + \omega_0^2\tau^2}q\mathsf{E}_y\right] \qquad (4.79)$$

The second form of (4.79) was chosen so that the first term is identical to the $F\tau$ product we used in our initial evaluation of (4.56). The rest of the terms, therefore, result from the presence of the magnetic field.

Doing the same thing for the $y$ velocity:

$$Im(\overline{V''}) = \frac{1}{1 + \omega_0\tau^2}\left(\frac{q\tau}{m^*}\right)(-\mathsf{E}_y - \omega_0\tau\mathsf{E}_x)$$

$$= \frac{1}{m^*}\left[-q\mathsf{E}_y\tau + \frac{\omega_0^2\tau^3}{1 + \omega_0^2\tau^2}q\mathsf{E}_y - \frac{\omega_0\tau^2}{1 + \omega_0^2\tau^2}q\mathsf{E}_x\right] \qquad (4.80)$$

Putting (4.78), (4.79), and (4.80) into (4.57) with the appropriate component of velocity for each case introduces no new problems in evaluating the integrals. We find ourselves averaging $v^2 e^{-m^*v^2/2kT}$ times each of the terms in (4.79) and (4.80) and the last term in (4.78) over all $v$ and multiplying the result by $m^*/3kT$. [See (4.59).] Using the angle bracket notation [see 4.59)] to indicate such an average, we have

$$\overline{v_x} = -\frac{q\mathsf{E}_x}{3kT}\left[\langle v^2\tau\rangle - \left\langle\frac{\omega_0^2\tau^3 v^2}{1 + \omega_0^2\tau^2}\right\rangle\right] + \frac{q\mathsf{E}_y}{3kT}\left\langle\frac{\omega_0\tau^2 v^2}{1 + \omega_0^2\tau^2}\right\rangle \qquad (4.81)$$

$$\overline{v_y} = -\frac{q\mathsf{E}_y}{3kT}\left[\langle v^2\tau\rangle - \left\langle\frac{\omega_0^2\tau^3 v^2}{1 + \omega_0^2\tau^2}\right\rangle\right] - \frac{q\mathsf{E}_x}{3kT}\left\langle\frac{\omega_0\tau^2 v^2}{1 + \omega_0^2\tau^2}\right\rangle \qquad (4.82)$$

$$\overline{v_z} = -\frac{q\mathsf{E}_z}{3kT}\langle v^2\tau\rangle \qquad (4.83)$$

## A MORE ACCURATE APPRAISAL OF THE HALL EXPERIMENT

The last three equations allow us a more realistic interpretation of the Hall measurements. In the typical Hall measurement, as illustrated in Figure 4.15, the $y$ and $z$ components of the current are zero. Note that this is a boundary condition on a particular experiment, so that the results we are about to obtain apply to that experiment only.

From (4.83) we have that $I_z = 0$ implies that $E_z = 0$, not a very surprising or exciting result. Equation 4.82 has more to say. Since $E_x$ is not zero and $\overline{v_y} = 0$, (4.82) requires

$$E_y = \frac{- E_x \langle \dfrac{\omega_0 \tau^2 v^2}{1 + \omega_0^2 \tau^2} \rangle}{\langle v^2 \tau \rangle - \langle \dfrac{\omega_0^2 \tau^3 v^2}{1 + \omega_0^2 \tau^2} \rangle} = \frac{- E_x \langle \dfrac{\omega_0 \tau^2 v^2}{1 + \omega_0^2 \tau^2} \rangle}{\langle \dfrac{v^2 \tau}{1 + \omega_0^2 \tau^2} \rangle} \qquad (4.84)$$

The simplification of the denominator takes advantage of the fact that averaging is distributive across addition (i.e., the sum of the averages is the average of the sum). Equation 4.84 put into (4.81) gives $\overline{v_x}$ completely in terms of $E_x$:

$$\overline{v_x} = - \frac{q E_x}{3kT} \left\{ \langle v^2 \tau \rangle - \langle \frac{\omega_0^2 \tau^3 v^2}{1 + \omega_0^2 \tau^2} \rangle + \frac{\langle \dfrac{\omega_0 \tau^2 v^2}{1 + \omega_0^2 \tau^2} \rangle^2}{\langle \dfrac{v^2 \tau}{1 + \omega_0^2 \tau^2} \rangle} \right\} \qquad (4.85)$$

Compare (4.85) to the drift term in (4.59′). The only way that the drift in the $x$-direction can be the same with and without the magnetic field is to have the second and third terms cancel. There is but one situation in which that can occur: $\tau$ must be independent of $v$! Such an improbable relationship just does not occur. Therefore, we must conclude that the magnetic field perpendicular to the flow of current changes the conductivity of the sample. Such a result should make good physical sense when you consider the trajectory of an electron in the combined electric and magnetic fields. Since there is no *net* flow in the $y$ direction, the $E_y$ Hall field must exactly balance the magnetic force for the *average* electron. Of course, if the forces balance for the average electron, they cannot balance for any other electron. The electrons travel in trajectories that curve away from the $x$-direction. Accordingly, the $x$-components must be less than they would be without the magnetic field.

It is customary to describe this change in terms of the percentage change of resistivity and to evaluate it under the usual experimental conditions where $(\omega_0\tau)^2 \ll 1$. It is not hard to get magnetic fields big enough to violate this approximation—2 webers/m$^2$ (20 kilogauss) will usually be enough at room temperature—but the run-of-the-mill Hall measurement is taken at much lower fields. Under that approximation, (4.84) and (4.85) become

$$\frac{E_y}{E_x\omega_0} = \frac{E_y}{E_xB}\left(\frac{m^*}{q}\right) = -\frac{\langle v^2\tau^2\rangle}{\langle v^2\tau\rangle} \tag{4.84'}$$

$$v_x = -\frac{qE_x}{3kT}\langle v^2\tau\rangle\left\{1 - \frac{\omega_0^2\langle v^2\tau^3\rangle}{\langle v^2\tau\rangle} + \frac{\omega_0^2\langle v^2\tau^2\rangle^2}{\langle v^2\tau\rangle^2}\right\} \tag{4.85'}$$

Since the conductivity is $\sigma = nq\bar{v}_x/E_x$, we have from (4.85'):

$$\sigma = \sigma_0\left\{1 - \frac{\omega_0^2\langle v^2\tau^3\rangle}{\langle v^2\tau\rangle} + \frac{\omega_0^2\langle v^2\tau^2\rangle^2}{\langle v^2\tau^2\rangle}\right\} \tag{4.86}$$

where $\sigma_0 = (q^2n/3kT)\langle v^2\tau\rangle$ is the conductivity of the sample without the magnetic field. Writing this in terms of the resistivity:

$$\rho = \rho_0 + \Delta\rho(B) = \frac{1}{\sigma} \cong \frac{1}{\sigma_0}\left(1 + \frac{\omega_0^2\langle v^2\tau^3\rangle}{\langle v^2\tau\rangle} - \frac{\omega_0^2\langle v^2\tau^2\rangle^2}{\langle v^2\tau^2\rangle^2}\right) \tag{4.87}$$

The last step in (4.87) used the usual Taylor series approximation to $1/(1-x) \simeq 1 + x$. This approximation is valid when $x$ is small, which will be true in (4.87) as long as our original approximation, $\omega_0^2\tau^2 \ll 1$, holds. We can now write the percentage change in resistivity:

$$\frac{\Delta\rho}{\rho_0} = B^2\left[\left(\frac{q}{m^*}\right)^2\frac{\langle v^2\tau^3\rangle\langle v^2\tau\rangle - \langle v^2\tau^2\rangle^2}{\langle v^2\tau^2\rangle^2}\right] \tag{4.88}$$

According to (4.90) the magnetoresistance should be proportional to the square of the magnetic field. Figure 4.17 shows some typical magnetoresistance data. Note that the change in resistance in the $n$-type material, especially for the 77°K data, is more than 100% at field strengths that are relatively easy to achieve. The low-field portion of the curve does follow a square law as (4.88) predicts. Why does the slope fall off at higher fields? [Answer: Going back to (4.85), before the small-field approximation was made, and allowing the $\omega_0^2\tau^2$ term to become the dominant part of the denominator (the large-field approximation), the velocity becomes

$$\bar{v}_x \cong -\frac{qE_x}{3kT}\langle v^2\rangle^2/\langle v^2/\tau\rangle \tag{4.89}$$

In other words, in the high-field approximation, *the resistance is independent of the magnetic field*. This saturation effect takes place when the electrons make several revolutions before scattering ($\omega_0\tau \gg 1$); once the field has be-

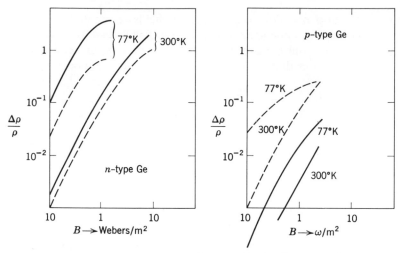

**Figure 4.17** Magnetoresistance data for Ge at liquid nitrogen and room temperatures. The current is always flowing in the [100] direction. The dashed lines are for the magnetic field perpendicular to the current (transverse magnetoresistance). $B$ was in the [010] direction. The solid lines are for the $B$ field in the [100] direction (longitudinal magnetoresistance). Data from G. L. Pearson and H. Suhl, *Phys. Rev.* **83**, 768 (1951).

come large enough so that the average electron makes several revolutions between collisions, it makes little difference whether they make one more or one less.]

Why does the saturation occur for lower fields at lower temperatures? (Answer: As long as the mobility is phonon-limited, the mean time between collisions for any given velocity decreases as the temperature increases. Accordingly, for phonon-limited conduction, $\omega_0^2\tau^2$ will be larger for lower temperatures. Note that at 77°K, the $p$-type sample is already starting to saturate at 0.1 webers/m².)

Now let's look at the interpretation of (4.69). According to the second form of (4.69), $\mu = -E_y/(E_xB_z)$. In the low-field limit, we have from (4.84'): $E_y/(E_xB_z) = -(q/m^*)\langle v^2\tau^2\rangle/\langle v^2\tau\rangle$. But from (4.59'), the electron mobility was given by $\mu_e = (q/3kT)\langle v^2\tau\rangle$. Noting that $\langle m^*v^2\rangle = 3kT$ (4.33'), we can see the difference between the Hall and drift mobilities:

$$\mu_{eH} = -\frac{q}{m^*}\frac{\langle v^2\tau^2\rangle}{\langle v^2\tau\rangle} \tag{4.90}$$

for the Hall mobility whereas

$$\mu_{eD} = -\frac{q}{m^*}\frac{\langle v^2\tau\rangle}{\langle v^2\rangle} \tag{4.91}$$

**Table 4.3** Properties of Several Semiconductors

| Material | Band Gap eV | | Hall Mobility[a] cm²/Vsec (300°K) | | Effective Mass Ratio $m^*/m_0$ | | Relative Dielectric Constant $\varepsilon/\varepsilon_0 \,[\varepsilon_0 \simeq (1/36\pi) \times 10^{-11} f/cm]$ |
|---|---|---|---|---|---|---|---|
| | 300°K | 0°K | Electrons | Holes | Electrons | Holes | |
| Ge  (1) | 0.67 | 0.75 | 3950 (3900) | 3400 (1900) | $m_{\parallel}/m_0 = 1.6$ $m_{\perp}/m_0 = 0.08$ | 0.3,  0.04 | 16 |
| Si  (1) | 1.11 | 1.15 | 1900 (1350) | 425 (480) | $m_{\parallel}/m_0 = 0.97$ $m_{\perp}/m_0 = 0.19$ | 0.5,  0.16 | 12 |
| GaAs  (4) | 1.43 | 1.52 | 8500 | 400 | 0.068 | 0.5 | 10.9 |
| GaP  (2) (4) | 2.24 | 2.40 | 110 | 75 | 0.5 | 0.5 | 10 |
| InSb  (2) | 0.16 | 0.26 | 78,000 | 750 | 0.013 | 0.6 | 17 |
| AlSb  (2) | 1.63 | 1.75 | 200–450 | 200–450 | 0.3 | 1.8 | 11.5 |
| Se  (3) | | 1.6 | | $\left\{\begin{array}{l}0.4^b\\0.1\end{array}\right.$ | | | |
| PbS  (4) | 0.4 | 0.34 | 600 | 700 | 0.66 | 0.5 | 17 |
| CdS  (4) | 2.42 | 2.56 | 300 | 50 | 0.17 | 0.6 | 10 |
| C(diamond)  (4) | 5.47 | 5.51 | 1800 | 1600 | 0.2 | 0.25 | 5.5 |

SOURCES: (1) E. M. Conwell, *Proc. IRE* **46**, 76 (1958); (2) C. Hilsum and A. C. Rose-Innes, *Semiconducting III–V Compounds*, Pergamon Press, N.Y.; (3) *American Inst. of Physics Handbook*, Second Edition, McGraw-Hill, N.Y.; and (4) S. M. SZE, *Physics of Semiconductor Devices*, Wiley, 1969.

[a] Values listed in parentheses are drift mobilities.
[b] Selinium is noncubic and has a distinctly anisotropic conductivity.

for the drift mobility. Once again, unless the scattering time is independent of velocity, the Hall mobility is not the same as the drift mobility.

If we try to evaluate (4.90) using (4.42) for the lifetime, we run into a bit of an embarrassment—the integral containing $\tau^2$ diverges for low values of $v$. This divergence arises because phonons seldom scatter low-velocity electrons, so the time between collisions becomes infinite as $v \to 0$. Physically, of course, such a result is utter nonsense. What we have left out is the impurity scattering that should dominate at low velocities. This makes an accurate evaluation of $\mu_H/\mu_D$ very difficult. We might guess that the ratio is greater than 1 because the Hall mobility "tends to diverge." In the few materials in which both the drift and the Hall mobilities have been measured, $\mu_H/\mu_D$ is usually larger than 1, sometimes by quite a bit (see Table 4.3).

The first expression in (4.69)—for the Hall constant, $R_H$—is supposed to give us information about the carrier concentration. To see what it really indicates, let us consider the low-field case in which $\sigma \simeq \sigma_0$ and (4.84′) hold. Then

$$R_H = \frac{-E_y}{J_x B_z} = \frac{-E_y}{\sigma_0 E_x B_z} = \frac{-\mu_H}{\sigma_0} \tag{4.92}$$

Note that $R_H$ is the ratio of two easily measured parameters. Putting in $qn\mu_{eD}$ for $\sigma_0$ shows what we need for a useful interpretation:

$$R_H = \frac{-\mu_{eH}}{qn\mu_{eD}}$$

Thus, if we knew the ratio of $\mu_{eH}$ to $\mu_{eD}$, we would indeed be able to measure the carrier concentration in a simple way. Unfortunately, this ratio depends to some extent on the impurity concentration, so usually no exact and simple interpretation is possible. Most experimentalists assume that $R_H \simeq -1/qn$ just as (4.69) implies. It is not that they do not know better; it's just that there are no other practical alternatives. Besides, you seldom need to know the carrier concentration to better accuracy than (4.69) gives.

## A FEW PROBLEMS

As routine exercises, interpret the following pieces of data.

### Exercise 4.1

The Hall constant for a sample of Ge at room temperature is found to be $-3 \times 10^2$ cm$^3$/coulomb. What is the approximate net impurity concentration and is the material $n$ or $p$ type? [Answer: The sign of $R_H$ gives the sign of the dominant carrier. Using (4.69), the material is $n$-type (the $-$ sign) with a net impurity concentration of $1/(1.6 \times 10^{-19} \times 3 \times 10^2) = 2.08 \times 10^{16}$ cm$^{-3}$.]

**Exercise 4.2**

With proper note of the fact that the averages above should really be taken over a Fermi distribution rather than a Boltzmann distribution, we can use the results above for metals as well as semiconductors. Consider the following data for W and Cu.

|    | $\rho$ in $\Omega$ cm | Atoms/cm$^3$ | $R_H$ in cm$^3$/Coulomb |
|----|-----------------------|--------------|--------------------------|
| Cu | $1.69 \times 10^{-6}$ | $8.45 \times 10^{22}$ | $-0.55 \times 10^{-4}$ |
| W  | $5.33 \times 10^{-6}$ | $6.35 \times 10^{22}$ | $+1.18 \times 10^{-4}$ |

What is the dominant carrier in each of these commonly used metals? [Answer: From the sign of the Hall coefficient, electrons are the principal carrier in copper and holes are the principal carrier in tungsten. (Yes, shattering as it may be, the lamp filament by which you are reading this is heated by *holes!*)] What is the apparent carrier concentration for each case? (Answer: $1.14 \times 10^{23}$ cm$^{-3}$ for copper and $0.53 \times 10^{23}$ for tungsten.) What is the Hall mobility in each case? [Answer: From (4.92), $\mu_H = |\sigma_0 R_H|$. Accordingly, $\mu_H = 32.6$ cm$^2$/Vsec for copper and 22.2 cm$^2$/Vsec for tungsten. Compare these with the Hall mobilities measured for the semiconductors listed in Table 4.2. You can see that the high-conductivity characteristic of metals reflects their large number of carriers rather than a high mobility for the individual carriers.]

**Exercise 4.3**

As a final problem, consider the case in which there are both holes and electrons. Since the Hall constant changes sign if the carrier type is changed, there must be some "blend" of holes and electrons for which the Hall coefficient is zero. Find the ratio of the densities of holes to electrons that *in the low-field limit* corresponds to $R_H = 0$. What will the value of the Hall mobility be if $R_H = 0$? [Answer: The second question is easily answered from (4.93). If $R_H = 0$, $\mu_H = 0$. Since the drift mobility does not vanish for either species of carrier in any concentration, it is pretty evident that Hall mobility will distinctly misrepresent the drift mobility when there is no very dominant carrier. To answer the first part, proceed from (4.82) using the low-field limit. Both holes and electrons must be considered, and the boundary condition in the y-direction is that no net current flows. Thus $p v_{yh} = n v_{ye}$. For the Hall constant to vanish, $E_y$ must vanish. Equation 4.82 becomes

$$v_{yh} = -\frac{q E_x}{3kT} \omega_{0h} \langle v^2 \tau_h^2 \rangle$$

$$v_{ye} = -\frac{qE_x}{3kT}\,\omega_{0e}\,\langle v^2\tau_e^2\rangle$$

Setting the $y$ current equal to 0 yields $n\omega_{0e}\langle v^2\tau_e^2\rangle = p\omega_{0h}\langle v^2\tau_h^2\rangle$. Taking the ratio of $p$ to $n$ and eliminating common terms gives the answer:

$$p/n = \frac{m_h^*}{m_e^*}\,\frac{\langle v^2\tau_e\rangle}{\langle v^2\tau_h\rangle}\,]$$

Note that if the averages of the scattering time for holes and electrons are not grossly different, the Hall constant will have a null when the ratio of the carrier concentrations equals the ratio of the carrier masses. For a material like InSb (see Table 4.3), the holes are 46 times as heavy as the electrons. Thus, a moderately $p$-type sample of InSb will still give a negative Hall constant. That is fair enough when you remember that the Hall coefficient gives the sign of the *dominant* carrier, in the sense of the carrier transporting more of the current, not the one that is necessarily numerically dominant.

## HALL MEASUREMENTS WITH ELLIPSOIDAL ENERGY SURFACES

If an electric field is applied to an electron lying on an ellipsoidal constant-energy surface, the electron accelerates in a direction that, in general, is not in the direction of the field. In Chapter 2 we saw that for crystals with cubic symmetry, the sum of all such accelerations over *all* the equivalent ellipsoids gave a net acceleration in the direction of the field with the average effective mass given by $(1/m^*)_{av} = (1/m_\parallel + 2/m_\perp)/3$.

Taking the average in the presence of a magnetic field is not nearly so neat. For example, in the calculation that we just did, we found that a magnetic field in the $z$-direction did not influence the conductivity in the $z$-direction (cf. equation 4.83). This was a direct result of the fact that there is a magnetic force only for those components of the velocity that are normal to the magnetic field. With spherical constant-energy surfaces, an $E_z$ generates only a $v_z$, so $v \times B = 0$.

With ellipsoidal energy surfaces, the velocity and the electric field will be parallel only if the field is parallel to a major or minor axis of the ellipse. For Si, it is possible to align the electric field with a major or minor axis of all six ellipsoids simultaneously (i.e., in the $\{100\}$ directions). In Ge, since the major axes are not normal to each other, no such alignment is possible.

When the electric field is not aligned with one of these special directions, we must rewrite (4.70), (4.71) and (4.72), taking into account the anisotropy in the effective mass. For cubic crystals and very small magnetic fields we would expect the conductivity and the associated Hall coefficient to remain isotropic. (On the other hand, since the magnetic field introduces an inherent anisotropy—a preferred direction—we should expect both the conductivity and the Hall constant to be functions of the crystallographic direction for large magnetic fields.) If this assumption is valid (it's not too difficult to show that it is), we can pick a convenient direction and simplify the analysis. Let us consider a silicon crystal (see Figure 2.25) with the electric field along the [100] direction and the magnetic field along the [001] direction. Then both the electric and magnetic forces are aligned with a major or minor axis for each of ellipsoids. Consider first a [100] ellipsoid. Equations 4.70 through 4.72 change only slightly to

$$\frac{dv_x}{dt} = - \frac{qE_x}{m_\parallel} - \left( \frac{qB_z}{m_\parallel} \right) v_y = - \frac{qE_x}{m_\parallel} - \omega_\parallel v_y \qquad (4.93)$$

$$\frac{dv_y}{dt} = - \frac{qE_y}{m_\perp} - \left( \frac{qB_z}{m_\perp} \right) v_x = - \frac{qE_y}{m_\perp} + \omega_\perp v_x \qquad (4.94)$$

$$\frac{dv_z}{dt} = - \frac{qE_z}{m_\perp} \qquad (4.95)$$

Our procedure is pretty much the same as before, but we must account for the two different cyclotron frequencies. Accordingly, multiply (4.93) by $\sqrt{\omega_\perp}$ and (4.94) by $j\sqrt{\omega_\parallel}$. Then add the two together. Instead of the complex velocity, $V'$, we get products of velocities and square roots of cyclotron frequencies, but that really adds nothing new. Defining $W' \equiv \sqrt{\omega_\perp}\, v_x + j\sqrt{\omega_\parallel} v_y$, the new version of (4.73) is

$$\frac{dW'}{dt} + j\sqrt{\omega_\parallel \omega_\perp}\, W' = - q \left[ \frac{\sqrt{\omega_\perp}}{m_\parallel} E_x + j \frac{\sqrt{\omega_\parallel}}{m_\perp} E_y \right] \qquad (4.96)$$

The solution proceeds identically with (4.75) including the averaging over all possible scattering times as in (4.76). The solution yields a new version of (4.80):

$$v_{y[100]} = \frac{1}{\sqrt{\omega_\parallel}} Im(W'') = \left( \frac{- q\tau}{1 + \omega_\perp \omega_\parallel \tau^2} \right) \left( \frac{E_y}{m_\perp} - \frac{E_x \omega_\perp \tau}{m_\parallel} \right) \qquad (4.97)$$

This covers only one minimum, [100] and its mirror image [$\bar{1}$00]. For the [010] and [0$\bar{1}$0] minima, the result is the same as (4.97) with the interchange of $m_\perp$ with $m_\parallel$ and $\omega_\perp$ with $\omega_\parallel$. For the [001] and [00$\bar{1}$] minima, the result is obtained from (4.97) by replacing all ( $\parallel$ ) terms by ( $\perp$ ) terms. Averaging the contribution from the three different ellipsoids gives us $v_y$ which we then must average over a Maxwellian distribution as in (4.82). In this case, we must take proper account of the effective mass anisotropy. $v_y$ is given by:

$$v_y = \frac{1}{3}\left[ \frac{-q\tau}{1+\omega_\perp\omega_\parallel\tau^2}\left\{ \mathsf{E}_y\left(\frac{1}{m_\perp}+\frac{1}{m_\parallel}\right) - \mathsf{E}_x\tau\left(\frac{\omega_\perp}{m_\parallel}+\frac{\omega_\parallel}{m_\perp}\right)\right\} - \right.$$

$$\left. - \frac{q\tau}{1+\omega_\perp^2\tau^2}\left\{ \frac{\mathsf{E}_y}{m_\perp} - \mathsf{E}_x\frac{\tau\omega_\perp}{m_\perp}\right\}\right] \tag{4.98}$$

Since we are dealing with the small field case, we may ignore $\omega_\perp\omega_\parallel\tau^2$ and $\omega_\perp^2\tau^2$ compared to 1. Furthermore, in our standard Hall experiment, $v_y = 0$. Solving for the ratio of $\mathsf{E}_y$ to $\mathsf{E}_x$:

$$\frac{\mathsf{E}_y}{\mathsf{E}_x} = \frac{qB_z\left[\dfrac{2}{m_\parallel m_\perp}+\dfrac{1}{m_\perp^2}\right]\langle v^2\tau^2\rangle}{\left(\dfrac{2}{m_\perp}+\dfrac{1}{m_\parallel}\right)\langle v^2\tau\rangle}$$

Dividing by $B_z$ and rearranging gives $\mu_H$ in a very convenient form:

$$\mu_H = \frac{\mathsf{E}_y}{\mathsf{E}_x B_z} = \frac{q\left[2+\dfrac{m_\parallel}{m_\perp}\right]\langle v^2\tau^2\rangle}{m_\perp\left(2\dfrac{m_\parallel}{m_\perp}+1\right)\langle v^2\tau\rangle} \tag{4.99}$$

With ellipsoidal energy surfaces, the drift mobility is given by

$$\mu_D = \frac{1}{3}\left(\frac{2}{m_\perp}+\frac{1}{m_\parallel}\right)q\frac{\langle v^2\tau\rangle}{\langle v^2\rangle}$$

Computing the ratio of $\mu_H$ to $\mu_D$:

$$\frac{\mu_H}{\mu_D} = \frac{\langle v^2\tau^2\rangle\langle v^2\rangle}{\langle v^2\tau\rangle^2}\frac{3\dfrac{m_\parallel}{m_\perp}\left(\dfrac{m_\parallel}{m_\perp}+2\right)}{\left(2\dfrac{m_\parallel}{m_\perp}+1\right)^2} \tag{4.100}$$

Equation 4.100 applies equally well to any cubic crystal in the low-field limit. It should be compared with the result for a spherical surface, (4.90) and (4.91). Since $m_\parallel > m_\perp$, the effect of the ellipsoidal energy surface is to reduce the ratio of the Hall to drift mobility. If the ellipsoids are very prolate —as in germanium, for example—the ratio reduces to three fourths of the value for a spherical surface. Note that the Hall mobility for Ge in Table 4.3 is almost the same as the drift mobility, while for Si, which is much less prolate, the Hall mobility is much larger than the drift mobility.

The Hall coefficient is similarly affected, since $R_H = \mu_H/qn\mu_D$. This implies still more uncertainty in interpreting the Hall coefficient as a measure of the carrier concentration. Once again, the saving feature is that the Hall coefficient usually gives the carrier concentration within a factor of 2, which is usually good enough. Obviously, however, one cannot make an accurate determination of the drift mobility using Hall measurements in any way. In the next chapter, we will examine a method for determining *minority* carrier drift mobility with good accuracy (the Haynes-Shockley experiment). Unfortunately, there is no particularly good way to measure the majority density or mobility with high accuracy. We are more or less stuck with the easy-to-perform but hard-to-interpret Hall measurement.

As a final and historically interesting note on ellipsoidal energy surfaces, consider the possibility of a change in conductivity along the direction of the magnetic field, called *longitudinal* magnetoresistance. If we put the magnetic field in some arbitrary direction, equations 4.93, 4.94, and 4.95 must have all the other cross-product terms added. If we then solved for the current from each ellipsoid in the direction of the magnetic field, we would find that the magnetic contributions did not cancel unless the energy surfaces were spherical. As a matter of fact, in such spectacularly ellipsoidal materials as Ge, the longitudinal magnetoresistance term is greater than the transverse one. Figure 4.17 shows the longitudinal and transverse magnetoresistance for both *n*- and *p*-type germanium with current flowing in the $\{100\}$ direction. Note how much smaller the longitudinal magnetoresistance is in the almost spherical bands of the *p*-type material. It was in an effort to explain this enormous difference that the ellipsoidal energy surfaces were first postulated (B. Abeles and S. Meiboom, *Phys. Rev.* **95**, 31 (1954)]. It was not until a year later that the first cyclotron resonance experiments were reported (see Chapter 2), confirming in an unambiguous way the elliptical shape and the many-valleyed nature of the bottom of the germanium conduction band.

## CONDUCTIVITY IN MODERATE ELECTRIC FIELDS

In a text concerned with device analysis, we may ignore the behavior of semiconductors in high magnetic fields because such fields play almost no role at all in device technology. High electric fields, on the other hand, play a central role in the behavior of almost all semiconductor devices. Even the lowly resistor can occasionally show some startling activity when subjected to high electric fields. A piece of lightly doped *n*-type GaAs between two ohmic contacts acts like a simple resistor at low fields. But subject it to high fields in an appropriately resonant structure, and it starts to emit microwaves coherently and in copious amounts. Although this Gunn effect is limited to a few *n*-type semiconductors, a number of high-field effects are exhibited by all semiconductors and find very important application even in very routine devices. For example, drift velocity saturation, avalanche multiplication, and interband tunneling are all quite important to transistor operation.

What we are prepared to do here is to examine the transition from the ordinary ohmic conductance that is evident at low fields to the region where the drift current is quite independent of the applied field. What we will find is that as the temperature of the carriers begins to differ noticeably from the temperature of the lattice, the differential mobility $(\partial v/\partial E)$ falls well below the low field mobility. We will be using results derived from our assumptions of a Maxwellian distribution and a constant-phonon cross-section, so some cautionary information for the consumer is definitely in order. The constant cross-section assumption gives a mobility-versus-temperature relationship that is reasonably close to what is observed only in certain cases. It is not out of the question that this very arbitrary assumption might represent the net result of the several scattering processes, at least for those cases in which it properly predicts the temperature dependence. If this were the case, using the same cross-section to estimate the field dependence of the drift velocity should give moderately good results. As it turns out, it does, but don't regard this as a proof of anything.

The use of the word "temperature," implying the existence of an equilibrium distribution of kinetic energy, is equally suspect. There have been several direct measurements of the carrier distribution under high electric fields, and not too surprisingly, they do not reveal Maxwellian distributions. Once you remove the constraints of the law of detailed balancing, there is absolutely no reason why an equilibrium distribution should remain. (The word *temperature* gets about as badly bent in common scientific parlance as the word *democracy* in the political sphere. $kT$ is often used to describe the average energy of an ensemble whether or not the ensemble has approached a state of maximum entropy. Don't do it! Temperature is a useful concept and should be left that way.) With these two caveats, let us proceed.

If you reexamine the derivation of the temperature dependence of the mobility that gave equation 4.65 $[\mu \propto (kT)^{-5/2}]$, you will see that the temperature dependence evolves from two independent steps. The first was the derivation of the scattering rate for an electron with velocity, $v$. This is a function only of the phonon distribution, so the temperature that enters there is the lattice temperature. The second step involved averaging the scattering time over the electron velocity distribution. If one may assume a Maxwellian distribution, another $T^{-3/2}$ enters at this second step. This temperature describes the electron ensemble. Accordingly, if the electrons and the lattice are not quite at the same temperature, we should write the mobility as

$$\mu \propto (kT_L)^{-1}(kT_e)^{-3/2} \qquad (4.101)$$

What we now need to do is calculate the rate at which the field delivers energy to the electrons and the rate at which the electrons deliver energy to the lattice. Both of these will be functions of the electron temperature. In the steady state, the two rates will be equal. Requiring equality gives the relationship between the electron temperature and the electric field.

It is easy to write the input power from $P_{in} = \sigma E^2$. According to (4.101), this should be

$$P_{in} = q\mu_0(T_L/T_e)^{3/2} E^2 \tag{4.102}$$

That was easy, but to write $P_{out}$ will not be so trivial.

A not implausible approach to finding $P_{out}$ is to consider the emission of phonons as an example of blackbody radiation. Certainly the statistics are the same—both are bosons—and the same premises that underlie the photon equilibrium apply to phonons. We can begin directly from the Planck radiation law (3.63) with only a little modification. Equation 3.63 was derived for photons in a cavity. Photons are always transverse and travel through space at the speed of light, $c$. Phonons, on the other hand, can be either transverse or longitudinal, and they travel through the crystal with a speed that depends on both their mode and their energy. To simplify things a bit, let us assume constant transverse and longitudinal phonon velocities $c_t$ and $c_l$. With that, we may write a Planck phonon radiation law for the energy per unit volume between $v$ and $v + dv$:

$$dE(v) = \left(\frac{2}{c_t^3} + \frac{1}{c_l^3}\right)4\pi\ \frac{hv^3 dv}{e^{hv/kT} - 1} \qquad \text{in joules/cm}^3 \tag{4.103}$$

Two things will eventually present some problems with (4.103). There is no energy in the lattice above the frequency of the most energetic optical phonon, and there is a gap between the acoustic and optical branches. However, long before we get into too much difficulty with (4.103), we will come a cropper with the problem of specifying the "absorptivity" of the medium. In blackbody theory, you normally assume either that you have a lossless medium in a cavity with ideally black walls or that you have an equilibrium situation in the cavity that you can sample with a little port into the volume. From the first, you can get the radiation from the surface by setting the easily calculated flux incident upon the surface equal to the emitted flux. From the second you get (3.63). What we have in our phonon case is a cavity that has perfectly reflecting walls with a bulk absorptivity throughout the volume. Since we want to get the power delivered to the lattice, we must be able to specify the rate at which the energy given by (4.103) would be absorbed under conditions of thermal equilibrium. Total "blackness" would be absurd. The absorptivity must be quite finite. This brings us right back to the same problem that has dogged this whole chapter. The absorption (or emission) rate is determined by the details of the scattering process that we have been unable to specify. Lacking these details that would give us the absorption spectrum, the most straightforward thing to do is to assume that the electron ensemble is an absorbing medium of uniform grayness. We may then describe this electron cloud by an absorptivity, $\alpha$, in $cm^{-1}$ (i.e., the intensity of a propagating wave falls by a factor of $1/e$ in a distance $1/\alpha$). The energy-absorption rate in equilibrium is then $dP = \alpha c dE$ where $c$ is the appropriate phonon velocity. Putting this into (4.103) gives

$$dP_{out} - 4\pi\alpha \left( \frac{2}{v_T^2} + \frac{1}{v_L^2} \right) \frac{h v^3 dv}{e^{hv/kT} - 1} \qquad \text{in watts/cm}^3 \qquad (4.104)$$

We should now integrate (4.104) to get the total power. A quick bout with the definite integral tables yields a "Stefan-Boltzmann law" for the total radiated flux:

$$P_{out} = A[T_e^4 - T_L^4] \qquad \text{in watts/cm}^3 \qquad (4.105)$$

where $A$ contains all of the constant factors from (4.104).

We may now set the input power (4.102) equal to the output power above to get

$$\mathsf{E}^2 = [A/q\mu_0 n_0] \frac{(kT_e)^4 - (kT_L)^4}{(T_L/T_e)^{3/2}} \qquad (4.106)$$

The coefficient in (4.106) appears to be a function of $n_0$, but since the absorptivity of the electron ensemble (contained in $A$) would undoubtedly be proportional to the electron density, the coefficient would be characteristic of the type of carrier and the lattice and independent of the carrier density. If the coefficient were known, it would be easy enough to select various values of $T_e$ to see what field would produce them. Once the temperature is known, the mobility may be determined from (4.101).

Since $A$ is not directly accessible, an alternative procedure is to work the problem backwards, proceeding from a direct measurement of the drift velocity at some elevated field toward a value of the entire coefficient in (4.106). Figure 4.18 shows the effectiveness of (4.106) and (4.101) in fitting some of the better drift-velocity data. The value of $A/q\mu_0 n_0$ was determined at an intermediate field point from the data, and then the velocity-versus-field curve developed for a wide range of fields. The resulting curves fit the data reasonably well up to the point where pronounced saturation is seen to set in. Since this pronounced saturation is quite clearly the result of the onset of a "new" scattering mechanism in which both cross-section and distribution assumptions are in error, it is not at all surprising that our theory fails to predict it. As you can see from the curves, we obtain a fit over a much wider range for the situations in which the mobility followed approximately a $T^{-5/2}$ dependence than for the case of electrons in Ge which follows a $T^{-3/2}$ law. This is also as expected.

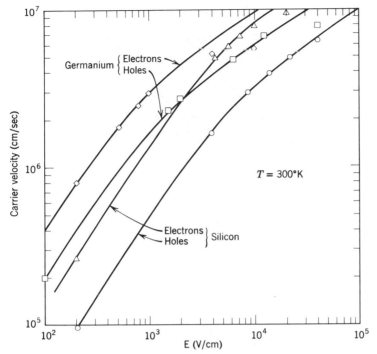

**Figure 4.18** Carrier drift velocity versus the electric field. The curves are derived from (4.109) and (4.104). The germanium data are from A. C. Prior, *Phys. Chem. Solids* **12**, 175 (1960). The silicon data are from C. B. Norris, Jr., and J. F. Gibbons, *IEEE Trans. Elec. Dev.* **ED-14**, 38 (1967).

One may also plot $T$ versus $E$ from the same calculations. These curves are shown in Figure 4.19. As you can see, the temperature increases rather slowly at first, but the rate continues to increase until a power law ($T \propto E^a$) develops at higher fields. A quick inspection of (4.106) shows that for $T_e^4 \gg T_L^4$, the relationship is $T_e \propto E^{4/11}$. At a lattice temperature of 300°K, an increase in electron temperature of 100° is enough to establish this power law. This relationship between $E$ and $T_e$ can be put into (4.101) to obtain the rule

$$v_{\text{drift}} \propto E^{5/11} \simeq E^{1/2} \tag{4.107}$$

which is what is observed for intermediate fields. There are a number of situations in semiconductor device analysis in which the explicit nonlinear relationship between the drift velocity and the field is required. The inter-mediate-field region is usually characterized by (4.107), while the high-field region is well represented by a field-independent velocity. For the three cases that are well represented by our *ad hoc* assumptions, the transition from the

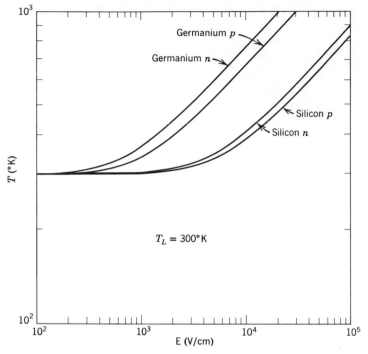

**Figure 4.19** Carrier temperature versus the applied electric field. The curves are those derived from (4.109) normalized to fit the data shown in Figure 4.18.

power law to the field-independent drift velocity occurs in the vicinity of $8 \times 10^6$ cm/sec. The saturated drift velocity for these three cases is just about $10^7$ cm/sec at 300°K.

## A PROBLEM

If an electron or hole has an energy in excess of 1.5 $E_g$, it can lose a large portion of its kinetic energy by generating a hole-electron pair. This process, called carrier multiplication, plays a very important role in many semi-conductor devices. We will deal with it in more detail in the next chapter, but for the moment we would like to be sure that it is not going on to any great extent.

To see whether multiplication might be a factor in scattering, let us find the electric field that, assuming (4.106) is correct, would allow one hole in $10^3$ to have enough energy to create a hole-electron pair in germanium. The solution proceeds in two steps. The first is to obtain an expression that gives the fraction of the carriers with energy greater than some certain amount. This must then be solved to find the temperature at which the fraction is $10^{-3}$ from $1.5 E_g = 0.99$ eV to $\infty$. Then we can use Figure 4.19 to obtain the electric field for that temperature.

To obtain the fraction we proceed as in the evaluation of (4.53), but we integrate from $E_1$ to $\infty$:

$$\frac{p_1}{p_0} = \frac{2}{\sqrt{\pi}} \int_{E_1}^{\infty} e^{-x} \sqrt{x}\, dx \qquad \text{where} \qquad x = E/kT \qquad (4.108)$$

The only easy form of (4.108) is with $E_1 = 0$. For all other ranges of integration, a numerical method must be employed. Table 4.4 presents the results of a Simpson's-rule integration of the compliment of (4.108),

$$\frac{p_1}{p_0} = 1 - \frac{2}{\sqrt{\pi}} \int_0^{E_1} e^{-x} \sqrt{x}\, dx$$

From the table it is easy to pick out the energy that has $10^{-3}$ of the hole population above it. It is just about $8.1\ kT$, which thus becomes $E_1$. Therefore, $0.99$ eV $= 8.1\ kT_h$. Since $kT_L = 0.026$ eV, we have $T_h/T_L = 4.7$ or

**Table 4.4** The Fraction of Electrons or Holes with Energy Greater than $E_1$ in Units of $kT^a$

| $E_1$ in $kT$ | 0.0 | 0.2 | 0.4 | 0.6 | 0.8 |
|---|---|---|---|---|---|
| 0. | 1.000000 | 0.940297 | 0.849530 | 0.753067 | 0.659449 |
| 1. | 0.572461 | 0.493683 | 0.423543 | 0.361843 | 0.308055 |
| 2. | 0.261493 | 0.221410 | 0.187063 | 0.157742 | 0.132794 |
| 3. | 0.111623 | 0.093702 | 0.078562 | 0.065797 | 0.055051 |
| 4. | 0.046017 | 0.038434 | 0.032076 | 0.026750 | 0.022294 |
| 5. | 0.018569 | 0.015457 | 0.012860 | 0.010694 | 0.008888 |
| 6. | 0.007384 | 0.006132 | 0.005091 | 0.004224 | 0.003504 |
| 7. | 0.002906 | 0.002409 | 0.001996 | 0.001654 | 0.001370 |
| 8. | 0.001134 | 0.000939 | 0.000777 | 0.000643 | 0.000532 |
| 9. | 0.000440 | 0.000364 | 0.000301 | 0.000249 | 0.000206 |
| 10. | 0.000170 | 0.000141 | 0.000116 | 0.000096 | 0.000079 |
| 11. | 0.000065 | 0.000054 | 0.000045 | 0.000037 | 0.000031 |
| 12. | 0.000025 | 0.000021 | 0.000017 | 0.000014 | 0.000012 |
| 13. | 0.000010 | 0.000008 | 0.000007 | 0.000006 | 0.000005 |
| 14. | 0.000004 | 0.000003 | 0.000003 | 0.000002 | 0.000002 |
| 15. | 0.000002 | 0.000001 | 0.000001 | 0.000001 | 0.000001 |
| 16. | 0.000001 | 0.000001 | 0.000001 | 0.000001 | 0.000000 |

$a$ The table is accurate to 1 part in the last place. The table is based on a Maxwellian distribution that is strictly true only for thermal equilibrium in a parabolic band.

$T_h = 1410°K$. This is far off-scale in Figure 4.19, so carrier multiplication has not been a problem in these calculations. Extrapolation to $T_h$ in that figure would give about 70,000 V/cm, a value that is quite clearly beyond whatever range of validity can be claimed for our calculations.

## CHAPTER SUMMARY

The central theme of this chapter has been the behavior of free carriers under the influence of electric and magnetic fields. We have confined our-selves to the effects that are of central importance to the most common semi-conductor devices. Of necessity, this leaves out much that is both interesting and valuable in the vast and very active field of carrier transport.

We began with a discussion of the legitimacy of using the solutions of Chapter 2 for the solution of real problems. Chapter 2 had dealt with the perfectly periodic, time-independent crystal. Real crystals are filled with such a variety of fixed and time-varying aperiodicities that the utility of the Bloch wave solutions was certainly not obvious. Although a rigorous de-limitation of the degree of imperfection that could still be treated by the Bloch technique is beyond the scope of this book, by an analogy with the solution of a damped violin string we were able to conclude that the use of Bloch waves was a sensible first approximation.

We then turned our attention to the scattering of Bloch waves by two of the most important imperfections: ionized impurities and phonons. The ionized impurities were treated using the Conwell-Weisskopf technique, which is an extension of Rutherford scattering. In this description, the par-ticlelike behavior of the electron was paramount. The solution for an individual scattering event was an open, hyperbolic orbit that was characterized by the angle through which the electron was turned. This angle was a function of electron's initial speed and its moment arm with respect to the scattering center. When we averaged the loss of initial velocity over all moment arms, we obtained (4.15), which gave the mean scattering time for an electron with velocity, $v_0$, as

$$\tau_R \propto \frac{v_0^3}{\log\left(1 + v_0^4 b_m^2 / a^2\right)}$$

We next turned to scattering by phonons. At normal temperatures in pure semiconductors, this is the dominant scattering mode. We first had to estab-lish the character of phonons. A one-dimensional crystal model was devel-oped that predicted transverse and longitudinal modes of oscillation and gave their $E:k$ plot. Inequivalent atoms or lattice sites introduced two (or more) sets of such modes. We labeled the lowest branches the acoustic modes and upper ones optical modes. The principal difference between them is that an optical phonon always has substantial amounts of energy for all wave numbers including no crystal momentum, while an acoustic phonon has zero energy at zero wave number. We then extended our one-dimensional model to three dimensions (in a very superficial manner) and found the density of phonon states in $k$-space. With this density-of-states function, we proceeded to find the number of phonons per energy interval and summed this over the acoustic branch to find the total number of acoustic phonons. To do this, we had to take advantage of the fact that the phonon energy is $\ll kT$ at room temperature. For energies much less than $kT$, the Bose factor becomes $f_B(E) \approx kT/E$. By approximating the Brillouin zone with a sphere and the $E:k$ plot with straight lines (constant velocity of sound), we were able to obtain the total number of phonons and the energy of the average phonon. At room temperature this average phonon had an energy of only $\sim kT/250$, a very small amount compared to the kinetic energy of the average conduction electron $(3kT/2)$.

Since the average electron moves much more rapidly than do the phonons, the lattice vibrations could be treated as a spatially periodic function wherein the electron-phonon interaction is very similar to the Bragg scattering of Chapter 2. In fact, if the electron's $E:k$ plot has spherical constant-energy surfaces, the typical low-energy phonon gives what is conventionally called Bragg's law: $2\lambda_p \sin \theta = n\lambda_e$ or $2k_e \sin \theta = nk_p$. For elliptical constant-energy surfaces or surfaces of greater complexity, the small energy of the phonon still requires the electron to scatter about from one state to another on almost the same surface. Nothing quite so neat as Bragg's law results, but the vector scattering diagram that characterized conservation of momentum (Figure 4.10, for example) is quite similar to the one for the spherical surface.

At this point we were ready, finally, to derive the relationship between the average scattering time for an electron and the electron's initial speed. We made the ad hoc assumption that the scattering cross-section for electron-phonon collisions was independent of the phonon's energy as long as the collision satisfied conservation of energy and momentum. The uncertainty principle gave us the energy-momentum range for phonons that would satisfy a particular transition. Summing over all possible collisions gave us the collision rate, which is the inverse of the mean collision time. Finally, we found that the mean collision time is the same as the mean randomization time. Our result was $1/\tau_R \propto kTv_e^3$.

We next determined the relationship between the randomization time and the drift and diffusion of the electrons. The first step was to establish the Boltzmann transport equation, which described the time rate of change of the Boltzmann or Fermi factor for each point in phase space as a function of the applied fields, density variations, and collisions. To use the Boltzmann transport equation, we had to assume a velocity distribution very close to equilibrium form which is a Maxwellian distribution. We were then able to average over all velocities to find the net flow of electrons that resulted from applied fields and density variations. The average velocity in the presence of an electric field but no magnetic field is

$$v_z = \left[ -\frac{1}{n}\frac{\partial n}{\partial z} - \frac{q}{kT}\,\mathsf{E}_z \right]\frac{1}{3}\langle v^2\tau \rangle$$

We defined the mobility and diffusion coefficients as

$$\mu_e \equiv -\frac{v_z}{\mathsf{E}_z} = \frac{q}{3kT}\langle v^2\tau \rangle$$

$$D_e \equiv -v_z \left/ \left( \frac{1}{n}\frac{\partial n}{\partial z} \right) \right. = \frac{1}{3}\langle v^2\tau \rangle$$

Using these definitions, the expressions for current density for holes and electrons become

$$J_h = qp\mu_h\mathsf{E} - qD_h\nabla p$$

$$J_e = qn\mu_e\mathsf{E} + qD_e\nabla n$$

From the definitions and the Boltzmann transport equation, it was evident that $\mu/D = q/kT$. This equation is called the Einstein relationship. We also derived it from the law of detailed balancing and the expressions for the hole and the electron currents. Finally, we performed the averages demanded by the Boltzmann equation to determine the temperature variation of the mobility. We found that our ad hoc phonon scattering theory predicted that $\mu \propto T^{-5/2}$, while for impurity scattering over reasonably narrow temperature swings, the prediction was $\mu \propto T^{3/2}$. Thus, at low temperatures, impurity scattering should dominate, while at higher temperatures, except in the most impure or imperfect samples, phonon scattering should dominate.

We next turned to combined magnetic and electric fields and explored the Hall effect and magnetoresistance. We concluded from a very rudimentary analysis that we might determine the mobility and carrier concentration from Hall measurements, but on more detailed analysis, we found that the Hall mobility was given by

$$\mu_H = \mu_D\frac{\langle v^2\tau^2 \rangle \langle v^2 \rangle}{\langle v^2\tau \rangle^2}$$

where $\mu_D$ is the true drift mobility. In general, these mobilities would not be the same. A similar difference was found between the Hall carrier concentration and the true carrier concentration.

Our final efforts were devoted to the variation of the mobility with the electric field. This phenomenon results from the heating of the electrons to temperatures substantially in excess of that of the lattice in which they reside. By separating the electron and lattice temperatures in the drift velocity analysis and computing the power transferred to the electrons by the field and the power transferred to the lattice by the electrons, we were able to find the relationship between the applied field and the electron temperature. From this we were able to compute the average velocity of the electrons as a function of the electric field for moderately high electric fields. For several important cases the results derived from our ad hoc model fitted the experimental data over a surprisingly wide range. The model did not predict, however, the drift velocity saturation that is always observed at sufficiently high fields.

In this chapter we have been dealing with charge carriers moving within a band. In the next chapter, we will consider how they may move from one band to another. That will complete the general physics that we need to analyze a large group of semiconductor devices. The remainder of the book will be devoted to such analysis.

### Problems

1. Phonons do not travel through a lattice unimpeded any more than electrons do. Otherwise, heat would travel through solids at the speed of sound. (In some very curious low-temperature experiments in liquid helium, heat does just that. However, being a boson makes He truly the "horse of another color" in physics.) It is not obvious from our derivation of the eigenfunctions of a lattice where this scattering would come from. None the less, if phonons are to come to equilibrium with each other and their surroundings, some interaction must be possible. (An interaction through another species of particle is one possibility. For photons, other particles are absolutely necessary to achieve equilibrium. Since Maxwell's equations are perfectly linear and time-invariant in vacuum, photons do not interact with each other unless there are other particles present.) A particularly interesting interaction is phonon-phonon scattering.

   (*a*) What physically reasonable change(s) must be made in equation 4.16 to allow phonon-phonon scattering to take place?

   *Answer:* The problem with (4.16) is that it is a linear time-invariant equation. Time invariance is more or less obligatory, but a Hook's law, $F = -Kx$, relationship is clearly only a first approximation. As soon as higher-order terms are added (e.g., $F = -K_1x - K_3x^3$), different frequencies interact, and scattering is permitted.

   (*b*) If the phonon's probability of scattering were simply proportional to the number of phonons that were available for scattering interactions, what sort of temperature dependence would you expect to find for the thermal conductivity of nonmetallic materials (that is, no electronic heat conduction) at room temperature and above?

*Answer:* With this very simple-minded assumption, a simple solution is possible. At room temperature and above, the number of phonons in the lattice is proportional to $kT$. Thus, one should expect the conductivity to go as $1/T$. At these temperatures, $\sigma \propto 1/T$ is indeed what is observed.

**2.** A world without phonons would be strange indeed! Consider the following example: Assume that you had a perfectly rigid lattice in which no vibrations were possible. The only scattering mode for the free carriers is ionized impurity scattering. (This is not a bad description for scattering in a very cold lattice.) Noting that the denominator in (4.15′) is a very slowly varying function of the velocity, let us take it as constant so that we may write $\tau_R = Av^3$. For the purposes of this problem, let us take $A = 10^{-33}$ sec$^4$/cm$^3$, a not unreasonable value. Let the crystal be a semiconductor with $10^{16}$ majority carriers/cm$^3$. Let these carriers always be in the Maxwellian distribution appropriate to their mean kinetic energy. Place across the sample an electric field of 1 V/cm from your best ideal voltage source. Now consider the following rather intriguing questions.

**(a)** Find the rate of change of temperature as a function of temperature. (*Hint:* Use the Dulong-Petit expression, $C = 3nk/2$, for the specific heat of the free carriers.)

*Answer:* $dT/dt = 2.62 \; T^{3/2}$°C/sec.

**(b)** From you answer in (*a*), determine the time it would take for the temperature of the carriers to diverge if the initial temperature is $T_0 = 1$°K. Diverge it will in finite time, since the mobility and hence the current and power run away with increasing temperature.

*Answer:* By integration of the answer in (*a*), we find that $1/\sqrt{T_0} + 1/\sqrt{T} = = 1.31 \; t$. $T$ diverges for $t = 0.764$ sec, not a very long time.

**(c)** Even though phonon scattering may not be important in the calculation of the low-temperature mobility, the result above is quite obvious nonsense. What keeps the electrons cool in such low-temperature experiments in real life?

*Answer:* The collisions with the lattice defects do transfer some energy to the lattice; phonons are generated, taking energy away from the hot carriers. In calculating the bounce of ping-pong balls off a trailer truck, one normally ignores the reaction of the truck. However, given *enough* ping-pong balls . . . .

3. All of the derivations within the chapter assumed that the temperature of the entire sample was the same. It is clear, however, that if there were a flux of phonons diffusing down a bar, the flow would tend to carry electrons and holes along with it. Similarly, if there were a flow of carriers, the phonons would tend to be carried along. These effects are not only readily observable, they also provide valuable thermoelectric transducers for measuring temperature and pumping heat and even generating useful amounts of electricity. They can also prove to be a bothersome nuisance in the measurement of small potentials from nonthermal sources where thermal gradients are difficult to avoid. These effects are readily derived from the work already completed in this chapter. Notice that in the presence of a thermal gradient:

$$\frac{df}{dx} = \frac{\partial f}{\partial T}\frac{\partial T}{\partial x} + \frac{\partial f}{\partial x}$$

and

$$\frac{\partial f}{\partial T} = \left(\frac{\partial \alpha}{\partial T} + \alpha\frac{m^*v^2}{2kT^2}\right)e^{-m^*v^2/2kT}$$

(a) Show that the average velocity in the $z$-direction in the presence of a $z$-directed electric field, thermal gradient, and concentration gradient is given by

$$\bar{v}_z = \left(-\frac{1}{\alpha}\frac{\partial \alpha}{\partial T}\frac{\partial T}{\partial z} - \frac{1}{\alpha}\frac{\partial \alpha}{\partial z} - \frac{q\mathrm{E}_z}{kT}\right)\langle v^2\tau\rangle + \frac{\partial T}{\partial z}\left(\frac{-m^*}{2kT^2}\right)\langle v^4\tau\rangle$$

(b) In the absence of a concentration gradient, show that the open-circuit electric field is proportional to the temperature gradient. Find the coefficient of proportionality. (This is the Seebeck coefficient. The effect is utilized in thermocouples.)

*Answer:* The coefficient is

$$\left[\left(\frac{m^*}{2qT}\right)\frac{\langle v^4\tau\rangle}{\langle v^2\tau\rangle} - \frac{kT}{q\alpha}\frac{\partial \alpha}{\partial T}\right]$$

(c) Find the thermal conductivity resulting from the free electrons under open circuit conditions.

(*Hint:* the heat flux is $Q = n\overline{Ev_z}$.)

*Answer:* The conductivity is $-Q/(dT/dz)$. Proceeding in the same fashion as in (a), one takes the average value of $Ev_z$ and then, since $v_z = 0$ for the open-circuit case, takes advantage of the result in (a) to eliminate the electric field and all of the other terms in the same bracket with the field.

The result is

$$\sigma_T = \left( \frac{nm^{*2}}{4kT^2} \right) \left[ \frac{\langle v^4\tau \rangle^2}{\langle v^2\tau \rangle} - \langle v^6\tau \rangle \right]$$

(d) If there were a current flowing but no temperature gradient, would there still be a flow of heat?

*Answer:* Yes, a phenomenon known as Thomson heating. The heat flux under conditions of no thermal gradient is given, for holes, by

$$Q_z = \left( -\frac{1}{p}\frac{dp}{dz} - \frac{q\mathsf{E}}{kT} \right) \frac{pm^*}{2} \langle v^4\tau \rangle$$

(e) The result in (d) suggests that the definition of thermal conductivity given in (c) may need some reform. Using that definition, find the thermal conductivity in terms of the zero current thermal conductivity and terms that depend on current.

*Answer:* Going through the same steps in (c), but this time not setting $\bar{v}_z$ to zero gives, for electrons

$$\sigma_T\big|_{I \neq 0} = \frac{nm^*\bar{v}_z}{2\dfrac{\partial T}{\partial z}} \frac{\langle v^4\tau \rangle}{\langle v^2\tau \rangle} + \sigma_T\big|_{I=0}$$

Since this result blows up for zero gradient if $\bar{v}_z \neq 0$, the only sensible definition for $\sigma_T$ is at zero electrical current.

## CHAPTER 5

# INTERBAND TRANSITIONS AND EXCESS CARRIERS

The last chapter dealt at some length with the interactions that tend to restore equilibrium to the spatial and velocity distributions within a single band. The average motion of the carriers represents a dynamic balance between the external disturbance and the tendency of the thermal reshuffling to restore equilibrium. When the disturbance is not too great, the average motion of the carriers is linearly dependent on the magnitude of the disturbance.

A similar situation occurs when the electron distribution among the bonds is forced away from its equilibrium value. For example, if a large number of hole-electron pairs are generated by shining light on a sample, it is found that the excess carriers disappear with time by recombining with carriers of the opposite type. Since a very large number of semiconductor devices operate by injecting excess carriers into the several bands, it is rather vital to know how long these carriers will remain around. In a number of important cases, we will find that the average excess carrier will recombine in a time that is relatively independent of the number of excess carriers. For that very simple case, we will get the commonly observed exponential decay of the excess number of carriers. Device analysis is much easier when such a rule obtains. Unfortunately, for many cases, the recombination rate depends rather strongly on the carrier concentration, and the analysis of device behavior is much more difficult. Although we will deal almost entirely with the simple exponential case in subsequent chapters, it is important to know when this rule can be expected not to hold. Accordingly, in this chapter we will take a reasonably close look at several of the processes that have been shown to be responsible for the recombination and generation of free carrier pairs.

One thing should be clear at the outset. Both the theory and the experimental determination of carrier generation and recombination are still subjects of current research. It is a very difficult problem because there are so many different ways in which a hole and an electron can recombine. Only a few of them are experimentally distinguishable, and in only a few cases is one mode of recombination completely dominant over the others. Some progress can be made through the law of detailed balancing. It will be progress enough for our needs, but there is much work yet to be done in this field.

## GENERATION AND RECOMBINATION PROCESSES

To describe all of the processes that can be responsible for the creation of a hole-electron pair, we must denumerate each of the ways in which a proper amount of momentum and energy can be absorbed or emitted. These processes can be extremely complicated. For example, a valence electron might get to the conduction band by sequentially absorbing seven low-energy phonons, a photon and the kinetic energy of an energetic hole. Needless to say, we are not about to launch into an analysis of that interaction. However, it is important to note that stepwise processes are quite possible as long as there are intermediate states along the way. In perfect crystals, there are no intermediate states within the band gap, but real crystals have both impurity and imperfection states within the gap. Our first move is to divide the recombination modes into those that do and those that do not require an intermediate state in order to proceed.

To have any importance, a one-step band-to-band recombination process must be simple. This follows directly from the fact that the probability of $N$ independent events occurring simultaneously is the product of the probabilities that each might occur separately. None of the processes is very likely to occur in the times of interest. (The "time of interest" is essentially the definition of simultaneity. In a wave process, it is the time to go half a wavelength or so. In semiconductors, this is typically a number of the order of $10^{-13}$ seconds.) Hence, the improbability of the complete event grows very rapidly with increasing complexity.

There are only a few simple ways that a one-step transition can take place. These are direct photorecombination (the emission of a photon to carry the energy away), indirect photorecombination (the emission of a photon to carry the energy away and the simultaneous emission of a phonon to satisfy conservation of momentum), the Auger (ō zhā′) process (the energy and momentum released by the recombination given to one or more free carriers), and tunneling (described below). The first two we examined in Chapter 2. Actually, we only discussed the stimulated absorption of photons, which is a carrier-pair generation process. However, we have certainly done enough with detailed balancing to see that knowing the absorption rate is sufficient to find the emission rate. As we saw in Chapter 2, the direct process is the more likely (i.e., it has the higher cross-section) of the two by several orders of magnitude. Direct photorecombination is a very important part of the recombination process in any direct material (e.g., GaAs). On the other hand, indirect photorecombination is usually so unimportant that it is rather hard to observe.

The band-to-band Auger process is difficult to observe in normal device operation, but it becomes very obvious in avalanche breakdown. This makes good sense when you consider that an electron (or hole) must have considerably more kinetic energy than $E_g$ if it is to be able to create a hole-electron pair. Since all useful semiconductors have $E_g \gg kT$, the number of free carriers capable of creating hole-electron pairs is normally vanishingly small. From detailed balancing it follows that in thermal equilibrium the number of pairs that can recombine by an Auger process is also vanishingly small. On the other hand, a large electric field can so heat the free carriers that $kT_e$ is no longer much, much less than $E_g$. These hot electrons do have sufficient energy to create hole-electron pairs, and they do so in great numbers. Although the one-step Auger process is unimportant for most device considerations, it puts a very definite upper limit on the electric field that may be applied to any device. We shall look at it in reasonable detail when we discuss avalanche multiplication later in this chapter.

The final one-step process is strictly a wave phenomenon. As we saw in Chapter 1, a wave contained in a box (i.e., the finite potential well problem) will penetrate into the walls of the box. We ran across this again in Chapter 2 in the discussion of the Kronig-Penney model. There we found that if the classically forbidden region of space had a finite thickness, the wave penetrated all the way through the forbidden region. A wave passing through such a barrier is said to *tunnel* through the forbidden region.

If a *p-n* junction is made between two degenerately doped semiconductors, the valence band in the *p*-material is approximately aligned with the conduction band in the *n*-material (see Figures 3.7 and 5.12). If the junction is very abrupt, the forbidden region (essentially, the depletion region) is very narrow and the electrons can easily *tunnel* in either direction. Although seldom done in the literature, we can consider this tunneling as a pair generation and recombination process. If the material is direct, the tunneling automatically satisfies conservation of energy and momentum. If the material is indirect, tunneling requires the emission or absorption of a phonon. Tunneling finds its principal application in the analysis of the breakdown characteristics of heavily doped junctions and the field emission of electrons in Schottky barriers and from hot metals in a vacuum tube. A brief analysis of tunneling will be made after we discuss avalanche breakdown.

## MULTISTEP RECOMBINATION PROCESSES

Most of the recombination in indirect semiconductors takes place at somewhat mysterious sites called *recombination centers*. Some of these are moderately well understood and other are not, but they all seem to have the property of allowing the recombination to take place in several steps that occur sequentially rather than simultaneously. For example, it is well known that the introduction of a very small amount of gold into silicon drastically shortens the time it takes an electron-hole pair to recombine. The gold atoms act as recombination centers in roughly the following way. Figure 3.4*b* shows that gold has two levels deep within the forbidden band. Consider the case for a moderately *n*-type sample. The Fermi level would lie reasonably well up in the forbidden band, substantially above the upper (acceptor) gold state. Is the state full or empty? (Answer: Full. States well above the Fermi level are empty and those below are filled.) The gold site is, therefore, negatively charged in its equilibrium state. If some holes are now injected (by shining light on the sample, for instance), the holes will be attracted to the negatively charged gold atoms. This coulombic attraction extends a considerable distance from the gold site. As the hole is drawn in toward the gold atom, it may suffer a collision with a phonon or another free carrier. If it loses enough energy in this collision, it will go into orbit around the gold side. The hole orbiting the gold atom is in many respects like a hydrogen atom, so we should expect to find a series of excited states. It is quite likely that the hole will first go into one of these excited states. Will the excited states be closer or further from the valence band than the ground state? (Answer: Nearer the valence band. Hole states are more energetic the *lower* they are on the $E:k$ plot.) The hole can drop into less energetic states by emitting phonons or photons in much the same way as an excited atom radiates its energy. It might also make an Auger transition. Eventually, by one route or another, the gold atom is in its neutral ground state and a hole has disappeared from the valence band. The neutral gold atom in the silicon site shows an attractive potential to *both* holes and electrons. This should not seem at all obvious, but if you examine Figure 3.4*b* you will see that gold can act as either a donor or an acceptor. (Such an impurity is called *amphoteric*.) However, since $n \gg p$, the gold is much, much more likely to pick up an electron. Accepting the electron may also be a multistep process, with the electron cascading down a series of excited states. The gold atom is once again negatively charged and ready to capture another hole. The net effect of this interaction is that a hole-electron pair has vanished, with the energy and momentum of the pair being carried off by a number of phonons, free carriers, or photons.

It is worthwhile to take a moment to see why impurities should be so effective in catalyzing recombination. From the wave mechanics of Chapters

1 and 2, recall that a Bloch wave—$u_k(x)\, e^{jkx}$—represents a plane wave whose expectation value of momentum is $\hbar k$. It should be obvious that an electron in orbit around a nucleus is not very much like a plane wave. What is the expectation value of its linear momentum? (Answer: Anything in a closed orbit must have an average momentum of zero.) Now comes the critical (and not very easy) question. If the linear momentum of the orbiting electron is measured, what values will be obtained? (Hint: Ask the same question about some more familiar classical orbit or take a look at the one-dimensional "orbit" problem—the particle in a box— done in Chapter 1.) [Answer: If the particle were a satellite in a circular orbit, the momentum would have a constant magnitude and the distribution of observed momenta would form a circle in momentum space. On the atomic scale, we should expect to find some spreading in the distribution of measured $k$ values. The problem is to expand a particular eigenfunction for the electron orbiting the ion in terms of a complete set of plane waves (e.g., the Bloch waves from the valence and conduction bands). This is, in principle, a straightforward Fourier expansion. However, since neither the eigenfunction nor the Bloch waves are really known, the expansion cannot be performed. What seems pretty reasonable, however, is to guess that eigenstates with energies just below the conduction band will be "made up" mostly of conduction-band wave functions from the conduction-band minima; states that lie very close to the valence band would be represented mostly by valence-band maxima wave functions. When we sample the linear momentum of one of these states, we will usually find those values whose Fourier amplitudes are largest. Note carefully that the actual wave function, which is a sum of the eigenfunctions of the wave equation describing that impurity in its particular crystal site, in no way resembles a plane wave. There is no requirement that the wave look like any of the terms in its Fourier expansion. For example, full-wave rectified 60-cycle AC has little resemblance to a 120-cycle sine wave, yet a measurement of the rectified wave's 120-cycle content shows that about 18% of the AC power in the wave will pass through a unity gain 120-cycle band-pass filter.]

For a transition to be probable, the event(s) that cause(s) it must occur rather frequently. Events that can be stimulated by low-energy phonons are likely to happen because there are so many low-energy phonons. Thus if the impurity allows the electron to make transitions that correspond to low-energy phonon emission, recombination will be more probable. One way to describe the fact that the bound electron's Fourier expansion contains more than one component is to say that the electron "spends part of its time in each of several different states." Presumably, some of these states come from each conduction-band minimum. If a free electron in one minimum of the conduction band were near an impurity site, and if it were stimulated to emit a phonon, it might well drop into orbit with its state at that moment matching the state that an orbiting electron would have. This is much the same situation that occurs when a satellite on an open orbit approaching the moon fires its retrorockets and matches its state to a closed orbit. It should be reasonably clear that most of the time the electron will be in the wrong place or the phonon will be of the wrong magnitude; most collisions just do not result in a bound orbit. Thus, although the average carrier collides about every $10^{-13}$ seconds, the shortest recombination times that are found in indirect materials are of the order of $10^{-8}$ seconds, and $10^{-5}$ or $10^{-6}$ is much more common.

Let us now take a look at the statistics of recombination to see how these times will vary with such obvious variables as the impurity concentration and the carrier concentration.

## GENERATION AND RECOMBINATION RATES THROUGH DEEP TRAPS

The model we are about to use was first proposed by Shockley and Read [*Phys. Rev.* **87**, 835 (1952)] and Hall [*Phys. Rev.* **83**, 228 (1951)] and seems to fit the facts moderately well. We consider only the simplest two-step process in which the impurities or imperfections have a single donor or acceptor level somewhere in the middle of the forbidden band. (I.e., we are discriminating against amphoteric and multidonor or multiacceptor impurities such as gold or mercury. The single donor can give up only one electron to become ionized or accept one electron to become neutral. It may do so, however, through a series of excited states that terminate in the ground state that we call the donor state.) The impurity is referred to as a *trap* or a *recombination center*. The trap can capture an electron from either the valence band or the conduction band, although the probabilities for the two different captures may be vastly different. However, we normally keep track of the holes in the valence band, so it is customary to refer to electron capture from the valence band as hole emission. Thus, as shown in Figure 5.1, the trap is capable of four actions: the emission and absorption of holes and the emission and absorption of electrons.

From the discussion of how a trap captures an electron, it should be fairly obvious that the probability that a particular electron will be captured as it passes the trap depends on the electron's kinetic energy and its wave number. Let us assume for simplicity that the probability for capture is dependent only on the electron's velocity. If the electron velocity distribution is Maxwellian, we can find the total rate of electron capture by the ionized donor traps, $N_T^+$, by taking our usual Maxwellian average:

$$R_e = \langle v_e \sigma_e(v) \rangle N_T^+ n \tag{5.1}$$

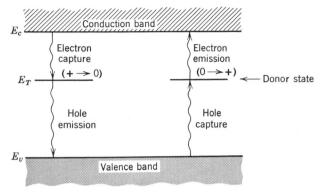

$E_c$

$E_T$

$E_v$

Conduction band

Electron
capture

$(+ \rightarrow 0)$

Hole
emission

Electron
emission

$(0 \rightarrow +)$  ← Donor state

Hole
capture

Valence band

**Figure 5.1** The four possible transitions that a simple donor trap may make. Those on the left change the donor from an ion to a neutral atom. Those on the right change it from neutral to a positively charged ion.

$\sigma_e$ is called the *capture cross-section*, which is the coefficient that relates the product of the electron's velocity and the density of ionized traps, $v_e N_T^+$, to the probability of the electron's being captured. It has very much the same meaning as the collision cross-section that we employed so liberally in the last chapter. It is customary to define a new velocity-cross-section product:

$$v_t \sigma_e \equiv \langle v_e \sigma_e(v) \rangle \qquad \text{where} \qquad v_t \equiv \langle v \rangle \qquad (5.2)$$

$\sigma_e$ does not necessarily correspond to any real cross-section, but the notation is convenient and common, so we might as well use it. Accordingly, (5.1) becomes $R_e = v_t \sigma_e N_T^+ n$. If we were to measure the trapping rate and if we took $v_t$ as the mean thermal velocity, we would obtain a numerical value for the capture cross-section $\sigma_e$. This is the number that is usually quoted in the literature (values between $10^{-15}$ and $10^{-16}$ cm$^2$ being common).

We can write an equation similar to (5.1) for hole capture:

$$R_h = v_t \sigma_h N_T^0 p \qquad (5.3)$$

where $N_T^0$ is the density of neutral traps and $\sigma_h$ the average hole-capture cross-section. Since we know the capture rate, we would be able to write the mean time for capture of a free electron. What is it? Answer: The electron capture rate is $R_e = -\partial n/\partial t = v_t \sigma_e N_T^+ n$. Solving for $n(t)$ (assuming that $N_T^+$ is independent of $t$) gives

$$n = n_0 e^{-t/\tau_e}$$

where

$$\tau_e = \frac{1}{v_t \sigma_e N_T^+} = \frac{n}{R_e}$$

The mean value for the capture time is

$$\bar{t} = \frac{\displaystyle\int_{t=0}^{\infty} t\,dn(t)}{\displaystyle\int_{t=0}^{\infty} dn(t)} = \tau_e$$

Since equilibrium demands that capture be balanced by reemission, we must define four times, one for each process that is shown in Figure 5.1. These are

$$\tau_{ec} = \frac{n}{R_e} = \frac{1}{v_t \sigma_e N_T^+} \qquad \text{(electron capture)}$$

$$\tau_{ee} \qquad \text{(electron emission)}$$

$$\tau_{hc} = \frac{p}{R_h} = \frac{1}{v_t \sigma_h N_T^0} \qquad \text{(hole capture)}$$

$$\tau_{he} \qquad \text{(hole emission)}$$

(5.4)

Detailed balancing requires that the emission and absorption rates be equal in thermal equilibrium, so with the equilibrium number of holes, electrons, and ionized traps we have

$$\frac{n_0}{\tau_{eco}} = \frac{N_{T_0}^0}{\tau_{ee}} \tag{5.5}$$

$$\frac{p_0}{\tau_{hco}} = \frac{N_{T_0}^+}{\tau_{he}} \tag{5.6}$$

which serve to determine the mean time for emission in equilibrium.

We now make the obvious but not necessarily true assumption that the emission time is independent of the carrier concentration and the number of ionized traps. Such an assumption presumes that the ionization of the traps is not caused by collisions of free carriers with the trap and also that the traps are sufficiently sparse so that they do not interact. There are several very interesting violations of the last assumption that have some practical consequences. For example, it is found that an electron captured by a donor trap can recombine with a hole captured by a not-too-distant acceptor trap, leading to the emission of a photon. The efficiency of this process in producing photons that are not absorbed by the material itself has led to improvements in light-emitting diodes. At the same time, the efficiency in light-emitting diodes is limited by the Auger process, which depends on the number of free carriers available for inelastic collisions. However, for most purposes, the independent emission assumption seems accurate enough.

If $N_T$ is the total number of donor traps, what is the equilibrium number of neutral traps, $N_{T_0}^0$? Answer: A trap is a deep impurity, so traps obey impurity statistics. Thus,

$$N_{T_0}^0 = f'(E_T) N_T = N_T \left[ \frac{1}{1 + \frac{1}{2} e^{(E_T - E_f)/kT}} \right]$$

There is no reason to assume that the trap level is not near the Fermi level, so the Boltzmann approximation is not appropriate. Since the total number of traps is fixed,

$$N_{T_0}^+ = N_T - N_{T_0}^0 \tag{5.7}$$

From (5.5) and (5.4),

$$\tau_{ee} = \frac{N_{T_0}^+ \tau_{ec0}}{n_0} = \frac{N_{T_0}^0}{n_0 v_t \sigma_e N_{T_0}^+} \tag{5.8}$$

From (5.7),

$$\frac{N_{T_0}^+}{N_{T_0}^0} = \frac{N_T}{N_{T_0}^0} - 1 = \frac{1}{f'(E_T)} - 1 \tag{5.9}$$

As usual, $n_0 = N_c f(E_c)$. Substituting that and (5.9) into (5.8) yields

$$\tau_{ee} = \frac{f'(E_T)}{n_0 v_t \sigma_e [1 - f'(E_T)]} = \frac{2 e^{-(E_T - E_f)/kT}}{v_t \sigma_e N_c e^{-(E_c - E_f)/kT}} \tag{5.10}$$

Note that in obtaining (5.10), the Boltzmann approximation was used for $f(E_c)$ but not for $f'(E_T)$. Taking the exponential in the numerator down to the denominator, we obtain the desired form for the mean electron emission time:

$$\tau_{ee} = \frac{2}{v_t \sigma_e N_c e^{-(E_c - E_T)/kT}} \equiv \frac{2}{v_t \sigma_e n_T} \tag{5.10'}$$

$n_T$ is the number of electrons that there would be in the conduction band if the Fermi level were at the trap level (which it might well be, but that is not relevant). An essentially identical argument leads to the relationship for $\tau_{he}$, the mean hole-emission time:

$$\tau_{he} = \frac{1/2}{v_t \sigma_h p_T} \tag{5.11}$$

$p_T$ is the number of holes that would be in the valence band if the Fermi level were at the trap level. The difference between (5.10) and (5.11) points up again that $f'(0) = 2/3 \neq 1/2$; the impurity statistics are not quite the same as the free-carrier statistics.

We now have the four capture and emission times that we need. Having assumed that the capture cross-sections are independent of whether the occupation numbers are the equilibrium values or not, we can write the general equations for the net rate at which trapping and reemission make electrons appear in the conduction band and holes in the valence band:

$$\frac{\partial n}{\partial t} = -\frac{n}{\tau_{ec}} + \frac{N_T^0}{\tau_{ee}} = -v_t \sigma_e \left( n N_T^+ - \frac{1}{2} n_T N_T^0 \right) \tag{5.12}$$

$$\frac{\partial p}{\partial t} = -\frac{p}{\tau_{hc}} + \frac{N_T^+}{\tau_{he}} = -v_t \sigma_h (p N_T^0 - 2 p_T N_T^+) \tag{5.13}$$

Note that if equations 5.12 and 5.13 are both equal to zero, we may obtain the satisfying result $np = n_T p_T = n_i^2$. This shows that we have found at least one mechanism by which that mass-action law, obtained in Chapter 3 from purely statistical considerations, would be brought about.

We may write an equation for the rate at which the number of neutral traps is increasing. In the steady state, this equation will have to equal zero and (5.12) will have to equal (5.13).

$$\frac{\partial N_T^0}{\partial t} = -\frac{N_T^0}{\tau_{ee}} - \frac{p}{\tau_{hc}} + \frac{N_T^+}{\tau_{he}} + \frac{n}{\tau_{ec}}$$

$$= -v_t \sigma_e \left[ \frac{n_T N_T^0}{2} - n N_T^+ \right] + v_t \sigma_h \left[ 2 p_T N_T^+ - p N_T^0 \right] \tag{5.14}$$

Since (5.14) is zero in the steady state, we can use (5.14) with (5.7) to obtain $N_T^0$ and $N_T^+$:

$$N_T^0 = N_T \frac{\sigma_e n + \sigma_h 2 p_T}{\sigma_e \left( \dfrac{n_T}{2} + n \right) + \sigma_h (2 p_T + p)} \tag{5.15}$$

$$N_T^+ = N_T \frac{\sigma_h p + \sigma_e (n_T/2)}{\sigma_e \left( \dfrac{n_T}{2} + n \right) + \sigma_h (2 p_T + p)} \tag{5.16}$$

Finally, returning to (5.12) and (5.13), which are equal in the steady state, we may obtain the steady-state recombination rates:

$$\frac{\partial n}{\partial t} = \frac{\partial p}{\partial t} = -v_t \sigma_e N_T \frac{\left[ np\sigma_h + nn_T(\sigma_e/2) - \frac{1}{2} nn_T\sigma_e - n_T p_T \sigma_h \right]}{\sigma_e \left( \frac{n_T}{2} + n \right) + \sigma_h(2p_T + p)}$$

$$= -v_t \sigma_e \sigma_h N_T \frac{\left[ np - n_i^2 \right]}{\sigma_e \left( \frac{n_T}{2} + n \right) + \sigma_h(2p_T + p)}$$

$$= -\frac{\left[ np - n_i^2 \right]}{\left( \frac{n_T}{2} + n \right) \tau_{h0} + (2p_T + p)\tau_{e0}} \qquad (5.17)$$

where $\tau_{h0} \equiv 1/N_t \sigma_h N_T$ and $\tau_{e0} \equiv 1/v_t \sigma_e N_T$.

Equation 5.17 clearly vanishes when $np = n_i^2$, and as we shall see later, that will occur only when $n = n_0$ and $p = p_0$. If we write (5.17) in terms of the *excess* carriers ($n' \equiv n - n_0$ and $p' \equiv p - p_0$) we obtain

$$\frac{\partial n'}{\partial t} = \frac{\partial p'}{\partial t} = -\frac{\left[ n_0 p' + p_0 n' + n'p' \right]}{\left[ (n_T/2) + n_0 + n' \right] \tau_{h0} + (2p_T + p_0 + p')\tau_{e0}} \qquad (5.17')$$

[At this point we might as well get in step with the rest of the world by dropping the factors of 2 and 1/2 in (5.17) and (5.17'). This is not really as bad as it sounds. The recombination centers rarely have a very well-defined energy, often being somewhat variable agglomerations of impurities and other lattice defects. The uncertainty in the trap energy carries over to $n_T$ and $p_T$. Furthermore, one is seldom knowledgeable about the trap center involved unless the traps were deliberately introduced. Thus, in most cases, you will not be sure whether the center is a donor or an acceptor. Since the acceptor sites interchange the 2 and the 1/2 in (5.17), since we normally are ignorant of the exact nature of the center, and since factors of 2 are not very significant for the exponentially varying elements of (5.17), it is reasonable to drop the factors of 2 and make the rate equations for donors and acceptors look exactly the same. This will be done throughout the remainder of the book.]

## LIFETIME OF EXCESS CARRIERS

It is usually the case that $n' \simeq p'$, since charge neutrality is difficult to upset in a conductive medium. We will discuss this at some length below, but what we want to do now is to see how $n' = p'$ affects (5.17'). The situation that is met with most often in device analysis is the strongly extrinsic material in which the injected excess-carrier density is much less than the majority carrier density. For example, if the material were $p$-type, we would have $n' = p' \ll p_0$. Note that since $p_0 \gg n_0$, it is quite reasonable to have $n' \gg n_0$. Since $p_0$ is so much greater than any of the other densities, we may greatly simplify (5.17') by dropping all of the terms that are not significant. With the trap level presumably very deep, the insignificant terms should include $n_T$ and $p_T$. Thus:

$$\frac{1}{n'} \frac{\partial n'}{\partial t} = \frac{1}{p'} \frac{\partial p'}{\partial t} = -\frac{1}{\tau_{e0}} \tag{5.18}$$

If there is no other recombination process, and if the injection of excess carriers is stopped at $t = 0$ with $n'_0$ excess electrons per cm³, (5.18) may be integrated to give the exponential-decay law that is so often observed:

$$n'(t) = p'(t) = n'(0)e^{-t/\tau_{e0}} \tag{5.19}$$

Obtaining the average time to recombine from (5.19) yields

$$\bar{t} = \frac{\displaystyle\int_{t=0}^{\infty} t\, dn(t)}{\displaystyle\int_{t=0}^{\infty} dn(t)} = \tau_{e0} \tag{5.20}$$

Thus, the average excess carrier recombines in a time, $\tau_{e0}$. If $n'(0) \gg n_0$, essentially all of the electrons are "excess" electrons, so $\tau_{e0}$ is referred to as the *minority-carrier lifetime*. As you can see from (5.17'), it is the minority-carrier lifetime only in a strongly $p$-type material. If the material is strongly $n$-type the minority-carrier lifetime is $\tau_{h0}$.

Using $n' = p'$ and $n_0 p_0 = n_i^2$, the general form of (5.18) can be reduced to a function of one variable—$n_0$ or $p_0$—provided that the excess-carrier concentration is always quite insignificant. What we obtain is

$$\frac{1}{n'} \frac{\partial n'}{\partial t} = -\frac{1}{\tau} = \frac{n_i^2 + p_0^2}{p_0^2 \tau_{e0} + p_0 \left[ n_t \tau_{h0} + p_T \tau_{e0} \right] + n_i^2 \tau_{h0}} \tag{5.21}$$

Equation 5.21 is a smoothly varying function of the equilibrium-hole concentration. A typical plot is shown in Figure 5.2. Note that the excess-carrier lifetime achieves a maximum value greater than either $\tau_{e0}$ or $\tau_{h0}$. If $\tau_{e0} = \tau_{h0}$, the maximum occurs when $p_0 = n_i$. Why should the lifetime of the excess carrier be longer with the Fermi level in the middle of the gap than it is with the Fermi level near the band edges? (Answer: When the sample is highly

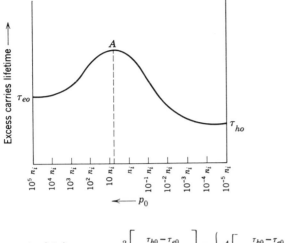

at point $A$:  $p_0 = -n_i^2 \left[ \dfrac{\tau_{h0} - \tau_{e0}}{\frac{n_T}{2}\tau_{h0} + 2p_T\,\tau_{e0}} \right] + \left\{ n_i^4 \left[ \dfrac{\tau_{h0} - \tau_{e0}}{\frac{n_T}{2}\tau_{h0} + 2p_T\,\tau_{e0}} \right]^2 + n_i^2 \right\}^{\frac{1}{2}}$

**Figure 5.2** Excess-carrier lifetime for the case in which the excess-carrier concentration is insignificant compared to the majority-carrier concentration.

extrinsic, the traps are completely filled with the majority carriers. In essence, all of the traps are waiting to capture minority carriers, and they will do so as rapidly as their capture cross-section permits. For the more intrinsic case, the traps are only partially full of majority carriers, so there are fewer traps trapping at any moment. With fewer effective traps, the lifetime of the excess carriers is longer.)

There is another extreme at which (5.17′) simplifies once again. That is the high-injection case where $n' = p' \gg$ either $p_0$ or $n_0$. Such a state is not difficult to achieve if the sample is not very extrinsic, and it frequently assumes some importance in device analysis. For this high-injection-level case, we have

$$\frac{1}{n'}\frac{dn'}{dt} = -\frac{1}{\tau_{h0} + \tau_{e0}} \tag{5.22}$$

This result makes sense if you imagine that at $t = 0$ all of the traps are filled with electrons. In a time $p'/\tau_{h0}$ the average trap would fill with a hole. Then, in a time $n'/\tau_{e0}$, it would flip back again and be ready to capture another hole. Since $n' = p'$, (5.22) results. In the low-injection case—for example $n' = p' \ll p_0$—the capture of an electron took $n'/\tau_{e0}$ but the capture of the hole required $p_0/\tau_{h0} \ll p'/\tau_{h0}$. Thus, (5.18) resulted.

In a case of intermediate injection—an extremely common case in the operation of real devices—the recombination rate is not linearly dependent on the excess-carrier concentration. Accordingly, there is no legitimate way to define an excess-carrier lifetime. The usual procedure is to use the low-injection value and either hope or pretend that it won't lead to excessive errors. With the exact-rate equations, analytic solutions become too cumbersome and one must resort to computor simulations.

One must also be somewhat cautious about applying the concept of lifetime to those situations in which processes other than trapping may affect the recombination rate (e.g., photo-recombination in direct materials and Auger recombination). We have handled only the simplest possible situation —the single-level trap whose emission times are independent of the concentration of holes, electrons, and filled traps. The results we have obtained fit Ge and Si pretty well but may be extremely inaccurate for many other semiconductors, particularly the compounds of period II and period VI such as CdS. Since our interest is primarily in the analysis of the common semiconductor devices, which are usually constructed from Si or Ge, the results that we have obtained are sufficient. But do not conclude that because we stop here, the entire tale has been told.

## Problem 5.1

A sample of Cu-doped Ge with $10^{15}$ Cu and $10^{17}$ Sb/cm$^3$ shows a minority carrier lifetime at low injection of $10^{-7}$ seconds.

1. Which ionization level of Cu in Ge is active? (See Figure 3.4a.) [Answer: From Figure 3.4a the Cu$^=$ level at 0.26 eV below the conduction-band edge must be the active one, since $E_c - E_f = (kT) \log (N_c/n) = = (0.026) \cdot \log (1.02 \times 10^{19}/10^{17}) = 0.12$ eV. Thus in thermal equilibrium, all the copper levels are filled with electrons.]

2. The effective mass for holes in Ge is 0.30 $m_0$. What is the capture cross-section for holes? [Answer:

$$v_t = \sqrt{\frac{3kT}{m^*}} = \sqrt{\frac{3 \times 0.26 \text{ eV}}{0.3 \times 0.567 \times 10^{-15} \text{ eV sec}^2/\text{cm}^2}} = 2.2 \times 10^7 \text{ cm/sec}$$

Then, from (5.17) we get $\sigma_h = 1/(v_t \tau_{ho} N_t) = 1/(2.2 \times 10^7 \times 10^{-7} \times 10^{15}) = = 4.6 \times 10^{-16}$ cm$^2$. Notice that $\sigma_h$ gives a "radius" for copper atoms of about 1.45 Å, which is roughly the size of an atom by any measurement. You should not draw any great significance from this apparent agreement in sizes, but it is emotionally satisfying to have the particle have pretty much the same size no matter how you measure it.]

**3.** Would you expect the lifetime to increase or decrease with increasing temperature? [Answer: Barring any pathological behavior for $\sigma_T$, the two factors that would be expected to change significantly with temperature are $v_t$ and, if the level is not very deep, the ratio of $n_0$ to $n_t$. The copper level is certainly deep enough so that in this case $n_t$ is insignificant for all temperatures of interest. This leaves only $v_t$, which will increase as the square root of the temperature. Accordingly, we would expect the lifetime to go roughly as $T^{-1/2}$. If, on the other hand, we had a trap that was not very deep compared to $E_f$, we would have to include the $n_t$ term. This goes up with temperature and tends to counteract the velocity term. Where $n_t$ becomes important, the lifetime can easily increase with increasing temperature. Such is frequently the case in silicon.]

## CHARGE NEUTRALITY

Back in Chapter 3, we found it convenient to assume that charge neutrality was a good approximation in all regions in which the doping gradient was reasonably low. At the time we were concerned with the equilibrium distribution of carriers. Since most devices operate with substantial numbers of excess carriers, it behooves us to reexamine that assumption. If you will recall, our motivation in assuming charge neutrality was to avoid solving (3.45′), which was horribly transcendental. Our present needs are similar. In the next section we will derive the equations that describe the motion of excess carriers using (4.61), (4.62), and the concept of excess-carrier lifetime. Equations 4.61 and 4.62 both contain the product of the electric field and a carrier concentration. If the electric field must be determined from the net charge density, (4.61) and (4.62) become coupled in a most unfortunate way through Poisson's equation. What must be done first is to show that charge neutrality is a good approximation for the materials we must consider in device applications. As we shall see, this will permit us to determine both the currents and the field without undue labor.

The argument is based on the relatively high conductivity of the usual device material. What is there about a piece of metal, for example, that proscribes charge imbalance *within* the metal? [Answer: From Maxwell's equations and Ohm's law for a metal, we have

$$\nabla \cdot \mathbf{E} = \frac{\rho}{\varepsilon}$$

$$\nabla \cdot \boldsymbol{J} = -\frac{\partial \rho}{\partial t}$$

$$\sigma \mathbf{E} = \boldsymbol{J}$$

Solving for the charge density yields

$$\frac{1}{\rho} \partial\rho = -\frac{\sigma}{\varepsilon} \partial t$$

or, upon integration:

$$\rho = \rho_0 e^{-(\sigma/\varepsilon)t} \tag{5.23}$$

Equation 5.23 says that any net charge imbalance must decay with a time constant of $\varepsilon/\sigma$. $\varepsilon$ varies over a rather limited range, being $0.885 \times 10^{-13}$ farads/cm for vacuum, slightly larger for metals, and about an order of magnitude larger for semiconductors. $\sigma$, on the other hand, covers at least 20 orders of magnitude, being about $10^6$ mhos/cm for copper and $10^{-14}$ mhos/cm for typical ceramic insulators at room temperature. Thus, for metals, the excess-charge relaxation time is the ridiculously low value of $10^{-19}$ seconds (ridiculous because what meaning does $\sigma$ have if we consider times that are six orders of magnitude less than a collision time!) while for insulators, the relaxation times range from seconds to hours. Without looking too closely at the specter raised by the overly short metallic relaxation time, we still must conclude that any charge imbalance in a metal vanishes with atmost haste.]

We must not be too hasty in applying (5.23) to an inhomogeneous sample, however. When there is a gradient in the concentration of mobile carriers— even if there is no gradient in the charge concentration—thermal reshuffling generates a diffusion current that is brought to a halt by the diffusion-generated charge buildup. This was the situation that we treated in some detail in Chapter 3. We found that a field could (and would) exist in an inhomogeneously doped semiconductor. Although the general solution for this field was not obtainable, a simple solution was found for two cases—the small concentration gradient (neutral approximation) and the large con-concentration gradient (depletion approximation). We can readily quantify how big or small the gradients may be to satisfy these approximations (cf. A. K. Jonscher, *Principles of Semiconductor Device Operation*, Appendix E, Wiley, 1960). However, since our principal interest at the moment is excess carriers, the thermal-equilibrium considerations that governed our reasoning in Chapter 3 are not pertinent. What we wish to determine is $p' - n'$.

When excess charge is injected into a semiconductor, if the charge remains mobile, a steady-state situation can come about in one of three ways:

1. If majority carriers are injected into a moderately extrinsic sample, the material will behave in much the same fashion as a metal. Equation 5.23 will apply, so the excess charge will distribute itself on (or pass through) the surface. This process will occur so rapidly that no significant amount of excess charge will ever be accumulated.

**2.** If minority carriers are injected into a moderately extrinsic sample, the sample behaves like a reasonably good conductor, but this situation is unique to conductors with both positive and negative mobile charges. The minority carriers do not repel the principal current carriers; they attract them! Even though (5.23) applies, the excess minority carriers do not rush to the surface. Rather, they attract majority carriers toward them and reduce the excess charge by drawing in *excess majority carriers.* Once again, the net excess charge distributes itself on the surface or passes through the surface of the sample, but in this case the flow of particles is in the opposite direction from (1). Case (2) is the one that most concerns us, and it leads to the conclusion that $n' \simeq p'$.

**3.** If a single species of carrier is injected into a semiconductor that has so few free carriers that it is essentially an insulator, (5.23) nominally applies but is generally not the determining influence. The excess carriers are free to move and would normally repel each other, traveling to the surface as in (1). However, in the usual situation, the carriers are injected from a highly conductive region into the insulating region. The insulating region acts like a dielectric with the bounding conductive region serving as a ground plane. The uncompensated charge in the insulating region induces attractive image charges in the conducting region. These attractive forces bind the injected charge into a region close to the conducting surface. The presence of the space charge in the insulating region strongly inhibits the injection of more carriers. When this situation obtains, the current flow, if any, is referred to as *space-charge limited.* We shall run across space-charge limits in Chapter 6 in the discussion of high-frequency transistor performance. Keep in mind that is is also possible for the excess minority carriers to become bound to traps. Since the symbols $n'$ and $p'$ mean excess *free* carriers, for the case of significant trapping, $n' \neq p'$ even though charge neutrality will still hold.

## THE CONTINUITY EQUATION

For most of the remainder of this book, we will be dealing with the situation in which $n' \simeq p'$. This in itself puts no limitation on $n'$ or $p'$; they simply must be equal. Our immediate problem is to determine how the excess-carrier concentration varies with time. This will lead directly to the differential equation used to describe *bipolar devices* (i.e., those that depend on the flow of both kinds of carriers for their operation. Examples include the junction diode and the bipolar junction transistor).

Consider the small volume of semiconductor shown in Figure 5.3. Let the current be flowing in the $x$-direction. We wish to compute the time rate of change of the hole concentration in the small volume. Carriers may enter

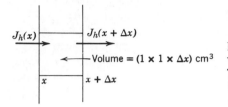

**Figure 5.3** Current flow through a small volume somewhere in a semiconductor. The net rate of hole storage from the flow through the volume is $\partial p/\partial t = -1/q\,(\partial J_h/\partial x)$.

(or leave) the volume by only two processes. Either they flow in (or out) or they are generated (or recombine) within the volume. The number of holes flowing in per $cm^2$ at $x$ is $J_h(x)/q$. At $x + dx$, the number flowing out per $cm^2$ is $J_h(x + \Delta x)/q$. Accordingly, the net number of holes per $cm^2$ flowing into the volume is $[J_h(x) - J_h(x + \Delta x)]/q$. To get the net influx per unit volume we must divide by $\Delta x$. Thus:

$$\partial p/\partial t|_{\text{flow}} = (1/q)\,[\,J_h(x) - J_h(x + \Delta x)\,]/\Delta x \Rightarrow -\,(1/q)\partial J_h/\partial x$$

From (5.18) we have the rate of change due to recombination and thermal generation. Putting the two parts together yields what is usually called the *continuity equation*. In one dimension:

$$\frac{\partial p'}{\partial t} = -\frac{1}{q}\frac{\partial J_h}{\partial x} - \frac{p'}{\tau_h} \tag{5.24}$$

If there is a non-thermal source of generation, G, this should be added to the right side of (5.24). Generalizing to three dimension (although many problems can be worked fairly legitimately in one dimension):

$$\frac{\partial p'}{\partial t} = -\frac{1}{q}\nabla \cdot \mathbf{J}_h - \frac{p'}{\tau_h} \tag{5.24'}$$

We may convert (5.24) and (5.24') into an equation in $p$ and $\mathbf{E}$ by using (4.62). The result is

$$\frac{\partial p'}{\partial t} = -\mu_h\nabla \cdot (p\mathbf{E}) + D_h\nabla^2 p - p'/\tau \tag{5.25}$$

Similarly, for electrons,

$$\partial n'/\partial t = +\mu_e\nabla \cdot (n\mathbf{E}) + D_e\nabla^2 n' - n'/\tau . \tag{5.25'}$$

Now for an approximation that greatly simplifies finding solutions to (5.25): The typical place in which we will want to apply (5.25) is a strongly extrinsic region, a region where the conductivity is relatively high. If the current is to be a reasonable value, the $\sigma E$ product must also be reasonable. What is $\sigma$ in terms of the carrier densities? [Answer: $\sigma = q(n\mu_e + p\mu_h)$.] Let us say we are dealing with an $n$-type piece of silicon where $n_0 = 10^{16}/cm^3$. What is $p_0$? (Answer: $p_0 = n_i^2/n_0 \simeq 10^4/cm^3$.) If $10^{14}$ holes/cm$^3$ were injected, $p = p' + p_0 \simeq p'$. On the other hand, since $n' = p'$, $n = n' + n_0 \simeq n_0$. Since $\mu_h < \mu_e$ and $p \ll n$, the drift term for the minority current must be very small compared to the drift term for the majority current. It is reasonable to ignore *relatively* minute currents, so for the large number of cases in which the minority density is much less than the majority density (i.e., the *low-injection case*), the minority current is described quite well by the diffusion term alone. Thus:

$$J_h \simeq - qD_h\nabla p \simeq - qD_h\nabla p' \tag{5.26}$$

and

$$\frac{\partial p'}{\partial t} = + D_h\nabla^2 p' - \frac{p'}{\tau_h} \tag{5.27}$$

Except for very high frequencies, the analysis of bipolar semiconductor devices can be considered a steady-state problem. Accordingly, equation 5.27 reduces even further to:

$$\frac{\partial p'}{\partial t} = 0 = D_h\nabla^2 p' - \frac{p'}{\tau_h} \tag{5.27'}$$

The most convenient solution (for one dimension) is usually of the form

$$p' = A \cosh\frac{x}{L_h} + B \sinh\frac{x}{L_h} \quad \text{where} \quad L_h = \sqrt{D_h\tau_h} \tag{5.28}$$

$L_h$ is called the *hole diffusion length*. It is the distance the average excess hole will diffuse before recombining.

Our next move is to examine in some detail exactly when one can and cannot use (5.27). We will find some interesting implications of $p' = n'$. But before proceeding, try your hand at the following two problems. The first is a simple turn-the-crank exercise. The second requires a greater stretch, but it should give you a pretty good idea of what one would have to do to solve an actual device problem.

## Problem 5.2

A uniformly doped square die of Ge has a cross-section of 1 cm$^2$ and a thickness of 200 $\mu$m. It is doped with $10^{17}$ Ga/cm$^3$ and has the following coefficients: $n_i^2 = 6.25 \times 10^{26}$

$$\mu_h = 1600 \text{ cm}^2/\text{Vsec} \qquad \tau_e = 1.1 \times 10^{-4} \sec$$
$$\mu_e = 3600 \text{ cm}^2/\text{Vsec} \qquad kT/q \simeq \tfrac{1}{40}\text{V}$$

By edict of the author, the following boundary conditions apply: On one side (call it $x = 0$), the excess carrier concentration is 0. On the other side ($x = w$), the electron concentration is $10^{15}$/cm$^3$. Find the minority current as a function of $x$.

*Solution.* From (5.26) [or, more exactly, from (4.61) with the same approximation about the minority carrier density that was used for (5.26)], we have $J_e = qD_e\nabla n'$. Since the sample is so wide compared to its thickness, we may certainly treat this as a one-dimensional problem. Accordingly, $J_e = = qD_e dn'/dx$. In the same way that (5.27') was obtained, we obtain

$$D_e \frac{d^2 n'}{dx^2} - \frac{n'}{\tau_e} = 0 \tag{5.29}$$

with solution:

$$n' = A \cosh \frac{x}{L_e} + B \sinh \frac{x}{L_e}, \qquad L_e = \sqrt{D_e \tau_e} \tag{5.30}$$

The boundary condition at $x = 0$ requires $A = 0$. The boundary condition at $x = w = 200$ $\mu$m leads to $B = 10^{15}/\sinh(w/L_e)$. Solving for the current as a function of position:

$$J_e = qD_e \frac{dn'}{dx} = \frac{10^{15} qD_e}{L_e} \frac{\cosh \dfrac{x}{L_e}}{\sinh \dfrac{w}{L_e}}$$

Finally, we must use the Einstein relationship between $\mu$ and $D$ to find $D_e$: $D_e = \mu_e(q/kT) = 90$ cm$^2$/sec. Thus,

$$J_e = 0.144 \frac{\cosh \dfrac{x}{0.1}}{\sinh 0.2} \text{ amps/cm}^2 \tag{5.31}$$

Notice how much current one can obtain by diffusion alone. $J_e$ is at its maximum at $x = w$, having the substantial value of 0.73 amps/cm$^2$. This may seem to conflict with your observation that diffusion is an extremely slow process. However, the speed of diffusion depends on what is diffusing.

In (4.60) and (4.95), the diffusion coefficient was found to be $D = \langle v^2\tau \rangle/3$. In other words, to have a high diffusion coefficient, the particles must have a high mean square velocity, and the mean free time between collisions should also be large. Electrons are very light, typically four to five orders of magnitude lighter than a molecule. Accordingly, electrons should have $v_e^2 = 3kT/m$ four to five orders of magnitude higher than mobile molecules. Electrons moving through a crystal lattice will suffer a collision with a phonon in something on the order of $10^{-12}$ seconds (i.e., a mean square velocity of $3 \times 10^{14}$ cm$^2$/sec$^2$ and a diffusion coefficient of 10 cm$^2$/sec), corresponding to a mean distance between collisions of the order of $10^{-5}$ cm. The density of molecules in a liquid is such that a mean distance between collisions is of the order of 5 or 10 Å—$10^{-7}$ cm. Thus, if electrons in a solid have diffusion coefficients of the order of 1 to 10 cm$^2$/sec, molecules in a liquid are going to have diffusion coefficients of the order of $10^{-6}$ cm$^2$/sec. With diffusion coefficients like that, if you want that martini to get mixed, you had better use the stirring rod!

If (5.31) is evaluated at $x = 0$, we find an electron current of only 0.97 $I_{max}$. What happened to the 3% that vanished? [Answer: The electrons carrying the current vanished by recombining with holes. However, the current did not vanish. Kirchhoff would never approve of a vanishing current! The hole that recombined with the electron was an excess hole ($p' = n'$), and it must have gotten to the recombination site somehow. If it was carried along with the electron from $x = w$, the currents of the electron and hole cancel; there is no current to vanish. If, on the other hand, the hole was injected at $x = 0$, traveling toward the electron, the motion of the electron to the right of the recombination site is the *same* current as the motion of the hole to the left of the recombination site. In other words, if the total current is observed, it adheres strictly to Kirchhoff's current law; individual components of the current, on the other hand, may grow or decline as the boundary conditions and the continuity equation demand.]

## Problem 5.3

Any time the equilibrium state is disturbed, the thermal reshuffling will produce a flow of one sort or another—whatever sort is appropriate for reestablishing equilibrium. Frequently, flows or gradients of one sort (e.g., thermal gradients and heat flow) will produce flows and gradients of another sort (e.g., electric fields and current). Consider the following example of how energy is transported into a crystal by the motion of free carriers.

A uniformly doped slice of germanium (see Figure 5.4) has an ohmic contact on one side; the other side of the slice is illuminated by an intense beam of visible light. Visible light lies in the energy range of 1.7 to 4.1 eV. From Figure 2.30, the attenuation factor for visible light in Ge is of the order

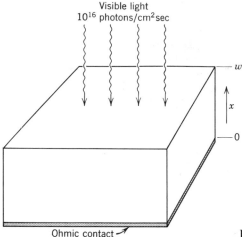

Ohmic contact ⤴

**Figure 5.4** Structure for problem 5.3.

of $10^4$/cm. For all practical purposes, then, the light will be totally absorbed within the first few microns of the surface. Thus, each photon that penetrates the surface generates a hole-electron pair essentially at the surface.

The ohmic contact is a boundary where the excess carrier concentration is held at zero regardless of the flux of carriers into or out of the small volume near the surface. How such a contact is created physically will be discussed in the next chapter, but for the moment, let $p'(0) = n'(0) = 0$. Let the illuminated surface be at $x = w$. If we assume that there is no significant recombination at the illuminated surface (probably a poor assumption), the steady-state situation must have the number of carriers leaving the $x = w$ surface equal to the number generated there. Furthermore, if the slice is left as an open circuit, the TOTAL current must be zero. With a few facts on the material, this is enough to determine the electric potential between 0 and $w$. You are asked to find this $V$ and also to determine the net energy flow that results from the motion of the free carriers (under the assumption that no thermal gradient exists).

*Data.* Material: Ge doped with $10^{17}$ Ga atoms/$cm^3$

$\mu_e = 2000 \text{ cm}^2/\text{Vsec}$     slice thickness $= w = 200 \ \mu M$
$\mu_h = 1000 \text{ cm}^2/\text{Vsec}$
$\tau_e = 2 \times 10^{-6} \text{ sec}$

Let:

$$kT/q = 1/40 \text{ V}$$
the illumination level $= 10^{16}$ photons/sec $cm^2$
(about the level of normal unfocused sunlight)

*Hints.* Although $J_T = J_e + J_h = 0$, neither $J_e$ nor $J_h$ must be zero by itself. The energy is transported principally in the form of hole-electron pairs, which can recombine and release their energy to the lattice. The key to the solution is the fact that the minority carriers satisfy an equation of the form (5.27'), while for the majority carriers the electric field must be included.

*Solution.* The outline of our procedure should go something like this: We may find the minority-carrier density from the continuity equation and the boundary conditions. From the minority-carrier density, it it easy to obtain the minority current, as in the last problem. Once $n'$ has been found, $p'$ is also known, since $n' = p'$. From $J_T = 0$, $J_e = -J_h$. Substituting from (4.61) and (4.62), the appropriate expressions for current densities, the only unknown in the equation is E. Integrating then gives $V$. The energy flow is the flow of excess minority (or majority) carriers times the gap energy.

Proceeding with these steps, we obtain

$$n'(x) = 10^{16} \frac{L_e}{D_e} \frac{\sinh(x/L_e)}{\cosh(w/L_e)} = 5.32 \times 10^{11} \sinh(x/L_e) < 1.93 \times 10^{12}/\mathrm{cm}^3$$

$$J_e = qD_e \frac{dn'}{dx} = 10^{16}q \frac{\cosh(x/L_e)}{\cosh(w/L_e)} = 4.25 \times 10^{-4} \cosh(x/L_e) \leq 1.6 \ \mathrm{mA/cm}^2$$

$$J_h = q\left[\mu_h pE - D_h \frac{dp}{dy}\right] = q\left[\mu_h(p_0 + n')E - D_h \frac{dn'}{dx}\right] = -qD_e \frac{dn'}{dx}$$

Solving for E (note that $n' \ll p_0$):

$$E = \frac{D_h - D_e}{\mu_h p_0} \frac{dn'}{dx} = \frac{\left(\frac{D_h}{D_e} - 1\right)J_e}{q\mu_h p_0} = -\frac{4.25 \times 10^{-4} \cosh(x/L_e)}{1.6 \times 10^{-2} \times 1000}$$

$$\lesssim -5 \times 10^{-5} \ \mathrm{V/cm}$$

Integrating to obtain $V$:

$$V = \frac{D_h - D_e}{\mu_n p_0}[n'(0) - n'(w)] = 0.483 \times 10^{-6} \ \mathrm{V}$$

The energy flux is

$$P = E_g J_e/q = 0.66 \times 4.25 \times 10^{-4} \cosh(x/L_e) \lesssim 1.05 \ \mathrm{mW/cm}^2$$

(Note that you must keep your units consistent. If you use $E_g$ in eV, the charge of the electron is 1. If you use $q = 1.6 \times 10^{-19}$ coulombs, the proper value of $E_g$ is $1.06 \times 10^{-19}$ joules.)

Some of these results are really quite significant. For example, the relatively large flux of photons generates a huge number of carriers in the little volume near the illuminated surface, but these carriers diffuse away and recombine so rapidly that the excess-carrier concentration is never very large. Note that the maximum excess-carrier density is still five orders of magnitude less than the majority-carrier density and a bit more than two orders of magnitude bigger than the minority equilibrium density. Thus, the usual "low-level" assumptions are quite accurate (i.e., ignoring $n$E compared to $p$E). Probably the most significant result is the tiny voltage necessary to keep the two currents in step. Kirchhoff's current law will always impose some restraint on the system, but as you can see, in a *reasonably extrinsic* semiconductor, the field required to satisfy Kirchhoff's law is very small.

## AMBIPOLAR DIFFUSION

In the normal operation of a large fraction of practical devices, the "low-injection" assumption is not very well satisfied. $n'$ still equals $p'$, but the minority-carriers density in certain portions of the device may be comparable to the majority density. Equations 5.25 and 5.25' still hold, but the electric-field terms in the minority-carrier currents may no longer be neglected. Indeed, even in the low injection case, it is possible to use the electric fields resulting from doping concentration gradients to substantially increase the minority-carrier transport and improve high-frequency device performance. What is to be done now is to examine the transport of excess carriers without any restriction on the injection level. The completely general case is too unwieldy to be very helpful, so the discussion will be limited to uniform doping. There are other special cases that are also amenable to solution and that yield useful results, but these will be saved for application to particular devices.

The development we are about to follow was first presented by van Roosbroeck [*Phys. Rev.* **91**, 282 (1953)]. What we are going to do is solve (5.25) and (5.25') for the motion of the *excess* carriers under the assumptions that $n' = p'$ and that the impurity concentration is not a function of position. Since $n' = p'$, the excess carriers must move in concert. The force that holds the two excess-carrier distributions together is nothing more elaborate than the attraction of oppositely charged particles. This force will not show up in our analysis; it is hidden in $n' = p'$. But its presence will lead to some reasonably unexpected and very observable effects.

We start by expanding (5.25) in terms of the equilibrium and excess-carrier concentrations:

$$\frac{\partial p'}{\partial t} = - \mu_h \left[ (\mathbf{E}' + \mathbf{E}_0) \cdot \nabla (p' + p_0) + (p' + p_0) \nabla \cdot (\mathbf{E}' + \mathbf{E}_0) \right] +$$

$$+ D_h \nabla^2 (p' + p_0) - \frac{p'}{\tau} \tag{5.32}$$

In thermal equilibrium, $p'$ and $\mathbf{E}'$ are zero everywhere, so (5.32) says the terms containing only equilibrium terms must cancel. Thus, before making any assumptions at all, we have

$$\frac{\partial p'}{\partial t} = - \mu_h \left[ \mathbf{E} \cdot \nabla p' + \mathbf{E}' \cdot \nabla p_0 + p \nabla \cdot \mathbf{E}' + p' \nabla \cdot \mathbf{E}_0 \right] + D_h \nabla^2 p' - \frac{p'}{\tau} \tag{5.32'}$$

If we now assume that the impurity concentration is uniform, the equilibrium electric field and the derivatives of the equilibrium carrier concentrations both vanish. What remains is

$$\frac{\partial p'}{\partial t} = - \mu_h \left[ \mathbf{E}' \cdot \nabla p' + p \nabla \cdot \mathbf{E}' \right] + D_h \nabla^2 p' - \frac{p'}{\tau} \tag{5.33}$$

Doing the same thing for (5.25') and substituting $p'$ for $n'$:

$$\frac{\partial p'}{\partial t} = + \mu_e \left[ \mathbf{E}' \cdot \nabla p' + n \nabla \cdot \mathbf{E}' \right] + D_e \nabla^2 p' - \frac{p'}{\tau} \tag{5.34}$$

The only unfortunate element in the two equations is the excess space-charge concentration—the $\nabla \cdot \mathbf{E}$ term. We may remove this untidy term in two steps. First, add $\mu_e n$ times equation 5.33 to $\mu_h p$ times (5.34). The result is

$$(\mu_e n + \mu_h p) \left( \frac{\partial p'}{\partial t} + \frac{p'}{\tau} \right) = \mu_h \mu_e (p - n) \mathbf{E} \cdot \nabla p' + (p \mu_e D_h + n \mu_h D_e) \nabla^2 p' \tag{5.35}$$

Dividing through by $(\mu_e n + \mu_h p)$ and taking advantage of the fact that $\mu_e D_h = \mu_h D_e$ (from the Einstein relationship), we obtain

$$\frac{\partial p'}{\partial t} = - \frac{p'}{\tau} + \frac{\mu_h \mu_e}{\mu_e n + \mu_h p} (p - n) \mathbf{E} \cdot \nabla p' + \frac{(p + n) \mu_h D_e}{\mu_e n + \mu_h p} \nabla^2 p' \tag{5.35'}$$

The strong similarity between (5.35') and (5.25) suggests defining new *ambipolar* mobility and diffusion coefficients. They are

$$\mu^* \equiv (p - n)/(n/\mu_h + p/\mu_e) = \frac{\mu_h \mu_e (p - n)}{\mu_e n + \mu_h p} \tag{5.36}$$

and

$$D^* \equiv (p + n)/(n/D_h + p/D_e) = \frac{D_h D_e (p + n)}{D_e n + D_h p} \tag{5.37}$$

Putting these into (5.35′) yields the ambipolar diffusion equation:

$$\frac{\partial p'}{\partial t} = - \frac{p'}{\tau} + \mu^* \mathbf{E} \cdot \nabla p' + D^* \nabla^2 p' \tag{5.38}$$

Now compare (5.38) with (5.25). The equations differ in only one important respect—the electric field is to the left of the $\nabla$ operator in the ambipolar equation but on the right in the normal diffusion equation. Under what circumstances would the two positions be equivalent? (Answer: If and ONLY if $\nabla \cdot \mathbf{E}' = 0$.) Since $\nabla \cdot \mathbf{E} = \rho/\varepsilon$, the ambipolar diffusion equation would appear to be based upon the assumption that the charge density is everywhere 0. Although we did not assume $\nabla \cdot \mathbf{E}' = 0$, if $p' = n'$ and the sample is uniformly doped, $\nabla \cdot \mathbf{E}'$ does equal 0. The difficulty with assuming $\nabla \cdot \mathbf{E}' = 0$ directly is seen if we write Poisson's equation in terms of $p'$ and $n'$: $\nabla \cdot \mathbf{E}' = q(p' - n')/\varepsilon$. Since $q/\varepsilon$ is of the order of $10^{-7}$ V cm for all semiconductors, we can compare the term containing $\nabla \cdot \mathbf{E}'$ with the others in (5.33). Consider, for example, the term $p'/\tau$. A typical excess-carrier lifetime in Si or Ge is $10^{-6}$ sec. A reasonably representative value of excess-carrier concentration in an operation device is $10^{15}/cm^3$. Thus, one reasonably important term in the equation has a magnitude of $10^{21}/cm^3 sec$. The term containing the space charge is $\mu_h p \nabla \cdot \mathbf{E}' = = \mu_h p q (p' - n')/\varepsilon$. $p'$ and $n'$ are both of the order of $10^{15}/cm^3$, but that tells us nothing about their difference. If $p$ is the majority density—say $10^{17}/cm^3$—the term $\mu_h p \nabla \cdot \mathbf{E}'$ will be of the same order as $p'/\tau$ for an excess-carrier difference of only about $10^8/cm^3$! That is, if $p'$ differs from $n'$ by 1 part in $10^7$, for the example given above, the rate of change of carrier concentration due to space charge, will be as significant as that due to recombination. Since (5.38) was obtained by algebraic elimination of the $\nabla \cdot \mathbf{E}$ term, it should be fairly evident that it is not based on $\nabla \cdot \mathbf{E}' = 0$. It is based on $p' \simeq n'$ being a reasonably good approximation. The very fact that $\mu_h p \nabla \cdot \mathbf{E}'$ would be so large if $p'$ was not very close to $n'$ assures us that our approximation is good. The effects of the small space charge that "glues" the excess holes to the excess electrons are still in (5.38). They are included within the definitions of $\mu^*$ and $D^*$.


## THE HAYNES-SHOCKLEY EXPERIMENT

The most direct demonstration of all of the effects predicted by (5.38) is the Haynes-Shockley experiment [*Phys. Rev.* **81**, 835 (1951)]. There are several

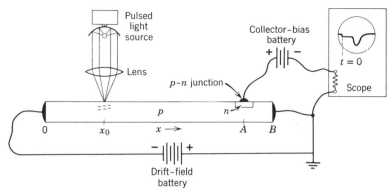

**Figure 5.5** Arrangement of the Haynes-Shockley apparatus using optical injection.

equivalent ways of performing the experiment, one of which is illustrated in Figure 5.5. A thin, rectangular semiconductor bar has ohmic contacts (ones that allow free passage of majority and minority carriers in either direction) fixed to each end (0 and $B$). A third contact is placed near one end (at $A$), but this contact is a $p$-$n$ junction. Since there is a built-in field in the junction that prevents the majority carriers from diffusing across the junction (i.e., the holes stay on the $p$-side and the electrons on the $n$-side), excess minority carriers that reach the junction will be swept across the depletion region by the field. Thus, an unbiased or reverse-biased junction (i.e., the applied field enhances the built-up field) is said to *collect* excess minority carriers.

The light source that is shown in the figure puts out a very brief ($\sim$ 1 $\mu$sec), intense pulse of light. It can be a xenon strobe light or, even better, a GaAs laser. The light pulse is focused on the bar at $x_0$ some distance from the collecting contact. Electrical connections are made as shown. The figure shows the polarities appropriate for a $p$-type bar.

The experiment proceeds as follows. The drift-field battery causes a DC current to flow through the bar, setting up a uniform electric field throughout the bar. The light pulse produces a large local density of excess holes and electrons at the focus of the lens. These are excess carriers, so they must obey (5.38). Figure 5.6 illustrates what is happening within the bar. At $t = 0^+$, the excess carriers are concentrated in a very narrow region about $X_0$. They are subject to the electric field, E, so they start to drift. They are also in a region with a high-density gradient, so they start to diffuse. The holes would "like" to drift to the left and the electrons to the right, but the attractive forces between them prevent the excess-carrier densities from separating by more than a very small amount. The dipole field formed by the slightly separated excess-charge densities runs counter to the applied field. Within the dipole-field region, the field is less than the applied field but the total carrier concen-

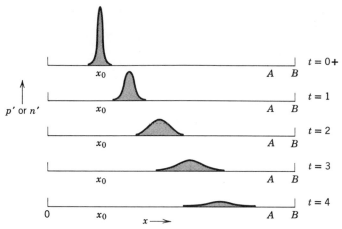

**Figure 5.6** The distribution of excess carriers in the Haynes-Shockley experiment as a function of time. The center of gravity of the distribution drifts with velocity $\mu^* E$. The carriers diffuse away from the center of gravity with diffusion coefficient $D^*$. The excess carriers vanish with a time constant $\tau$.

tration is higher than the equilibrium value. What this resembles is a traffic jam on a busy highway. Conservation of current requires the total flow everywhere to be the same, but in the region of the jam the flow is made up of a greater number of carriers going more slowly than what is observed in the undisturbed regions on either side. Just as the traffic jam (i.e., the high concentration of cars) can move in a direction opposite to the traffic (or can remain stationary), the field-contained grouping of excess carriers is a stable entity in spite of the fact that the individual carriers may enter and leave the region of higher density. The motion of the traffic jam rather than its component parts is what is predicted by (5.36). For the strongly extrinsic case, where the large equilibrium majority-carrier density can give a very large response to small changes in the field, all of the adaptation is provided by the majority carriers and the pulse of excess carriers (of both kinds) drifts with the velocity of the minority carriers. Notice that within the jam, the total majority flux is less than outside the jam. For the "intrinsic" case ($n = p$), the prevention of charge separation requires a field within the dipole region that keeps both carrier fluxes equal to their values outside the region of excess density. Accordingly, the jam is stationary ($\mu^* = 0$).

Regardless of how the pulse drifts along, the excess carriers will diffuse about their aggregate center of mass. In other words, the pulse spreads out. At the same time, the excess carriers are recombining, so the number of carriers contained in the pulse is decreasing with time.

When (and if) the pulse reaches the junction at $A$, the field in the depletion region collects a large fraction of the minority carriers that still remain within the pulse. The collector battery pumps these carriers around the collector circuit, producing a signal on the oscilloscope that mimics the shape of the pulse reaching $A$. Without too much difficulty, one may obtain approximate values of $\mu^*$, $D^*$, and $\tau$ from this experiment. By synchronizing the scope with the light pulse, one can measure the time it takes the pulse to drift to $A$. Knowing the distance from $X_0$ to $A$ gives the apparent drift velocity, $v_d$. The mobility is then $v_d/E$.

$D^*$ is a bit tougher to measure. In principle, one can solve the transient diffusion equation [e.g., (5.27)] with any initial pulse shape. However, the solution is usually in the form of a not-too-helpful infinite series of exponential or complementary error functions. Even for the conceptually simple $\delta$-function pulse, the solution about the center of gravity of the pulse is given by

$$p' = \frac{p_T}{\sqrt{\pi D^* t}} e^{-[(x^2/4D^*t) - t/\tau]} \tag{5.39}$$

where $p_T$ is the total number of holes in the pulse at $t = 0$. The way to obtain a value of $D^*$ from a Haynes-Shockley experiment is to generate a table of pulse widths at half height as a function of $D^*t$ from (5.39). Since $t$ and the pulse width at half height can be read from the scope, the table allows one to find $D^*$.

In principle, $\tau$ can be obtained by integrating the pulse to find the number of excess carriers remaining after a time $t$. In practice, this is far from the easiest way of doing it. However, even though we might hesitate to measure $D^*$ and $\tau$ using Haynes-Shockley apparatus, it is important to notice that you can very easily see the effects of diffusion and recombination right on the scope face. For example, if you double the drift field, the pulse you will observe will arrive twice as fast, be narrower, and have greater area (more excess carriers) than the one at half the field.

## CONDUCTIVITY MODULATION

If you look carefully at the pattern drawn on the scope face in Figure 5.5, you will note that there is a step in voltage at $t = 0$. Had we observed the current through the drift field battery, we would have seen it increase abruptly as the light struck the sample. This increase in current implies that the light is lowering the resistance of the bar. The voltage appearing on the scope face shows the increased amount as a voltage drop from $B$ to $A$. Since no light strikes the sample near $B$ or $A$, we must conclude that the conductivity is unchanged there. Thus, the increase in current gives a bigger IR drop, which shows up on the oscilloscope.

What we wish to do next is to show that the increase in the bar's conductance is proportional to the sum of the conductances of the *excess* carriers. Conductance is not to be found in (5.38). We must return to (4.61) and (4.62) to find the current and the voltage:

$$J_t = J_e + J_h = q[(p_0 + n')\mu_h + (n_0 + n')\mu_e]E + q(D_e - D_h)\frac{dn'}{dx} \tag{5.40}$$

Since the current is constant throughout the bar (ignoring the collecting contact), we may integrate (5.40) to obtain the voltage across the bar:

$$J_t B = q(p_0\mu_h + n_0\mu_e)V_{BO} + q\int_0^B (u_h + \mu_e)n'E\,dx +$$

$$+ q(D_e - D_h)(n'(B) - n'(0)) \tag{5.41}$$

The first term on the right is nothing more than the equilibrium conductance times the applied voltage. The last term goes to zero unless the excess-carrier concentration becomes substantial at one of the ends. The ohmic contacts prevent any substantial excess-carrier concentration from building up at the ends, so the last term (for *this* experiment, please note) is negligible. Finally, the only field whose integral amounts to anything is the applied field. If the applied E is approximately a constant and can be removed from within the integral sign, (5.41) reduces to

$$J_t B = q(p_0 u_h + n_0 u_e)E_{DC}B + qE_{DC}(u_h + u_e)\int_0^B n'\,dx \tag{5.41'}$$

The conductivity is $J_t/E$, giving

$$\sigma = \sigma_0 + q(\mu_h + \mu_e)\bar{n}' \tag{5.42}$$

where $\bar{n} \equiv (1/B)\int_0^B n\,dx$ is simply the average excess-minority-carrier density. This simple result—showing that the excess carriers behave like any other carrier as far as conductivity is concerned—has a rather important influence on a number of common devices. In particular, devices that require rather lightly doped regions, for one of several design reasons, can still carry large currents without excessive voltage drops if a sufficient number of excess carriers are injected into the lightly doped region.

Conductivity modulation is also the principal effect used to measure excess-carrier lifetime. One measures the conductivity as a function of time after the light has been turned off. If the surface were inactive and if all the traps behaved themselves, a simple exponential decay of the excess carriers would be mirrored by the exponential decay of the bar's conductance. Unfortunately, although conductivity modulation is the method most often employed for lifetime measurements, the results are usually far from simple to interpret. The most common problem is surface recombination. Surfaces are very complicated affairs that will provide some entertainment in later chapters. Here all you need know is that the recombination rate in the vicinity of the surface is usually very much greater than it is within the volume. The disappearance of excess carriers at the surface leads to transverse gradients in the carrier concentration; thus, the flow is no longer one-dimensional. Worse then that, the surface-recombination rate is unknown and probably not very stable. Thus the problem of extracting the right number from the exponential decay on the oscilloscope is reasonably complicated.

As a final blow to the man who would measure lifetimes, sooner or later he will run into excess-carrier conductivities that do not decay exponentially at all and may show ridiculously long lifetimes. Generally, this effect results from traps that are not recombination centers. A minority carrier gets gobbled up by such a trap, leaving it charged. This requires a compensating majority carrier, but for reasons that are not at all obvious, the two cannot recombine as long as the minority carrier is trapped. The excess majority carrier remains in the sample, increasing the sample's conductivity until the minority carrier escapes from the trap and recombines. This can take days in a cold sample sitting in the dark. Such effects are a nuisance in the active field of photoconductive light measurement (e.g., CdS light meters). What is wanted is a resistor whose conductance is proportional to the amount of light falling on it, not one whose resistance is proportional to how much light there was last week. Normally, at room temperature, this is not too much of a problem, unless you want to measure fast light changes. However, much infrared photodetection is done at extremely low temperatures where trap times of a week are not entirely out of the question.

## SURFACE-RECOMBINATION VELOCITY

All semiconductor devices are influenced to some degree by the nature of their surfaces. Not only does the surface represent an abrupt discontinuity in the periodic lattice, but it is also of necessity badly contaminated with a wide variety of impurities. Frequently, even well-prepared surfaces harbor large numbers of crystal imperfections generated in the very process of making a surface. If the sample's temperature is not raised to values close to the crystal's melting point, the surface contamination and imperfection can penetrate into the crystal only a very short distance—a few atomic layers. The thickness of the surface layer is roughly an order of magnitude less than any other dimension that is normally of concern in device analysis. It is also a region in which extraordinary impurity gradients exist, with their associated electric fields and charge concentrations. Rather than trying to analyze this tiny but complicated volume of material that forms the crystal surface, it is customary to treat the surface as if it were a truly two-dimensional affair that imposes some sort of boundary conditions on the volume that it encloses. This is much the same sort of thing that is done in electromagnetic field theory, where the penetration of an electromagnetic wave into a conductive material is replaced by some equivalent "surface conductance." Such a step certainly makes the problem of analysis much easier, and you only get into trouble when the thickness of the boundary layer is of explicit importance.

To describe the recombination (or generation) of carriers that takes place so rapidly in the tiny surface volume, we assume that the recombination rate is proportional to the number of excess carriers "at the surface." As long as we are dealing with dimensions that are large compared to the surface layer's thickness, the meaning of "at the surface" is clear; within the *analysis* there will be no noticeable changes over the dimensions characteristic of the surface thickness, so, in effect, the surface is two-dimensional. In actuality, within the surface layer there are very rapid changes, but at the scale that is normally employed, all of the surface activity can be considered to be taking place within zero volume. Since we are ignoring the volume, we cannot write an equation of the form of (5.18). It isn't dimensionally correct.

What we wish to do becomes clear when you consider conservation of particles. For example, in the steady state, if electrons are recombining within the surface volume at a rate proportional to the excess electron density in the vicinity of that volume, the number of electrons entering that surface volume per unit time must also be proportional to the excess electron density. At zero volume, that statement would be phrased: The electron flow into the surface is proportional to the excess electron density at the surface. This is a perfectly reasonable boundary condition of the Sturm-Liouville type:

$$- J_e\big|_{\text{surface}} = qsn'\big|_{\text{surface}} \tag{5.43}$$

where $s$ is the constant of proportionality. Dimensionally, $s$ has the units of cm/sec, so it has received the rather misleading name of *surface-recombination velocity*.

To see how the introduction of surface recombination modifies the solution to a problem you have already met, reconsider problem 5.3. A typical figure for the surface-recombination velocity is 3000 cm/sec. This will affect only the illuminated surface, since the other side is terminated in an ohmic contact. How does the introduction of a finite surface-recombination rate change the answer to problem 5.3? [Answer: Let the electron flux $J_e$ at the surface, $w$, be comprised of the flow out of the surface due to the generation of $10^{16}$

electrons/cm$^2$ sec minus the flow into the surface due to the recombination of electrons within the surface:

$$(1/q) J_e(w) = 10^{16} - sn'(w) \tag{5.44}$$

In view of the fact that $n'(0) = 0$, we may write the solution to (5.29) in the form

$$n'(x) = n'(w) \left[\sinh (x/L_e)\right]/\sinh (w/L_e) \tag{5.45}$$

The minority current is dominated by diffusion, so (5.44) gives us

$$(D_e/L_e) n'(w) \coth (w/L_e) = 10^{16} - sn'(w) \tag{5.46}$$

or, solving for $n'$:

$$n'(w) = \cfrac{10^{16}}{\cfrac{D_e}{L_e} \coth \cfrac{w}{L_e} + s}$$

For the original problem, $s = 0$, but everything else was the same. Accordingly, all of the answers given to problem 5.3 are reduced by the ratio of $n'(w)$ with $s = 3000$ to $n'(w)$ with $s = 0$. That is:

$$\left. \boxed{ \frac{V|_{s=3000}}{V|_{s=0}} = \cfrac{1}{1 + \cfrac{s}{\cfrac{D_e}{L_e} \coth \cfrac{w}{L_e}}} = \frac{1}{1 + 0.578} = 0.634 } \right.$$

Note how large a difference the surface recombination makes. This suggests a (not very good) way of measuring bulk and surface lifetime. Since one can remove material from the back surface before making the ohmic contact, one could, in principle, prepare a set of samples from the same wafer, all of which had the same properties except for the sample thickness, $w$. By performing the experiment described in problem 5.3 for several of the samples, it is a relatively simple matter to unravel the equations to get $s$ and $\tau$ from the measured voltages. The problem with the experiment is that it requires an accurate knowledge of a number of different factors—the number of photons absorbed, the majority-carrier concentration, and the minority-carrier mobility. One must also measure rather small voltages very accurately, and one is obliged to make a transparent electrical contact to the upper surface that does not interfere with the surface properties. There is a substantial conflict between making a contact that does not interfere with microvolt measurements while at the same time preserving surface properties. "At this point, the experiment begins in earnest...."

## THE GENERATION OF EXCESS CARRIERS BY THE APPLIED FIELD

In the last three chapters, we have always treated thermal reshuffling as a process that moves all measureables simultaneously toward their equilibrium values. It is possible to have major disruptions of this scheme if the relaxation of one disturbance creates another. Such chainlike relaxation takes the whole system toward equilibrium, but, on a transient basis, various elements of the system may be driven far away from their equilibrium values.

Consider a particularly important example of this effect called *avalanche multipication*. If the free carriers (say electrons) in a semiconductor (or a gas, for that matter) are subjected to a sufficiently large field, a noticeable fraction of them will have energy in excess of the band gap (or ionization threshold in the case of a gas). If collisions are sufficiently frequent so that the carriers are thermalized, and if we can calculate the carrier temperature, $T_e$, we can obtain this fraction quite directly as

$$K_1(E_g, T_e) = \frac{1}{n_0} \int_{E_g}^{\infty} f(E) \, g(E) \, dE \qquad (5.47)$$

Evaluating the integral will be rather difficult for most materials, since $g(E)$ becomes quite nonparabolic for energies of the order of $E_g$. We have already treated the parabolic case in Chapter 4 [cf. (4.108) and Table 4.3]. Evaluating the fraction does not really concern us, however. What is important to note is that if there are carriers with energy in excess of $E_g$, these carriers could give up their kinetic energy by creating a hole-electron pair. The process is the inverse of the Auger recombination process that was discussed briefly at the beginning of this chapter. A rather simplistic picture of the interaction would show a high-energy carrier colliding with an electron in the valence band and, in a billiard-ball fashion, imparting enough energy and momentum to the electron so that it is in a conduction-band state. This leaves a hole behind in the valence band, so a pair of carriers is created. (A similar collision between a high-energy electron and a gas atom can tear an electron off the gas atom, resulting similarly in a carrier pair: the electron and the ion.) The minimum energy required of the incident carrier can be found from conservation of momentum and energy, once the initial and final states are specified. In the simplest case, where the particles all have the same mass and the transition is direct, the minimum energy is $3E_g/2$. Consider the energy available in the center-of-mass coordinate system and this result will be obvious. One particle with momentum $mv$ comes in and three with momentum $mv/3$

come out. The energy available for the collision is

$$E_{\text{coll}} = \tfrac{1}{2} m \left[ v^2 - 3 \left( \frac{v}{3} \right)^2 \right] = \frac{mv^2}{3} = \tfrac{2}{3} E_{\text{initial}}$$

With different masses or indirect transitions, the threshold energy will be different. Thus, holes and electrons with the same energy may very well have different probabilities for creating hole-electron pairs.

To see why this pair-creation process should be important, consider what might happen as a group of $n$ electrons traversed a region of length $L$. The field is quite large, so that a moderate fraction of the electrons have energy in excess of $E_g$. In traversing the distance $L$, let us say that a fraction, $\gamma$, of the electrons generate new pairs. Thus, the first group of electrons to reach $L$ has $(1 + \gamma)n$ electrons. (If you are worried about conservation of current— and you should be—that will be attended to in a moment. Right now we are counting particles, which is not the same thing as measuring current, since the element of time has not been included.) There are also $\gamma n$ holes that must be heading back to the origin. To keep our arithmetic simple (and for no other reason), let us assume that the same fraction, $\gamma$, of the holes generate new carrier pairs. Thus, there will be $\gamma^2 n$ electrons that will arrive at $L$ because of the holes. These electrons are no different from any others, so they too will generate some new pairs (namely $\gamma^3 n$ of them). This infinite series can be easily summed, since it is the standard geometric series. The total number of electrons that will arrive at $L$ as a result of the $n$ electrons that left the origin is given by

$$n_f = n(1 + \gamma + \gamma^2 + \gamma^3 \ldots) = \frac{n}{1 - \gamma} \tag{5.48}$$

If the distance, $L$, and the field are sufficiently large, $\gamma$ will approach unity and the number of electrons arriving at $L$ will diverge! The rapid growth and runaway characteristics of this pair production in high fields has earned it the name "avalanche multiplication." It is an important process in most high-field breakdowns in semiconductors and gases, and it is responsible for a host of useful devices. (It is also responsible in part for a much larger number of very unhelpful phenomena—lightning bolts, destructive breakdown of insulators, neon signs, etc.)

Although (5.48) contains the essence of avalanche multiplication, to make any use of it for device applications we must know how to predict $\gamma$ given E and the boundary conditions of the device. More directly, we would like to be able to predict the current, given the voltage across the device. This involves relating $\gamma$ to the field, carefully accounting for both electron and hole multiplication rates, conserving the total current, keeping track of ambipolar forces and attending to boundary conditions when these determine the availability of carriers of either type. In this chapter, it is appropriate to look at $\gamma$ versus E, although the complexity of pair production will limit us to an overly simple model. The analysis of some practical devices, especially avalanching in diodes and junction transistors, is properly postponed until the next chapter. However, to give some feeling for how the simplest of avalanching devices might act, we will very briefly consider the terminal characteristics of a germanium resistor.

## A SIMPLIFIED MODEL OF PAIR PRODUCTION

To accurately predict the rate of generation of carrier pairs, we need two pieces of data. First, we must have the probability of pair generation as a function of the initial carrier's energy and momentum. Then we need the *actual* distribution function for the carriers in the system under study. If monochromatic (i.e., single frequency, which by $E = hv$, implies single energy) beams of carriers are available, as they often are for electrons or ions in a rarified atmosphere, the first piece of data may be obtained. For solids, on the other hand, the best information available is educated guesswork.

To obtain the distribution function, one must be able to set up and solve the Boltzmann transport equation. One must include such large energy-loss mechanisms as pair production, optical phonon emission, and intervalley scattering. These processes are effective only for the most energetic electrons, but it is only these that can cause pair production. There are many difficulties blocking a solution to this complex problem, and this is not the place to pursue them. We have open to us two alternatives. We can pretend that the applied field versus temperature relationship developed in the last chapter remains valid all the way up to avalanche. A good guess on $T_e$ versus E should give a relationship between the pair-production rate and the electric field that is roughly correct.

Alternatively, we could assume a very simple collision model, one in which the electrons (or holes) were subject to two types of collisions—ionizing and nonionizing—each of which is represented by a fixed mean free path. Furthermore, we could assume that each collision results in a total loss of kinetic energy. With this drastic simplification of the scattering processes, it is quite easy to obtain a relationship between the applied field and the multiplication rate. The result fits the experimentally observed law rather well, although that may well be fortuitous.

The premise behind the latter model is that the energy gained from the field is quite large compared to the mean thermal energy. When we are talking about minimum ionization energies of an eV or greater, that assumption is certainly reasonable. To have $kT = 1$ eV requires a temperature of $11,600°$K. Thus, for electron temperatures well below $10^4°$K, we may treat each carrier as if it starts from rest. Consider a group of $p$ holes that start at $x = 0$. We assume that until the hole reaches some energy, $E_1 > E_g$, it suffers only one kind of collision (usually assumed to be the emission of an optical phonon) and that collision can be characterized by a constant mean free path, $l_1$. (There is always a mean free path, but we are assuming that it is independent of E.) If a hole suffers a collision, we assume that it loses most of its kinetic energy. Thus, if $p$ holes start at $x = 0$, the number that reach the ionization threshold, $E_1$, is given by

$$p^* = pe^{-d/l_1} \tag{5.49}$$

where $d$ is the distance it takes the field to accelerate a hole to energy $E_1$: $d = E_1/q\mathsf{E}$. As soon as the holes have reached $E_1$, they may start generating carrier pairs. The rate of generation will be some moderately complicated function of energy, but since we do not know what it is, we might as well assume the simplest form we can. Once again, let the collision rate depend on energy in the proper way so that the mean free path for pair creation is $l_2$. Since the scattering rates are proportional to the inverses of the mean free paths, the mean distance that the $p^*$ holes will go before suffering either sort of collision is given by

$$l^* = d + \cfrac{1}{\cfrac{1}{l_1} + \cfrac{1}{l_2}} = d + \frac{l_1 l_2}{l_1 + l_2} \tag{5.50}$$

The next question is: What fraction of the original group of holes created a hole-electron pair when they scattered? Answer: Only those holes that reach $E_1$ can create pairs. Of the $p^*$ holes that reach $E_1$, some generate phonons and some pairs of carriers. The probabilities for both processes are inversely proportional to the mean free paths, so the number of pairs generated is given by

$$\Delta p = p^* \cfrac{\cfrac{1}{l_2}}{\cfrac{1}{l_1} + \cfrac{1}{l_2}} = p^* \frac{l_1}{l_1 + l_2} \tag{5.51}$$

From (5.50) and (5.51) we may obtain the carrier-pair generation rate per unit length:

$$\frac{\Delta p}{l^*} = \cfrac{p^* \cfrac{l_1}{l_1 + l_2}}{d + \cfrac{l_1 l_2}{l_1 + l_2}} = \frac{p^* l_1}{d l_1 + d l_2 + l_1 l_2} \tag{5.52}$$

Since the generation rate is proportional to the number of holes present, it is customary to define ionization coefficients, $\alpha_h$ and $\alpha_e$, such that

$$\frac{dp}{dx} = \alpha_h p + \alpha_e n = -\frac{dn}{dx} \tag{5.53}$$

[Note that the $(-)$ sign reflects the direction of propagation of the electrons.] If the various distances that appear in (5.42) are small enough so we can ignore the fact that the ionization occurring at $x$ is proportional to the number of holes that started some distance away from $x$, we may make (5.52) into the differential form of (5.53), and by substituting for $p^*$ and $d$ we obtain

$$\left.\frac{dp}{dx}\right|_{\text{holes}} = pe^{-E_1/qEl_1} \frac{1}{l_2 + \dfrac{E_1}{qE}\left(1 + \dfrac{l_2}{l_1}\right)} \tag{5.52'}$$

Thus, $\alpha_h$ is given by

$$\alpha_h = e^{-E_1/qEl_1}\left[\frac{1}{l_2 + \dfrac{E_1}{qE}\left(1 + \dfrac{l_2}{l_1}\right)}\right] \tag{5.54}$$

Note how sensitive the ionization coefficient is to variation of the electric field. It is hard to observe generation rates much less than $10^3$ pairs per cm per carrier, while rates in excess of $10^6$ would be hard to sustain. According to (5.54), this entire range—three orders of magnitude—amounts to a mere factor of 2 for the field. You must expect very small changes in the field to produce very large changes in the carrier concentration once the threshold for observable ionization is achieved.

The "temperature" argument is even simpler, although no more convincing. It certainly makes sense that a fraction of the electrons with energy greater than $E_1$ will create a pair in a unit time. If we assume that the fraction is independent of the distribution of energies (essentially the same assumption we made above), then the generation rate will be proportional to the number of carriers with energy above $E_1$. Once we assume that the distribution is Maxwellian, we are back to Table 4.3, which is plotted in Figure 5.7. Note that the curve is essentially a straight line on semilogarithmic paper for $E_1 > 3kT_e$. This suggests an exponential approximation, and a good fit is given by

$$\frac{n}{n_0} \simeq 3.5e^{-E_1/kT_e} \tag{5.55}$$

To make (5.55) resemble (5.54) we need only assume that $T_e$ is proportional to E. Remember, of course, that we have no physical reason for doing this, but knowing that it gives the right answer gives it an aura of respectability.

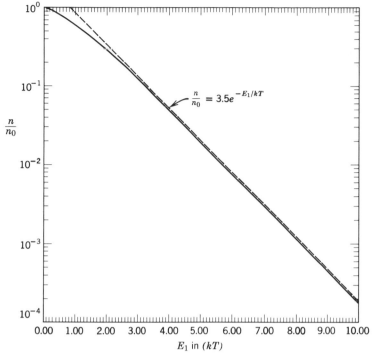

**Figure 5.7** Fraction of the electrons with energy in excess of $E_1$ for a Maxwellian distribution. The solid curve is a plot of the data in Table 4.3.

To measure $\alpha$ is nontrivial at best. It is normally done by measuring the multiplication factor for a reverse-biased junction diode, a technique that is full of pitfalls. We'll discuss that experiment briefly in Chapter 6. In the meanwhile, let us accept the measurements and compare them with (5.54). Over reasonable range, $\log \alpha$ should be proportional to $1/E$. This is plotted in Figure 5.8. Notice that the curves are indeed reasonably straight. Notice also that holes and electrons have different $\alpha$'s, with holes being the more efficient pair producer in Ge while electrons are more efficient in Si.

Why is the ionization rate at a given field higher in Ge than Si? [Answer: Whether or not (5.54) is an exact representation of the ionization coefficient, it must be true that a larger band gap implies a higher ionization threshold. Thus, for a given rate of ionization, a larger band gap requires a higher field. We might also point to Figure 4.19, which shows the effect of mobility on the carrier temperature. The higher the mobility, the higher the temperature at a given field. The higher the temperature, the higher the multiplication rate. According to this argument, electrons should have higher $\alpha$'s than holes— true for Si but false for Ge.]

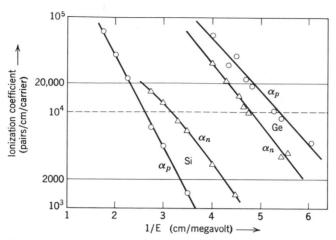

**Figure 5.8** Measured values of the ionization coefficient versus the inverse of the applied field. The curves prove to be reasonably straight lines on a semilogarithmic plot as equation 5.44 predicts. The data for germanium are from S. L. Miller, *Phys. Rev.* **99**, 1234 (1955). The silicon data are from A. G. Chynoweth, *Phys. Rev.* **109**, 1537 (1958).

## THE INCREMENTAL RESISTANCE OF AVALANCHING DIODES

The word *diode* is sometimes used as a synonym for rectifier, but that is a provincial use of the word. A precise definition of *diode* is any device that is connected to the outside world by precisely two leads. From a circuit point of view it is a one-port. The word is normally used to describe nonlinear one-ports, but the nonlinearity need not be asymmetric as in a rectifier.

What we want to look at here are diodes that are operating at fields sufficient to result in avalanche multiplication. The diodes we will examine are particularly simple structures—a bar of high-resistivity semiconductor terminated at each end by an ohmic contact. From a practical standpoint, such a structure is difficult to use because the high fields involved imply very large power densities. This makes it necessary to employ very short pulses to avoid liquifying your sample. Nonetheless, the simplicity of the structure recommends it for discussion purposes if not for experiment.

The characteristics of the diode are shown in Figure 5.9. At currents below a few amperes, the diode is a simple, ohmic resistor. As the field and current are raised, the electron temperature begins to rise significantly, causing the mobility to fall. By the time the voltage has reached 200 V, the mobility is falling so fast that the current is almost independent of the voltage. As the current approaches point $F$, avalanche multiplication is becoming evident. The remarkably rapid increase in carrier concentration for small increases in

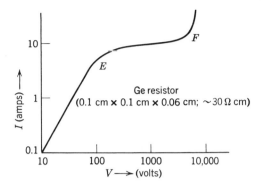

Figure 5.9 *V-I* characteristics of a germanium resistor operating at fields well above the ohmic region.

the electric field means that the differential conductance is enormous; small increments in the voltage yield very large increments in the current. This all makes sense, but what about the boundary conditions? What are the limits on the carrier concentration at the two terminals?

Truly ohmic contacts should act as infinite sources or sinks for carriers, so there would never be any lack of available carriers nor any limit to the number that may flow into the contact. Real ohmic contacts are actually rather complicated, but they act reasonably ideally. A good first approximation to the action at the ohmic contacts is to say that the excess-carrier concentration at the contacts is zero.

Within the bulk is a more complicated situation. The currents are, as usual, the products of the carrier drift velocity (almost field-independent in these high fields) and the carrier charge densities, but the densities now depend on position and field through (5.53). If the carrier concentrations are at all high, ambipolar forces are important. These will couple into (5.53) in a most unfortunate way.

In the regions near the contacts, the very large carrier gradients contribute large components of diffusion current. As a matter of fact, if you try to take everything into account at once, you will obtain a most spectacular set of equations. Approximate solutions are easily obtained only for the two extremes—high conductivity where $p' \simeq n'$ is dominant and very low conductivity where the ambipolar forces may be ignored.

If the carrier concentration is large enough that $p' \simeq n'$, the excess carriers cannot pull apart. If, as is likely to be the case, the excess-carrier concentrations vastly exceed the equilibrium concentrations, the ambipolar mobility goes to 0 (see equation 5.36), and the excess concentration is unable to move.

What we are dealing with in this case is an example of contagious conductivity modulation. Since the current is constant throughout the resistor, a region of high conductivity (i.e., high carrier concentration) will have a relatively small field within it, while a region of low conductivity will have a large field. A large field implies a high ionization coefficient and vice versa. Thus, the high-resistance regions "break down" by avalanching until the sample is fairly uniformly "ionized." The steady state is shown in Figure 5.10. Note that the ohmic contacts clamp the concentration of excess carriers at the edges and that this leads to substantial diffusion currents. Except for the ends, the resistor looks pretty uniform.

The other extreme—essentially the insulator, where $p$ and $n$ are both so small that $p'$ need not equal $n'$ and the electric field may be moderately independent of the current—is actually a case of greater practical interest. The situation of almost no free carriers and a very large electric field is just what one finds in the depletion region of a $p$-$n$ junction under reverse bias. In fact, avalanching in the bulk of a semiconductor resistor is almost never observed in the materials usually employed for device fabrication. On the other hand, it is relatively easy to observe avalanching in $p$-$n$ junctions, since this is the common cause of a sudden breakdown in a junction device.

In the semiinsulating case, the dominant equation within the bulk is (5.53). The solution to (5.53) is easy to evaluate only when the $\alpha$'s are independent of position. What must hold true if the $\alpha$'s are to be independent of position? [Answer: From (5.54), the electric field must be constant within the bulk. This requirement must be met strictly since the ionization coefficient depends so strongly on the electric field.]

If the $\alpha$'s are constant, (5.53) may be easily solved to yield a solution of the form:

$$n = c_1 + c_2 e^{(\alpha_h - \alpha_e) x}$$

$$p = - c_1 \left( \frac{\alpha_h}{\alpha_e} \right) - c_2 e^{(\alpha_h - \alpha_e) x}$$

It would seem that all that we need to do to complete (5.56) is to specify the carrier concentrations at the contact where they are injected. Unfortunately, we have ignored a couple of unignorables that preclude so simple an answer. First, we have ignored the inescapable fact that the total current through the diode is constant. Second, we have ignored the diffusion currents, which must become important near the ohmic contacts. We must conclude that the electric field does vary and (5.56) does not represent the answer. Since this semiinsulating problem is of some practical significance, let us take a quick look at a more general solution.

When the field is sufficiently large to cause ionization, the carriers are so hot that the carrier-drift velocity is almost independent of the field. Accordingly, the current is simply proportional to the carrier concentration—independent of the field. If we write (5.53) in terms of the currents, we have

$$\frac{dI_h}{dx} = \alpha_h I_h + \alpha_e I_e \tag{5.57}$$

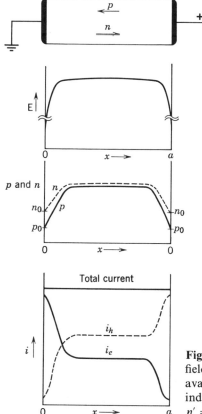

**Figure 5.10** Spatial distribution of the electric field, carrier concentration, and currents in an avalanching semiconductor resistor in which the induced conductivity is sufficient to insure that $n' = p'$ everywhere.

The total current, $I_t$, is constant, so we may eliminate $I_e$ from (5.57) to obtain

$$\frac{dI_h}{dx} = \alpha_e I_t + (\alpha_h - \alpha_e) I_h \qquad (5.57')$$

To obtain a solution we must choose some boundary conditions. Let us choose a case in which the material is sufficiently $p$-type so that the electron current in the absence of avalanching is negligible. Figure 5.11 illustrates the example. The hole current initiated at $x = 0$ grows as it traverses from 0 to $a$. The electron current starts at a negligible level at $x = a$ and grows to substantial proportions at $x = 0$.

To get the analytical form of the solution to (5.57'), note that the differential equation becomes a perfect differential on multiplication by the proper

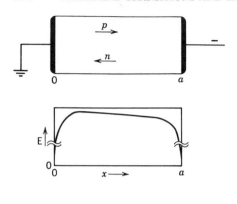

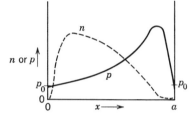

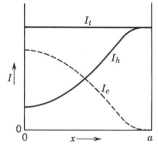

**Figure 5.11** Spatial distribution of the electric field, carrier concentration, and currents in an avalanching semiconductor diode in which the carrier concentrations are sufficiently small that ambipolar forces may be ignored. The sample is $p$-type. At the ohmic contacts the current is predominantly diffusive, so the field is much smaller near the contacts. There is also a shift in the field that results from the variation in conductivity throughout the sample.

integrating factor. In this case, the integrating factor is

$$\exp\left[\int_0^x (\alpha_e - \alpha_h)\, dx\right] \equiv e^{U(x)}$$

Multiplying through by the integrating factor yields

$$e^U \frac{d}{dx}\left(\frac{I_h}{I_t}\right) + e^U (\alpha_e - \alpha_h)\left(\frac{I_h}{I_t}\right) = \frac{d}{dx}\left(e^U \frac{I_h}{I_t}\right) = \alpha_e e^U$$

which may be integrated directly to give

$$I_h(x) = \left[\frac{1}{M_h} + \int_0^x \alpha_e e^U dx\right] I_t e^{-U} \tag{5.58}$$

The constant of integration, $I_t/M_h$, is the initial hole current. Thus, $M_h$ is the ratio of the initial hole current to the final hole current. It is called the *hole-multiplication factor*.

In principle, all we need to do to use (5.58) is to specify the electric-field distribution, determine the $\alpha$'s corresponding to this field, and then perform the the required integrations. However, if you will reexamine (5.54), I think you will agree that this is not a task that one would undertake lightly. In fact, it is only by gross approximation that even the simplest distribution can be bent into a tractable yet plausible solution. Even numerical integration would provide a modicum of entertaiment.

To obtain a curve such as those in Figure 5.8, one measures the multiplication factor in a suitable structure (e.g., a *p-n* junction of fairly accurately known form). Then (5.58) is solved for the multiplication factor, determining in principle what the integral of the difference in ionization coefficients is. Finally, with a certain amount of hand waving and a best guess at the electric field distribution, the definite integrals for the multiplication factor are turned inside out to give $\alpha$ versus E. The approximations used to evaluate $\alpha$ from $M$ appear to be consistent within a factor of 20 to 30% using a number of different structures.

We will have occasion to reexamine avalanche breakdown in depleted regions when we look at the maximum voltage that junction diodes and transistors can support. For the moment, however, we have completed our study of interband transitions except for one rather intriguing possibility. Is it possible to achieve such large fields that an electron could move directly from the valence band into the conduction band?

## THE CASE OF REALLY BIG FIELDS

The answer is yes; in fact, our work in Chapter 1 contains all that is needed to predict such an effect. Such large fields would be difficult to obtain in an avalanche situation because the currents would be astronomical. What we need is a structure that has a very large built-in field. The obvious one is an abrupt, heavily doped *p-n* junction. It is relatively easy to build a junction diode with a field in the junction in excess of $10^6$ V/cm. If one dopes both the *n* and *p* sides so that they are degenerate (i.e., so that the Fermi level is within the allowed bands), the carriers can make transitions between the valence and conduction bands with no applied fields whatsoever. The band diagram for such a junction is illustrated in Figure 5.12.

The essence of the physics behind these transitions can be seen by answering one question: If the electron can make a transition from the valence to the conduction band without absorbing energy from an applied field (It certainly cannot absorb energy from the thermal-equilibrium field!), what is the energy

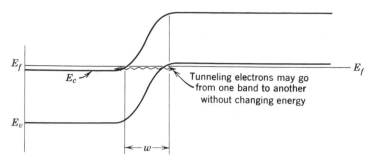

**Figure 5.12** Band diagram for a degenerately doped *p-n* junction. The depletion width, *w*, must be kept small if there is to be any significant amount of tunneling.

difference between the valence and conduction band? (Answer: Although this may look like a "Who's buried in Grant's tomb?" question, it is not. The answer, by inspection of Figure 5.12, is that at any point in space, the difference between the valence and conduction bands is the same—$E_g$. However, the valence band on the *p* side is at the same potential as the conduction band on the *n* side. Thus, if an electron could go from the valence band on the *p* side to the conduction band on the *n* side, the transition would require no energy. It might still require some momentum and it certainly requires that the electron traverse a finite region of the crystal with an energy that is "not allowed.") It is important to note that the depletion region in a heavily doped, abrupt junction is extremely narrow. Thus the region in space where the electron's wave function does not represent an allowed state is very narrow.

In exercise 2 and Figure 1.9 of Chapter 1, we found the wave function for a particle in a finite potential well. We found that the wave penetrated into the forbidden region, its amplitude falling exponentially with the distance from the edge of the allowed region. We then looked at wave propagation in a periodic structure and found that when the wave was propagating in an unallowed direction, the intensity at each layer was reduced by reflection into one of the allowed orders. This again illustrated that the "unallowed" wave could propagate a considerable distance through the crystal before "vanishing." If the crystal was not large compared to the attenuation length, a substantial portion of the initial wave could penetrate completely through the forbidden region. From the particle point of view, the photon would have a reasonable probability of "tunneling" through the potential barrier.

Note carefully what the criteria for tunneling are: The forbidden region must be of finite length; that is, there must be allowed states for the particle with its particular energy on both sides of the forbidden region. Second, the region itself must not be very wide compared to the attenuation length for the incident wave. What does the attenuation length depend on? [Answer: The attenuation length $L$ (the inverse of the propagation constant $\alpha$) is dependent on the difference between the total energy of the wave and the local potential energy. If the region is forbidden, the potential energy required exceeds the total energy of the particle. If the potential function does not change too rapidly with position, we may write the propagation factor as we did in Chapter 1:

$$\alpha = 1/L = \sqrt{(V - E_i)\, 2m/h^2} \tag{5.59}$$

There is a very real question about what to use for $V(x)$ in (5.49). It is difficult to answer for two reasons. First, the distance $L$ is short enough that we might well question whether Bloch waves were a good approximation to the solution. Second, even if we employ a Bloch-wave solution, during the transition from one band to the other, the solution appropriate to one band must change to the solution appropriate to the other. This may involve a phonon for $k$ conservation.]

(If you are familiar with electromagnetic wave guides, the problem we are discussing here has an almost exact analog. If one connects two geometrically dissimilar wave guides, both of which may conduct waves with frequency $v_1$, by a short section of guide that has its cutoff frequency well above $v_1$, one has two electrically different allowed regions connected by a short forbidden region. A fraction of the photons with energy $hv_1$ that are incident on the "junction" from the left will be transmitted to the guide on the right. Since the guides are different, mode conversion must take place for the transmitted waves. If the middle guide is to connect two disssimilar guides, its cross-section must vary along its length. As you can see, this would make for a most unattractive though rather important field-theory problem. The solid-state problem is beset with most of the same difficulties.)

Although all of the difficulties with the tunneling problem have not been successfully resolved, approximate solutions for most cases have been generated and appear to fit the measured data fairly well. What we want to do here is to obtain an order-of-magnitude estimate for the transmission probability for an electron incident upon the junction. To make life easy, we will assume the simplest possible model that still resembles the real problem. We will assume that the transition is direct. If we chose our material properly (e.g., (GaAs), the transition would be direct, and this obviates the necessity of considering mode conversion (i.e., change of $k$). Next, we will replace the curved section of the band-edge-versus-position plot (Figure 5.12) with straight lines. This is equivalent to assuming that the electric field in the depletion region is a constant. Obviously this is not true, but the exact form is really not too critical. Finally, we must choose the proper form of $V$ to insert into (5.59). This choice is not very obvious, but the following observation can guide us. The bending of the bands in the process of thermal equilibration is what causes the middle of the depletion region to be forbidden to waves that are allowed on either side of it. This bending is a continuous function of position, so no abrupt changes of potential are permitted. Since the wave is allowed on either side of the depletion region, the function $(V - E_i)$ must start at zero on one side and finish at zero on the other. The simplest function that we could choose with two zeroes is a parabola. Putting the maximum at the origin:

$$V - E_i = A - Bx^2 \qquad (5.60)$$

$A$ must represent some fraction of the gap energy, $E_g$ and $B$ must be chosen so that $V - E_i$ passes through 0 at the right places. With our constant-field assumption, the forbidden-zone width is given by $w = E_g/q\,\mathsf{E}$. To have $V - E_i$ vanish at $\pm\, w/2$, the ratio of $A/B$ must be given by $A/B = E_g^2/(4\mathsf{E}^2 q^2)$. It is reasonable to attribute the $x^2$ term to the square of the potential shift caused by the field: $q\mathsf{E}x$. To keep the dimensions correct, we must have $B$ in the form $q^2\mathsf{E}^2/E_g$, although we have no a priori reason to pick $E_g$ rather than some other arbitrary fraction such as $(5/17)\, E_g$. Thus, (5.60) has become

$$V - E = \frac{(E_g/2)^2 - (q\mathsf{E}x)^2}{E_g} \qquad (5.60')$$

If we want our solution to be applicable to three-dimensional crystals, we must allow for the fact that only that portion of the energy associated with motion normal to the plane of the junction may interact with the potential in the forbidden zone. In other words, only the component of momentum perpendicular to the plane of a mirror is affected by the mirror. If we define $E_\parallel$ and $E_\perp$ as the components of energy associated with motion in the direction of tunneling and normal to the direction of tunneling, respectively, we get the three-dimensional form of (5.60'):

$$V - E_\parallel = \frac{(E_g/2)^2 - (q\mathsf{E}x)^2 + E_g E_\perp}{E_g} \qquad (5.61)$$

The final step is to average the attenuation coefficient over the depletion region—in effect, taking the product of all the infinitesimal attenuations $e^{-\alpha dx}$. Since we have deducted $E_\perp$ from the total energy available to the particle, we must find the new zeroes of (5.61). This will give us the effective width of the forbidden region, $w'$, as a function of $E_\perp$, $E_g$, and the electric field.

$$w'/2 = \frac{1}{q\mathsf{E}} \sqrt{\left(\frac{E_g}{2}\right)^2 + E_g E_\perp} \qquad (5.62)$$

The total attenuation is given by $e^{-a}$ where $a$ is [from (5.59) and (5.61)]:

$$a = \int_{-w'/2}^{w'/2} \alpha \, dx = \int_{-w'/2}^{w'/2} \sqrt{\frac{2m}{\hbar^2}(V - E_{\parallel})} \, dx \qquad (5.63)$$

Considering the form of (5.61) and (5.62), a convenient change of variable is $y = qEx/\sqrt{(E_g/2)^2 + E_g E_{\perp}}$. Putting this into (5.63) gives

$$a = \frac{(2m)^{1/2}}{q \, E\hbar} \frac{(E_g/2)^2 + E_g E_{\perp}}{E_g^{1/2}} \int_{-1}^{1} \sqrt{1 - y^2} \, dy$$

$$= \frac{\pi}{2} \frac{(2m)^{1/2}}{q \, E\hbar} \left[ \frac{(E_g/2)^2 + E_g E_{\perp}}{E_g^{1/2}} \right] \qquad (5.63')$$

If you are surprised that $E_{\parallel}$ does not appear in (5.63'), note that in the simplest form (5.60') the electron's energy does not enter at all. This is because we have replaced the curved portions of Figure 5.12 with straight lines. Thus, the distance across the forbidden region (and the height of the barrier) is independent of $E_{\perp}$. The only reason $E_{\perp}$ enters is that, for any electron energy, the electron's ability to penetrate the barrier depends on the height of the barrier compared to the energy that carries the electron through the barrier.

Another question should be asked: What value of $m$ should be used, since it is highly unlikely that the effective mass will be the same in both bands? Since $m$ came from equation 5.59, we should look there for our answer. Consider the energy balance in bending the bands to align the band edges. Figure 5.13 shows an exaggerated diagram of Figure 5.12. The electron enters the forbidden band at point $D$ and arrives at the conduction band at point $F$. (Note the meaning of the diagram. All points along a line represent the same state. At different points, $x_1$ and $x_2$, the same state can represent a different absolute energy, but a given state always has the same value of $k$ and the same energy with respect to the band edge.) As the electron approaches or leaves the forbidden zone on a line of constant absolute energy, it passes through many allowed states, exchanging some kinetic energy with its supply of potential energy. The two states that are eventually connected by the tunneling event have, at any point $x$, an energy difference given by

$$\Delta E = E_g + \frac{\hbar^2 k_c^2}{2m_c} + \frac{\hbar^2 k_v^2}{2m_v} \qquad (5.64)$$

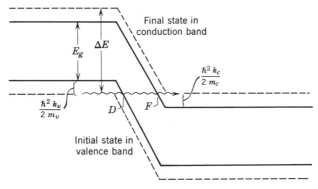

**Figure 5.13** An exaggerated version of Figure 5.12 for illustration of the energy shift in the bands required to align the initial and final states.

This energy difference must be the amount of band bending provided by the field: $qEw$. Furthermore, we have required that the transition be direct, so $k_v = k_c = k$. Putting these two into (5.64) gives

$$qEw = E_g + \frac{\hbar^2 k^2}{2}\left(\frac{1}{m_c} + \frac{1}{m_v}\right) \tag{5.65}$$

or

$$k = d\sqrt{\frac{2}{\hbar^2}\left(\frac{m_c m_v}{m_c + m_v}\right)(E_g - qEw)}$$

Note that (5.65) gives us (5.59) if we replace $w$ by $x$ running from $-w/2$ to $w/2$ and $m$ by $m_c m_v/(m_c + m_v)$. Thus, internal consistency requires that we employ $m = m_c m_v/(m_c + m_v)$. Note, however, that (5.59) and the whole idea of using Bloch waves is somewhat suspect, so we may be being consistently wrong.

The final question is: How does one obtain the transmission probability, $P$, given the attenuation coefficient, $a$? (Answer: $e^{-a}$ is the attenuation factor for the wave amplitude. The particle density is proportional to the square of the wave amplitude, so the transmission probability is

$$P = e^{-2a}) \tag{5.66}$$

To get some idea of what sort of fields it takes to get a reasonable tunneling probability, we should evaluate a specific example. Note that the field can be the combined built-in and applied field. It also goes without saying that if there is a substantial tunneling current going one way at thermal equilibrium, there is an equal and opposite one coming back.

Since we have considered tunneling only in direct materials, an example must be selected from a material such as GaAs. (The results will fit Ge reasonably well, since there is a direct valley only 0.14 eV above the indirect valley.) Let us say that we had a very heavily doped GaAs junction diode that had $5 \times 10^{19}$ majority carriers/cm$^3$ on both sides of the junction. Such heavy doping would lead to many difficult problems if we wanted to do things very rigorously (e.g., the band gap becomes a function of the local doping; the complete Fermi function must be used instead of the Boltzmann approximation; etc.), but since a crude estimate of the tunneling probability will suit our needs, let us pretend that no complications would arise. Using 1.3 eV as both the band gap and $qV_D$, and with $\varepsilon = 1.1 \times 10^{-12}$ farads/cm, the depletion model gives a total depletion length of 0.085 $\mu$m and an average field of $1.54 \times 10^6$ V/cm. The reduced mass is 0.062 $m_0$. Plugging these into (5.63') and ignoring $E_\perp E_g$ gives

$$P \simeq e^{-11.2} = 10^{-4.87}$$

A transmission probability of the order of $10^{-5}$ may not seem very large to you, but before you dismiss it as insignificant, consider what it is $10^{-5}$ of. Very crudely, the particle flux incident upon the junction is of the order of $n v_t \simeq 10^{27}$/cm$^2$ sec. Since both sides of the junction are degenerately doped, we can say that the states on both sides of the junction are half filled. Thus, our transmission probability suggests that a flux of the order of $10^{22}$ particles/cm$^2$ sec ($10^3$ amperes/cm$^2$) is flowing in both directions. These are very large currents indeed. If we apply a small amount of reverse bias to unbalance the flows, we should expect to see very large currents flowing. We would not be disappointed. The reverse conductance of such heavily doped junctions is enormous. Since the usual junction diode hardly conducts at all in the reverse direction, the junction we have just considered is said to have "broken down" by tunneling. This effect is also called Zener breakdown after Clarence Zener, who first examined the possibility of tunneling transitions between bands [C. Zener, *Proc. Roy. Soc.*, London **145**, 523 (1934)].

It is important to note what a huge difference in $P$ a small change in doping can produce. Noting that $P$ is of the form:

$$P = e^{-C/E} \qquad (5.66')$$

we see from the evaluation of $P$ above that a 20% decrease in the average field produces an order-of-magnitude decrease in the tunneling probability. For fields very much less than $10^6$ V/cm, the tunneling probability will be essentially zero.

## CHAPTER SUMMARY

The central theme of this chapter was interband transitions — the creation and annihilation of hole-electron pairs. Both low- and high-field cases were included since both play important roles in common semiconductor-device characteristics.

We began by defining the principal one-step, band-to-band transition processes. These comprised the direct phototransition in which a photon accounts for the pair's energy and there is no change in momentum; the indirect phototransition in which a photon must be emitted or absorbed to account for the pair's energy and a phonon for their momentum difference; the Auger process in which the energy of the pair and the momentum difference, if any, is accounted for by the change in kinetic energy and momentum of another free carrier; and the tunneling process, in which an electron goes from an allowed state in one band to an allowed state of the same absolute energy in the other band, passing through a portion of the crystal that is "forbidden" to electrons of that energy. The first process is important mainly in direct materials such as GaAs; the second is seldom important, although it is observable; the third is important principally as the generation process in avalanche multiplication; and the fourth is dominant in very heavily doped, very abrupt p-n junctions.

Having concluded that one-step transitions were not usually important in indirect materials such as silicon and germanium, we turned to the analysis of the simplest multistep generation-recombination process. These transitions took place at recombination centers such as donor or acceptor states that were deep within the forbidden band. The recombination centers or traps catalyze transitions by providing a staircase of allowed states between which transitions are reasonably probable.

We analyzed one particular type of trap — the single donor — using kinetic equations (rate equations) that could be evaluated by equilibrium considerations. The four possible events that could occur at a trap were the capture or emission of a hole and the capture or emission of an electron. A mean time for each event was defined. The capture times were presumed to be inversely proportional to the product of the appropriate carrier's mean thermal velocity, the capture cross-section for the carrier, and the number of traps in the state appropriate for that transition:

$$\left.\begin{aligned}\tau_{ec} &= \frac{1}{v_t \sigma_e N_T^+} \\[2mm] \tau_{hc} &= \frac{1}{v_t \sigma_h N_T^0}\end{aligned}\right\} \quad \text{for a donor trap}$$

We next assumed that the emission of holes or electrons from traps was not influenced by the density of free carriers or by the state of nearby traps. This assumption does not always hold, but it appears to be valid for most situations. Then, by the application of equilibrium arguments, the values for the mean electron and hole emission times were obtained:

$$\tau_{ee} = \frac{2}{v_t\sigma_e n_T} \qquad \text{where} \qquad n_T = N_c e^{-(E_c - E_T)/kT}$$

$$\tau_{he} = \frac{\frac{1}{2}}{v_t\sigma_h p_T} \qquad \text{where} \qquad p_T = N_v e^{-(E_T - E_0)/kT}$$

Finally, by some algebraic manipulation of the rate equations, the steady-state recombination rates were obtained. The factor of 2 and $\frac{1}{2}$ in those equations were dropped. Then in terms of the excess-carrier concentrations, the steady-state equation became:

$$\frac{\partial n'}{\partial t} = \frac{\partial p'}{\partial t} = - [n_0 p' + p_0 n' + n'p']/[(n_T + n_0 + n')\tau_{h0} + (p_T + p_0 + p')\tau_{e0}]$$

where $\tau_{h0} = 1/v_t\sigma_h N_T$ and $\tau_{e0} = 1/v_t\sigma_e N_T$.

There were two situations in which the rate equations could be written in the form of $(1/n')(\partial n'/\partial t) = \text{const}$. They were labeled low or high level depending on whether the minority-carrier density was much less than or much more than the equilibrium majority density. For these two cases, the excess-carrier density decayed exponentially with a characteristic time, $\tau$, given by $\tau = \tau_{e0}$ for a $p$-type sample, $\tau = \tau_{h0}$ for an $n$-type sample, low injection, and $\tau = \tau_{e0} + \tau_{h0}$ for either case at high-injection levels. For the intermediate case—neither high nor low injection—we did not obtain an exponential decay, so there is no constant decay time.

We next established a rule for excess carrier concentrations that we had already invoked in the equilibrium situation: that no substantial charge imbalance can exist for very long in a region of moderate conductivity. This rule followed directly from Maxwell's equations. In the simplest of terms, if a current can flow to relieve a charge imbalance, it will. Thus, $p' = = n'$. (There are exceptions to $p' = n'$ that still yield charge balance. For example, if a portion of the excess carriers of one type get trapped, their charge is still present but they are no longer free carriers.)

Using the concept of excess-carrier lifetime and the equations from Chapter 4 for determining current flow, the *continuity equations* were established:

$$\frac{\partial p'}{\partial t} = - \mu_h \nabla \cdot (p\mathbf{E}) + D_h \nabla^2 p - p'/\tau$$

$$\frac{\partial n'}{\partial t} = \mu_e \nabla \cdot (n\mathbf{E}) + D_e \nabla^2 n - n'/\tau$$

For minority carriers under conditions of low injection, these further reduced to

$$\frac{\partial p'}{\partial t} \cong D_h \nabla^2 p' - p'/\tau$$

$$\frac{\partial n'}{\partial t} \cong D_e \nabla^2 n' - n'/\tau$$

In the usual one-dimensional problem, the general solution to the simpler form is

$$p' \text{ or } n' = A \cosh \frac{x}{L} + B \sinh \frac{x}{L} \quad \text{where} \quad L = \sqrt{D_h \tau_h} \quad \text{or} \quad \sqrt{D_e \tau_e}$$

When a case of intermediate injection must be analyzed or wherever we want to keep track of the excess majority carriers, one must be careful to include the ambipolar force that keeps $p' = n'$. Including $p' = n'$ and eliminating the charge-imbalance terms ($\nabla \cdot \mathbf{E}$ terms) between the two equations led to the ambipolar diffusion equation:

$$\frac{\partial p'}{\partial t} = -\frac{p'}{\tau} + \mu^* \mathbf{E} \cdot \nabla p' + D^* \nabla^2 p'$$

where

$$\mu^* = \frac{p - n}{n/\mu_h + p/\mu_e} \qquad D^* = \frac{p + n}{n/D_h + p/D_e}$$

To illustrate the effects predicted by the ambipolar diffusion equation, one of the several versions of the Haynes-Shockley experiment was described. A measurement of $\mu^*$ was simple. Although not very easily, values of $D^*$ and $\tau$ could also be obtained from this classic experiment that allows you to "see" the minority carriers.

The Haynes-Shockley experiment also pointed up another important effect of the injection of excess carriers—conductivity modulation. It was shown that although the excess carriers might be so tied together that the pulse of carriers could not drift at all, the effect of the extra carriers on the conductivity was identical to the effect of the equilibrium carriers.

The shape of the Haynes-Shockley sample suggested a problem we had not yet considered. The high ratio of surface to volume made us wonder whether the observed recombination was going on in the bulk or in the surface layer. The analysis of charge flow with surface recombination was handled using the concept of surface-recombination velocity, $s$. This coefficient relates the current flowing into the surface to the excess-charge concentration at the surface:

$$-J_e\big|_{\text{surface}} = qsn'\big|_{\text{surface}}$$

A reexamination of a problem we had done ignoring surface recombination showed that the observed values of surface-recombination velocity can make a substantial difference in the total observed recombination.

We completed the chapter by considering interband transitions that occur in the presence of high or extremely high electric fields. For the high-field case, hot carriers create new carrier pairs by an Auger process. When the field is high enough, the new carrier pair will also multiply, leading to a runaway situation called avalanche multiplication. The rate at which carriers were generated was developed using an extremely simplistic model. Ionization coefficients $\alpha_h$ and $\alpha_e$ were defined by the equations $dp/dx = = \alpha_h p + \alpha_e n = - (dn/dx)$. After all awkward facts were assumed away, we obtained a relationship between the ionization coefficients and the field of the form:

$$\alpha_h = e^{-(E_1)/qEl_1} \left[ \dfrac{1}{l_2 + \dfrac{E_1}{qE}} \right] \quad \text{where} \quad \begin{aligned} E_1 &= \text{ionization threshold energy} \\ l_1 &= \text{mean free path for nonionizing} \\ &\quad \text{scattering} \\ l_2 &= \text{mean free path for ionizing} \\ &\quad \text{scattering} \end{aligned}$$

This rule seems to fit the observed data moderately well. The extreme sensitivity of $\alpha$ to variations in $E$ is quite evident, a fact that can make the analysis of avalanching in varying fields quite complicated. We examined two cases, avalanching in a region of high conductivity and in a region of very low conductivity. The first case was done mostly by hand waving and impassioned appeals to reason, but some analytic analysis was possible in the second and more important case. Since the high field implies a field-independent drift velocity, we were able to write the rate of change of the hole current as

$$\frac{dI_h}{dx} = \alpha_e I_e + \alpha_h I_h$$

The total current was constant, so the electron current could be eliminated. This led to a differential equation that could be solved to give

$$I_h(x) = \left[ \frac{1}{M_h} + \int_0^x \alpha_e e^{u(x)} \, dx \right] I_t e^{-u(x)} \quad \text{where} \quad u(x) = \int_0^x (\alpha_e - \alpha_h) \, dx$$

The quantity $M_h$, the ratio of the final hole current to the initial hole current, is called the hole-multiplication ratio. Although the equation above is more or less complete, using it can be far from trivial.

The extremely-high-field case brought up the subject of tunneling. When the conduction band in one portion of the crystal is brought into alignment with the valence band in another part of the crystal, electrons may tunnel through the forbidden region to available states on the other side. To have very much probability of tunneling, the forbidden region must be very narrow. To align the bands requires band bending, and to bend them rapidly requires very large electric fields—either built in or applied. If a sufficiently high field is not built in, one obtains avalanche breakdown before tunneling becomes apparent. Thus, tunneling is dominant only in very heavily doped, very abrupt junctions.

A simple quantitative model of tunneling in direct materials was developed. The derivation followed directly from the "particle-in-a-well" problem of Chapter 1. We assumed a Bloch wave solution (after expressing some doubts) and a parabolic form for the potential energy in the forbidden zone. We then assumed that the reflection coefficient for the whole forbidden zone was the integral of the reflections from each differential element in the zone. This integral performed, we obtained for a transmission probability:

$$p(\mathsf{E}, E_\perp) = \exp\left\{ (\sqrt{2m}\,\pi/q\mathsf{E}\hbar) \left[ \frac{(E_g/2)^2 + E_g E_\perp}{E_g^{1/2}} \right] \right\}$$

$E_\perp$ is the energy tied up in motion perpendicular to the direction of tunneling. This energy is unavailable for tunneling, so it effectively raises the barrier energy and increases the barrier width. Just as in avalanche multiplication, the tunneling rate is extremely sensitive to the field.

At this point in the book, we have examined in greater or lesser detail all of the physics necessary to understand the behavior of most semiconductor devices. What remains to be done is to apply this extensive background to a number of common devices to see what makes them function. The next two chapters will be a blend of applied physics, current semiconductor technology, and electronic modeling. Only a few devices will be considered—various types of diodes and transistors—and you might well wonder whether all the physics was really necessary. Unfortunately, the answer is yes. Certainly several topics could have been eliminated, but you will be suprised at how much physics you will use in following an electron in one lead and out another. The physics will not be hard to use, but it certainly was not easy to learn. What lies ahead should be more obvious.

## Problems

1. A thin, very lightly doped, uniform slab of $n$-type silicon containing $10^{14}$ donors/cm$^2$ is placed in a beam of very-high-energy electrons. Each electron produces many hole-electron pairs even though the beam is almost unaffected by its passage through the slab. A uniform generation rate $\Gamma = 10^{18}$ hole-electron pairs per cm$^3$ per sec is expected. Let us say that

the slab has the following properties in its bulk and has its surface terminated by ohmic contacts:

$$\tau_h = 10^{-5}\,\text{sec} \qquad\qquad\qquad \mu_e = 1200\,\text{cm}^2/\text{Vsec}$$

$$kT = 1/40\,\text{eV} \qquad\qquad\qquad \mu_h = 400\,\text{cm}^2/\text{Vsec}$$

Let the slab be 200 $\mu$m thick.

(a) Find the ratio of the conductance with the beam on to that with the beam off.

*Answer:* Even with no diffusion, $p = 0.1n_0$, so low-level analysis is applicable. The solution for $p$ with a uniform generation rate, $\Gamma$, and ohmic contacts at 0 and $a$ is

$$n' = p' = \Gamma\tau_h \left[ 1 - \frac{\cosh\dfrac{x}{\sqrt{D_h\tau_h}}}{\cosh\dfrac{a}{\sqrt{D_h\tau_h}}} \right]$$

By integration over the slab to find the average carrier density,

$$G_0 = 4.8\,\text{mhos/cm}^2$$

$$G' = 0.64\,\text{mhos/cm}^2$$

so

$$\frac{G_0 + G'}{G_0} = 111\%$$

(b) What fraction of the holes recombine within the slab?
*Answer:* Directly from the arithmetic in (a), 84.2%.

2. Let us say that we know a lot more about a recombination center in silicon than we usually do. Say the center is a donor lying 0.25 eV below the intrinsic level. Its electron and hole capture cross-sections are both $10^{-16}$ cm$^2$, independent of temperature. The maximum solid solubility of the impurity causing the center is $10^{15}/\text{cm}^3$.

(a) What lifetime would be observed at 300°K in a sample with $10^{17}$ As atoms/cm$^3$ as well as $10^{15}$ of these recombination centers?
*Answer:* All other factors are small, leaving only $\tau_{e0}$. With the average effective mass of electrons in Si and T = 300°K, $\tau_{e0}$ turns out to be 0.52 × $10^{-6}$ sec.

(b) What lifetime would be observed at 300°K in a similar sample that contained only 3 × $10^{15}$ As/cm$^3$?

*Answer:* For this example, one must observe that the ionized donors partially compensate the acceptors. A quick calculation shows that, at $2 \times 10^{15} = p$, the centers are about 90% ionized; accordingly $p \simeq 2 \times 10^{15}$. Then, from (5.17):

$$\tau = \tau_{eo}\left(1 + \frac{2p_T}{p_0}\right) = 0.52 \times 10^{-6}\left(1 + \frac{0.45 \times 10^{15}}{2 \times 10^{15}}\right) = 0.64 \times 10^{-6} \text{ sec}$$

(*c*) If the temperature were increased to $400°K$, what lifetime would then be observed with the sample from (*b*)?

*Answer:* One must account both for the increases in $v_t$ (which reduces $\tau_{eo}$) and the increase in $p_T$ (which raises $\tau$). The increase is by a factor of $\sim 47$ while the reduction is by a factor of $\sim 0.87$, so the net is an increase by a factor of 40.8.

(*d*) What is the physical change (e.g., which process changes) that brings about the increase in lifetime with increasing temperature?

*Answer:* There are four rates to be considered. However, given $\sigma \neq \sigma(T)$, the capture rates increase with $T$, lowering $\tau$. What are left are the emission rates. Since the recombination centers are much closer to the valence than the conduction band, it is clear that hole emission is much more likely than electron emission. There are several equivalent ways to describe what goes on. The equilibrium statistical argument is that the Fermi level moves up with increases in $T$, so the occupation number for the traps increases. Thus, there are less traps available to capture conduction electrons. The rate-equation argument is that electrons jumping up out of the valence band compete for the limited number of recombination centers available with electrons falling down out of the conduction band. From either viewpoint, the conduction electrons have less chance to recombine, so the lifetime increases.

3. One of the best high-speed, high-sensitivity middle-infrared detectors is a very cold piece of *p*-type germanium doped with a moderately deep acceptor impurity such as Cu, Hg, Cd, or Zn (see Figure 3.4*a*). These *photoresistive* detectors operate by sensing the generation of majority carriers caused by the photoionization of the relatively deep acceptor states. The detectors must be operated at temperatures sufficiently low to ensure almost complete *freeze out* (deionization) of the deep impurities. In this problem you are asked to compute some of the properties of such a detector.

   Let the sample be "optically thin" (i.e., the intensity of the light does not vary much in traversing the whole sample), even though this is not what one would want for best low-level detection. Let the sample have $10^{16}$

Cu atoms/cm³, each with an optical absoption cross-section of $10^{-14}$ cm² for 10 $\mu$m (0.1 eV) radiation. Let the hole-capture cross-section for ionized acceptors be $10^{-14}$ cm² at 22°K and let the hole mobility at 22°K be 5000 cm²/Vsec. At 22°K the band gap in Ge is about 0.75 eV and the thermal velocity about $10^6$ cm/sec.

(a) First, find the conductivity of the detector material at 22°K without illumination.

*Answer:* At 22°K, $N_v = 1.13 \times 10^{17}$ cm³. Since the acceptors are presumably neutral at this temperature, the Boltzmann approximation is valid for both the acceptors and the band itself. The results are

$p = 2 \times 10^{12}$/cm³ (note: only 0.02% of the Cu atoms are ionized)
$E_f - E_v = 0.023$ eV $= 12.2\, kT$
$E_A - E_v = 0.04$ (from Figure 3.4a)
$\sigma = 1.6 \times 10^{-3}$ $\Omega$cm

(b) Second, find the general expression for the steady-state conductivity of the bulk material in a uniform light flux of 0.1 eV photons.

*Hint:* Take advantage of the fact that the only source of holes at this temperature is the acceptor population. Accordingly $p = N_T^-$ and $\sigma = q\mu_h p$.

*Answer:* The governing rate equation is

$$\frac{\partial p}{\partial t} = -v_t \sigma_h \left[ p^2 \left( 1 + \frac{p_0}{N_{T_0}^0} \right) - \frac{p N_T p_0}{N_{T_0}^0} \right] + G$$

where $G = \sigma_L N_T^0 \mathscr{L}$ is the generation term due to the light flux $\mathscr{L}$ (photons/cm²sec). Since $N_T^0 = N_T - p$, in the steady state we have

$$p^2 \left( 1 + \frac{p_0}{N_{T_0}^0} \right) - p \left[ \frac{N_T p_0}{N_{T_0}^0} + \frac{\sigma_L \mathscr{L}}{v_t \sigma_h} \right] + \frac{\sigma_L N_T \mathscr{L}}{v_t \sigma_h} = 0$$

(c) The result for (b) is a nonlinear $\mathscr{L}$ versus $p$ relationship. However, under most conditions of practical interest, the light flux is pretty small, so that $N_T^0 \simeq N_T$. Under these small-signal conditions, find the linear and quadratic terms in the relationship of $p$ versus $\mathscr{L}$. For the reasonably strong signal of $10^{-9}$ watts/cm² ($\sim 10^{11}$ photons/cm² sec), show that the response is essentially linear.

*Answer:* Under the low-level assumption, the equation for $p$ versus $\mathscr{L}$ is approximately

$$p^2 - p p_0 + \frac{\sigma_L N_T \mathscr{L}}{v_t \sigma_h} = 0$$

from which

$$p = p_0 + \frac{\sigma_L N_T \mathscr{L}}{p_0 v_t \sigma_h} - \frac{2\sigma_L^2 N_T^2 \mathscr{L}^2}{p_0^3 v_t^2 \sigma_h^2} + \dots$$

The linear term is approximately $10^{-3}p_0$ for a signal of $10^{11}$ photons/cm$^2$ sec. The quadratic term is $10^{-6}p_0$. Thus, the signal is detected quite linearly. (N.B. Actually, the response to the *electric field* is quadratic, since the photon flux is proportional to the square of the electric field. However, if the intensity is what is conveying the information, the detection is linear.) Observe, however, that at 1 $\mu$W/cm$^2$, substantial nonlinearity would be observed.

(*d*) Under the conditions of part (*c*), what is the excess-carrier lifetime? Does it get longer or shorter at higher light intensities?

*Answer:* 500 $\mu$sec. Increasing the light flux increases $N_T^-$ and thus decreases the average lifetime for the excess carriers.

4. Tunneling junctions of various kinds are finding many practical applications in the processing of low-voltage, high-frequency signals. A particularly spectacular example is the very simple MOM (metal-oxide-metal) junction made by placing a sharply pointed tungsten wire against a nickel stud. The native oxide layer on the nickel ($\sim$ 10Å thick) forms the barrier through which the electrons must tunnel. Since the tungsten wire can be made very small, the capacity of the MOM junction can be very small, and the junction can respond coherently to oscillations in the infrared region ($\sim 4 \times 10^{13}$ Hz). Consider a very simple model of such a junction in equilibrium. The alignment of the Fermi levels for the two metals causes a uniform field in the oxide of $4 \times 10^6$ V/cm, running from the tungsten to the nickel. The oxide conduction band edge is 3 eV above the Fermi level at the tungsten interface. Let $kT = 0.026$ eV and $m = m_0$ for both metals.

(*a*) Find the transmission probability as a function of the applied voltage for an electron with energy $E = \frac{3}{2}(kT)$. Show by evaluation that the transmission probability does not change very rapidly with applied bias.

*Answer:* Letting the oxide thickness $= a$ and $V = V_0 - q\mathsf{E}_x$ in the oxide, we have

$$d\alpha = \left[\frac{2m}{\hbar^2}(V - E)\right]^{1/2} dx$$

from which

$$\alpha = -\left[\frac{2m}{\hbar^2}\right]^{1/2} \frac{2}{3q\mathsf{E}}[(V_0 - E)^{3/2} - (V_0 - E - q\mathsf{E}a)^{3/2}]$$

The total voltage across the oxide is that built in plus that applied: $V_T =$ $= V_0 + V = Ea$. The built-in voltage is 0.4 V. Since $P = e^{-2\alpha}$, one may evaluate $P$ for various $V$ to be [notice that for $V = -0.4$ (i.e., $E = 0$), one must reevaluate the integral]:

| $V$ | $\alpha$ | $P$ |
|---|---|---|
| 0 | 8.46 | $10^{-7.35}$ |
| $-0.1$ | 8.54 | $10^{-7.42}$ |
| $+0.1$ | 8.41 | $10^{-7.32}$ |
| $-0.4$ | 8.80 | $10^{-7.64}$ |

(b) Since something of the order of $10^{28}$ electrons/cm$^2$ are incident on the junction from each side each second, a substantial flux could be crossing the oxide. However, since the transmission probability is not a strong function of voltage and is symmetric in any case, describe the process by which the voltage produces a large net flux.

*Answer:* The applied potential shifts the bands so that electrons coming from one side see more empty states than electrons coming from the other side. It is this difference in the number of full initial states and empty final states that accounts for the principal changes in the net flux.

# 6

# DEVICES THAT WORK PRINCIPALLY BY DIFFUSION: (A) THE JUNCTION DIODE

The next five chapters will deal with four devices. The analysis for each will begin with a highly simplified, idealized model of the actual device. As your sophistication develops, the model will be progressively refined to make it a more accurate representation, but it will always be less complicated than the real device. Since the behavior of any semiconductor circuit element is the net result of a host of different interactions, both within the bulk and at the variety of different surfaces that bound a real device, such simplification is necessary. A blunt frontal assault that tried to account for everything at once would be essentially incomprehensible.

Unfortunately, the construction steps that lead to a proper (i.e., accurate) model are neither obvious nor necessarily unique. Although none of the modeling presented in the next chapters is new to the literature, it is presumably new to the reader. This will undoubtedly lead to the question: "Why in the world is he doing THAT?" Unfortunately, the answer is clear only in retrospect. Most readers find the material quite well directed the second time through, when the coherence of the whole analysis is maintained by a degree of familiarity. You may find that an occasional reference to the chapter summaries will be helpful the first time through.

There is a lesson to be learned from the fact that the material is quite clear and simple a posteriori. The original development of a good analytical model is considerably more difficult than it looks in retrospect. It is seldom obvious which of the many simple steps that could be taken will lead to important insights. Details of the model for a bipolar transistor were actively debated for almost a decade. New insights still appear sporadically. For none of the devices that we shall discuss is the model "finished."

The next three chapters will develop the analysis of the junction diode and the bipolar transistor. In these devices, the principal limit to current flow is the rate at which the available minority carriers can diffuse through a neutral region. The standard model represents the action of the applied potentials as modulating a boundary condition—namely, the number of minority carriers available at the boundaries of the neutral regions. The first-order analysis of such devices proceeds in very much the same fashion as in problem 5.3, with solutions for the minority-carrier concentrations being formally in the form of (5.28). The principal difficulty at this stage is in establishing (i.e., convincing yourself of) the validity of the assumed boundary conditions. Once these are accepted, the solutions to the I-V characteristics for the first-order model are obtained quite readily. Let us begin by taking a quick look at what the model is.

Figure 6.1 shows the structure of a typical *diffused planar* diode that might be used for moderate power—say about 1 to 5 amps forward and 100 V peak reverse voltage. The figure caption briefly describes the steps that would produce such devices. Note that a very large fraction of the junction area is a plane. Unless junction curvature is critically important—occasionally it is—we would certainly be tempted to represent the diode as a truly plane device, one in which all variation was along a single axis perpendicular to the plane of the junction. Such a one-dimensional representation is shown in Figure 6.2. The figure has a primitive symmetry about the depletion region. That narrow region joins two bulk regions that are each terminated at their remote ends by an ohmic metal-semiconductor junction. For reasons that will be presented later in this chapter, we assume that the *excess*-carrier concentration at the ohmic contacts is always negligible and that the minority-carrier concentration at the "edges" of the depletion region is proportional to $\exp[qV/kT]$ where $V$ is the applied potential. Neither of these assumptions should be obvious; in fact, neither is strictly true. But before arguing the validity of the assumptions, let us see where they would lead.

If you review problem 5.3 on page 243, you will see that having a minority-carrier concentration that is at its equilibrium value on one side (the ohmic contact) and proportional to $\exp(qV/kT)$ on the other will lead immediately to a diffusive current given by

$$I_{\text{minority}} = \text{constant} \cdot [\exp(qV/kT) - 1] \tag{6.1}$$

Since the minority current on one side of the depletion region is fed by the majority current on the other, all we need to do get a first approximation to the total current is to assume that negligible recombination takes place within the relatively narrow depletion region. Again, this is an assumption that is not necessarily warranted, but more of that anon. Since the total current throughout any diode must be constant, if I know both the majority and the minority current at one edge of the depletion region, I have found the total current. Transposing the minority current from one side of the

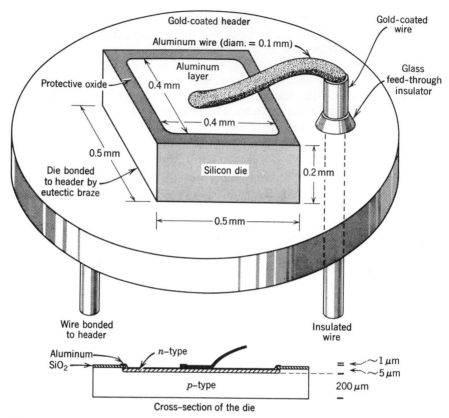

**Figure 6.1** A medium-sized planar diode mounted on a TO header. The final structure is fabricated in two groups of steps. In the first, many diodes are fabricated on a single wafer of silicon by diffusing boron through windows that have been etched in the protective oxide covering the surface. After diffusion, the windows are reopened and coated with aluminum. Then the wafer is scribed and broken into individual dice. The second group of operations involves packaging the diode. The die is bonded to the header by mechanically scrubbing it on the gold surface at a temperature in excess of 400°C. An Au-Si eutectic solution forms that makes a very effective braze between the header and the die. A wire is then bonded to the aluminum layer on top and to an insulated wire coming through the header. These are typically cold welds. The package is finished by sealing a top to the header in an inert atmosphere.

depletion region to the majority current on the other gives me this last piece of data, and I may write directly from (6.1):

$$I_{\text{total}} = I_{\text{sat}} \left[ \exp \left( qV/kT \right) - 1 \right] \tag{6.2}$$

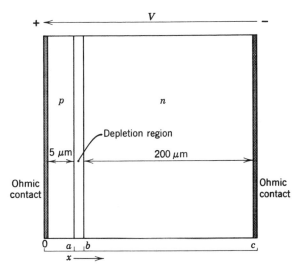

**Figure 6.2** A one-dimensional representation of the diode shown in Figure 6.1.

where $I_{sat}$ is an appropriate constant called the *saturation current*. [The name derives from the fact that *if* (6.2) holds, the current for negative voltages quickly saturates at $-I_{sat}$.]

It is important to recognize just where the various parts of (6.2) come from in order to be able to predict what will happen as the various physical parameters of the diode are varied. First, where did the exponential term enter? [Answer: There are two parts to this answer, one for the majority-carrier current and one for the minority-carrier current at the point of evaluation. For the minority-carrier current, the exponential term came directly from the assumed exponential voltage dependence of the minority-carrier concentration at the boundary of the depletion region. The majority current is simply the assumed exponentially-dependent minority current from the opposite side, translated under the assumption of negligible recombination within the depletion region. If either of these assumptions fails in an important way, we would expect (6.2) to fail.]

Which factors will determine $I_{sat}$? [Answer: $I_{sat}$ is the proportionality factor between the exponentially dependent minority-carrier concentration and the current. We should thus expect $I_{sat}$ to be directly proportional to the diffusion coefficients of the minority carriers and to the number of minority carriers there in equilibrium. It should also be inversely proportional to the length that the carriers must diffuse. If you throw in the charge per carrier and the area of the junction, you have the whole thing.]

Before going through the algebra that gives the details of (6.2), consider for the moment another possibility—the case in which the limit to the flow is the junction barrier itself. That is, if a carrier crosses the barrier, it goes immediately to the ohmic contact. If the barrier height is controlled by the voltage, the Boltzmann distribution of electron energies implies that the number of carriers that can (and will) cross the barrier is exponentially dependent on the voltage. Once again, we would expect the current to depend exponentially on the voltage, but we would also conclude that the proportionality factor for such diodes would be very different from $I_{sat}$ of (6.2). In particular, since $I_{sat}$ is proportional to the minority-carrier densities, we must expect the saturation current for the junction diode to be extremely temperature sensitive. If, on the other hand, the current is determined by the number of majority carriers that can cross the junction, the proportionality factor (in contrast to the current itself) need not be temperature sensitive (although it usually is). Examples of the latter kind of junction include the metal-semiconductor junction (Schottky barrier) and field emission (for low to moderate fields) from metals into vacuum. The lesson to be learned from this is that although the universality of the Boltzmann distribution makes the current in most diodes depend exponentially on the applied voltage to some power, all diodes do not behave in the same way. When you are making a diode, how it is supposed to function determines quite directly what you must do to obtain good devices.

## THE P-N JUNCTION OUT OF EQUILIBRIUM

As we found in Chapter 3, a *p-n* junction in thermal equilibrium has a potential across it given by

$$V_D = \frac{kT}{q} \ln (N_D N_A / n_i^2) \tag{3.49}$$

Since the diffusion potential is a perfectly legitimate voltage, it seems logical to assume that, in the presence of some applied voltage, the electrons would respond to the algebraic sum of the two local fields. Apart from possible ohmic drops in the bulk regions, the applied bias will appear across the junction, making the total junction voltage $V_j = V - V_D$. The sign of the applied voltage is taken as positive in the direction that reduces the potential step.

We must now agree on how to use this new nonequilibrium potential. All of the happy results that derive directly from the principle of detailed balancing no longer apply. In fact, we have a situation that should appear pretty intractable. We have introduced a nonequilibrium potential distribution into a region of very large and rapidly varying electric field and concentration gradient. You might well wonder whether anyone could determine even what to put into the Boltzmann transport equation, let alone do the integration itself. To the best of my knowledge, no one has ever published such a fundamental analysis for the depletion region. However, this apparent obstacle to the analysis of the junction can be circumvented if we can claim that two sides of the depletion region remain approximately in equilibrium with each other. If the carrier concentrations are in approximate equilibrium on opposites sides of the junction, the Boltzmann relationship that we are seeking would hold

$$p_n = p_p e^{qV_j/kT}$$

$$n_p = n_n e^{q\,V_j/kT}$$

(6.3)

(The subscripts indicate on which side of the depletion region the concentration in question is to be found.) Note that (6.3) is not quite yet in the form that will give us (6.1) The postulate that yielded (6.1) was that the minority-carrier density (e.g., $p_n$) is proportional to $\exp(qV/kT)$. Furthermore, I have not yet established the validity of (6.3); I have only said that *if* the carriers on either side of the depletion region are essentially in equilibrium with each other, (6.3) would hold. The heart of the matter is why anyone would suppose that this quasi-equilibrium situation would hold across the depletion region when it had an external potential impressed upon it.

The answer is a combination of two arguments. First, as you saw in Chapter 3, equilibrium requires the detailed balancing of all flows but puts no restriction on the magnitudes of the individual components of any flow. Equilibrium is usually maintained by large, completely balanced fluxes that serve to rapidly reduce any departures from the equilibrium distribution. If the net number of minority carriers flowing *away* from the junction is limited entirely by the rate at which the carriers can diffuse (or drift and diffuse) through the neutral region, then, within the depletion region itself, there can be two huge and almost equal fluxes going in opposite directions. This is just the sort of condition that would keep the two sides (and everywhere in between) distributed in the most likely random manner. All right, you say, IF there is a large flux in each direction to give all that stirring, you might buy (6.3). But what argument can I give to convince you that these fluxes exist? The answer, which is the second part of the demonstration, turns out to win only half of my case. I can argue successfully for (6.3) for conditions of forward bias, but not under reverse bias ($V < 0$). Fortunately, we don't really need (6.3) for the analysis of reverse-biased junctions. Let us begin with the forward-biased case.

Even though the large electric field within the depletion region might well make the mobility a function of position, let us take a look at the order of magnitude that we might expect for the oppositely directed drift and diffusion currents within the depletion region. For holes, for example, we would have

$$J_h = q[\mu_h \mathsf{E}p - D_h(dp/dx)]$$ (4.62)

Some typical values are $\mu_h = 400$ cm$^2$/Vsec, $D_h = 10$ cm$^2$/sec, and $\mathsf{E} = 10^4$ V/cm. Note that this field exists only within the depletion region. The particular values of these coefficients are not too critical to the argument, but the next ones are. In a typical *forward-biased* junction situation, we would have something like $10^{17}$ holes/cm$^3$ on the $p$ side of the depletion region with $10^{12}$ to $10^{15}$ on the $n$ side. [Such a statement is based on the supposition that (6.3) holds at least approximately, which for the moment can be considered an experimentally confirmable fact.] Let us compare the magnitude of the net hole current in the neutral $n$ region with the value of the two fluxes predicted by (4.62) within the depletion region. What I wish to demonstrate is that the individual fluxes predicted by (4.62) are enormous compared with the net flux of holes found in the neutral $n$ region.

Let us say that the distance the average hole must diffuse through the neutral region before recombining or reaching an ohmic contact is of the order of 10 $\mu$m. If the region is quite neutral, the electric field is probably negligible, so the current is given by the second term in (4.62). The flux in the neutral region is thus $- D_h dp/dx = 10 \times 10^{15}/10^{-3} = 10^{19}$ holes/cm$^2$sec. That's about an amp/cm$^2$, which is a moderate current density, neither very high nor very low. Now consider the fluxes in the depletion region. At the potential midpoint of the depletion region, assuming something of the form of (6.3), we would have $p = 10^{16}$/cm$^3$. With a field of the order of $10^4$ V/cm, the drift flux is given by $\mu_h \mathsf{E}p = 400 \times 10^4 \times 10^{16} = 4000 \times 10^{19}$ holes/cm$^2$sec. That's 4000 times the flux we found at the edge of the neutral region! Q.E.D. To have the small net flux, the diffusive component must be almost equal in magnitude to the very large drift component that we just determined. These are the large oppositely directed fluxes that establish (6.3).

However, before dashing out to celebrate this revelation, note some of the limitations of the demonstration. First, we once again run into the rather fuzzy "edge" of the depletion region. For the example just discussed, the two "huge" fluxes must taper to the net flux at this edge. Although the change from large field to negligible field can be rapid, it is not really abrupt.

Second, the demonstration depended on having not only a large electric field but also a substantial carrier population. Under reverse bias, the depletion region is almost completely devoid of carriers. Thus, we must conclude that there is almost no flux and no reason to assume an equilibrium-like distribution. However, if there are a negligible number of carriers, do you really care how they are distributed? The answer is usually no.

Finally, although we assumed a forward bias sufficient to give us a substantial carrier population throughout the depletion region, we still kept the voltage low enough so that a significant barrier $(V_D - V)$ remained. Under sufficient forward bias, both the electric field and the concentration gradient that typify the depletion region vanish. This makes the opposing-flux argument degrade to the comparison of two indeterminates; but once again, if the depletion region is not well defined, do you care whether you know the carrier concentrations on its undefined edges?

In sum then, we have established the accuracy and meaningfulness of (6.3) over the region in which that equation has great importance. This solves the principal problem of the first-order analysis of the junction diode. What remains is algebra.

## A FIRST-ORDER ANALYSIS OF THE P-N JUNCTION DIODE

Let us now return to the one-dimensional diode of Figure 6.2. At $x = 0$ and $x = c$ are ohmic contacts. By assumption (to be defended shortly), the excess-carrier concentrations at these contacts is negligible. The points $x = a$ and $x = b$ are the edges of the depletion region. The excess-carrier concentrations at these two points are to be found through (6.3). Equation 6.3 can be written in a somewhat more convenient form by noting that when $V_j = -V_D$, we should obtain the equilibrium values for the carrier concentrations $(p_{p0}, p_{n0}, n_{p0}, \text{ and } n_{n0})$. Thus, in terms of the applied potential, $V$, we have

$$
\left.
\begin{aligned}
\frac{p_n}{p_p} &= e^{q(V - V_D)/kT} = \frac{p_{n0}}{p_{p0}} e^{qV/kT} \\[2em]
\frac{n_p}{n_n} &= e^{q(V - V_D)/kT} = \frac{n_{p0}}{n_{n0}} e^{qV/kT}
\end{aligned}
\right\}
\tag{6.3'}
$$

Equation 6.3' is quite general. For our first walk through the analysis, we are aiming for a solution of the form of (6.2). As we shall see, that limits us to situations in which the excess-carrier density is small compared to the equilibrium majority density. With this restriction, we have

$$
\left.
\begin{aligned}
p_p &= p_{p0} + p_p' \simeq p_{p0} \\
n_n &= n_{n0} + n_n' \simeq n_{n0}
\end{aligned}
\right\}
\quad \text{i.e.} \quad
\left\{
\begin{aligned}
p_p' &\ll p_{p0} \\
n_n' &\ll n_{n0}
\end{aligned}
\right\}
\tag{6.4}
$$

The situation described by (6.4) is called *low injection-level*. It is important to note that low injection-levels are obtained when the excess-*majority*-carrier density is insignificant compared to the equilibrium *majority* density. Although we can readily operate beyond the low-injection-level condition, when

(6.4) holds, (6.3') takes on the particularly simple form that leads directly to (6.2):

$$p_n = p_{n0}e^{qV/kT}$$
$$n_p = n_{p0}e^{qV/kT} \quad\quad (6.5)$$

Take care to note the difference between (6.5) and (6.4). Equation 6.5 is a statement about the minority-carrier densities; (6.4) is a statement about the majority densities. In the typical extrinsic semiconductor, the minority-carrier density can change by many orders of magnitude before the excess majority density determined by the charge-neutrality requirement $p' = n'$ interferes with the low-injection-level assumption. Thus, enormous swings of the minority-carrier density can occur without significantly affecting the majority-carrier density.

We are now ready for the formal analysis. We proceed by finding the minority-carrier concentration throughout the neutral bulk regions by solving (5.27') subject to the boundary conditions just stated. The minority-current densities are then obtained from (5.26), which is the form of (4.62) appropriate under conditions of low-level injection and reasonably uniform doping throughout the bulk regions. (Nonuniform doping can lead to significant "built-in" fields that must be considered under certain circumstances. We will examine them for the bipolar transistor, but they do not add much insight into typical diode behavior. Accordingly, we assume uniform doping for simplicity even though most junction-formation processes lead to a doping gradient on at least one side of the junction.) The minority currents will be will be evaluated at $a$ and $b$. We then assume that there is insignificant recombination within the depletion region so that the minority-current density at $a$ is the same as the majority-current density at $b$. This assumption is frequently invalid, but the exceptions will be saved for later entertainment. Then, by adding the minority and majority components at the single point, $b$, we obtain the total current for the diode.

Beginning with the solution in the $p$ region, we are solving for the excess electron concentration, $n'(x)$, determined by

$$\tau_e D_e \frac{\partial^2 n'_p}{\partial x^2} + n'_p(x) = 0 \quad\quad (5.27')$$

subject to the boundary conditions:[1]

$$n'_p(0) = 0$$
$$n'_p(a) = n_{p0}\left[e^{qV/kT} - 1\right] \quad\quad (6.5')$$

---

[1] For $V < 0$ $n'_p(a) \Rightarrow -n_{p0}$, or in other words, $n_p(a) \Rightarrow 0$. We don't really require our arguments in defense of (6.3) to be valid for $V < 0$ (they aren't) as long as the carrier concentration is very small.

The solution may be written

$$n'_p(x) = n_{p0}[e^{qV/kT} - 1] \sinh(x/L_e)/\sinh(a/L_e) \qquad (6.6)$$

where $L_e = \sqrt{D_e \tau_e}$ is the electron diffusion length on the $p$ side. In the same way, on the $n$ side we obtain

$$p'_n(x) = p_{n0}[e^{qV/kT} - 1] \sinh[(c - x)/L_h]/\sinh[(c - b)/L_h] \qquad (6.7)$$

with $L_h = \sqrt{D_h \tau_h}$.

The minority currents are found using (5.26):

$$J_h(x) = - qD_h \frac{\partial p'_n}{\partial x} = \frac{qD_h p_{n0}}{L_h} [e^{qV/kT} - 1] \frac{\cosh[(c - x)/L_h]}{\sinh[(c - b)/L_h]} \qquad (6.8)$$

$$J_e(x) = + qD_e \frac{\partial n'_p}{\partial x} = \frac{qD_e n_{p0}}{L_e} [e^{qV/kT} - 1] \frac{\cosh(x/L_e)}{\sinh(a/L_e)} \qquad (6.9)$$

Under the assumption that there is negligible recombination in the depletion region, the total current may be written as the sum of the minority currents on either side of the depletion region:

$$I_T = [J_e(a) + J_h(b)] A$$

$$= [e^{qV/kT} - 1] \left\{ \frac{qD_h p_{n0}}{L_h} \coth\left(\frac{c - b}{L_h}\right) + \frac{qD_e n_{p0}}{L_e} \coth\left(\frac{a}{L_e}\right) \right\} A \qquad (6.10)$$

The results given in equations 6.6 through 6.9 are shown in the graphs in Figures 6.3 (forward bias) and 6.4 (reverse bias). The numerical values chosen for the coefficients are typical values for silicon, though even for a single material, a great deal of variation is both possible and useful. The $I$-$V$ characteristics predicted by (6.10) are shown in Figure 6.5.

The most conspicuous and important feature of this first-order analysis is the extreme asymmetry in the $V$-$I$ curve. Note that (6.10) is precisely in the form of equation 6.2, with $I_{sat}$ being the expression within the curly brackets times the area $A$. The coefficients in $I_{sat}$ are just what you would expect from the discussion associated with equation 6.2.

There are several features of equation 6.10 that should be examined rather carefully, since they underlie much of the thinking and the language used to describe diffusion-controlled devices. The two terms that comprise the saturation current are each composed of three variable factors: a geometric factor $[(1/L_e)\coth(a/L_e)]$, a material factor $(D_e)$, and a doping factor $(n_{p0} = n_i^2/N_A)$. Two of these factors, geometry and doping, are very clearly in the hands of the device designer. The diffusion coefficient is definitely affected by the device-manufacturing steps, but it is not a very easily controlled parameter.

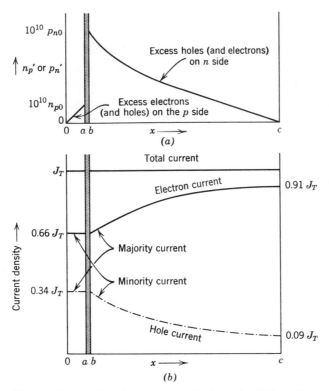

**Figure 6.3** Graphs of equations 6.6 through 6.9 for $V = +0.6$ V (forward bias). Note that although there are five times as many holes injected into the $n$ side as electrons into the $p$ side, the electron current is everywhere larger. Two things contribute to the lion's share of the current going to the electrons. First, $D_e > D_h$ (generally true) and second, the $p$ side is so much shorter than the $n$ side. The latter is usually the more important factor and is obviously a function of the diode's construction. (a) Excess-carrier density versus position according to (6.6) and (6.7) for $V > 0$. (b) Hole and electron-current densities as a function of position according to (6.8) and (6.9) for $V > 0$.

A given side of a diode is classified by its geometric factor as *long based*, *intermediate based*, or *short based* according to whether the width of the ohmic (bulk) region, $a$, is much longer, of the same order, or much shorter than the diffusion length, $L_e$. For the short-based case, $\coth(a/L_e) \Rightarrow L_e/a$, so the geometric factor reduces to $1/a$. This means that the diffusion length does not appear in (6.10) or any of the equations that preceded it. For example, (6.6) reduces to a first-order polynomal in $x$. The $p$ side of the diode is shown as short based ($a$ taken as $L_e/10$). The minority-carrier distribution is a straight

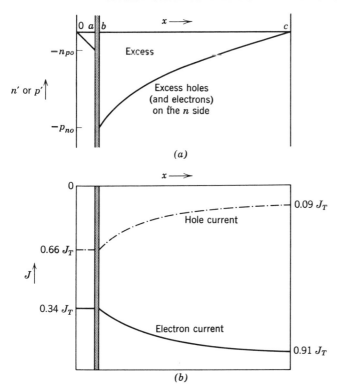

(a)

(b)

**Figure 6.4** Graphs of equations (6.6) through (6.9) for $V < -0.1$ V. Apart from the minus sign on the ordinate, these curves are the same as Figure 6.3, with one colossal exception. The ordinates in each case are $10^{-10}$ of the ordinates in (6.3). Also, note that this figure is essentially independent of $V$ as long as $V < -0.1$ V. (a) Excess-carrier density versus position according to (6.6) and (6.7) for $V < 0$. (b) Hole and electron-current densities as a function of position according to (6.8) and (6.9) for $V < 0$.

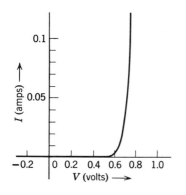

**Figure 6.5a** The $V$–$I$ curve for the diode on a linear scale. The curve is a drawing of equation 6.10 with the set of reasonably representative coefficients given in the text. Notice that the diode essentially "snaps on" at about 0.65 V.

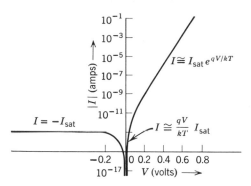

**Figure 6.5b** The absolute value of $I$ versus $V$ in a semilogarithmic plot. The curve is based on (6.10).

line on that side. (See Figures 6.3 and 6.4.) The gradient of the carrier density is a constant, so the minority-carrier current is also independent of position throughout the $p$ side. Physically, what short based means is that almost all of the carriers get across the bulk region without recombining. Conservation of current and negligible recombination imply that each component of the current is constant.

The $n$ side is an example of an intermediate-based case ($c - b = 2L_h$), so no algebraic simplification is possible. Note that, as the minority carriers disappear by recombination in going from $b$ to $c$, their share of the current is taken up by the majority carriers (which are flowing in to recombine with the minority carriers). Note also that we have never solved directly for the majority current. It was found from the fact that the total current is conserved.

How should we proceed to find the voltage drop, if any, across the ohmic regions? [Answer: Exactly as in problem 5.3. That is, since we know the majority current and the majority-carrier density ($p' = n'$), we can solve for the only unknown in the majority-current equation—the electric field. In other words, E is just right to insure conservation of current. Integrating the field gives the voltage drop. For the low-injection case, where $p_p = p_{p0}$, etc., the integration is particularly simple.]

The truly long-based case occurs in practice only with materials that have exceedingly short lifetimes. In most cases, this means direct materials such as GaAs ($\tau < 10^{-9}$ sec), although it is possible to introduce "lifetime killers," such as gold, copper or iron, to achieve a very short lifetime in the common indirect semiconductors. With gold-doped silicon or a direct material, diffusion lengths as short as 1 $\mu$m can be readily obtained. If we produced a planar structure such as that shown in Figures 6.1 and 6.2, using a typical 200 $\mu$m-thick wafer of short lifetime material, it is clear that we would indeed have a very long-based diode on the substrate side. Although we can obtain the long-based diode solution by limiting arguments on equations 6.6 through 6.10, it is easier to simply write the solution to (5.27′) in the exponential form and use the length to dismiss the increasing exponential. The result is to replace each of the hyperbolic functions with an exponential with the same argument; for example, the term $\sinh(x/L_e)/\sinh(a/L_e)$ goes to $\exp(x/L_e)/\exp(a/L_e) = \exp[-(a-x)/L_e]$. The exchange of the exponentials for the hyperbolic functions produces no great surprises; it merely decouples the ohmic contacts from the junction, and the length adds some series resistance to the diode.

To get some feeling for the behavior of real diodes operating according to our first-order analysis, some reasonably routine exercises are called for. Then we must consider how one might actually make a diode, including how to obtain an ohmic contact.

### Problem 6.1.

Let us consider some germanium diodes, since at low and moderate voltages they behave very much as the first-order analysis predicts. For all cases let $D_e = 90$ cm$^2$/sec, $D_h = 40$ cm$^2$/sec, and the excess-carrier lifetime $10^{-5}$ sec. Let the wafer be 110 $\mu$m thick with the bottom 100 $\mu$m p-type and the top 10 $\mu$m n-type. The area of the diode is 1 mm$^2$. Let three diodes be constructed, one with both sides uniformly doped at $5 \times 10^{15}$ impurities/cm$^3$; one with the p side at $5 \times 10^{15}$/cm$^3$ but the n side at $5 \times 10^{17}$/cm$^3$; and, finally, one with both sides at $5 \times 10^{17}$/cm$^3$. For each case, find the saturation current, the ratio of the hole to electron current at the junction, and the forward bias necessary, at $kT \simeq 1/40$ eV, to achieve a current of 10 mA.

*Answers:*

|          | $I_{sat}$      | $I_h/I_e$ | $V$ at 10 mA |
|----------|----------------|-----------|--------------|
| Case I   | 9.88 $\mu$A    | 4.25      | 0.173 V      |
| Case II  | 1.98 $\mu$A    | 0.0425    | 0.215 V      |
| Case III | 0.0988 $\mu$A  | 4.25      | 0.288 V      |

The results are noteworthy in several ways:
(a) Light doping leads to a higher saturation current and thus a higher current for a given bias.
(b) A large asymmetry in the doping can result in the current being carried principally by one kind of carrier. Similarly, making one length considerably longer than the other tends to make one carrier dominant. Both of these effects are of critical importance in transistor design.
(c) An enormous swing in the saturation current requires only a small change in bias to achieve the same current ($\sim 58$ mV per factor of 10 at room temperature).

## Problem 6.2.

For cases II and III in problem 6.1, find the voltage drop across the two bulk regions and find how this voltage depends on the current.

*Solution.* Writing the expression for the majority current on the $p$ side, and taking advantage of $p' = n'$ and $p_p \cong p_{p0}$, we have

$$J_h = J_T - J_e = q\mu_h p_{p0} E + qD_h dn'/dx$$

Eliminating the derivative using $qdn'/dx = J_e/D_e$ and solving for $E$:

$$E = \frac{J_T - \left(1 - \dfrac{D_h}{D_e}\right)J_e}{q\mu_h p_{p0}} = \frac{J_T - \left(1 - \dfrac{D_h}{D_e}\right)qD_e\dfrac{dn'}{dx}}{\sigma_{p0}} \tag{6.14}$$

The second form of (6.14) points out explicitly that the denominator is the equilibrium conductivity of the $p$ side. Thus, the first term in (6.14) $(J_T/\sigma_{p0})$ represents the normal ohmic drop in a resistive material. The second term is the ambipolar field required to keep $p' = n'$. To obtain the voltage drop across the $p$ region, we must integrate (6.14):

$$V_{cb} = -\int_b^c E dx = \frac{J_T(c - b)}{\sigma_{p0}} - \frac{\left(1 - \dfrac{D_h}{D_e}\right)qD_e n'_p(b)}{\sigma_{p0}} \tag{6.15}$$

A completely similar argument gives the drop on the $n$ side:

$$V_{ao} = -\int_0^a E dx = \frac{J_T(a)}{\sigma_{n0}} - \frac{\left(1 - \dfrac{D_e}{D_h}\right)qD_h p'_n(a)}{\sigma_{n0}} \tag{6.15'}$$

The dependence of the voltage on the current is quite clear: the voltage is proportional to the current—in other words, it is ohmic.

The ambipolar fields are in the proper direction to make the excess majority carriers stay with the excess minority carriers. Since the electron mobility is greater than the hole mobility, this means that the electric field must point in the direction the minority carriers are diffusing (i.e., "downhill" on a concentration-versus-position plot). This is in opposite directions on either side of the junction, so the ambipolar voltages tend to cancel each other while the resistive terms add. Note that the resistive terms are proportional to the diodes' thickness, while the ambipolar field is independent of the sample size. Accordingly, the ambipolar terms will be most important for thin-based regions. Neither term will be very important if the resistivity of the material is very low. To get some idea of what is big and small, you must compare the ohmic drops with the junction voltage. Evaluating (6.15) and (6.15′) for cases II and III gives

$$V_{a0} = \frac{10^{-3} + 0.05 \times 10^{-3}}{288} = 3.29 \, \mu\text{V}$$

$$V_{cb} = \frac{10^{-2} - 0.398 \times 10^{-2}}{1.28} = 4.7 \, \text{mV}$$

case II

$$V_{a0} = \frac{10^{-3} + 10^{-3}}{288} = 6.94 \, \mu\text{V}$$

$$V_{cb} = \frac{10^{-2} + 10^{-3}}{128} = 6.97 \, \mu\text{V}$$

case III

It is only for the lightly doped side that there is a voltage that is even perceptible when compared with the junction voltage drop. This is a very important conclusion: Under low-injection conditions, the full voltage drop across the diode is across the depletion layer.

Speaking of low-injection conditions, the analysis above required that $p_n(a)$ and $n_p(b)$ be calculated, so with no further labor we can see directly whether the low-injection condition has been met. The three cases are listed below, and, as you can see, all three are acceptable.

|          | $p_n(a)$              | $n_{n0}$           | $n_p(b)$               | $p_{p0}$           |
|----------|-----------------------|--------------------|------------------------|--------------------|
| Case I   | $1.25 \times 10^{14}$ | $5 \times 10^{15}$ | $1.25 \times 10^{14}$  | $5 \times 10^{15}$ |
| Case II  | $6.25 \times 10^{12}$ | $5 \times 10^{17}$ | $4.99 \times 10^{14}$  | $5 \times 10^{15}$ |
| Case III | $1.25 \times 10^{14}$ | $5 \times 10^{17}$ | $1.25 \times 10^{14}$  | $5 \times 10^{17}$ |

## PRODUCING P-N JUNCTIONS

Although one can produce a satisfactory *p-n* junction by very carefully cold-welding a *p* slice to an *n* slice—that is, forming a bond between the two solids without ever melting either—it is certainly not the normal way to proceed. There are two basic approaches to generating such junctions. In one class of junctions, a layer of one type is grown on a layer of the other type; in the other method, an impurity of one type is diffused through the surface into a solid layer of the other type. There are many variations on how one can accomplish each of these steps, but they are very easily classified as a grown junction or a diffused junction. Naturally, it is possible to confuse the issue by performing a diffusion after growth, but the steps are usually quite distinct. Grown junctions are usually very abrupt—that is, only a few atomic layers form the transition from the $p$ side to the $n$ side. This is the type of junction we have considered so far. By comparison, the diffused junction, as its name implies, goes more gradually from $p$ to $n$, with a narrow $i$ (intrinsic) region in the middle.

Both the oldest and newest methods for producing large-area junctions (i.e., not the so-called "point contacts," which are a bit mysterious at best) are grown-junction techniques that produces what could properly be called *epitaxial* junctions. Epitaxial growth refers to growth of single-crystal layers on a single-crystal substrate from material precipitated out of a gaseous or liquid carrier. The precipitating atoms align themselves with the lattice underneath by diffusing about on the surface until they find themselves a proper potential well to fall into. Since the single crystal has a much lower free energy than a polycrystalline mass, it is the more stable form. Accordingly, if the precipitation rate is not too fast and the temperature of the crystal substrate is high enough, excellent crystals can be grown by epitaxy.

The oldest of the epitaxial techniques produces what is called an *alloyed* junction. The process begins, like all of the other techniques, with a thin slice—a *wafer*—of the base crystal, say arsenic-doped germanium. On top of the wafer, as shown in Figure 6.6, is placed a lump of indium or gallium, $p$-type dopants that happen to have extremely low melting points. Let us say we are using Ga, which melts at 30°C (but boils a fantastic distance away at 2403°C!). If we heat the slice up to a few hundred degrees centigrade, we will form a solution of Ge in Ga with a concentration of several percent Ge. As the figure shows, in dissolving the Ge, the Ga etches its way down into the wafer, removing the contaminated or damaged upper layers and leaving a very clean surface sealed under the liquid. The temperature of the wafer is then lowered. This reduces the solubility of the Ge, so the solution becomes supersaturated. The excess Ge precipitates onto the crystal below, growing a new single crystal layer that must be, considering the circumstances, absolutely saturated with Ga. At these temperatures, the solid solubility of Ga in Ge is of the order of $10^{18}/cm^3$, so the regrown Ge is very strongly $p$-type. (Note that this "saturated" solid solution contains less than 1 atom of Ga per $10^4$ atoms of Ge.) On top of the regrown Ge is a mush of Ga and some polycrystalline Ge. The mush and most of the polycrystalline material is removed by etching, leaving an excellent surface for making an ohmic contact. This is a technique that is still much in use where one wants to obtain a very heavily doped, very abrupt junction.

It is frequently desirable to grow lightly doped layers, and it is almost always desirable to have complete control over the thickness and doping

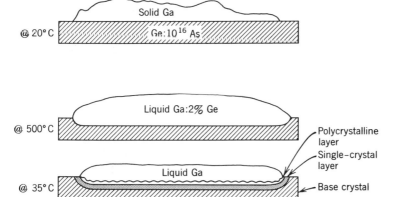

**Figure 6.6** The alloy technique for producing a *p-n* junction. The liquid gallium dissolves some of the base crystal, forming a saturated solution of gallium and germanium. Upon cooling, the Ge precipitates out of solution, growing first into a gallium-saturated single-crystal layer, and then into a polycrystalline layer. The Ga and polycrystalline Ge are removed, leaving an abrupt *p-n* junction with the regrown material heavily doped.

profile of the grown layer. With liquid epitaxy it is difficult to exercise these options, so most people have turned to gaseous epitaxy, where everything and anything can be turned on and off with a valve. The basic process is to decompose a silicon (or Ge if that's what you are growing) compound by heat and/or chemical reaction to liberate silicon atoms at the surface of the growing crystal. The most common compounds for growing silicon are $SiCl_4$ and $SiH_4$ (silicon tetrachloride and silane). Both are highly obnoxious poisons, the first a liquid at room temperature, the second a gas. A typical epitaxial system based on silane is shown in Figure 6.7. The silicon wafers are put on a carbon boat and placed in a quartz reaction vessel. The system is purged of air and filled with $H_2$ plus about $1/10\%$ $SiH_4$. The radio frequency generator is turned on, heating up the carbon susceptor and the silicon wafers sitting on top of it. The reaction going on at the surface of the silicon is fairly simple:

$$SiH_4 \rightleftharpoons Si + 2H_2 \qquad (6.16)$$

By varying the amount of silane and/or the temperature, the direction of the reaction may be driven either way. The first step, then, is to etch off the top surface to provide a clean surface for growing. Then the temperature is raised to drive the reaction into the growth regime ($\sim 1200°C$), and the flow rate is adjusted to obtain a growth rate of about 1 $\mu$m/minute. Doping of the growing layer is achieved simply by adding an exceedingly small

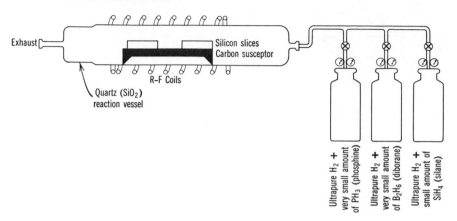

**Figure 6.7** An all gas epitaxial crystal-growth system. By adjusting the flow rate on the silane and the temperature of the silicon wafers, silicon may be added to or removed from wafers according to the reaction given in (6.16). Donors may be added to the growing crystal by adding a minute amount of phosphine; acceptors by a minute amount of diborane. Typical growth rates for high-quality results are about 1 $\mu m/min$.

amount of diborane ($B_2H_6$) for acceptors or phosphine ($PH_3$) for donors. By a reaction similar to (6.16), the phosphorus or boron is added to the growing crystal. The system is about as simple as it sounds, although there is a fair amount of knob twirling to get it working. Small differences in gas purity, temperature, and flow rates can be critically important and the cleanliness requirements are severe.

The material grown by a well-run epitaxial system is quite good. By simply adjusting the impurity gas flow rate, one can grow lightly doped layers on heavily doped substrates (one of epitaxy's unique and industrially important advantages) or even taper the doping to suit any particular design need. Although the growth is done at a relatively elevated temperature, the growing time is usually short enough so that very little diffusion takes place into or out of the substrate. Thus, the gaseous epitaxial process usually produces abrupt junctions, just as the liquid epitaxial system does.

The other method for producing $p$-$n$ junctions—solid-state diffusion—is the heart of the planar technique. Its principal advantage is that we may mask the surface so that the diffusion proceeds only in certain very well localized areas. In this way we can produce simultaneously several thousand transistors on one wafer with the same labor required for producing one. It is also possible to interconnect them if that is desired. The diffusion technique depends on the fact that atoms can diffuse—albeit very slowly—through even so rigid a structure as the lattice of a single crystal. As in all diffusions, the atoms move about through thermal reshuffling. However, if all of the atoms, both in the silicon and in the impurity, are in proper lattice sites, there would be no place for anything to move. Such a situation is the equivalent of a completely full band; no conduction is possible. If there are some vacancies in the lattice—missing atoms, the analog of the hole—there is the possibility of diffusive and conductive transport. Since it requires a relatively large amount of energy to create a vacancy in high-melting-point semiconductors, vacancies must be rather rare. Accordingly, we should expect the transport of atoms within the lattice to depend most strongly on the number of vacancies present. If the energy to create a vacancy, called the *activation energy*, is $E_a$, how many vacancies should there be at temperature $T$? Answer: Since we know that the number is small compared to the number of atoms, $N_0$, the Boltzmann factor must apply and the number of vacancies is given by

$$N_{VAC} = N_0 e^{-E_a/kT} \qquad (6.17)$$

If we were investigating how mobile the silicon atoms themselves were, we should expect that the transport coefficients would resemble (6.17). However, since we are observing the diffusion of impurity atoms, which are not so tightly bound as the host atoms, we must expect that the impurity and the vacancy will interact to lower the activation energy to $E_a'$. Thus, the diffusion coefficient should resemble

$$D = D_0 e^{-E_a'/kT} \qquad (6.18)$$

where the largest $E_a'$ is $E_a$ and that is for the host atoms themselves. Equation 6.18 does fit much of the measured data. When the process is studied in detail, complications arise because not all impurities substitute for host atoms; some fit in between and are called *interstitial* impurities. Interstitial diffusion is not vacancy-limited, so it should and does proceed much more rapidly than substitutional diffusion. (For example, the tiny He atom can diffuse surprisingly rapidly through most solids even at room temperature and below. If one keeps liquid helium in a thin-walled glass dewar [vacuum flask], after a number of months of use, one finds that the vacuum has decreased substantially. The culprit is He diffusing right through the glass walls.) It is also observed that different impurities interact with each other through the same ambipolar forces we have considered for holes and electrons. However, for almost all of the impurities deliberately diffused into Ge and Si, (6.18) holds very well. For Si, $E_a'$ is of the order of 4 eV and $D_0$ about 200 cm$^2$/sec. A specific example is As in Si: $E_a' \simeq 4.38$ eV and $D_0 \simeq 280$ cm$^2$/sec. Thus, at 1300°C (1573°K), the diffusion coefficient for As in Si is $2.8 \times 10^{-12}$ cm$^2$/sec (or in the units that one would think in for such work: 1 $\mu$m$^2$/hr).

(A good source for many of the analytical tools of the whole-planar process, including practical details and measured data, is A. S. Grove, *Physics and Technology of Semiconductor Devices*, Wiley, 1967.)

Once you have agreed that the diffusion could go on, we are back to an equation like (5.27) [N.B., not (5.27'), which is a steady-state equation] to describe the motion of the impurities through the lattice. Equation 5.27 includes terms for the rate of change in concentration resulting from recombination and from diffusion. Unless chemical reactions can go on, the impurities cannot "recombine." Such reactions are not usually very important, so we shall ignore them. Thus, (5.27) becomes

$$\partial N_D / \partial t = D \nabla N_D \Rightarrow D \partial^2 N_D / \partial x^2 \qquad \text{for one-dimensional problems} \qquad (6.19)$$

Two reasonably practical boundary conditions that lead to well-tabulated functions are diffusion into a semi-infinite plane slab with the surface concentration held at a constant value, $C_S$, and diffusion into a semi-infinite plane slab from an impulse ($\delta$-function) of impurities located in the surface. The "semi-infinite slab" bit comes from the obvious slowness of most solid-state diffusion. It would take so long to diffuse impurities from one side of a typical wafer to the other side that, for all practical purposes, the wafer is infinitely thick.

To see where the two boundary conditions come from, consider the usual two-step diffusion process: The wafer is placed in an open-ended quartz tube in the middle of a furnace whose temperature may be of the order of 1000°C. Let us say that the wafer is As-doped Si ($N_D \simeq 10^{16}/\text{cm}^3$) and that we wish to produce a $p$ layer doped with boron. The first step is to expose the hot silicon wafer to reasonably dilute diborane vapor (say 0.1 % diborane, 99.9 % inert gas). The solid solubility of boron in Si at this temperature is about $10^{21}/\text{cm}^3$. A surface concentration of $10^{21}/\text{cm}^3$ is very quickly achieved at this partial pressure of diborane, and once the limit of solubility is reached, no more may dissolve. Thus, we have an example of the first kind of boundary condition. The solution of (6.19) for constant surface concentration, $C_s$, is given by

$$N_A(x, t) = C_s \operatorname{erfc} \left[ \frac{x}{2\sqrt{Dt}} \right] \qquad (6.20)$$

Erfc $(x)$ is called the *complementary error function* and is defined by

$$\operatorname{erfc}(x) \equiv 1 - \operatorname{erf}(x) \equiv 1 - \frac{2}{\sqrt{\pi}} \int_0^x e^{-y^2} dy \qquad (6.21)$$

This is a well-tabulated function. Graphs appropriate to (6.20) are shown in Figure 6.8. Some useful relationships are given in Table 6.1.

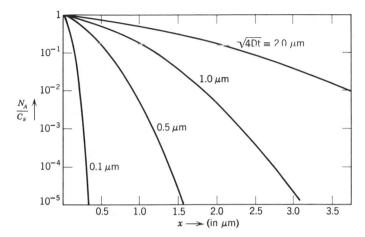

**Figure 6.8** Distribution of impurities after a single-step diffusion in which the surface concentration was held constant at $C_s$. The plot shows equation 6.20 for four different values of $\sqrt{4Dt}$ where $D$ is the diffusion constant and $t$ the time during which the diffusion was carried out. If the background of donor impurities was assumed to be at $10^{16}/\text{cm}^3$ and $C_s = 10^{21}/\text{cm}^3$, the junction would occur where the acceptor concentration crossed the $10^{-5}$ line.

**Table 6.1.** Error Function Approximations, Integrals, and Derivatives.

$$\text{erf}(x) \equiv \frac{2}{\sqrt{\pi}} \int_0^x e^{-y^2} dy$$

$$\text{erfc}(x) \equiv 1 - \text{erf}(x)$$

$$\text{erf}(x) \Rightarrow 2x/\sqrt{\pi} \qquad \text{for} \qquad x \ll 1$$

$$\text{erfc}(x) \Rightarrow \frac{1}{\sqrt{\pi}} \frac{e^{-x^2}}{x} \qquad \text{for} \qquad x \gg 1$$

$$\frac{d\,\text{erf}(x)}{dx} = -\frac{d\,\text{erfc}(x)}{dx} = \frac{2}{\sqrt{\pi}} e^{-x^2}$$

$$\int_0^x \text{erfc}(y)\,dy = x\,\text{erfc}(x) + \frac{1}{\sqrt{\pi}}(1 - e^{-x^2})$$

$$\int_0^\infty \text{erfc}(y)\,dy = 1/\sqrt{\pi}$$

Figure 6.8 is arranged to be a "universal" plot of (6.20). To see what has happened in our specific example, we must insert the appropriate value of $D$, $C_s$, and the length of time, $t$, that the diffusion was run. At 1000°C, the diffusion coefficient of boron is about $5 \times 10^{-3}$ $\mu m^2/hr$ and $C_s \simeq 10^{21}/$ cm$^3$. Thus, at this low a temperature, the shortest diffusion shown corresponds to half an hour and the junction (where $N_D = N_A$) would be at a depth of only 0.63 $\mu m$. Note that increasing the time to 2.5 hours would drive the junction in only another micrometer.

If getting a *p-n* junction were our sole aim, we would have completed the job with this first step. However, one frequently does not want to have such a high concentration of impurities in so large a percentage of the diffused side of the junction as (6.20) and Figure 6.8 require. The obvious answer is to let the boron continue diffusing while not adding any more to the surface. In other words, shut off the diborane and continue the diffusion. To speed up the *drive-in* diffusion, the temperature is raised to about 1200°C. To protect and seal the surface, the drive in is done in an oxidizing atmosphere (e.g., moist air). The oxygen very quickly seals the surface by forming $SiO_2$ (quartz). Most materials diffuse exceedingly slowly through quartz (there are important and even critical exceptions), so after a few minutes in the oxygen the total amount of impurity in the wafer is approximately constant; none can come in or leave.

This drive-in diffusion requires the solution of (6.19) with (6.20) as an initial condition. The closed-form solution to the exact problem does not exist, but if we are going to drive the junction in a few micrometers, the initial *deposition* diffusion looks very much like a $\delta$-function by comparison. For the $\delta$-function with a total number of impurities per cm$^2$, $Q$, the solution to (6.19) is

$$N_A(x, t) = \frac{Q}{\sqrt{\pi D t}} e^{-x^2/4Dt} \tag{6.21}$$

Equation 6.21 is called a *Gaussian distribution*. To find $Q$, (6.20) must be integrated, which is easily accomplished when we note from Table 6.1 that $\int_0^\infty \text{erfc}\,(x)\,dx = 1/\sqrt{\pi}$. Keeping track of the fact that there have been two diffusions, so there are two times and two diffusion coefficients, another form of (6.21) is

$$N_A(x, t_{\text{drive}}) = \frac{2}{\pi} C_s \frac{\sqrt{D_{\text{dep}}\,t_{\text{dep}}}}{\sqrt{D_{\text{drive}}\,t_{\text{drive}}}} e^{-x^2/4\,D_{\text{drive}}\,t_{\text{drive}}} \tag{6.21'}$$

Curves for (6.21) are shown in Figure 6.9. Be warned that Figure 6.9 shows an approximate solution for a distinctly *simplified* model. Two things may

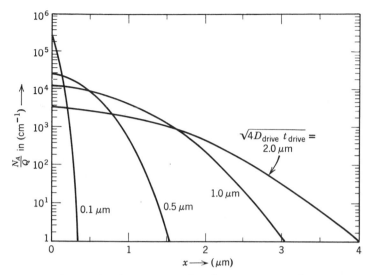

**Figure 6.9** Distribution of impurities after a two-step diffusion. The first step deposits a total impurity content of $Q$ impurities per cm$^2$ in a very thin surface layer. The surface is then sealed so that the total impurity content remains at $Q$. The ordinate gives the ratio of the impurity concentration to $Q$ as a function of position for four drive-in times.

seriously affect the accuracy of the model. If the doping level is much higher than $n_i$ (which is pretty large at 1200°C), the steep gradient of the impurity distribution will result in a relatively large built-in electric field. Since the impurities are mobile, the field will cause them to move in such a way that the distribution will become more uniform. Naturally, this yields a different impurity profile than that predicted by (6.21). Another source of considerable error is all of the activity going on at the oxidizing surface. The oxygen is chewing its way down through the very heavily doped upper layers of the crystal. (A typical oxide coating may be 0.2 to 0.5 $\mu$m thick.) The solubility of the impurity may be higher or lower in quartz than in silicon, so some re-distribution in the surface layer is to be expected. This redistribution will undoubtedly affect the final results. In real production work, diffusion depths must frequently be held to tolerances of the order of 0.1 $\mu$m or less. Our once-over-lightly is inadequate for such applications. What our simplified model gives is a good idea of the way things almost work.

**Problem 6.3.**

Take a moment to see how long the diffusion should run to drive the junction in to about 4 $\mu$m after the deposition diffusion described above. At 1200°C, the diffusion coefficient for B is about 0.42 $\mu m^2$/hr. First, what is $Q$ from the deposition? [*Answer:* From the integral of (6.20), we have

$$Q = \sqrt{4D_{dep}t_{dep}} \, C_s/\sqrt{\pi} \tag{6.22}$$

The deposition diffusion was run until $\sqrt{4Dt} = 0.1$ $\mu m$ and $C_s$ for boron is $10^{21}$/cm$^3$. Thus, $Q = (1/\sqrt{\pi}) \times 10^{16}$/cm$^2$. (Note that some units had to be adjusted!)]

What will the concentration of acceptors be at the junction? (*Answer:* At the junction $N_A = N_D = 10^{16}$/cm$^3$. In terms of the ordinate of Figure 6.10, $(N_A/Q) = \sqrt{\pi}$. It is evident on Figure 6.9 that the junction will be at a depth of about 4 $\mu m$ if $\sqrt{4D_{drive}t_{drive}} \simeq 2.0$ $\mu m$. For the diffusion coefficient given, that requires a drive in time of 2.38 hours.]

What is the surface concentration at the end of the drive-in diffusion? [*Answer:* From Figure 6.9, $N_A/Q$ at the surface is about $3.2 \times 10^3$. With the value of $Q$ found above, this gives a surface concentration of about $10^{19}$/ cm$^3$, about two orders of magnitude below $C_s$ during the deposition diffusion.]

We shall return to this technique when we discuss transistor structures, at which time the use of the oxide layer will become quite clear. For the moment, we have three plausible ways of generating *p-n* junctions. We must now make some ohmic contacts to connect these junctions to the outside world.

## THE METAL-SEMICONDUCTOR JUNCTION

If you allow two dissimilar materials to come to thermal equilibrium with each other, you may confidently state that their bands have bent sufficiently to bring their Fermi levels into alignment. Whether you could observe this distortion of the bands, however, depends entirely on what materials you happen to have chosen. In insulators, the Fermi level is almost indeterminate. (Remember the illustration about diamond with $E_g = 5.47$ eV and $n_i$ of the order of $10^{-27}$/cm$^3$? You could swing the carrier concentration $p$ or $n$ type by 40 orders of magnitude and not notice the change! In fact, a single donor or acceptor in a cm$^3$ of diamond would pull the Fermi level up or down by 1.55 eV.) Unless the position of the Fermi level is of some consequence, band bending has little meaning.

For metals, on the other hand, the Fermi level runs right through an allowed band, and the shift of a moderate fraction of an eV corresponds to a charge accumulation of the order of $10^{20}$ electrons/cm$^3$. As with heavily doped semiconductors, this implies a very short transition region. Furthermore, since the bands within a metal are usually rather wide with the Fermi

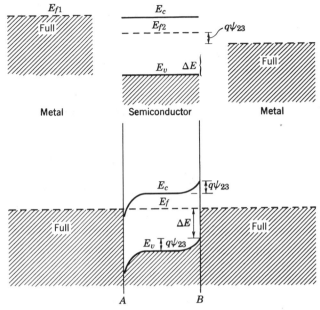

**Figure 6.10** Mythical band diagram for two metal-semiconductor junctions. The myth involves the assumption that the band diagram of the semiconductor itself would be flat all the way to the edges. The diagram shows an *n*-type semoconductor. The metal on the left has its Fermi level higher than the semiconductor's, the one on the right, lower. The junction on the left (at *A*) is thus an enhancement region in the semiconductor; the one on the right (at *B*) is a depletion region. The junction at *B* would rectify, while the one at *A* would not.

level more or less in the middle, a substantial amount of band bending can transpire without ever having a region where the Fermi level—the energy where the states are both half full and half empty—is not within an allowed band. Accordingly, metal-to-metal junctions are always ohmic. The carriers are free to flow in either direction. (The built-in field can be observed, however, if the junction is heated or cooled. The built-in field is a function of temperature, so that if a junction between dissimilar metals is heated, the built-in potential at the hot end will be different from that at the cold junction. The difference can be readily measured by inserting a sensitive voltmeter in the loop. This is the basis of the thermocouple.)

When one makes a junction between a metal and a semiconductor, some new options and a few surprises appear. The surprises are associated with the realities of the semiconductor surface, so let us save them for a moment. The options are illustrated in Figure 6.10, in which an *n*-type semiconductor is joined to two different metals. One metal has its bulk Fermi level above that of the semiconductor; the other, below it. Since we ignore the surface, it is easy to describe what happens when the three are joined. Some of the excess electrons from the metal on the left diffuse into the semiconductor, filling states in the conduction band and thus bending the band edge down toward the Fermi level. On the other side, electrons from the semiconductor spill into the metal, emptying some of the conduction-band states and pushing the bands upward.

Note that there is a barrier to electron flow at *B* but none at *A*. Thus, in terms of majority (electron) carrier flow, *B* would rectify but *A* would not. The holes, which are the bulk minority carriers, would also behave as if they were at a rectifying junction at *B*. If a hole came from the left, it would *"fall"* into the metal and float upward through the continuum of states within the metal until it vanished in the sea of empty states above the Fermi level. Thus, minority carriers would be collected from the semiconductor. Under forward bias, the barriers for hole injection into the semiconductor and electron injection into the metal would be lowered, and injection would take place. Note from the figure that the hole and electron barriers have different heights. This is almost always the case when dissimilar materials are joined. However, both holes and electrons would obey an equation somewhat similar to the injection in a *p-n* junction; in both cases we are dealing with a Boltzmann factor giving the number of carriers with sufficient energy to cross over a voltage-variable barrier. Thus, for a metal-semiconductor rectifying contact, we should expect an equation of the form $J = J_0(e^{qV/kT} - 1)$. Such asymmetry in the conduction across a metal-semiconductor junction is the physical phenomenon behind the oldest known solid-state rectifiers: copper-copper oxide and selenium. Both of these rectifiers had been discovered and were in active use many years before their behavior was even crudely understood. Such metal-semiconductor junctions are called *Schottky barrier diodes*.

Let us now take a look at point *A* in Figure 6.10. Here we have an intimation of an ohmic contact. Even though the conduction characteristics may not be entirely symmetrical, it should be fairly obvious that not very much stands in the way of a great many electrons going in either direction. Majority-carrier conduction across the boundary seems assured. But what about minority-carrier conduction? After all, we have used an "ohmic contact" boundary condition, $p' = n' = 0$. Does that fit the present case? The answer is no. There is a barrier to minority-carrier flow at *A* that almost guarantees accumulation of inflowing excess holes. What we would like to find at *A*— at least as far as minority carriers are concerned—is no barrier and a very short lifetime for excess carriers. If this could be obtained without spoiling the majority flow, the ohmic contact might be possible.

As it turns out, if you simply put a metal contact on a semiconductor surface, you almost always get something that behaves much more like a rectifier than an ohmic contact. This phenomenon is pretty independent of the choice of the metal. The secret, as suggested by J. Bardeen in 1947, lies in the special properties of the surface. When these are taken into account, it becomes more evident how to generate a metal-semiconductor transition that behaves in a very ohmic fashion.

The rather complicated discontinuity that constitutes a surface produces a large number of localized states with energies in the forbidden band. These may be either donor or acceptor states, and they arise from a number of different causes. First there are the *Tamm* states, which come about because the surface terminates the perfectly periodic potential, leaving all of the surface atoms with an unsatisfied bond. Then there are states that result from the almost inevitable accumulation of all sorts of impurities at the surface. The surface impurity layer represents a termination that is somewhat less abrupt than a truly clean surface, so a "dirty" surface generally has fewer surface states than a truly atomically clean one. All device surfaces are "dirty" in this sense. Whatever their source, the surface states seem to lie mostly near the middle of the gap (deep states), and they are pretty dense near the surface. These extra states require the bands to bend, just as more shallow donors or acceptors do. In most cases, these surface states are so dense that the Fermi level must pass very close to the center of the distribution. Figure 6.11 illustrates the expected state distribution and the band diagram. In the upper drawing, you can see how the surface-state distribution has created barriers to electron (majority carrier) flow at both sides with no metal contact at all. Because the density of surface states is so large, the Fermi level goes right through the center of the distribution. When the metals are brought in contact with the surface, it is the surface states that empty or fill. Since there are so many of them, it takes a very small band shift to align the Fermi levels. (Remember that "aligning the Fermi levels" is equivalent to having the charge separation, caused by the diffusion that empties or fills the states, create enough of a field to halt further net flow. If the density of states is high, getting enough empty or full states requires a small shift of the bands with respect to the Fermi level.) Thus, the addition of the metal contacts has very little effect on the band diagram near the surface. We ought to get rectifying contacts *all the time*!

Well, not quite. We could beat the system if:

**1.** The surface of states could be entirely filled or emptied by heavily doping the surface; or

**2.** The barrier could be made so thin that the carriers could tunnel right through it.

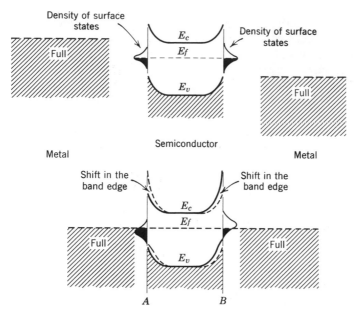

**Figure 6.11** A more realistic band diagram for two metal-semiconductor junctions. This diagram includes the surface states that were ignored in Figure 6.11. The accumulation of electrons in the surface states cause the bands to bend with no contact whatever. When the metal contact is made, it merely empties or fills a few more surface states to align the Fermi levels. Because of the high density-of-surface states, very little band bending is required to equilibrate the system. The difference in the surface-state filling, which depends on the metal chosen, can be seen in the lower figure. Note, however, that although differences in the bands are observable on comparison of the two sides, both $A$ and $B$ now represent rectifying contacts.

Both approaches are used to get satisfactory ohmic contacts.

The usual ohmic contact is achieved in two steps. First, a diffusion is performed guaranteeing that the surface is very heavily doped. Then an aluminum metalization layer is evaporated onto the surface and "baked" in (i.e., a microdiffusion is performed). Metal-to-metal contacts are then made to the aluminum to connect to the outside world. Gold may also be used on a prediffused surface. For pressure contacts, In or Ga is sometimes used instead of Al. Figure 6.12 illustrates the results of the aluminization scheme for both a $p$- and an $n$-type material.

The arguments for the ohmic properties of these junctions are different for majority and minority carriers and also for the $n$ and $p$ materials. For the minority carriers we are basically seeking an extremely fast recombination rate at the surface. If $\tau \to 0$, for all practical purposes the excess carriers

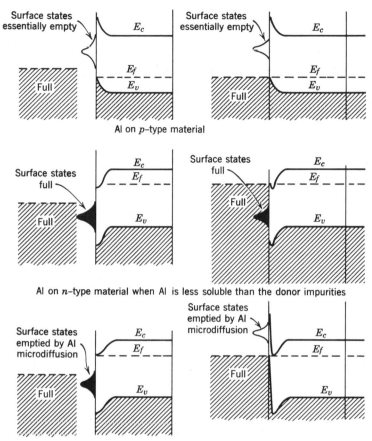

**Figure 6.12** The band diagram for three practical ohmic contacts.

would disappear as fast as they came in. Our study of recombination through traps would argue that one could lower the lifetime only a certain amount by doping (Figure 5.2). However, we did not consider the case of exceedingly heavy doping. The high percentage of Al, the other impurities, and the very large number of surface states lead to the introduction of many more traps and also to cooperative recombination through pairs of nearby traps. Furthermore, the heavy doping introduces so much lattice strain that the crystalline regularity that prohibits indirect transitions is effectively destroyed. $k$ would no longer be a "good" quantum number. Thus, as we get to the very heavily doped surface, excess carrier lifetimes become very, very short indeed. Since a very heavily doped surface will make an ohmic contact for minority carriers, both steps (1) and (2) will help make the contact ohmic.

For majority carriers, we must have an undrainable source and an un-fillable sink. A metal represents just what we are looking for, but the contact must be made in such a way that the carriers may travel freely from the metal to the semiconductor and back. We want no barrier formed. This is the same thing as saying that the edge of the majority-carrier band should be aligned with the Fermi level in the metal. At the Fermi level in a metal is an infinite (for all practical purposes) source of both full and empty states.

For the aluminum on p-type material, as Figure 6.12 shows, the aluminum simply adds to number of acceptors already there, assuring the degeneracy of the doping and the band-edge alignment that we sought. For the n-type material, the aluminum is a counter dope. If the Al is less soluble than the concentration of n-type dopant (not unlikely), the degenerately doped n-material is still degenerate (only slightly less so). If the Al is more soluble, one obtains a microscopically thin, degenerate to generate p-n junction. That makes an ideal, exceedingly thin, tunneling junction, so the carriers can whiz right through the barrier. Since this tunneling junction is generally back-to-back with a normal p-n junction, you can never get the tunneling junction biased far enough forward to observe any of the peculiar tunnel-diode be-havior. In the reverse direction (forward for the nondegenerate junction), the current flows freely by tunneling, so it looks just like an ideal ohmic con-tact. Note that, in the degenerately doped material, we may argue in the same fashion as with the metal and the surface states: the high density of states means that a very small shift in the bands is quite sufficient to align the Fermi levels.

Now that some practical manufacturing techniques have been introduced, we have a chance to compare our simple, first-order model with more realistic descriptions. The first-order model will not do too badly; the AC and DC characteristics of the model are not very drastically modified by practical considerations. If you wonder at the amount of effort we are spending on the diode, remember that it is the basic building block of all diffusion-con-trolled and some drift-controlled devices. Once the diode's operation is well established, the operation of the more complex structures is relatively easy to understand.

## SMALL-SIGNAL ANALYSIS OF THE JUNCTION DIODE

There are three different points of view in describing the nonlinear behavior of a diode. We may want to know the DC characteristics, such as those dis-played in Figure 6.5. Such plots give the value of $I$ for a fixed value of $V$. Alternately, we might want the *switching* characteristics of the diode. For this we need a dynamic display of $V$ versus $t$ after the application of a step in current. The analysis of switching requires the solution of a non-steady-

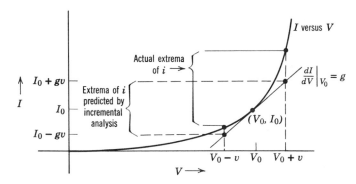

**Figure 6.13** Incremental analysis to find $i$ given $v$. The tangent to the curve is used as an approximation to the curve. The approximation is extended far beyond its reasonable range to illustrate the difference between the exact and incremental analyses. This difference between $i = gv$ and the true $i$ is called the distortion. It must be kept reasonably small for the incremental analysis to be useful.

state equation such as (5.27). Finally, an incremental or differential analysis can be performed to permit the representation of the diode by linear circuit elements. The introduction of linear elements makes circuit analysis immeasurably easier, but you must remember that the method is limited to incremental shifts about a fixed bias point. A junction diode is a very nonlinear element, and no amount of wishing will make it into a resistor and a couple of capacitors.

The incremental analysis proceeds from the assumption that all of the dependent electrical variables (e.g., current and charge) are smoothly varying continuous functions of the independent variable (e.g., voltage). Thus, if we select some DC bias point, $(V_0, I_0)$, and add to it a small AC signal, say $ve^{j\omega t}$, the current may be represented by $I_0$ plus a small AC component:

$$I(t) = I_0 + ie^{j\omega t} = I_0 + yve^{j\omega t} \tag{6.23}$$

where $y \equiv (dI/dV)|_{V_0}$ is called the *small-signal, differential,* or *incremental admittance.* Equation 6.23 is strictly true only in the limit as $v \to 0$, but for many applications, it is a very useful approximation to replace the actual V-I curve by its tangent at the bias point. This is illustrated in Figure 6.13. Note carefully that $y$ is defined for sinusoidal excitation, so charge storage and flux storage must be accounted for. In the best of all possible worlds, the physical model suggests an electrical model comprising only lumped, frequency-independent circuit elements—$R$'s, $L$s and $C$'s—from which $y$ may be obtained, but you should be prepared for a less satisfactory denouement. You can't win them all.

Before pursuing the arithmetic of the incremental model, its not-very-obvious utility should be discussed. Just how often does one get the chance to do something useful with a lumped-circuit approximation to a junction diode? If we stick to individual junction diodes, the answer is not very often. It is only in the microwave field that junction diodes are much employed as voltage-variable, quasi-linear circuit elements. However, when the diodes become part of more elaborate structures—in particular, junction or field-effect transistors—that are routinely employed for linear amplification, the incremental model enjoys extremely wide employment. Accordingly, we will take a brief dip into the incremental model here, waiting until the transistor appears for total immersion.

The small size of the typical junction diode precludes any large flux storage, so it is not unreasonable to dismiss all inductances as insignificant. Naturally, if you want to use a diode in UHF or microwave applications, don't begin by throwing out what may be the most important term of all, but for frequencies well up into the radio bands, typical diodes have negligible inductance. Our model must then contain only $r$'s and $c$'s. The configuration of the typical diode suggests that the simplest model would be that shown in Figure 6.14. In that circuit, $r_b$ is the incremental resistance associated with voltage drops in the bulk regions; $g_j$ is the incremental conductance associated with the voltage across the depletion region; and $c$ is the incremental capacity associated with charge storage in either the bulk or depletion regions. [Note the standard convention: All incremental voltages and currents and their ratios (e.g., $i/v = g$) are indicated by lower-case letters with lower-case subscripts.]

The $r$'s and $c$'s can be obtained very easily from our first-order model. To begin with, we found in problem 6.2 on page 302 that the voltage across the bulk regions was negligibly small under the low-injection-level requirement for the first-order analysis. Thus, in a rather trivial way, we obtain $r_b \simeq 0$.

Since charge storage will not enter at very low frequency, differentiating the DC $I$-$V$ relationship (6.10′) gives $g_j = 1/r_j$ as

$$g_j = \left.\frac{\partial I}{\partial V}\right|_{V_0} = \frac{q}{kT} I_{\text{sat}} e^{qV_0/kT} \tag{6.24}$$

In the forward direction, $I \simeq I_{\text{sat}} e^{qV/kT}$ and one obtains the very useful relationship:

$$g_j = (q/kT)I_0 \qquad I_0 > 0 \tag{6.25}$$

Equations 6.24 and 6.25 contain a most startling conclusion: As long as the low-level-injection assumption holds, all diodes carrying the same forward current have the SAME incremental conductance! Will they also have the

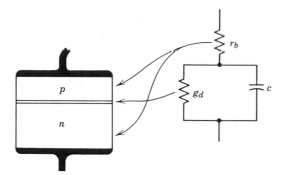

**Figure 6.14** Incremental model of the junction diode. The linear elements are functions of the bias point. $r_b$ is associated with the incremental voltage drop across the bulk regions; $r_j$ is associated with the incremental voltage drop across the depletion region; and $c$ is the capacity associated with charge storage in both the bulk and depletion regions.

same bias? [Answer: An emphatic no! According to (6.10), all of the differences from one junction diode to another are contained in $I_{sat}$, a quantity that can range over many orders of magnitude. For every order-of-magnitude difference in $I_{sat}$, the bias voltage must shift $2.3kT/q$ ($2.3 \cong \log_e 10$) or roughly 57 mV at room temperature. You have here the first example of the difference between a large- and small-signal analysis. The "large signal" is the bias voltage; to compute the bias current, we must introduce the principal non-linearity. To compute the increment in current given the small increment of voltage, only the slope of the $I$-$V$ curve at the bias point must be known. For junction diodes this happens to be proportional to the bias current. Such a relationship between the slope and the current is peculiar to junction diodes, but on any single-valued $V$-$I$ curve, specifying one of the bias coordinates is enough. For a junction diode, it just happens to be more convenient to specify $I_0$.]

## CHARGE STORAGE IN THE JUNCTION DIODE

The conventional picture of a capacitor is a pair of conductors separated by an insulating dielectric. The electric field in the dielectric and the charge stored on the conductors are intimately associated with each other. From an electrical point of view, however, capacitance may be associated with any pair of terminals that require that a change in voltage be associated with a change in the charge stored. That the charge associated with the change in voltage give rise to an electric field through a Gauss's-law relationship is not required at all. A question immediately arises: How is it possible to have charge stored without an accompanying electric field? (Answer: The charge may not be associated with the field if the charge is stored in the form of a neutral plasma—i.e., if $p' = n' = 0$.) As equations 6.6 and 6.7 show, there is a charge storage in the neutral regions of a diode associated with the applied voltage. Since these equations were derived from the diffusion equation (5.27'), the capacity associated with charge storage in the neutral regions is called *diffusion capacity*.

A second question now arises: If the voltage is not associated with a field from the stored charge, where is the field for this voltage to be found? (Answer: The voltage appears across the depletion region, and it IS associated with the charge within the depletion region. This charge storage within the depletion region is attributed to a second capacity, the *depletion* or *transition* capacity. Since it is precisely the same voltage that appears across both the diffusion and depletion capacities, the two capacities must be in parallel. Keeping them separate is convenient because they depend on the bias voltage in such very different ways.)

It is unfortunate but true that if we set about finding the diffusion capacity by finding the change in the charge stored for a given change in the applied voltage—certainly the obvious procedure—the result that we will get is incorrect. According to (6.23), the diode should be represented by an admittance, $y = i/v$, whose positive imaginary part is the capacitive susceptance. As you shall see, the distributed nature of the charge storage leads to a significant difference between the $c = \Delta Q/\Delta V$ calculated on a DC basis and the $\omega c =$ positive imaginary part of $y$ calculated from proper AC considerations. We begin with the simple DC calculation: To find the total charge stored in the bulk regions, we should integrate (6.6) and (6.7) and multiply the result by the area, $A$, and $q$ (see Figures 6.2 and 6.3):

$$Q = Aq \left[ \int_0^a n_p' \, dx + \int_b^{c-b} p_n' \, dx \right] = Aq \left[ e^{qV/kT} - 1 \right] \times$$

$$\times \left\{ n_{p0} L_e \left( \frac{\cosh\left(\dfrac{a}{L_e}\right) - 1}{\sinh\left(\dfrac{a}{L_e}\right)} \right) + p_{n0} L_h \left( \frac{\cosh\left(\dfrac{c-b}{L_h}\right) - 1}{\sinh\left(\dfrac{c-b}{L_h}\right)} \right) \right\} \quad (6.26)$$

For the important short-based case [i.e., $a \ll L_e$ and $(c - b) \ll L_h$], (6.26) reduces to

$$Q = Aq \left[ e^{qV/kT} - 1 \right] \left\{ n_{p0}(a/2) + p_{n0} \left( \frac{c-b}{2} \right) \right\} \quad (6.26')$$

Note well at this point—for here is the flaw in this derivation—that (6.26) and (6.26') give the total charge without any reference to the relationship between the current and voltage that brought in the charge. A proper derivation of capacitance must be based on the phase difference between current and voltage. Arithmetically, this comes down to the difference between integrating (6.6) and (6.7) to get $Q$ and differentiating the same equations to get $J$. Although the two operations yield *almost* the same result, the difference is both real and important.

To complete this DC differential analysis, we need only to differentiate either version of (6.26) to obtain the result:

$$c = dQ/dV = qQ/kT \qquad (6.27)$$

In the short-based case, comparing (6.26') with the equivalent version of (6.10) yields the simple result that

$$Q = Q_e + Q_h = (a^2/2D_e)J_e + ([c - b]^2/2D_h)J_h$$

In other words, the charge storage associated with each minority-carrier current is directly proportional to that current. Then, from (6.27) we may conclude that the capacity associated with each current is also proportional to the current. We will now derive a proper small-signal capacitive susceptance. It too will be directly proportional to the current, but the coefficient of proportionality will be noticeably different.

## THE AC DIFFUSION EQUATION

Up to this point, we have been employing solutions derived from (5.27'), the form of the diffusion equation appropriate to the completely time-invariant case. It is not correct for time-dependent driving functions. The introduction of sinusoidal driving functions brings about a small but important change in our results.

If we assume that the excess-carrier concentration varies sinusoidally:

$$p'(x) = p_1(x)e^{j\omega t} \qquad (6.28)$$

then the diffusion equation

$$D_h \partial^2 p'/\partial x^2 - p'/\tau = \partial p'/\partial t \qquad (5.27)$$

becomes

$$D_h \partial^2 p'/\partial x^2 - p'/\tau = j\omega p'.$$

Upon rearrangement, this AC version of (5.27) takes on the form of (5.27'):

$$\partial^2 p'/\partial x^2 - p'/L_h^{*2} = 0 \qquad L_h^* \equiv L_h/\sqrt{1 + j\omega\tau} = \sqrt{D_h\tau/(1 + j\omega\tau)}$$

$$(6.29)$$

$L^*$ is called the *AC diffusion length*. Note that it is a complex number and that it is always shorter in magnitude than the normal diffusion length. Of course, for frequencies small enough that $\omega\tau \ll 1$, $L^* \simeq L$ and the solutions to (6.29) and (5.27′) are the same. For the more general case, solutions to (6.29) can be obtained from solutions to (5.27′) simply by replacing $L$ by $L^*$. This is the small, but, as you shall see, important change.

In order for solutions to (6.29) to be very useful, we must show that sinusoidal driving functions are likely to occur. The only common case—but an important one—where (6.28) is encountered is for the incremental portion of the carrier distribution. If the voltage applied to the diode can be described by $V(t) = V_0 + ve^{j\omega t}$, then, from (6.5′):

$$n'_p \simeq n_{p0}e^{qV/kT} \simeq n_{p0}e^{qV_0/kT}\left(1 + \frac{qv}{kT}e^{j\omega t}\right) \tag{6.30}$$

To write (6.30), I have had to assume that $v \ll kT/q \simeq 1/40$th of a volt at room temperature. This is quite different from the usual smallness criterion of $v \ll V_0$. If (6.30) may be used, the first term satisfies (5.27′) and the second (6.29). Expressions equivalent to (6.8) and (6.9) may be written for $j$, the small-signal AC current density. For example, the small-signal electron current on the $p$ side is given by

$$j_e(x) = \frac{qD_e n_{p0}}{L^*_e}\frac{\cosh\dfrac{x}{L^*_e}}{\sinh\dfrac{a}{L^*_e}}\frac{qve^{j\omega t}}{kT}e^{qV_0/kT} \tag{6.31}$$

Without heeding your consternation over the complex cosh's and sinh's, we proceed to find the AC admittance of the diode:

$$j_T = \frac{q^2v}{kT}e^{qV_0/kT}\left[\frac{n_{p0}D_e}{a}\left(\frac{a}{L^*_e}\right)\coth\left(\frac{a}{L^*_e}\right) + \right.$$

$$\left. + \frac{p_{n0}D_h}{(c-b)}\left(\frac{c-b}{L^*_h}\right)\coth\left(\frac{c-b}{L^*_h}\right)\right] \tag{6.32}$$

By definition:

$$y = i/v = Aj/v \tag{6.33}$$

Splitting (6.33) into its real and imaginary parts may improve your trigonometry, but it is not very enlightening. Only two cases are easy and both are of some interest—the short-based and long-based diodes. The long-based case shows how the simple circuit model breaks down at sufficiently high frequencies; the short-based case is of practical importance, and it is one example in which the difference between (6.33) and (6.27) becomes quite clear.

For the short-based case, the arguments of both the hyperbolic functions are much less than 1 in magnitude. Accordingly, the functions may be expanded to the first significant real and imaginary terms. For small $z$,

$$(z) \coth (z) = \frac{(z) \cosh (z)}{\sinh (z)} \Rightarrow \frac{1 + \dfrac{z^2}{2!}}{1 + \dfrac{z^2}{3!}} \cong \left(1 + \frac{z^2}{2!}\right)\left(1 - \frac{z^2}{3!}\right) \cong 1 + \frac{z^2}{3}$$

(6.34)

Noting that the arguments in (6.32) are of the form

$$a/L_e^* = (a/L_e)(1 + j\omega t)^{1/2}$$

we may put (6.34) into (6.33) to get the admittance for the case where the diode is short based with respect to the AC diffusion length:

$$y = \frac{q^2}{kT} e^{qV_0/kT} \left\{ \frac{n_{p0}D_e}{a} + \frac{p_{n0}D_h}{(c - b)} + j\omega \left[ \frac{an_{p0} + (c - b)p_{n0}}{3} \right] \right\}$$

(6.35)

Not too surprisingly, the real part of (6.35) gives the same conductance as the DC analysis. Thus, $g = qI_0/kT$ whether we use (6.32) or the short-based version of (6.10). The imaginary part is not the same as $qQ/kt$ from (6.26'), however. It is, in fact, precisely 2/3 of the DC estimate. The physical reason why (6.35) and not (6.26') is the correct expression is that the charge is distributed through what is in fact an RC transmission line. Equation 6.35 correctly accounts for the phase shifts associated with the distributed line. A DC analysis obviously does not.

The long-based case is rather different from any capacitor you have probably met before in that the result of the analysis depends on the frequency. For the long-based case, the hyperbolic cotangents go to 1 (regardless of $\omega t$) to give a new form of (6.33):

$$y = \frac{q^2}{kT} e^{qV_0/kT} \left[ \frac{n_{p0}D_e}{L_e} \sqrt{1 + j\omega\tau_p} + \frac{p_{n0}D_h}{L_h} \sqrt{1 + j\omega\tau_n} \right]$$

(6.36)

Note that since the excess-carrier lifetimes on the two sides need not be the same, the two times must be specified. If the frequency-lifetime products are sufficiently small, the square roots can be expanded to give (recall that $D_e\tau_e = L_e^2$)

$$y = \frac{q^2}{kT} e^{qV_0/kT} \left\{ \frac{n_{p0}D_e}{L_e} + \frac{p_{n0}D_h}{L_h} + j\frac{\omega}{2}[n_{p0}L_e + p_{n0}L_h] \right\}$$

(6.37)

This corresponds to (6.35) for the short-based case. Once again, the imaginary part differs from what we get from a DC analysis, this time by a factor of $1/2$.

For intermediate values of $\omega\tau$ there is no simple expansion, but when $\omega\tau \gg 1$, (6.36) again takes on a convenient form:

$$y = g + js = \frac{q^2}{kT} e^{qV_0/kT} \left[ n_{p0}\sqrt{D_e} + p_{n0}\sqrt{D_h} \right] \sqrt{\omega} \left( \frac{1+j}{\sqrt{2}} \right) \quad (6.38)$$

Equation 6.38 definitely includes a capacitive susceptance—the current leads the voltage—but it is not possible to define a meaningful capacity. Why not? [Answer: The susceptance depends on $\sqrt{\omega}$ rather than $\omega$. For a legitimate capacity, $s = \omega c$ where $c$ is independent of the frequency. Frequency-independent capacities are contained in (6.37) and (6.35), but not in (6.38).] Since the conductance in (6.38) is also frequency dependent, it is not possible to represent the diode at very high frequencies by any conventional lumped-circuit parameters. Junction diodes find their way into detector circuits at the highest-microwave frequencies generated. {I would normally have used "submillimeter" to describe these "highest" wavelengths, but recent work by Javan and co-workers has demonstrated that genuine non-linear junctions can be created that retain their junction properties in the middle infrared. Using minute tungsten whiskers to form point-contact metal-(oxide?)-metal junctions on a nickel plate, they showed that appropriate junctions would satisfactorily mix the 5.2 $\mu$m emission from a CO laser with a combination of conventional microwave frequencies and the output of a $CO_2$ laser at 10.4 $\mu$m. By careful measurement of the easily detected beat frequencies, they were able to determine the absolute frequency of the CO laser line, and all of this was done at $5.8 \times 10^{13}$ Hz! [See D. R. Sokoloff et al., *Appl. Phys. Letters* **17**, 257 (1970)].}

Although applications of conventional diodes at $10^{13}$ Hz are still a bit esoteric, they find enormous utility in the microwave region. It is at these relatively high frequencies that the linear representation of the junction diode finds its principal applications. The large market for microwave diodes suggests that a knowledge of how to and how not to model the junction diode can be very useful.

## THE DEPLETION (OR TRANSITION) CAPACITY

The charge storage associated with the depletion layer has a more classically capacitive character. The voltage across the junction is found by an application of Gauss's law, certainly a classical way to begin. If the depletion approximation is accurate, the incremental capacity of the junction diode is well represented by a parallel-plate capacitor whose plate separation

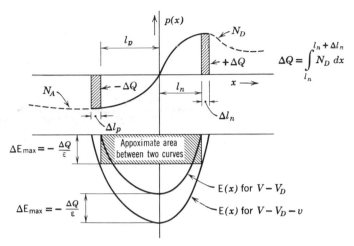

$$\Delta Q = \int_{l_n}^{l_n + \Delta l_n} N_D \, dx$$

**Figure 6.15** The charge concentration and the electric field as a function of position throughout the depletion region. The doping profile is entirely arbitrary. The hatched areas in both curves are quantities used in the calculation of the incremental capacity. Each of the hatched areas in the upper curve has an area of magnitude $Q$. The area of the lower curve is the incremental voltage, $v = \Delta Q / \varepsilon \, (l_p + l_n)$.

depends on the bias voltage. Even if the depletion approximation is not sufficiently accurate, the Gauss's-law integration yields a most conventional, easily observed capacity. The relationship between the carrier concentration and the easily measured depletion capacity is so direct that the principal tool for measuring concentration profiles is a $c$ versus $V$ plot. For the present, we will examine the depletion capacity only for the simple but common case in which the depletion approximation is quite adequate.

Under the low-level-injection assumption, essentially all of the applied voltage appears across the depletion region. To absorb this change in voltage, charge must be added or subtracted to the amount stored on each side of the junction. According to the depletion approximation, the charge must be added to or subtracted from the *edge* of the depletion region—that is, the depletion region must expand or contract to accommodate the extra voltage. There is no free charge to remove from within the depleted region, so there is no change in the charge distribution within the remaining portion of the depletion layer. For small changes in the total potential across the depletion region, the width of the depletion layer is approximately constant. Accordingly, on an incremental basis, we are adding or subtracting charges with a fixed separation, making the depletion region the dielectric that separates the parallel plates of the capacitor. Figure 6.15 illustrates the various steps for calculating the capacity from Poisson's equation. Not very surprisingly, we obtain

$$c_{dep} = \Delta Q / v = \frac{\varepsilon}{(l_n + l_p)} \qquad \text{(farads per unit area)} \qquad (6.39)$$

The only problem in using (6.39) is finding the depletion lengths as a function of $V_0$. We have already treated this problem for one particular case in Chapter 3. There the depletion widths were obtained for the uniformly doped abrupt junction under equilibrium conditions ($V_0 = 0$). For the non-equilibrium case—as long as we assume that the depletion model is accurate—the only change needed in the solution for the depletion lengths (equation 3.51) is the substitution of $V_D - V_0$ for $V_D$. Putting (3.51) so modified into (6.39) yields $c(V_0)$:

$$c = \frac{\varepsilon}{\left(\dfrac{2\varepsilon(V_D - V_0)}{q}\right)^{1/2}\left[\left(\dfrac{N_A}{N_D(N_A + N_D)}\right)^{1/2} + \left(\dfrac{N_D}{N_A(N_A + N_D)}\right)^{1/2}\right]} =$$

$$= \sqrt{\frac{q\varepsilon(N_A N_D)}{2(V_D - V_0)(N_A + N_D)}} \qquad (6.40)$$

Equation 6.40 is a reasonably good description of the observed capacity for uniform abrupt junctions such as those obtained by alloying. Diffused junctions tend to be *graded*. That is, the net impurity concentration is a continuous function of position throughout the depletion layer. The simplest such distribution is the linear one (see Figure 6.16):

$$\rho = q(N_D - N_A) = qKx \qquad (6.41)$$

If (6.41) is integrated twice, we obtain the voltage from Poisson's equation:

$$V_D - V_0 = \frac{2qKl^3}{3\varepsilon} \qquad (6.42)$$

If (6.42) is substituted in (6.39) (note that $C = \varepsilon/2l$), the depletion capacity is determined:

$$c_{DEP} = \varepsilon\left[\frac{qK}{12\varepsilon(V_D - V_0)}\right]^{1/3} \qquad (6.43)$$

Not only do (6.43) and (6.40) give the capacity as a function of voltage, but they also provide a reasonably direct way of determining the doping in the vicinity of the junction. You might well question whether one ever finds a distribution as simple as (6.41). Certainly the diffusion profiles we examined earlier in the chapter were far from linear. But if the diffusion is reasonably deep, the impurity profile will be changing slowly enough in the vicinity

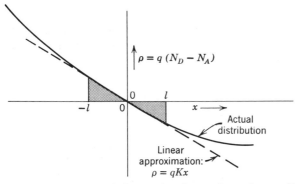

**Figure 6.16** The depletion region for an impurity profile that can be approximated by a straight line. The "junction" occurs where $\rho$ passes through 0. The linear approximation assures a symmetric charge distribution. $l$ is given by (6.42).

of the junction to be approximated by a straight line. If higher powers of $x$ are required, the double integration that led to (3.51) or (6.42) will yield

$$c \propto (V_D - V_0)^{-1/n}$$

where $n \geq 3$. In general, for any doping profile, (6.44) will apply with $n \geq 2$. Compare this with the exponential voltage dependence of the diffusion capacity. The comparison suggests that under reverse bias, the junction capacity is almost entirely the depletion capacity, while under sufficient forward bias, the reverse conclusion is true. The first conclusion is quite generally true, while the second is only usually true. For example, exceedingly narrow-based diodes will have very small diffusion capacities, so the depletion capacity under substantial forward bias could still be the dominant term.

Summing up our results for the incremental-circuit model under the assumptions of (1) low injection level and (2) moderate applied voltages, we have found that there are only two very significant terms: A conductance, $g_j$, that is proportional to the current, in parallel with a capacity, $c$, that is the sum of the depletion and diffusion capacities. The conductance is given by (6.24), the diffusion capacity by (6.26) and (6.27), and the depletion capacity by (6.40), (6.43), or (6.44), depending on the impurity distribution. Normally, under conditions of moderate forward bias, the diffusion capacity vastly exceeds the depletion capacity, while for moderate reverse bias, only the depletion capacity is important.

## Problem 6.4

To get some practice with the material developed so far and to develop some sense of the order of magnitude of realistic values, compute the incremental conductance and capacity for the silicon alloy junction diode described below.

Do it first at a forward bias of 0.5 V and then at a reverse bias of $-0.5$ V. Verify that the low-level approximation is reasonably valid, and find the bias at which the minority density on the lightly doped side equals half the majority density on that side.

*Data on the Diode.*
$A \equiv$ area $= 1$ mm$^2$ (that's a pretty large device)
Length of the $p$ side $= a = 10$ $\mu$m
Length of the $n$ side $= c - b = 100$ $\mu$m
For convenience, let $n_i^2 = 10^{20}$/cm$^3$ and $q/kT = 40$/V
On the $n$ side: $N_D = 10^{15}$/cm$^3$, $D_h = 10$ cm$^2$/sec, and $\tau = 10^{-5}$ sec
On the $p$ side: $N_A = 10^{17}$/cm$^3$, $D_e = 25$ cm$^2$/sec, and $\tau = 4 \times 10^{-6}$ sec

*Solution.* Starting with the forward bias case and assuming that the low-injection-level condition is met (to be checked later), we find $g$ from (6.24). $I_{sat}$ is obtained from (6.10) and the data above (Note that $L_e \gg a$ but $L_h = = c - b$. Thus, only the $p$ side is short based.):

$$I_{sat} = qA \left[ \frac{n_{e0}D_e}{a} + \frac{p_{n0}D_h}{L_h} \coth \frac{c-b}{L_h} \right]$$

$$= 1.6 \times 10^{-19} \times 10^{-2} \left[ \frac{10^3 \times 2.5}{10^{-3}} + \frac{10^5 \times 10}{10^{-2}} \coth 1 \right]$$

$$= 2.5 \times 10^{-13} \text{ amperes}$$

Then, at a forward voltage of 0.5 V, the current is

$$I \simeq I_{sat}e^{qV/kT} = 0.123 \text{ mA}$$

and the conductance is $g = qI/kT = 5 \times 10^{-3}$ mhos.
To find the diffusion capacity, $Q$ is obtained from (6.26):

$$Q = qA \left[ \frac{n_{e0}a}{2} + p_{n0}L_h \left\{ \frac{\cosh\left(\frac{c-b}{L_h}\right) - 1}{\sinh \frac{c-b}{L_h}} \right\} \right] \left[ e^{qV/kT} - 1 \right]$$

$$= 1.6 \times 10^{-19} \times 10^{-2} \left[ \frac{10^3 \times 10^{-3}}{2} + 10^5 \times 10^{-2} \left\{ \frac{0.543}{1.175} \right\} \right] e^{20} =$$

$$= 7.4 \times 10^{-19} \times 10^{8.7} = 3.7 \times 10^{-10} \text{ coul}$$

Then $c_{diff} = 2qQ/3kT = 0.0099$ $\mu$F, a moderately substantial capacity. Note that the capacity is almost all on the lightly doped, relatively long $n$ side. The diffusion capacity could easily be reduced (if that were important) by increasing the doping and shortening the base width. Increasing $N_D$, of course, would decrease $I$ and thus $g$, while decreasing $(c - b)$ would increase $I$ and $g$.

The depletion capacity is found from (6.40). First we must obtain $V_D = (kT/q) \log (N_A N_D / n_i^2) = 0.67$ V. Then, using (6.40):

$$c_{dep} = A \left[ \frac{q\varepsilon}{2(V_D - V_0)} \frac{(N_A N_D)}{(N_A + N_D)} \right]^{1/2} =$$

$$= 10^{-2} \left[ \frac{1.6 \times 10^{-19} \times 1.06 \times 10^{-12}}{2 \times 0.17} \times \frac{10^{32}}{1.01 \times 10^{17}} \right]^{1/2} = 232 \text{ pF}$$

Thus, in the forward direction, we find as expected $c_{diff} \gg c_{dep}$. [Note that increasing $N_D$ would decrease $c_{diff}$ but increase $c_{dep}$.]

In the reverse direction both $g$ and $c_{diff}$ go very rapidly to 0. Solving for $c_{dep}$ at $-0.5$ volts gives

$$c_{dep} = 85 \text{ pF}$$

To check the injection level we must examine the lightly doped side. Thus, $p_n = p_{n0} e^{qV/kT} = 10^5 e^{20} = 5 \times 10^{13}/\text{cm}^3$. This is sufficiently below $10^{15}$ to warrant using the low-level approximation. The minority-carrier density will be equal to half of the majority-carrier density when the minority density equals the equilibrium majority density (i.e., $n_n = n_{n0} + n_n' = n_{n0} + p_n' = 2p_n'$). This will occur on the lightly doped side while the heavily doped side is still well within the low-level criterion. Thus, we may use $p_n = p_{n0} e^{qV/kT} = n_{n0}$. Solving for $V$ gives

$$V = (kT/q) \log (n_{n0}/p_{n0}) = 0.575 \text{ V}$$

Since the current would be 2.5 mA at 0.575 V, you can see that the low-level condition can be exceeded rather easily in a lightly doped junction. A diode of this cross-section can easily handle amperes in the forward direction, so it is important to find out what effects moderate to high levels of injection have on the junction-diode characteristics.

## JUNCTION DIODES AT EXTREMES OF BIAS

The simple, low-level model that we have just considered does a pretty good job of predicting the forward current over a current range of several decades. However, if we go to relatively large or very small currents, it is not very surprising that some of the simplifications in the model break down. For example, if the current is very small, can we still ignore small amounts of generation or recombination within the depletion region? The $I_{sat}$ obtained in the last problem was only $2.5 \times 10^{-13}$ A. It certainly wouldn't take much generation or recombination to compete with a current that small. Similarly, although we can ignore voltage drops in the bulk regions at low currents, at high currents they may well absorb a large fraction of the applied voltage.

The failures of the simple model can be divided into four reasonably distinct effects: (1) At low current levels (which can include rather large values of reverse voltage), we may have to include recombination and generation in the space-charge layer. This will lead to larger currents than are predicted by the simple theory. (2) At moderately large forward currents, the low-level assumption is no longer correct. This will lead to a current dependence of $e^{qV/2kT}$ (as is also true in the very low-forward-current case, but for high currents, the current will be less than the simple theory predicts). (3) At very high currents, the $I$-$V$ curve departs completely from the exponential form, becoming either linear or quadratic, depending on the base widths of the diode. (4) Finally, at a large enough reverse voltage, the junction "breaks down" either by avalanche or tunneling (Zener breakdown). Each of these effects places very important restrictions on the performance of junction diodes and transistors, so they warrant at least a cursory examination.

Of the four effects, the easiest to analyze is the moderate-current case in which the low-injection-level assumption fails. For a highly asymmetric junction—say, $p_{p0} \gg n_{n0}$—a *moderate injection level* would be in the range where $n_{n0} \ll n_n \ll p_{p0}$. For example, if the acceptor density were $10^{18}/\text{cm}^3$ and the donor level were $10^{15}/\text{cm}^3$, moderate injection would cover the range $5 \times 10^{15} \lesssim n_n \simeq p_n \lesssim 5 \times 10^{17}$. Although (6.4) and (6.5) apply perfectly well for moderate injection, it is no longer reasonable to ignore the ambipolar fields within the lightly doped side when computing the voltage across the diode. Keep in mind that the reason the ambipolar field can be ignored for the *low-level case* is that a relatively small number of the majority carriers are sufficient to neutralize the excess minority charge. For the case we are now considering, it takes essentially all of the majority carriers to neutralize the minority charge.

Let us consider a specific problem. Under conditions of moderate injection, how much of a change in applied voltage is required to raise the minority-carrier density by a factor of ten on the $n$ side? (As long as $n_p \ll p_n$, this is equivalent to asking how much voltage it takes to raise the current by a factor of ten.) At low levels of injection, the answer comes directly from (6.4′):

$$10 = e^{q\Delta V'/kT} \Rightarrow \Delta V' = 2.3kT/q \simeq 58 \text{ mV at } 300°\text{K} \qquad (6.45)$$

By inspection of (6.4), it is evident that (6.4′) still holds for the intermediate injection case. Note, however, that (6.5′) is no longer valid, since $n_n \neq n_{n0}$.

Now for the crucial step. For the intermediate case, the total voltage does not appear across the junction; part of it is to be found across the neutral $n$ region. To find this potential, we proceed in essentially the same fashion that was used to find the Einstein relationship [cf. (4.62)–(4.66)]. The electron and hole currents are given by

$$J_e = q[\mu_e n\mathsf{E} + D_e dn/dx] \tag{4.63}$$

$$J_h = q[\mu_h p\mathsf{E} - D_h dp/dx] \tag{4.64}$$

By hypothesis, we are dealing with a moderate injection situation on the $n$ side, so $p \simeq n$. Also by assumption, $n_n \ll p_p \simeq p_{p0}$, so the hole current should be much larger than the electron current. In practice, $J_h \gg J_e$ must be tempered by the assorted constants that enter the current equations, such as base widths and diffusion constants. Unless the diode is extraordinarily asymmetric, there may not be much of a voltage span where both $p_n \simeq n_n$ and $J_e \ll J_h$ hold simultaneously. However, assuming that both hold, (4.64) above represents the *sum* of two roughly equal numbers and (4.63) represents their *difference*. Since $J_h \gg J_e$, it follows that the two terms are large compared to their difference. If we ignore the difference (i.e., set $J_e \simeq 0$), (4.63) may be converted to:

$$-\mathsf{E} = \frac{D_e}{\mu_e} \frac{1}{n} \frac{dn}{dx} = \frac{kT}{q} \frac{d \log n}{dx} \tag{6.46}$$

Or, upon integration from the ohmic contact (where $n = n_{n0}$) to the edge of the depletion region:

$$\Delta V'' = \int_{n=n_{n0}}^{n(a)} \frac{kT}{q} d \log n = \frac{kT}{q} \log \frac{n(a)}{n_{n0}} \quad \text{or} \quad n(a) = n_{n0} e^{q\Delta V''/kT}$$
$$\tag{6.47}$$

Equation 6.47 has exactly the same form as (6.4′), but $\Delta V''$ is the voltage across the neutral $n$ region. Since $p_n(a) = n_n(a)$, the increase by a factor of ten in $p_n(a)$, which required an increment of 58 mV across the depletion region, is accompanied by a factor-of-ten increase in $n_n(a)$, which requires another increment of 58 mV across the neutral $n$ region. Thus, in terms of the total voltage, $V$, at the terminals of the junction diode, the boundary condition is

$$\frac{p_n(V + \Delta V)}{p_n(V)} = e^{q\Delta V/2kT} \qquad (\Delta V = \Delta V' + \Delta V'') \tag{6.48}$$

Since the exponential variation in the carrier concentration will be the dominant change in (4.64), we can conclude from (6.48) that, at moderate injection levels, the slope of the logarithmic I-V curve has fallen by a factor of 2 from its low-level value. The high-current data shown in Figure 6.17 appears to support this conclusion.

When the hole and electron currents are individually observable, as in a bipolar transistor, it is important to note that only the principal current varies according to a relationship such as (6.48). The relatively high injection

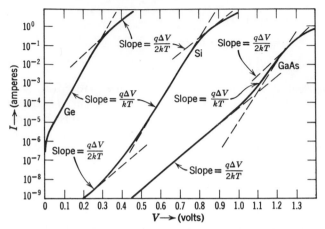

**Figure 6.17** The forward $I$-$V$ characteristics for three typical small-junction diodes at $300°K$. Note that GaAs never really follows "the law of the junction." Ge does follow the classical diode law until high injection sets in. Si shows the effects of generation in the depletion region at low currents, the normal diode curve at intermediate currents, and the typical high-injection response at high currents. (From Grove, *Physics and Technology of Semiconductor Devices*, Wiley, 1967.)

level on the lightly doped side of the junction effectively increases the doping on that side, increasing the share of the current carried by the majority carriers on the lightly doped side. Each component of current is proportional to the majority-carrier density at the depletion region edge reduced by $\exp\left[-\text{ barrier height}/kT\right]$. Thus each minority flux is proportional to the majority-carrier concentration on the opposite side. The barrier height for all carriers is going as $q\Delta V/2$ as in (6.48), while on the lightly doped side the number of majority carriers determined by (6.48) is also increasing as $e^{q\Delta V/2kT}$. This implies that the lesser of the two currents will still be increasing as $e^{q\Delta V/kT}$ and that the difference between the two current terms is rapidly vanishing. As we shall see, this will have a serious effect on transistor performance at moderate current levels.

At extremely high current levels, such as one might encounter in capacitor discharges or unintentional short circuits, voltage drops across the neutral regions can become the principal voltage across the diode. The recombination mechanisms that are essential to an ohmic contact may also saturate, leading to substantial minority-carrier accumulation with an attendant decrease in the diffusion current. The $I$-$V$ characteristics cease to be exponential, since the junction barrier is no longer the dominating element in current flow. Although this region of operation can be analyzed [cf. A. K. Jonscher, *J. of Electronics and Control*, **5**, 1 and **5**, 226 (1958)], it is not a region of normal operation nor one whose analysis is very illuminating. Accordingly, we will pursue it no further.

## DIODES AT VERY LOW LEVELS OF CURRENT

In problem 6.3, we obtained an $I_{sat}$ of $2.5 \times 10^{-13}$ amperes. If that diode were manufactured and $I_{sat}$ measured, one would obtain a reverse current of the order of $10^{-9}$ amperes. Furthermore, the current would be found to be noticeably voltage dependent. In contrast, if one manufactured a similar diode from germanium, both the measured and predicted $I_{sat}$ would be of the order of $10^{-6}$ amperes essentially independent of reverse voltage. Obviously, we have left out a term that seems small compared to a microamp but is large compared to a picoamp. This current arises from the carriers generated within the depletion region. Since we have ignored both recombination and generation with the depletion region, it is not surprising that they come back to haunt us. You might think that a current of $10^{-9}$ amperes is not likely to be very important, but it turns out that recombination within the depletion region places an important lower limit on the useful operating range of a transistor. Furthermore, from a rather practical standpoint, it is important to know whether the reverse current through a junction diode represents a fundamental limit or whether it is the net result of poor processing. It is rather easy to create diodes that leak around the junction rather than through it. Such leakage currents are inevitably unstable and frequently lead to catastrophic failure at voltages that the junction itself could endure. Thus, a knowledge of what to expect in the reverse direction is a valuable tool.

To evaluate the current under reverse bias is relatively simple. [We have already evaluated the current resulting from generation in the bulk regions. That is nothing more than $-I_{sat}$ from (6.10′).] The number of carrier pairs generated within the depletion region is given by (5.17) integrated over the volume of the depletion region. Under reverse bias, $n \simeq p \simeq 0$, so (5.17) reduces to the particularly simple form:

$$\frac{\partial n}{\partial t} = \frac{\partial p}{\partial t} = n_i^2/(n_T\tau_{h0} + p_T\tau_{e0}) = \text{constant} \tag{6.46}$$

The integral of the constant over the volume gives the total number of carriers generated within the depleted region. If $A$ is the junction area and $l_D$ the total depletion width, the current due to generation within the depletion region is

$$I = - I_{DEP} = -qAl_D n_i^2/(n_T\tau_{h0} + p_T\tau_{e0}) \tag{6.47}$$

Is $I_{DEP}$ voltage dependent? [Answer: Yes. $l_D$ increases with reverse bias according to (6.39), (6.42), or an equivalent expression. Thus, we would expect the reverse current to increase as the square root or cube root of the reverse voltage.]

To get some feeling for the importance of $I_{DEP}$, we should compare it with $I_{sat}$. Apart from variations in trapping times—which can be quite substantial from one material or sample to the next—the principal difference between $I_{DEP}$ and $I_{sat}$ is that $I_{DEP}$ is proportional to $n_i$ while $I_{sat}$ is proportional to $n_i^2$. [Remember that, for a given trap level, $E_T$, $n_T = n_i e^{(E_T - E_i)/kT}$ and $p_T = n_i^2/n_T$.] Thus, if the other factors in (6.47) and (6.10) are not grossly different from one material to the next, we should expect the ratio of the depletion and saturation currents to vary as [see (3.28) and (3.29)]:

$$I_{DEP}/I_{sat} \propto 1/n_i \propto T^{-3/2} e^{E_g/2kT} \tag{6.48}$$

Thus, for materials with large band gaps, the depletion current should dominate, while for smaller band gaps, the saturation current is dominant. This, of course, explains the difference between the reverse $I$-$V$ characteristics of Ge and Si. It is even more exaggerated for wider band-gap materials such as GaAs. Note, however, that in saying that $I_{DEP} \gg I_{sat}$ for GaAs, we are not saying that either term is very large. In fact, just the opposite is true. $I_{sat}$ is so small as to be unobservable. $I_{DEP}$ is pretty small itself, but it is the dominant current and thus the one that is observed.

Under forward bias, it follows that, if $I_{DEP}$ dominates the reverse current, it must initially dominate the forward current. This was pointed out by Sah et al. [C. T. Sah, R. N. Noyce, and W. Shockley, *Proc. IRE*, **45**, 1228 (1957)] in their classic paper on the influence of the depletion region on the $I$-$V$ characteristics of junction diodes. Unfortunately, under forward bias (6.46) does not hold and we must integrate (5.17) over a region in which $n$ and $p$ are functions of position. The integral is too complicated to obtain a convenient closed-form solution. We must bludgeon the integral into tractable form by approximating the integrand. The trick that is usually employed is to take advantage of the experimental fact that the most effective traps for recombination-generation processes are those that lie rather close to the center of the forbidden gap. Efficient generation centers are not necessarily at the center, but they are almost inevitably quite deep compared to the normal donors or acceptors. For example, Fe, Au, and Cu are effective "lifetime killers" in both Ge and Si. An examination of Figure 3.4 will show that these atoms all produce very deep levels. As you shall see below, if we may assume that the recombination centers are at or near the middle of the band, the integration of (5.17) may be greatly simplified.

To evaluate (5.17) under conditions of forward bias, we must be able to specify $n$ and $p$ throughout the depletion region. At low forward bias, we may assume that low injection conditions will obtain everywhere. Following the same logic that led to (6.4') and (6.5'), we should assume that the carrier concentrations are related by an equilibriumlike Boltzmann relationship:

$$n(x) = n_i e^{q[\psi(x) + V(x)]/kT} - n_0(x) e^{qV(x)/kT} \qquad (6.49)^*$$

$$p(x) = n_i e^{q[-\psi(x) - V(x) + V_1]/kT} = p_0(x) e^{q[V_1 - V(x)]/kT} \qquad (6.50)^*$$

$V(x)$ is measured from the $n$ side of the depletion region and $V_1$ is the total applied voltage across the depletion region. These two expressions are the extensions of (3.41') and (3.42') to the nonequilibrium case of an applied voltage.

If you take the product of (6.49) and (6.50), you get the extension of the $np$ product rule (3.28):

$$np = n_i^2 e^{qV_1/kT} \qquad (6.51)$$

Applying (6.51), (6.50), and (6.49) to the generation-rate equation (5.17) yields

$$\frac{\partial n}{\partial t} = - \frac{n_i^2 (e^{qV_1/kT} - 1)}{[n_T + n_0(x) e^{qV(x)/kT}] \tau_{h0} + [p_T + p_0(x) e^{q[V_1 - V(x)]/kT}] \tau_{e0}} \qquad (6.52)$$

Equation 6.52 is, of course, completely equivalent to (5.17), but it shows the route to a very convenient integration approximation.

The total current from recombination in the depletion region is given by

$$I_{DEP} = - qA \int_0^{L_D} \frac{\partial n}{\partial t} dx \qquad (6.53)$$

where 0 is at the edge on the $n$ side, $A$ is the junction area, and the thickness of the depletion region is $L_D$. Only the denominator of the integrand is a function of position, but its positional dependence may be very strong indeed. If it is true that the effective traps are those with energies far from the band edges, then $n_T$ and $p_T$ will be very small numbers. Making the reasonable assumption that neither $\tau_{e0}$ nor $\tau_{h0}$ is an extraordinarily large number, it is evident that (6.52) will take on a rather sharply peaked maximum approximately where $n\tau_{h0} + p\tau_{e0}$ is a minimum. That is, where

$$\tau_{h0} = \tau_{e0} e^{-q/kT[2\psi(x) + 2V(x) - V_1]} \qquad (6.54)$$

The sharpness of the maximum is evident in the powerful exponential dependence of the denominator.

---

* The quantities $\psi(x) + V(x) \equiv \phi_e(x)$ and $\psi(x) + V(x) - V_1 \equiv \phi_h(x)$ are often referred to as *quasi-Fermi potentials* or *imrefs* in analogy with the Fermi potential, $\psi(x)$. When imrefs are used, (6.49) and (6.50) look exactly like (3.41') and (3.42') with the exception that $\phi_e$ need not equal $\phi_h$. The $\phi$'s are nothing more than an algebraic convenience, but as such, they are rather popular.

Although we could proceed to evaluate (6.53) at this point, it will serve our purposes just as well if we make two mildly restrictive assumptions to simplify the arithmetic. First, we will assume that the traps are located at the intrinsic level so $n_T \simeq p_T \simeq n_i$. Second, we will assume that $\tau_{e0} = \tau_{h0} = \tau$. Neither of these alters the result in any fundamental way and the arithmetic is much simplified. Under the assumptions given, (6.52) becomes

$$\frac{\partial n}{\partial t} \simeq - \frac{n_i^2 \left( e^{qV_1/kT} - 1 \right)}{\tau n_i \left[ e^{q(\psi + V)/kT} + e^{q(V_1 - \psi - V)/kT} \right]} =$$

$$= - \frac{n_i}{\tau} \frac{\sinh \dfrac{qV_1}{2kT}}{\cosh \left[ \dfrac{q}{kT} \left( \psi + V - \dfrac{V_1}{2} \right) \right]} \tag{6.52'}$$

The evaluation of (6.53) now follows quite directly. Approximating the hyperbolic cosine by an exponential and replacing $[\psi(x) + V(x) - V/2]$ by $[- \mathsf{E}_{max} x]$ (the sharply peaked character of the integrand assures us that the integrand has become vanishingly small before the field has fallen much below its maximum value), the integral becomes simply

$$I_{DEP} = \frac{qAn_i}{\tau} \frac{kT}{q\mathsf{E}_{max}} \sinh \frac{qV_1}{2kT} \simeq \frac{qAn_i}{2\tau} \frac{kT}{q\mathsf{E}_{max}} e^{qV_1/2kT} \tag{6.55}$$

Once again, we want to know how $I_{DEP}$ compares with the "normal" diode current, $I_{sat} e^{qV_1/kT}$. Since $I_{sat}$ is proportional to $n_i^2$ and $I_{DEP}$ to $n_i$, we have the same situation that we saw in the reverse-bias case (6.48). Materials with large band gaps will show relatively much more recombination current than narrow-band-gap materials. However, since $I_{DEP}$ is proportional to $e^{qV_1/2kT}$ while the normal current is increasing as $e^{qV_1/kT}$, it follows that the normal current will eventually exceed the depletion-region recombination current. Thus, we should expect the semilogarithmic I-V curve first to rise as $qV/2kT$ and then to double its slope to $qV/kT$. (And then, from our previous analysis, we should expect the slope to fall back to $qV/2kT$ as high injection sets in.) Figure 6.17 shows that Si follows this prediction quite well. For Ge, $I_{DEP}$ is too small to be observed under forward bias, while for GaAs, $I_{DEP}$ dominates to such a degree that one reaches a high injection-level situation before any extensive portion of normal I-V curve is observed.

## JUNCTION DIODES AT LARGE REVERSE BIAS—BREAKDOWN

It is not too gross a description to say that a junction diode does not conduct in the reverse direction. Certainly, the incremental conductance of the typical silicon or germanium diode under reverse bias is extremely small. But such a description can hold only for a limited applied voltage, since all "insulators" will "break down" under sufficient electrical stress. *Breakdown* is an ill-defined term implying a more or less sudden and drastic increase in the incremental conductance as the result of a change in the essential character of the "insulator." Although a breakdown can be accompanied by irreversible damage to the insulator, many devices spend their entire useful lives in a state of perpetual breakdown.

For junction diodes, we must discriminate between two classes of breakdown—essential ones that represent a change in the conductance of the depletion region, and processing or packaging breakdowns that arise from phenomena unrelated to *p-n* junctions in the bulk material. In this latter class, we must include all breakdowns that result from the condition of the diode's surfaces or the finite extent of the diode.

Controlled breakdown of the junction diode is generally nondestructive. In fact, it is possible (and useful) to build junction diodes that can be quite properly described as being broken down when they are in a state of thermal equilibrium. Such diodes are classified as either backward or tunnel diodes, depending on their forward characteristics.

The two causes of essential breakdown are tunneling (Zener breakdown) and avalanche multiplication. Either or both may be active in any one diode, but it is generally safe to assume that breakdown at very low voltage—roughly of the order of a volt or less—is Zener breakdown, while those at a few volts or higher are avalanche breakdowns. That tunneling occurs only for very low reverse voltages follows from the requirement that the tunneling barrier be very thin. Since the depletion region spreads out with increasing reverse bias, it follows that the junctions most likely to show appreciable tunneling are those that are very thin to begin with. To make a narrow depletion region, one forms a junction that is heavily doped on both sides. Such junctions will have very large internal fields, so they will avalanche at low reverse bias. Thus, if tunneling is going to be significant, it will have to appear at very, very low reverse bias.

Figure 6.18*a* shows the equilibrium band diagram for a relatively heavily doped abrupt junction. The depletion region is very narrow and the electric field in the depletion region very large. If the diode were reverse biased by a relatively small amount, the edges of the bands would be shifted as shown in Figure 6.18*b*. Note that the application of the reverse bias causes the upper

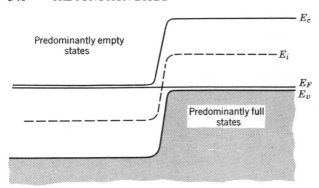

**Figure 6.18a** Band diagram for an abrupt, heavily doped *p-n* junction in thermal equilibrium. The edge of the valence band on the *p* side is close to the edge of the conduction band on the *n* side, but there is no overlap in equilibrium. Note that the transition region is very narrow.

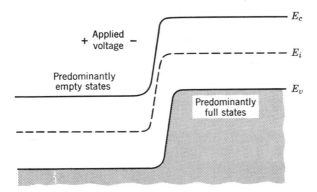

**Figure 6.18b** The same diode with some reverse bias applied. Note now that some of the full states in the valence band align with empty states in the conduction band. If the physical distance separating the full and empty states is not too large, substantial tunneling currents could now flow.

valence-band states on the *p* side to align with conduction-band states on the *n* side. It takes a very small voltage to achieve this alignment. In terms of $E_g$ and $V_D$, how much does it take? [Answer: The difference between the energy at the edge of the valence band on the *p* side and its energy on the *n* side is just $q(V_D - V)$ where $V$ is the applied voltage (positive for forward bias). Alignment of the conduction-band edge on the *n* side and the valence-band edge on the *p* side occurs when $V_D - V = E_g/q$. For any bias more negative than that, some empty states on the *n* side will be aligned with some full states on the *p* side. Note that the required back bias to achieve such an alignment is always less than $E_g/q$—substantially less than a volt even in a relatively lightly doped junction.]

If it is so easy to align the valence and the conduction bands across the depletion region, why don't all junctions show Zener breakdown at a fraction of a volt? (The answer, of course, is that alignment of full and empty states is a necessary but insufficient condition for current flow. There must also be a reasonably large transmission probability.) The transmission probability, at least for direct transitions, was given by (5.56) and (5.53') and was of the form:

$$P = e^{-C/\bar{E}} \tag{5.56'}$$

where $C$ is a function of the effective masses and the energy and the momentum of the transition. $\bar{E}$ is the *average* electric field in the depletion region. As you saw in Chapter 5, a field of the order of $10^6$ V/cm was required to achieve a substantial tunneling probability. Assuming an abrupt junction between two uniformly doped regions, we may find the depletion length and average field from (3.51) and (3.52):

$$l_t = l_n + l_p = \sqrt{\frac{2\varepsilon}{q}(V_D - V)\frac{(N_A + N_D)}{N_A N_D}} \tag{6.56}$$

and

$$\bar{E} = (V_D - V)/l_t = \sqrt{\frac{q(V_D - V)N_A N_D}{2\varepsilon(N_A + N_D)}} \tag{6.57}$$

Note that the average field increases only as $-V^{1/2}$. Why should that be so? (Answer: The average field is the voltage across the depletion region divided by the length of the depletion region. Since an increase in reverse voltage is accompanied by an increase in the width of the depletion region, the field must increase less rapidly than the voltage.)

By taking advantage of the fact that $\varepsilon$ is about the same for most semiconductors of practical interest ($\sim 10^{-12}$ F/cm), we may evaluate the critical voltage for tunneling for any particular doping. For the asymmetrically doped junction, setting (6.57) equal to $10^6$ V/cm gives

$$V_B = V_D - V \approx 1.25 \times 10^{19}/N \tag{6.58}$$

For a critical total voltage of 1 V, (6.58) says that the doping on the "lightly doped" side of the junction must be about $1.25 \times 10^{19}/cm^3$. As we shall see immediately below, if tunneling does not occur by the time the voltage is of the order of 1 V, the onset of avalanche multiplication prevents the voltage from ever reaching the critical value. We must thus conclude that for junctions whose more lightly doped side has much less than $10^{19}$ carriers/cm$^3$, Zener breakdown will not occur.

## AVALANCHE BREAKDOWN

At this point you might be curious about how one can discriminate between avalanche and Zener breakdown. Apart from the typical low breakdown voltage of the Zener process, the *I-V* characteristics are not especially different, and it might seem difficult to observe anything else. (Actually Zener breakdown does not "turn on" as abruptly as avalanche breakdown does, but "soft" breakdown is not an easily interpreted diagnostic tool.) However, we are not limited to observing breakdown alone. If we provide ourselves with an independent source of carriers that we may inject into the junction (e.g., by shining light on the junction), we may observe the fraction of the carriers collected as a function of the reverse voltage. If we collect more than we inject, we are undoubtedly observing avalanche multiplication. It also happens that avalanche breakdown is a very erratic process, yielding a rather noisy current. Chynoweth and McKay found that the noisiness of the avalanche current was distinct enough to differentiate it from Zener current [*Phys. Rev.*, **106**, 418 (1957)].

Consider a junction diode at moderate reverse bias. The injected current is $- I_{sat}$. If we examine the usual case of the reasonably asymmetric diode, $I_{sat}$ will be of predominantly one type—say, hole current. If we raise the bias toward the avalanche voltage, we will have the situation discussed in Chapter 5: a current of one dominant type traversing a region with very large electric fields and no requirement that $p' = n'$. For such a case, we found that the hole current could be described by

$$I_h(x) = \left[ \frac{1}{M_h} + \int_0^x \alpha_e e^{U(y)} \, dy \right] I_t e^{-U(x)} \tag{5.48}$$

We now wish to find the voltage at which the diode breaks down by the avalanche process. That is, we wish to find the voltage at which the output current would diverge were it not limited by some external process. This will be the point at which $M_h \to \infty$.

We may take a big step toward our goal by considering a particular diode structure in which we know the field versus the position. Let us choose the uniformly doped step junction with one side very heavily doped, since this is the junction we have considered most thoroughly. This structure is easily realized by alloying a junction on a uniformly doped substrate.

Even though we know the field distribution for such a junction, we are still confronted with some obnoxious integrals of the form:

$$\int \frac{A}{E(x)} \exp \left[ \int \frac{B}{E(x)} \, dx \right] dx$$

If the electric field were in the numerator of the integrand, the analysis would be much easier. To achieve that desirable arrangement, we take advantage of the fact that over a limited range, we may obtain a good match to (5.44) with a much simpler expression:*

$$\alpha_h = aE^n \qquad (6.59)$$

The constants $a$ and $n$ are obtained by fitting (6.59) to experimental data such as that in Figure 5.8.

What remains to be done is simply some equation shuffling with a judicious amount of fudging where necessary. The first step is to evaluate (5.48) at $w$, the far end of the depletion region (i.e., the $n$ side). At that point, $I_h = I_t$ (the total current), so (5.48) becomes

$$\frac{1}{M_h} = e^{U(w)} - \int_0^w \alpha_e(x)e^{U(x)}dx =$$

$$= \exp\left[\int_0^w (\alpha_e - \alpha_h)\,dx\right] - \int_0^w \alpha_e(x)\exp\left[\int_0^x (\alpha_e - \alpha_h)\,dy\right]dx \quad (6.60)$$

Next note that

$$\frac{d}{dx}e^{U(x)} = \frac{dU}{dx}e^U$$

so

$$\int_0^w \frac{dU}{dx}e^{U(x)}dx = e^{U(w)} - 1 \qquad (6.61)$$

Equation 6.61 may be used to reduce the number of integrals in (6.60) in the following two steps:

$$\frac{1}{M_h} = e^{U(w)} - \int_0^w (\alpha_e - \alpha_h)e^{U(x)}dx - \int_0^w \alpha_h e^{U(x)}dx =$$

$$= e^{U(w)} - e^{U(w)} + 1 - \int_0^w \alpha_h e^{U(x)}dx$$

or finally:

$$1 - \frac{1}{M_h} = \int_0^w \alpha_h e^{U(x)}\,dx \qquad (\Rightarrow 1 \text{ at breakdown}) \qquad (6.62)$$

---

* This approximation was first used by S. L. Miller, *Phys. Rev.* **99**, 1234 (1955).

Even with our less-than-admirable approximation for $\alpha_h$, (6.62) presents a bit of a problem because the ratio of $\alpha_h$ to $\alpha_e$ is a function of the electric field. The analysis becomes much easier if we assume that the associated pairs of lines in Figure 5.8 are parallel. This is not too bad an approximation for the Ge data, but it is obviously absurd for the Si data. In spite of the clumsy data-bending here, the results we will obtain are not too bad because most of the multiplication takes place within a short distance of the maximum field. Over such a short interval, it does not seem too unreasonable to assume that the ratio of the multiplication coefficients is a constant:*

$$\frac{\alpha_e}{\alpha_h} = \gamma \tag{6.63}$$

Under the assumption of an abrupt step junction, the field becomes

$$E = E_{max} \, x/w \tag{6.64}$$

We may relate $E_{max}$ and $w$ to the total potential across the depletion region. Since we are evaluating these values at breakdown, the total voltage is the breakdown voltage $V_B$, and

$$E_{max} = 2V_B/w \tag{6.65}$$

where

$$w = \sqrt{\frac{2\varepsilon}{q} V_B \frac{N_A + N_D}{N_A N_D}}$$

Using (6.63) in (6.62) allows us to perform one of the integrals:

$$1 - \frac{1}{M_h} = 1 = \frac{1}{\gamma - 1} \int_0^w (\gamma - 1)\,\alpha_h \exp\left[ \int_0^x (\gamma - 1)\,\alpha_h dy \right] dx =$$

$$= \exp\left[ (\gamma - 1) \int_0^x \alpha_h dy \right] - 1 \tag{6.66}$$

Putting (6.64) into (6.59) and then the result of that operation into (6.66) allows us to write, with some rearrangement:

$$\log \gamma = (\gamma - 1) \int_0^w a E_{max}^n \left( \frac{x}{w} \right)^n dx = \frac{\gamma - 1}{n + 1} a E_{max}^n \, w \tag{6.67}$$

---

* A less realistic but mathematically simpler assumption can be used with surprising success. J. Maserjian [*J. Appl. Phys.* **30**, 1613 (1959)] and D. P. Kennedy and R. R. O'Brien [*IBM J. Res.* **9**, 412 (1965)] have shown that they can fit the data quite well using $\alpha_h = \alpha_e = \alpha_i$ and $\alpha_i = -\,ae^{-bE}$ with $a$ and $b$ to be determined by matching experimental data for diode breakdown.

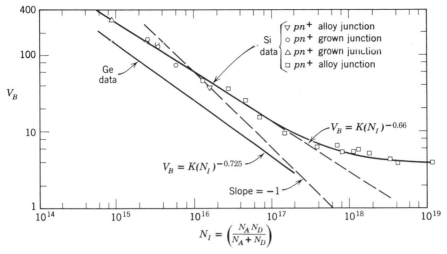

**Figure 6.19** Breakdown voltage versus doping for step junctions in silicon and germanium. The germanium data from S. L. Miller, *Phys. Rev.* **99**, 1234 (1955) and the silicon from S. L. Miller, *Phys. Rev.* **105**, 1246 (1957).

Now, insert (6.65) into (6.67) and the final result drops out:

$$\frac{n+1}{(\gamma-1)}\frac{\log\gamma}{a} = \frac{(2V_B)^n}{w^{n-1}} = \frac{(2V_B)^{(n+1)/2}}{\left[\frac{\varepsilon}{q}\left(\frac{N_A + N_D}{N_A N_D}\right)\right]^{(n-1)/2}}$$

Or, upon solving for $V_B$ and putting the sundry constants together into a single constant, $K$:

$$V_B = K\left(\frac{N_A + N_D}{N_A N_D}\right)^{(n-1)/(n+1)} \tag{6.68}$$

In matching (6.59) to the more exact exponential form (5.44), $n$ inevitably comes out $\gg 1$. Thus, when we compare (6.68) with the equivalent expression for Zener breakdown, (6.58), it is easy to see that avalanche breakdown must dominate at moderate doping levels while tunneling must predominate at very high doping levels.

It is relatively easy to measure the breakdown voltage as a function of doping from which $K$ and $n$ above may be determined. Figure 6.19 shows such data for both Si and Ge. Over a substantial range of impurity concentrations, both the curves in Figure 6.19 follow the straight-line log-log plot predicted by (6.68). The change in slope of the silicon curve as $N_I$ begins to get quite large represents a failure of our oversimplified theory. That the theory does not cover all possible doping levels should neither surprise nor

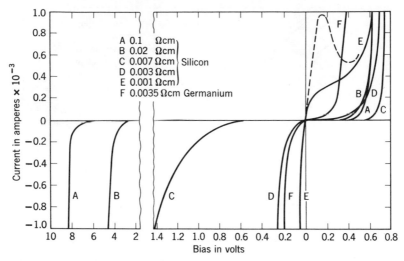

**Figure 6.20** The *I-V* characteristics of some heavily and extremely heavily doped abrupt junctions in silicon and germanium. The measured carrier concentrations for *C, D, E,* and *F* are given below. *A* would correspond to something of the order of $3 \times 10^{17}/cm^3$ and *B* to about $3 \times 10^{18}/cm^3$. All of the substrates were *n*-type and all of the *p* material was alloyed from aluminum.

| | | Silicon | | Germanium |
|---|---|---|---|---|
| Impurity, $n^+$ | As | As | As | As |
| Resistivity, ohm-cm | 0.001 | 0.003 | 0.007 | 0.0035 |
| $n$ concentration cm$^{-3}$ | $7 \times 10^{19}$ | $2.5 \times 10^{19}$ | $8 \times 10^{18}$ | $2.7 \times 10^{18}$ |
| $p^+$ alloying impurity | Al | Al | Al | Al |
| $p$ concentration cm$^{-3}$ | $1.3 \times 10^{19}$ | $1.3 \times 10^{19}$ | $1.3 \times 10^{19}$ | $4 \times 10^{20}$ |

The data and figure are from Chynoweth et al., *Phys. Rev.* **118**, 425 (1960).

alarm us. The curves predicted by (6.68) give a very good account of themselves over the entire range of normal device applications.

What is interesting about the leveling off of the Si curve in Figure 6.19 is that it suggests that tunneling might not occur in silicon. In fact, neither the germanium nor the silicon curves show any tendency to change to a slope of $-1$ as (6.58) suggests they must if tunneling is dominant. Does tunneling ever take over? The answer is an emphatic yes. Figure 6.19 is not carried far enough to show the change. Figure 6.20 corrects this omission by showing the *I-V* characteristics for some heavily and extremely heavily doped junctions in Si and Ge. Note that the breakdown voltage drops rather abruptly to values much less than 1 volt when the doping exceeds about $10^{19}/cm^3$. (The last point in Miller's data may indicate a carrier concentration in excess of what was actually there. A fairly large error in measuring the concentrations at the junction is not at all unlikely at these very high doping levels.)

That tunneling is indeed the dominant process is shown quite spectacularly by the forward characteristics of diode $E$. Unlike the others, which show the typical $I_{sat}(e^{qV/kT} - 1)$ behavior, diode $E$ shows a very large incremental conductance at the origin. This diode is, in effect, broken down at zero applied bias—in thermal equilibrium! Had the diode been constructed from a direct material such as GaAs, the conductance would have been substantially larger. In reverse bias, diode $E$ behaves like any other diode biased beyond breakdown. In the forward direction, however, the applied bias has a curious effect. Exactly at 0 bias, the tunneling currents must cancel in detail as equilibrium demands. As the bias increases, the band overlap is decreased so that the currents are unbalanced in favor of forward current. If the transition between bands is direct and the equilibrium overlap substantial, the net current flow can be quite large. However, as the bias is increased still more, the bands are completely separated (no overlap) and the tunneling current must cease. If the ratio of the tunneling current to the normal current at these low voltages is sufficiently great, an $I$-$V$ curve such as the dashed line in Figure 6.20 is observed. The current first increases rapidly with forward bias, then decreases with forward bias, and finally starts up again. Such diodes are called *tunnel diodes* or *Esaki diodes* after the man who first explained their behavior. (Previously they had been known as "splits" in the industry because the swing from the tunneling mode to the normal forward mode was much too fast to be observed on the oscilloscopes of the day. They were considered a nuisance, but it was known that one could eliminate this inexplicable $I$-$V$ characteristic by heating and diffusing the junction [and hence reducing the tunneling probability]. Then Leo Esaki explained* the curve and the industry suddenly found that they could sell these funny diodes.) That diode $E$ did not display quite so spectacular a swing from tunnel to normal forward current as the dashed line shows is an indication of how much less efficient it is to have a tunneling event that requires a phonon to conserve momentum. At sufficiently low temperature, where the Fermi function looks like a step and the number of phonons available is very small, we may actually observe structure in the $I$-$V$ characteristics that shows the tunnel current changing as the energy to generate a particular kind of phonon is reached. This is shown quite beautifully in Figures 6.21 and 6.22. Note that all of this structure disappears as soon as the temperature comes up enough to provide

* L. Esaki, *Phys. Rev.* **109**, 603 (1958).

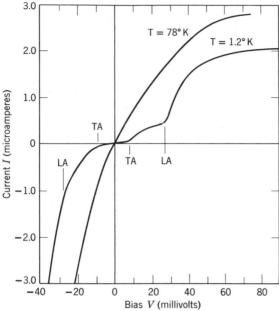

**Figure 6.21** Tunnel current measurements in two germanium diodes. The figure shows the *I-V* curves at very low values of current for low and extremely low temperatures. When the temperature is extremely low, the absence of phonons implies that indirect tunneling cannot take place until the electron has enough energy to create its own phonon. The threshold energies for transverse acoustic and longitudinal acoustic phonons show up very clearly. At 78 °K, there are ample supplies of phonons available from the thermal energy, so no phonon structure is visible. TA = transverse acoustical branch; LA = longitudinal acoustical branch. The data is from J. J. Tiemann and H. Fritzsche, *Phys. Rev.* **132**, 2506 (1963).

an ample supply of phonons. Note also, in Figure 6.22, the enormous increase in incremental conductance when the direct transition becomes energetically possible.

What we may conclude at this point is that tunneling can indeed be an important process, but that it won't be in most devices. Breakdown in most diodes will be by avalanche or by some limitation imposed by the surfaces or finite extent of the junction.

## A REVIEW PROBLEM

### Problem 6.5.

The typical way to make a silicon power rectifier is to alloy an aluminum wire into a thin *n*-type substrate. Consider such a diode with a cross-section

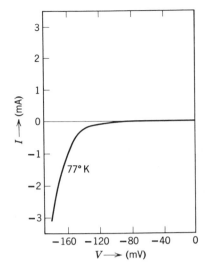

**Figure 6.22** Tunnel current measurements in two germanium diodes. The figure shows the onset of direct tunneling at 77°K. Note the enormous change from the vertical scale in the previous figure. In normal parlance, the diode in this figure would be said to be breaking down at about 140 mV. However, as you can see in Figure 6.21, tunneling is present at all voltages. This very clearly illustrates how much less efficient indirect-transition processes are compared to the direct ones. See Figures 2.30 and 2.31 for the equivalent comparison for optical transitions.

of 4 mm² on a substrate of 1 ohm-cm resistivity. The substrate thickness is 100 $\mu$m and the regrown $p$ layer is 6 $\mu$m thick. The $p$ layer has a carrier concentration of about $1.3 \times 10^{19}/\text{cm}^3$ (essentially degenerate). For the $n$-type material, an electron mobility of about 1200 cm²/Vsec is a probable value, while for the heavily doped $p$ material, something on the order of 100 cm²/Vsec is a likely value for the hole mobility. The mobility ratio, $\mu_e/\mu_h$, is 3 in both materials. The excess-carrier lifetime is $10^{-5}$ sec on the $n$ side but only $10^{-8}$ sec in the $p$ material.

You are asked to predict the following seven things [let $kT/q \simeq (1/40)\ V$]:

1. The reverse breakdown voltage.

2. The incremental capacity at $-50$ V.

3. The voltage across the diode for a forward current of 1 mA.

4. The current and voltage at which high-level injection effects begin to be evident (i.e., when $p_n' \simeq n_{no}$).

5. The incremental conductance and capacity at 1 mA. (Are there any limits on the frequency at which these incremental quantities may be used?)

6. If the $n$ side were short based—so you could ignore the resistive drops and the decay of the minority current—what would the voltage drop be at a forward current of 1 amp? Break this voltage into the component across the junction and the portion across the bulk material.

7. Under the short-base assumption of problem 6, determine the incremental conductance at a current of 1 ampere.

*Answers:*

(1) About 100 V (Figure 6.19); (2) $c = 118$ pF; (3) 0.552 V; (4) 0.641 V and 43.3 mA; (5) $g = 0.04$ mhos, $c = 0.14$ $\mu$F with the requirement that $\omega\tau \ll 1$ or $\omega \ll 10^5$ sec$^{-1}$ [see (6.33)]; (6) $V_T = V' + V'' = 0.718 + 0.077 = 0.795$ V; and (7) $g = (q/2kT) I = 20$ mhos.

## CHAPTER SUMMARY

This chapter has been devoted to the principal building block in diffusion-controlled semiconductor devices—the junction diode. The development proceeded at two levels. The initial effort was to establish a first-order physical model that would predict the basic *I-V* characteristics for the junction diode. To achieve this goal, it was necessary to postulate two types of boundary conditions: the ohmic contact and the law of the junction. At the postulated ohmic contact, the excess-carrier concentration was always zero, while at the edge of the depletion region at the *p-n* junction, the carrier concentration was given by

$$\frac{p_n}{p_p} = \frac{p_{n0}}{p_{p0}} e^{qV/kT}$$

$$\frac{n_p}{n_n} = \frac{n_{p0}}{n_{n0}} e^{qV/kT}$$

With these boundary conditions, we were able to find the carrier concentrations everywhere within the diode.

To keep this first model simple, we restricted our attention to situations in which there was negligible recombination or generation in the depletion region and in which the minority-carrier density was always much less than the majority-carrier density (i.e., the *low-level-injection* assumption). With the no-recombination assumption, we were able to determine the total current by simply adding the easily determined minority currents on either side of the depletion region. The minority currents are easy to find in the low-injection case because they are determined entirely by diffusion. The electric field may be ignored.

The DC current-voltage relationship was found to be

$$I = I_{sat}(e^{qV/kT} - 1)$$

where the saturation current, $I_{sat}$, is given by (see Figure 6.2):

$$I_{sat} = qA \left[ \frac{D_h p_{n0}}{L_h} \coth\left(\frac{c-b}{L_h}\right) + \frac{D_e n_{p0}}{L_e} \coth\left(\frac{a}{L_e}\right) \right]$$

The exponential *I-V* relationship is a direct result of the boundary value assumption. To reinforce the rationality of making that assumption, we examined the extent to which detailed balance would be upset by an applied potential. The flux argument supported the assumption of a quasi-equilibrium distribution of carriers throughout the depletion region.

Three different "base lengths" were defined in $I_{sat}$ above; the bulk region was called *short based* if its width were much less than a diffusion length, *long based* if the converse held true, and *intermediate based* if the width were of the order of a diffusion length. For the short-based case, very few of the excess carriers will recombine before reaching the ohmic contact; conversely, for the long-based case, very few of the minority carriers can reach the ohmic contact.

The voltage that appears in the exponential $I$-$V$ relationship is the voltage across the depletion region. To see whether there were any other voltages across the diode, we examined the voltage drops across the ohmic (bulk) regions (problem 6.2) on page 302. We found that the drop across the bulk regions was ohmic and that it comprised an ambipolar term and a simple resistive term. The drop tended to be pretty insignificant for all but very long-based cases, although it could be perceptible in very lightly doped regions.

Having established a simplified model for junction-diode behavior, we next turned to a brief examination of junction-manufacturing techniques. Both liquid and gaseous epitaxy were discussed. Each gave a technique for growing reasonably thick layers of one type on a substrate of another type. In liquid epitaxy, if the solvent is a dopant (e.g., gallium or aluminum), one obtains an extremely heavily doped layer. Alloy junctions are generally of this type. Lightly doped layers may be obtained by using inactive or insoluble (in the solid state) solvents [e.g., tin (inactive) or lead (insoluble and inactive)]. In gaseous epitaxy, one may easily control the doping throughout the growth process, producing tailor-made doping profiles.

The section on manufacturing techniques concluded with a brief development of the diffusion process. Both a one-step diffusion from a fixed surface concentration and the two-step predeposition-with-drive-in methods were examined. The reasons that this type of process is so popular will become evident in the next chapter, where it will prove to give the designer control over many of the factors essential to good transistors. Our results in this chapter showed that the diffusion technique led to graded impurity profiles, with steep gradients being obtained by short, shallow diffusions and shallow gradients by long diffusion times. The rate at which the doping material is transported into the wafer is proportional to the diffusion coefficient, which in turn depends exponentially on $1/T$. Thus, the diffusion process is quite dependent on temperature. For silicon, typical diffusion times are a half hour to several hours at temperatures between 1000°C and 1300°C, giving diffusion depths of several microns.

Once we had several manufacturing techniques at our disposal, we could examine the ohmic contact and the validity of the boundary condition we have been using to describe it. The usual metal-to-semiconductor junction is a rectifying one because of the barrier created by the large number of deep surface states. To achieve an ohmic contact, we found that one must dope the surface heavily enough to fill or empty these surface states completely, driving the Fermi level into either the valence or conduction bands. This degenerate doping at the surface allows the Fermi levels in the metal and semiconductor to line up without creating a barrier to majority-carrier flow. In this sense, the junction becomes ohmic; majority carriers can flow readily in either direction. Minority carriers, on the other hand, find a barrier at the junction because of the graded doping within the semiconductor itself. The free flow of minority carriers could occur only if the recombination rate within the heavily doped region were exceedingly short. Such short lifetimes are certainly not surprising in degenerately doped material, but the ohmic character of the junction then depends on the current flowing out of the junction being a small fraction of the generation rate. If such is the case, it is a reasonable assumption that the excess-carrier concentration at the ohmic contact is zero. On the other hand, it should be possible to make such contacts appear nonohmic by driving enough minority current into the junction. Such effects can be observed, but they normally fall well outside the range of practical device operation.

Having established the plausibility of our first-order model, we next developed a small-signal representation of the junction diode based on that model. The exponential $I$-$V$ characteristics led directly to a small-signal forward conductance: $g \simeq (q/kT)I$. Under reverse bias, $g \simeq 0$. Parallel with this conductance was an incremental capacity comprising two terms. The principal capacity under reverse bias arose from charge storage in the depletion region of the diode. This depletion or transition capacity was given by $c_{\mathrm{DEP}} = \varepsilon A / l$ where $A$ is the junction area and $l$ the depletion width. $l$ is a function of the doping and the applied bias. For the abrupt, uniformly doped junction it is given by (6.40); for the linearly graded junction, by (6.43).

A second capacity became important under forward bias. This capacity arose from charge storage in the neutral regions of the diode. We found that although the capacitive susceptance arising from this charge storage was readily obtained, it did not appear to be a traditional $s = \omega c$ relationship except at relatively low frequency. At high frequencies, the susceptance varied as $\omega^{1/2}$. It is possible to keep this diffusion susceptance linear over an extended range by making the diode very short based. However, this has the effect of making the susceptance negligibly small compared to $g$. Thus, we must conclude that the diffusion susceptance for the diode is either very small or unfortunately nonlinear in its frequency dependence. (This conclusion will not hold for the transistor, since the rather small capacity in the narrow base region occurs in a place where a small capacity is important. The reasons will become clear in the next chapter.)

Having a large and small signal representation of the first-order junction-diode model, we returned to our original assumptions to see where they would begin to fail in important ways. We considered three cases where one or more of the assumptions required for the first-order model were no longer valid. These included the case of low current-levels in moderately large band-gap materials where we found that we could not ignore the recombination and generation in the space-charge region; the moderately high-level injection situation where the excess-carrier concentration on one side of the junction exceeded the equilibrium majority-carrier density on that side; and the junction at reverse voltages high enough to break it down. For forward bias the first two situations led to identical incremental $I$-$V$ curves with

$$d \log (I)/dV = q/2kT$$

For the low-level situation, the $q/2kT$ slope arose from the dependence of the recombination-generation rate on the $pn$ product in the depletion region. For the high-level case, the slope was determined by the fact that the applied-voltage increments were equally divided between the junction and the lightly doped neutral region.

For reverse bias, generation within the space-charge region gave a reverse current substantially above the saturation current expected for the large-band-gap materials. Furthermore, the reverse current was proportional to the width of the depletion region, so the current was not independent of the voltage. We found that germanium, with its band gap of 0.66 eV, showed almost no depletion-region current at room temperature. Silicon, on the other hand, with its band gap of 1.1 eV, showed a substantial range in which the current was determined by the depletion region. For a material like gallium arsenide, the diffusion component barely becomes important before high-level injection sets in. Since the depletion-region currents are proportional to $n_i$ while the diffusion terms are proportional to $n_i^2$, all of these conclusions about which current is important may be altered by changing the temperature.

Our final topic was junction breakdown. We considered both essential causes of breakdown: Zener breakdown and avalanche breakdown. Our analysis showed that Zener breakdown was not likely to be observed except in junctions that are degenerately doped. Such junctions will break down at voltages below 1 V and can even be made so that they are broken down at 0 V. For voltages much above 1 V, the mechanism appears to be avalanche breakdown. The analysis of the avalanche process led to a number of uncomfortable equations that were solved with judicious approximations. The net result was an expression for the breakdown voltage as a function of the doping on each side of the junction:

$$V_B = K \left( \frac{N_A + N_D}{N_A N_D} \right)^{(n-1)/(n+1)}$$

This expression was appropriate to the abrupt, uniformly doped junction such as those one gets from an alloy process. The constants $K$ and $n$ must be determined for each material.

This completes the analysis of the junction diode, one of the two principal building blocks of semiconductor technology. Our next step is to consider a device made up of two such diode blocks—the junction transistor. This is the topic for the next two chapters. The chapter after that will deal with another junction-diode transistor—the JFET. Then we must turn to the other building block, the semiconductor oxide interface. This will lead us to the MOSFET, the final topic in the book.

### Problems

1. In the review problem at the end of the chapter (problem 6.5 on page 348), you were asked to compute a number of parameters for a silicon power rectifier. If the diode were made on an 0.3 ohm-cm germanium substrate instead (so that the lifetimes would be about 10 times longer, the mobilities about 3 times larger, the dielectric constant 1.33 times larger, and the band gap 40% smaller) what difference in performance would you expect? You need not limit yourself to the items considered in problem 6.5.

2. Some of the properties of a silicon $p$-$n$ junction are given below.

   (*a*) Determine the maximum bias at which the low injection assumption is useful.
   (*b*) At what frequency would the absolute value of the AC diffusion length on the $n$ side be $1/\sqrt{2}$ of the DC diffusion length?
   (*c*) What would the forward current be at 0.3 V? (Careful!)
   (*d*) Calculate the diode small-signal admittance at 5 GHz and 0.6 V DC bias.
   Parameters of the diode:

   | | *p side* | *n side* | |
   |---|---|---|---|
   | Doping | $10^{18}$ | $10^{16}$ | $cm^{-3}$ |
   | Width | 3 | 200 | $\mu m$ |
   | $D_e$ | 20 | 30 | $cm^2/sec$ |
   | $D_h$ | 7 | 10 | $cm^2/sec$ |
   | $\tau$ | $10^{-7}$ | $10^{-6}$ | sec |

   Junction area: $10^{-5}$ $cm^2$
   Let $kT = (1/40)$ eV

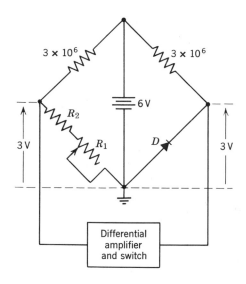

3. A germanium junction diode is to be used as the sensing element in an "all-solid-state" thermostat for household application. The proposed circuit is simply a Wheatstone bridge with the reverse-biased diode in one of the arms and an adjustable resistor in the other arm to set the temperature. The bridge is shown below with the setting appropriate to 70°F. You are asked to specify $R_2$ and the taper and the value of the wire-wound variable resistor $R_1$, so that a linear scale will result from 55°F to 90°F. What is the sensitivity of the bridge $(dV/dT)$ at 70°F?

4. One may use a junction diode operated in the forward mode as a voltage-variable attenuator. The diagram below shows a typical circuit. Assume that $I_{sat}$ is $10^{-6}$ amperes (i.e., a Ge device) and that $\omega\tau \ll 1$. What DC power would be required to give an AC attenuation of 1/2?

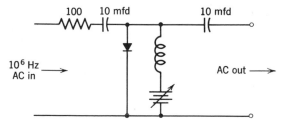

5. Diodes find much employment in computer circuitry, where they serve as voltage-sensing switches connecting successive amplifiers. At the speeds employed in modern computers, the time that it takes a diode to switch

from "on" to "off" and back is an important fraction of the total delay time. To turn the diode "off," one is obliged to remove the minority-carrier charge stored in the "on" state. The trick in making a fast diode is to minimize the amount of charge that must be moved. One of the standard procedures is to minimize the bulk minority-carrier storage by diffusing in a lifetime killer (usually gold) to make the diffusion length exceedingly short. To see what this buys you, compare the speed of the two diodes described below that are identical in all ways except for the addition of the gold.

Let the diodes be alloyed $p^+n$ structures on a 200 $\mu$m thick die. Let $D_h = 10$ cm$^2$/sec on the $n$ side. One diode has minority lifetime of $10^{-7}$ sec, the other of $10^{-9}$ sec. As you can see, both are certainly long-based. Prior to switching, each diode is carrying 1 mA. Switching is achieved by reversing the current until 90% of the total charge has been removed. By that time, the diode's voltage drop is large enough to make the current fall, and switching is complete.

(**a**) How much time would it take each of the diodes to switch if one may ignore the depletion capacity?

(**b**) If the junction area were $10^{-5}$ cm$^2$ and the $n$-side doping $10^{16}$/cm$^3$, would the change in the depletion charge make a significant contribution to the delay time?

*Answers:* $0.9 \times 10^{-7}$ sec and $0.9 \times 10^{-9}$ sec. The contribution of the depletion charge is small compared to the diffusion charge in the slow diode. For the fast one, the maximum junction voltage must be specified. At 10 V reverse bias, the depletion charge has changed by about as much as the diffusion charge.

6. In a germanium junction diode at room temperature, the dominant current is the diffusion current. At lower temperatures, we might expect the de-

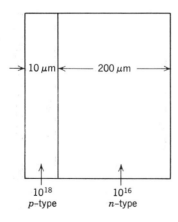

10 $\mu$m | 200 $\mu$m

$10^{18}$
$p$-type

$10^{16}$
$n$-type

pletion-region currents to become more important. An exact analysis is complicated by the fact that the capture cross-sections and the mobilities are rather obscure functions of temperature. On the other hand, the variations in $\tau_{e0}$, $\tau_{h0}$, $\mu_e$, and $\mu_h$ are likely to be relatively small compared to the variations in the carrier concentrations that depend exponentially on $1/T$. Thus, a preliminary analysis might well ignore all the temperature dependencies other than the exponential ones. On that basis, consider the following problem.

A one-dimensional representation of a germanium junction diode is shown below. Calculate the ratio of the depletion to the diffusion current at $-5$ V bias and $300°$K. Then determine the temperature at which the two terms would be equal to each other.

Diode area $= 10^{-4}$ cm$^2$

$\tau_{e0} = \tau_{h0} = 10^{-5}$ sec

$\mu_p$ on $n$ side $= 2000$ cm$^2$/Vsec

$\mu_n$ on $p$ side $= 3200$ cm$^2$/Vsec

# DEVICES THAT WORK PRINCIPALLY BY DIFFUSION: (B) A PHYSICAL MODEL OF THE BIPOLAR JUNCTION TRANSISTOR (BJT)

In the last chapter, we explored in considerable detail the various useful and nonuseful ways in which an individual junction diode could behave. It was in the pursuit of some of that information—the nature of the surface states—that J. Bardeen and W. H. Brattain first placed two diodes very close together on the same germanium substrate. They found that the current through one diode could be modulated by the current through the other, resulting in a rather large voltage gain ($\sim 20$ dB). They had, in fact constructed the first *bipolar transistor* [J. Bardeen and W. H. Brattain, *Phys. Rev.* **75**, 1208 (1949)].

From a historical point of view, it is interesting to note that the bipolar transistor was discovered in the effort to realize another rather similar valve— the *field-effect transistor* (FET). Shockley and Pearson had shown that such devices *should* be reasonably effective, but the first ones were not. In experiments to find out why, the bipolar transistor was discovered. Everyone jumped on the bipolar bandwagon; the FET was almost—but not quite— forgotten. It is not obvious why. The problems that had to be solved to make a good FET were both fewer and simpler than those that lay before the newborn bipolar transistor. It may have been that the large voltage gains that had been demonstrated for the bipolar transistor made it seem much more attractive than the FET, which appeared to provide only current and power gain. By the time the first effective FET was made by [G. C. Dacey and I. M. Ross, *Proc. IRE* **41**, 970 (1953)], it was known that FET's would also yield substantial voltage gains [W. Shockley, *Proc. IRE* **40**, 1365 (1952)], but by that time, the bipolar transistor was off and running. There seemed little chance of the FET ever challenging its supremacy.

Then, about 1960, a new element was added to the unequal contest. With the development of planar technology, people began to probe the possibility of using semiconductor processing techniques to manufacture whole functional blocks instead of discrete components. The economics of manufacturing strongly favor the FET in integrated circuits, so if the bipolar transistor does not have substantial electrical advantages, the FET will be the device selected.

Today both devices are effective, competitive, and reliable. For a good many years from now, the device and circuit engineer (more and more that's getting to be the selfsame person) will have to be thoroughly conversant with both devices. We will begin with the bipolar transistor, not for historical reasons, but because you are now primed for diode-diode interactions from the last chapter. In Chapter 9 we will turn to the junction FET (the one proposed by Shockley in his 1952 paper cited above) for our first look at a drift-controlled device. Finally, in Chapter 10 we will return to the original FET—the MOSFET—which was the element under study at the time of the invention of the bipolar transistor and the one that is most seriously challenging bipolar supremacy.

The analysis of the junction transistor proceeds in almost complete parallelism to the junction diode. In fact, with the background of Chapter 6, it is possible to do a rather complete analysis of the transistor in a relatively short space. Only two topics will require extensive development beyond the diode analysis we have already done. These are the small-signal analysis of the transistor and some of the limitations on the high-frequency performance of transistors. We will take a second look at planar-manufacturing techniques, since the *raison d'être* for the planar process is the control it provides over critical transistor parameters.

Just as in the analysis of the junction diode, a simplified first-order analysis provides a remarkably good general picture of transistor performance. From this analysis, two very useful circuit models may be constructed, one for linear (small-signal) analysis of transistor behavior, the other for large variations in bias. The models are quite durable, being equally useful when the transistor is driven into those regions of operation where the first-order analysis must be modified.

## THE BIPOLAR TRANSISTOR

Any three-terminal, solid-state device capable of linear amplification at zero frequency can be legitimately called a *transistor*. There are a large number of things that might be so labeled. The group that we are going to discuss in this chapter is called *bipolar* because these transistors require the flow of carriers of both polarities. The essential elements of the bipolar transistor are two diodes *in very close proximity to each other on the same substrate*. These diodes have all of the characteristics of individual junction diodes discussed in the last chapter plus one all-important new property. With an individual diode under reverse bias, the only source of minority carriers is generation within the volume or at the ohmic contact. In the bipolar transistor the second diode, since it is so close to the first, provides another and potentially very large source of minority carriers. The important feature of this second source is that it is a *controlled* source: The current flowing through the first diode is determined in large measure by the potential across the second diode. Note carefully that this current *collected* by the first diode is almost completely independent of the degree to which the first diode is reverse biased; a reverse-biased diode collects all of the minority carriers within a diffusion length of the depletion region no matter how they got there. Since the current through the first diode is independent of the voltage across it, that diode acts like a current source—one that is controlled by a voltage across another diode. This is the essence of transistor action.

Keep in mind that statements such as "the diode acts like a controlled current source" are really untrue. A current source is also an energy source, but a transistor is nothing but a valve. Although you certainly can get plenty of action in the shower when you turn on the valve, it is definitely not true that the valve by itself is a water source. It would be much more proper to say that the transistor and the *bias source* (e.g., the battery) together act like a current source. Nonetheless, I will follow the custom of describing valves as sources without making specific reference to their bias supplies. The only time this practice is likely to lead you astray is when the valve is fully open— that is, when the flow is limited not by the valve but by the bias supply. When such a state obtains, it is usually said that the transistor is "saturated." This is a doubly unfortunate choice of words. First, it is the power supply and not the transistor that has run out of reserves. Second, the word "saturated" also implies a mode of operation in which the current is rather independent of the applied voltage (as in $I_{sat}$) which is the opposite of the situation just described where the voltage of the power supply is precisely what is determining the current.

We now turn to a quantitative analysis of transistor action by applying the methods of Chapter 6 to the structure shown in Figure 7.1.

## THE PLANAR BIPOLAR TRANSISTOR AND A SIMPLIFIED MODEL

Figure 7.1 shows the layout of a small-signal transistor manufactured by the planar technique discussed in the previous chapter. The two diodes are created by diffusion from the same surface. First, a moderately large window is opened in the oxide layer on the *n*-type substrate. A *p*-type diffusion (frequently boron) through the window results in a *p-n* junction at a depth of the order of 3 to 4 $\mu$m. During this diffusion, an oxide layer is grown over the

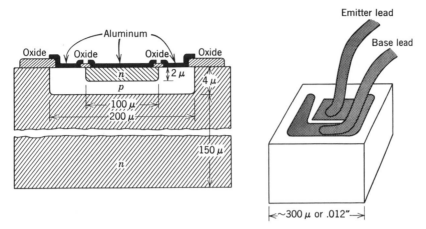

**Figure 7.1** Dimensions of a typical small-signal planar transistor. Note that the vertical scale is expanded by a factor of 10 over the horizontal scale. It is difficult, at first, to get used to the dimensions of real devices. The actual transistor, before it is put on a header or in another package, is roughly the size of a period, an object readily picked up by delicate tweezers but even more easily dropped. The wires shown in the drawing on the right are 25 $\mu$m in diameter (0.001"), about the limit of visual resolution.

window. A second, smaller window is opened in the center of the oxide layer covering the first diffusion. Through this second window, an $n$-type diffusion is performed (typically phosphorus), forming another junction within the volume of the first diffusion. At the end of the second diffusion, one has a layered structure—$n$-$p$-$n$—with very small spacing between the two junctions. The section in Figure 7.1 gives reasonably representative values for the dimensions of a typical transistor. Note that the vertical and horizontal scales differ by a factor of 10. Spacings between the layers are of the order of microns, while the windows are of the order of 100 microns wide. Thus, an undistorted drawing must be either gigantic or very hard to read. However, given Figure 7.1 to see how things fit together, it is a good idea to see just what an undistorted section of a transistor would look like. Figure 7.2 shows such a section at a magnification of 200 $\times$. Although the whole chip is pretty thin by conventional standards (of the order of the thickness of a few sheets of tablet paper), the active device is contained in a layer that is but a third of the thickness of a sheet of Saran wrap. Such a layer is not thin by atomic standards, however. To be 4 $\mu$m under the surface is to have 10,000 atoms between you and the surface layer.

The rather large aspect ratio that is evident in Figure 7.2 suggests that a reasonable analysis of the action of the transistor might be achieved by replacing the two-dimensional structure with a plane (rather than planar)

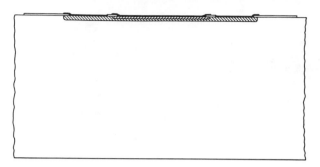

**Figure 7.2** An undistorted enlargement of the planar transistor in Figure 7.1. The magnification is 200 ×.

structure that is essentially one-dimensional. In effect, we would be taking the same approximation that is usually employed in analyzing the parallel-plate capacitor; we would be ignoring the fringing that occurs at the edges. As long as the effects that we wish to analyze do not depend critically on the local electric field, such an approximation should be rather accurate. However, before you accept the plane approximation as obvious and universal, think of at least one circumstance of practical importance in which the device's behavior depends critically on the local electric field. (An answer: The one that occurs to me is junction breakdown. If avalanching occurs at the corners, it will more than likely swamp out any other current flow that you may have been seeking.)

There is another problem with the one-dimensional approximation that is inherent in the bipolar transistor: any current that flows in or out of the base lead must run at right angles to our "one dimension." This is a fact that we will have to live with, since it turns out to have some first-order effects on transistor performance. However, we can still have an "almost" one-dimensional structure that includes transverse flows. Such a structure is shown in Figure 7.3. This is the figure upon which we will build all of our theory.

In comparing Figure 7.3 with Figure 7.1, notice that the active portion of the transistor is taken to be the area directly under the emitter. The portions of the base-collector junction that lie outside this area are represented by an essentially independent diode that shunts the active base-collector junction. (*Active* should be interpreted as meaning directly involved in the transistor action. The shunting diode is no less effective than any other junction diode.) The shunting diode is usually referred to as the *overlap diode.*

It should be stressed that the one-dimensional approximation, which is based on the dimensions of typical transistors, is not a critical assumption. Although the coefficients that we will obtain depend on the geometry, the

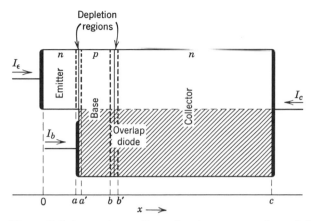

**Figure 7.3** A quasi one-dimensional representation of the bipolar transistor. The emitter, base, and collector are represented as having the same cross-section, with a second diode (the *overlap diode*) shunting the base-collector junction to represent those portions of the base-collector junction that are relatively remote from the emitter junction. Ohmic contacts are indicated by the solid bars with wires coming out; the edges of the depletion regions are indicated by dashed lines.

general results do not. The relationships between the terminal currents and voltages are really quite independent of the particular transistor's configuration.

## FLUXES AND CURRENTS WITHIN A BIPOLAR TRANSISTOR

The terminal currents of a transistor, which are defined by the standard convention as being positive going into the transistor, may be broken up into components that correspond to the physical processes going on within the transistor. This is a useful step in obtaining insight into the behavior of a transistor; however, when circuit analysis is the object, it is obvious that the terminal currents are the terms to work with. We begin with the components.

Consider the things that happen to the flow of holes in a junction diode as they go from the *p*-side ohmic contact to the *n*-side ohmic contact. The holes are generated at the ohmic contact and flow, let us say, to the right. The electrons generated at the same time flow away to the left through the metal lead attached to the ohmic contact. (Gustav Kirchhoff would have it no other way!) As the holes traverse the *p* region, they encounter excess electrons injected from the *n* side. On occasion, when circumstances are favorable, some of these excess carriers recombine, reducing the net flux of holes to the right and electrons to the left. Those holes that reach the junction traverse the depletion region. (N.B. We are dealing here only with the net flux—

the difference between the incident and reflected flux at the junction barrier. In the analysis of currents, it is only those carriers that get through that count.) Within the depletion region, recombination is much more probable (more of the traps are effective), so a noticeable fraction of holes may not make it across the entire depletion region. Those that do, diffuse across the n side toward the ohmic contact. A fraction of these recombine before reaching the contact; the remainder recombine at the ohmic contact.

A completely similar set of statements could be made for the electron flow from right to left. (The only exception is at the contacts. The flow of electrons through the contact is continuous, since the same type of carrier operates both in the metal and in the semiconductor. There is nothing particularly significant about this fact, however. If you prefer to have holes flow through the contact, use tungsten-wire leads. Hall measurements indicate that tungsten is a hole conductor.) If we wish to describe the same junction under reverse bias, we need only interchange the direction of the flows and substitute the word *generation* for *recombination*.

The principal reason for separating the flux into components according to where the recombination is taking place is analytical convenience. It is relatively easy to write algebraic expressions for the minority-carrier fluxes and not much more difficult to include recombination within the depletion region. Then, by an application of Kirchhoff's current law, the terminal currents are determined.

A similar analysis is quite appropriate to the bipolar transistor. There are two principal differences: Carriers can go from one terminal to either of two others, and the base region is bounded on both sides by a *p-n* junction. Thus, for an *n-p-n* transistor, such as that illustrated in Figure 7.3, we can have an electron flux going from left to right and another one, *entirely independent*, going from right to left. This bilateral symmetry—*n-p-n* spelled backwards is *n-p-n*—is so fundamental to the operation of the transistor that we should take a moment to look at the physical foundations on which it rests.

For majority carriers, a *p-n* junction represents a barrier. Only those carriers with enough energy can surmount the barrier. For minority carriers, on the other hand, the junction is a cliff to fall off. Thus, any electrons that are injected into the *p*-type base from one side and diffuse to the other side will fall out of the base on that other side. Another way of saying this that has direct applicability to the analysis of the flow rates is: For carriers injected into one side of the base, the other side looks like an ideal sink for excess carriers (just like an ideal ohmic contact). Implicit in such a statement is the independence of the two flows; it is as if we could label the electrons from one side and keep them separate from those from the other side. Of course this is not possible, but if the equation describing the flows is *linear*, any flow is the simple sum of its parts no matter how they are divided. There is certainly no

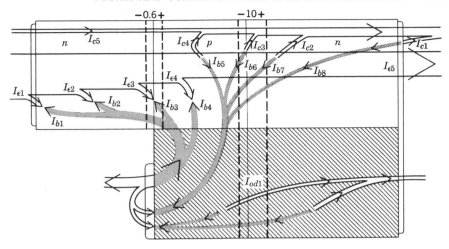

**Figure 7.4** Carrier flows in an *n-p-n* transistor in the normal active mode. Electrons are shown as clear arrows; holes are shown as gray arrows. The arrows indicate the flow of a particular component. If the arrows meet, recombination is taking place. If the arrows are in opposition, generation is occurring. At the base ohmic contact, both generation and recombination are shown as occurring. The currents—as opposed to the carrier flows—may be visualized in the above drawing by reversing the arrows on the electron flows. Some effort has been made to show which elements are likely to be large and which small, but the widths of the lines should not be taken to be even remotely quantitative.

doubt that the diffusion equation (5.27′) is a linear differential equation. Accordingly, our division of the total electron flow in the base into a flow to the right and a flow to the left is both physically sensible and mathematically acceptable.

Let us now disassemble the currents in a transistor as we just did for the single diode. Let the emitter junction be forward biased and the collector junction reverse biased. This choice will permit us to look at the effects of both types of bias at once. It also corresponds to the regime in which most transistors are operated. Figure 7.4 shows the carrier fluxes that would be flowing and Table 7.1 presents a list of the currents according to their destinations. It is important to keep in mind that under conditions of reverse bias, recombination becomes generation (and vice versa) and the flow reverses. Thus, the upper electron flux in Figure 7.4 is, in a peculiar sort of way, a mirror image of the lower electron flux. Each has five components that arise from recombination or generation in equivalent locations; each contributes to the terminal currents at both the emitter and collector; and each is responsible for a portion of the base current. Note that the base current for the

**Table 7.1.** A List of the Currents Flowing in a Transistor Delineated According to the Source and Destination of the Carriers Involved[a]

| Current Component | Current Carrier and Its Destination | Related Component |
|---|---|---|
| $I_{\epsilon 1}$ | Electrons that recombine at the emitter ohmic contact. | $I_{b1}$ |
| $I_{\epsilon 2}$ | Electrons that recombine within the emitter bulk. | $I_{b2}$ |
| $I_{\epsilon 3}$ | Electrons that recombine within the emitter-base depletion region. | $I_{b3}$ |
| $I_{\epsilon 4}$ | Electrons that are injected into the base and recombine within the base bulk region. | $I_{b4}$ |
| $I_{\epsilon 5}$ | Electrons that are injected into the base and diffuse to the collector junction where they fall into the collector and flow out the collector lead. | |
| $I_{b1}$ | Holes injected into the emitter that reach the emitter ohmic contact. | $I_{\epsilon 1}$ |
| $I_{b2}$ | Holes injected into the emitter that recombine within the emitter bulk region. | $I_{\epsilon 2}$ |
| $I_{b3}$ | Holes injected toward the emitter that recombine within the emitter-base depletion region. | $I_{\epsilon 3}$ |
| $I_{b4}$ | Holes that recombine with excess electrons in the base. | $I_{\epsilon 4}$ |
| $I_{od1}$ | Diffusion current in the overlap diode. | |
| $I_{od2}$ | Overlap diode current resulting from depletion region recombination-generation. (This component is not shown in Figure 7.4 to reduce the confusion.) | |
| $I_{c1} - I_{c5}$ | Same as $I_{\epsilon 1} - I_{\epsilon 5}$ but interchanging the words collector and emitter. (Note that if the junction is reverse biased, as the collector junction is in Figure 7.4, *recombination* becomes *generation* and the direction of the flow reverses.) | $I_{b5} - I_{b8}$ |

The leftmost brace groups the first nine rows as "Currents controlled primarily by $V_{be}$" and the last three rows as "Currents controlled primarily by $V_{bc}$".

[a] The current listed as *related component* on the right is the one that completes the flow of charge through the transistor. All of the currents except $I_{od2}$ are shown schematically in Figure 7.4.

bias given could conceivably have either sign depending on the relative sizes of the various components. The emitter and collector currents, on the other hand, are clearly $-$ and $+$ respectively.

Although I have stressed the bilateral symmetry of the active part of the transistor, keep in mind that there are usually some very significant asymmetries in the doping and other coefficients applicable to the three layers of the transistor. In fact, although it is possible to make an almost perfectly symmetric transistor, it would be far inferior to the very asymmetric ones that are normally manufactured. Reasons for this will become quite clear as soon as we have analytical expressions for the terminal currents at our disposal. These we now proceed to obtain.

## THE TERMINAL CURRENTS AS FUNCTIONS OF THE APPLIED BIAS

There are certainly an awesome number of current components illustrated in Figure 7.4, yet in their analysis, we will need to recognize only two different types of current. The most important class comprises minority-carrier currents in the bulk regions; the other class consists of the currents arising in the depletion regions. For the first class, the steps leading to (6.10) will give us the current as a function of the voltage across the injecting junction. For the second class, an application of (6.55) gives the current as a function of the applied bias. Our first task is to write the terminal currents, grouping together those terms that require a single application of one of the two analysis techniques. Note that, as usual, we will be finding the current by evaluating the minority-carrier currents at the edges of a $p$-$n$ junction depletion region.

$$I_\epsilon = -(I_{b1} + I_{b2}) + (I_{\epsilon 4} + I_{\epsilon 5}) - I_{c5} \qquad\qquad - I_{b3} \qquad (7.1)$$

diffusion terms     depletion terms

$$I_c = -(I_{b5} + I_{b6}) + (I_{c4} + I_{c5}) - I_{\epsilon 5} - I_{od1} \qquad - I_{b7} - I_{od2} \qquad (7.2)$$

[The sign convention employed in (7.1) and (7.2) is that a current is positive if it goes *into* the lead whose label it bears. Thus, $I_{b3}$ is positive if it makes the *base* current more positive. The overlap diode current is chosen as positive *into* the base.] The base current in Figure 7.4 could be evaluated in a similar manner by adding up the components in the figure, but a much more utilitarian approach is to invoke Kirchhoff's current law, by which

$$I_\epsilon + I_b + I_c = 0 \qquad (7.3)$$

All that remains is to tabulate the results. With surprisingly little experience, they can be written down as "obvious." You must keep several points in mind. The diffusion currents are found, as before, by solving the diffusion equation subject to proper boundary conditions. However, since we have the same boundary conditions as in the simple diode case, we just get the diode solution for each individual group of terms. The only place where this equivalence to the diode case is not obvious is the base region. Notice carefully that the excess minority carrier density at $a'$ in Figure 7.3 is given by

$$n'_b(a') = n_{bo}[e^{qV_{be}/kT} - 1] = (n_i^2/N_{Ab})[e^{qV_{be}/kT} - 1] \qquad (7.4)$$

while at the other side $(b)$ it is

$$n'_b(b) = (n_i^2/N_{Ab})[e^{qV_{bc}/kT} - 1] \qquad (7.5)$$

However, if you accept the idea that any minority carrier will pass right through the depletion region, then the excess-minority-carrier density of carriers *going to the right* at $b$ or *to the left* at $a'$ is 0. Equations 7.4 and 7.5 represent the excess density of carriers being injected *into* the base. Thus, for carriers going to the right, the boundary conditions are (7.4) on the left and $n' = 0$ on the right. These are the same boundary conditions that were used for the diode, so we get the same solutions. (Notice that the *sum* of the solutions at $a'$ and $b$ add up to the actual boundary conditions, as indeed they must!)

Before actually trying to write down the solutions, we must adopt some conventions and make a few noncritical assumptions, or the notation and equation length will be overwhelming. According to Figure 7.3, there are two areas of importance: the emitter area, $A_e$, and the overlap-diode area, $A_{od}$. Let the lengths of each of the regions by symbolized by $w$ with an appropriate subscript [e.g., $w_b = (b - a')$]. Since each layer of the transistor will be characterized by its own doping, minority diffusion length, and minority diffusion constant, it is very cumbersome to carry two subscripts to indicate the carrier type and the location of the carriers. Accordingly, we will keep only the location subscript (e.g., $D_b$ for the minority diffusion constant in the base). Let us assume that each layer is uniformly doped. (We'll correct for this later when it becomes important.) In evaluating the depletion currents, let us condense all of those really experimentally inaccessible coefficients into a single lifetime by making the same assumptions that we did to get (6.55). That is: $\tau_{ho} = \tau_{eo} = \tau$ and $n_T = p_T = n_i$. Finally, let us assume that all of the voltage differences between terminals appear across the junctions. (This is another condition we must later relax.)

Using these conventions and approximations, we may now write down explicit expressions for the components of the terminal currents. These are presented in Table 7.2 without further compromise. Examine each of the terms, referring to Figures 7.3 and 7.4 as needed, until you see where each one came from and agree with the result. Then let us turn to some interpretations of the results.

To get a good feeling for what Table 7.2 implies, consider the behavior of a transistor in the *common-base* circuit shown in Figure 7.5. The figure contains all of the elements required to make a simple transistor amplifier with $v_{in}$ being the signal voltage and $v_{out}$ being that portion of the voltage

**Table 7.2.** Evaluation of the Components of the Terminal Currents in Terms of the Voltages Applied to the Junctions[a]

| | Current | Value in Terms of the Junction Voltage | Comments |
|---|---|---|---|
| **Elements of the emitter current** — Controlled by $V_{b\epsilon}$ | $I_{b1} + I_{b2}$ | $\dfrac{qA_\epsilon D_\epsilon n_i^2}{L_\epsilon N_{D\epsilon}} \coth(w_\epsilon/L_\epsilon)\left[e^{qV_{b\epsilon}/kT} - 1\right]$ | Evaluated at $a$ |
| | $I_{\epsilon4} + I_{\epsilon5}$ | $-\dfrac{qA_\epsilon D_b n_i^2}{L_b N_{Ab}} \coth(w_b/L_b)\left[e^{qV_{b\epsilon}/kT} - 1\right]$ | Evaluated at $a'$ |
| | $I_{b3}$ | $\dfrac{qA_\epsilon n_i}{\tau_\epsilon} \dfrac{kT}{q\,\mathsf{E}_{\epsilon\,\max}}\left[e^{qV_{b\epsilon}/2kT} - 1\right]$ | Forward bias on emitter junction |
| | | $-\dfrac{qA_\epsilon n_i(a' - a)}{\tau_\epsilon}$ | Reverse bias on emitter junction; $(a' - a)$ is voltage dependent. |
| | $I_{c5}$ | $-\dfrac{qA_\epsilon D_b n_i^2}{L_b N_{Ab}} \operatorname{csch}(w_b/L_b)\left[e^{qV_{bc}/kT} - 1\right]$ | Evaluated at $a'$ |
| **Elements of the collector current** — Controlled by $V_{bc}$ | $I_{b5} + I_{b6}$ | $\dfrac{qA_c D_c n_i^2}{L_c N_{Dc}} \coth(w_c/L_c)\left[e^{qV_{bc}/kT} - 1\right]$ | Evaluated at $b'$ |
| | $I_{c4} + I_{c5}$ | the same as $(I_{\epsilon4} + I_{\epsilon5})$ but with the replacement of $V_{b\epsilon}$ with $V_{bc}$ | |
| | $I_{b7}$ | $\dfrac{qA_\epsilon n_i}{\tau_c} \dfrac{kT}{q\,\mathsf{E}_{c\,\max}}\left[e^{qV_{bc}/kT} - 1\right]$ | Forward bias on collector junction |
| | | $-\dfrac{qA_\epsilon n_i(b' - b)}{\tau_C}$ | Reverse bias on collector junction; $(b' - b)$ is voltage dependent. |
| | $I_{0d2}$ | Same as $I_{b7}$ but replace $A_\epsilon$ with $A_{od}$ | |
| | $I_{od1}$ | Same as $\left[(I_{b5} + I_{b6}) + (I_{c4} + I_{c5})\right]$ except that $A_{od}$ replaces $E_\epsilon$ | |
| | $I_{\epsilon5}$ | Same as $I_{c5}$ but $V_{b\epsilon}$ replaces $V_{bc}$ | Evaluated at $a'$ |

[a] The diffusion terms are determined by evaluating the gradient in the minority carrier density at the boundary indicated in the comment on the right. The locations refer to the scale in Figure 7.3.

across $R_L$ that is the response to $v_{in}$. Let us first look at the quiescent operation (i.e., $v_{in} = 0$). Note that the polarity of the two bias batteries is such that the emitter is forward biased while the collector is reverse biased. We may take

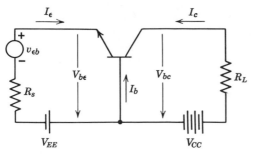

**Figure 7.5** Schematic of the transistor in a simple common-base circuit. The polarities of the applied biases are shown for the forward-active mode. The conventional definitions of the terminal currents and voltages are also shown.

advantage of that fact when we write down the terminal currents from Table 7.2. If the current through the emitter junction is large enough, we should be able to ignore the recombination within the depletion region. After all, with the diode we found that the depletion terms dominate only at low currents, while the diffusion terms are the important ones at medium to high currents. With this approximation, the emitter current is

$$I_\epsilon = - I_{b1} - I_{b2} + I_{\epsilon4} + I_{\epsilon5} - I_{c5} \tag{7.6}$$

From Table 7.2, you can see that the first terms are proportional to $[e^{qV_{b\epsilon}/kT} - 1]$. The coefficients for each of the four terms are accessible only to the designer of the transistor, so for electrical-circuit analysis it makes sense to compact the four into one *measurable* (positive) coefficient called $I_{\epsilon s}$.

$$I_\epsilon = - I_{\epsilon s} \left[e^{qV_{b\epsilon}/kT} - 1\right] - I_{c5} \tag{7.6'}$$

Keep in mind that we obtained (7.6′) as a useful approximation when we can ignore the currents arising from recombination (or generation) in the depletion region. As you discovered in the last chapter, this was a pretty satisfactory approximation at all current levels for germanium diodes at room temperature or for silicon diodes at more elevated temperatures. The simplicity of (7.6′) is certainly admirable, so let us pursue this approximation for a while. Then, in the same way that (7.1) leads to (7.6′), (7.2) gives us

$$I_c = - I_{cs} \left[e^{qV_{bc}/kT} - 1\right] - I_{\epsilon5} \tag{7.7}$$

$I_{cs}$ contains the set of coefficients equivalent to $I_{\epsilon s}$ plus the coefficient for the diffusion current in the overlap diode. Our final step here is to take advantage of the fact that $I_{\epsilon5}$ and $I_{c5}$ have the same voltage dependence as the other emitter and collector terms we are considering. Since these two terms represent the current transferred from one junction to the other, it makes sense

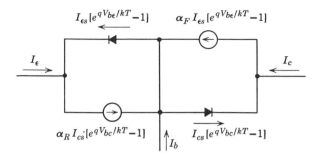

**Figure 7.6** The Ebers-Moll model for an *n-p-n* transistor. Keep in mind that the minority carriers in the base are electrons so that the particle flux and the current are flowing in opposite directions. In keeping with the standard two-port notation, all of the terminal currents are defined as positive into the device. Thus, in an *n-p-n* transistor (the most common type in silicon technology), the principal flux flows from emitter to collector while the current goes from collector to emitter.

to define a forward and reverse current transfer ratio, $\alpha_F$ and $\alpha_R$:

$$\alpha_F = \frac{I_{\epsilon 5}}{I_{\epsilon 1} + I_{\epsilon 2} + I_{\epsilon 4} + I_{\epsilon 5}} \tag{7.8}$$

$$\alpha_R = \frac{I_{c5}}{I_{c1} + I_{c2} + I_{c4} + I_{c5} + I_{od1}} \tag{7.9}$$

With these definitions, we may write the two terminal currents in a particularly simple and useful form:

$$I_\epsilon = - I_{es} \left[ e^{qV_{be}/kT} - 1 \right] + \alpha_R I_{cs} \left[ e^{qV_{bc}/kT} - 1 \right] \tag{7.10}$$

$$I_c = - I_{cs} \left[ e^{qV_{bc}/kT} - 1 \right] + \alpha_F I_{es} \left[ e^{qV_{be}/kT} - 1 \right] \tag{7.11}$$

Each of the terminal currents is seen to have a diodelike term that depends on the local junction voltage and a second term that is independent of the local junction voltage. This suggests a very simple electrical model of the transistor known as the Ebers-Moll model [J. J. Ebers and J. L. Moll, *Proc. IRE* **42**, 1761 (1954)]. The model is shown in Figure 7.6. It is a representation of equations 7.10 and 7.11 with several endearing attributes. First, its electrical behavior in a circuit is reasonably obvious. Second, all of its coefficients are, at least in principle, directly measurable. And finally, it is a model that is easily remembered.

Looking at the figure or the equations, it is fairly easy to see what $I_{es}$ and

$I_{cs}$ stand for electrically. If we make a diode from a transistor by tying either the emitter or collector to the base, $V_{be}$ or $V_{bc}$ will be 0. If, for example, the collector is tied to the base, equation 7.10 becomes the classical diode equation, with $I_{es}$ being the diode saturation current. Thus, a measurement of the saturation current with the base alternately tied to the emitter and collector should determine two of the four coefficients.

A small but important problem arises in performing this shorting between the leads. A glance at Figure 7.3 should convince you that shorting the base lead to the collector or emitter lead does not necessarily reduce the *junction* voltage to zero. The base and collector leads are too far from the junction to be sure that there are no voltage drops between the junction and the ohmic contacts. On the other hand, it is always easy to open a circuit. This guarantees that the total lead current is zero. Since the total lead current is a variable that we can easily measure and manipulate, it is frequently desirable to have the Ebers-Moll equations in terms of the total lead current. This is readily achieved by eliminating either one of the voltage-dependent terms between equations 7.10 and 7.11:

$$I_\epsilon = -I_{es}\left[e^{qV_{be}/kT} - 1\right] - \alpha_R\{I_c - \alpha_F I_{es}\left[e^{qV_{be}/kT} - 1\right]\}$$

$$= -\alpha_R I_c - I_{\epsilon 0}\left[e^{qV_{be}/kT} - 1\right] \tag{7.12}$$

where $I_{\epsilon 0} = (1 - \alpha_R\alpha_F)I_{es}$. In exactly the same way we obtain

$$I_c = -\alpha_F I_\epsilon - I_{co}\left[e^{qV_{bc}/kT} - 1\right] \tag{7.13}$$

withe $I_{co} = (1 - \alpha_R\alpha_F)I_{cs}$. (N.B. $I_{cs}$ is also a positive coefficient.) $I_{\epsilon 0}$ and $I_{co}$ are the saturation currents of the diode made by *open circuiting* the collector or emitter leads respectively.

Equations 7.12 and 7.13 are very positive statements about the behavior of transistors. The question is: How accurate are they for real devices? Since our principal compromise in getting (7.10) through (7.13) was ignoring the recombination and generation in the depletion regions, we should expect these equations to be accurate whenever the depletion-region terms are unimportant. Let us look at (7.13), for example. It predicts that for $qV_{bc}/kT \ll 0$, the collector current will be independent of the collector junction voltage and incrementally proportional to the emitter current. As $V_{bc}$ becomes greater than zero, the collector current becomes more and more negative until at about 0.6 V (for silicon at room temperature) the collector current is dominated by its exponential dependence on the collector voltage. Figure 7.7 is a plot of (7.13) generally referred to as the *collector characteristies*. In this plot, the collector current is plotted against the voltage from the collector to the common terminal (the base in this case). The other terminal's current is treated as an independent parameter. Thus, each curve in the

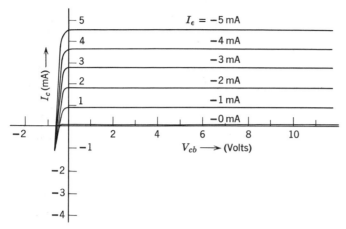

**Figure 7.7** The common-base collector characteristic curves for an *n-p-n* silicon transistor according to (7.13). On this scale $I_{co}$ is too small to be observed. $\alpha_F$ is taken to be only 0.9 to make the difference between $I_c$ and $I_e$ clearly evident.

family represent $I_c$ versus $V_{cb}$ for some particular value of $I_e$. Can we determine the Ebers-Moll parameters from such a plot? (Answer: In principle, yes. The only limitation is the accuracy with which the plot can be read. $\alpha_F = = \Delta I_c / \Delta I_e$, so measuring the separation between adjacent curves should yield $\alpha_F$. $I_{co}$ should be the collector current for zero emitter current. If the scale is sufficiently enlarged, we should be able to read this directly from the graph.)

Now comes the moment of truth. Figure 7.8 presents the measured data for a 2N707A *n-p-n* silicon high-frequency transitor. Figure 7.8a is the family of curves appropriate to (7.13); 7.8b is the set for (7.12). It is immediately clear how remarkably well we have done in 7.8a. It looks so much like Figure 7.7 that you have to look pretty carefully to note any differences at all. Figure 7.8b appears to be a lot less satisfactory. However, as you will soon see, Figure 7.8b merely exaggerates the deviations from the Ebers-Moll model; the same things go on in 7.8a. We will be able to draw a most satisfying result from Figure 7.8, for as it turns out, all of what you see is readily analyzed in terms of the physical model that we have developed so far.

## SUCCESSES AND SHORTCOMINGS OF THE EBERS-MOLL MODEL

The very close similarity of Figures 7.7 and 7.8a shows that equation 7.13 is an effective description of the behavior of a transistor. Equation 7.13 describes a transistor operating as the designer intended, with the principal flux flowing from emitter to collector. This is the so-called *forward active*

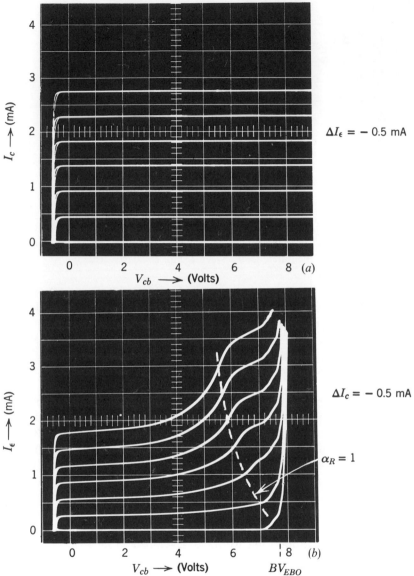

**Figure 7.8** (*a*) The common-base collector characteristic curves for a 2N707A *n-p-n* silicon transistor operating in the forward-active mode. (*b*) The emitter characteristic curves for the same transistor operated in the reverse-active mode. For both displays, the horizontal scale is 1 V/division and the vertical scale is 0.5 mA/division. The input current (emitter or collector respectively) is stepped 0.5 mA between adjacent curves, starting at the bottom with 0.

*mode.* If one runs the transistor backwards—*the reverse active mode*—, equation 7.12 should describe the situation. It is obvious from Figure 7.8*b* that (7.12) is not as accurate as (7.13). Since the only differences between the two equations are in the values of the physical parameters that go into the alpha's and the saturation currents, and since the theory that we used to derive equations 7.12 and 7.13 is general and quite effective, the only conclusion that can be drawn is that some of the physical parameters are not independent of the bias. Before searching for the culprits, let us detail the differences between what our equations predicted and what we have found in practice.

Since the saturation currents are too small to be seen on the scale of Figure 7.8, our two equations predict that the output current should be proportional to the input current and independent of the voltage on the output terminal. It is fairly clear that the output current in both figures is proportional to the input current with $\alpha_F = 0.93 \pm 0.05$ and $\alpha_R = 0.60 \pm 0.05$. However, it is equally clear that the output current is also somewhat dependent on the output terminal's voltage. It takes some imagination to see the effect in Figure 7.8*a*, but it is definitely in both sets of curves.

[By the way, if you are suprised at the low value of $\alpha_F$, note the expected accuracy of the measurement. It is difficult to get a very precise measurement of $\alpha$ from a curve tracer. If a curve tracer is all you have, a better approach is to measure $I_c$ versus $I_b$ and then obtain $\alpha$ from $\alpha = 1/(1 + I_b/I_c) \cong \cong 1 - I_b/I_c$.]

Since the power one may obtain from a current generator—including a dependent generator such as those in the Ebers-Moll model—is determined by the generator's shunt conductance and current strength, it is evident that one wants to have $\alpha_F$ as close to 1 as possible and the output conductance of the dependent generator as small as possible. This raises an interesting point: What is meant by the output conductance of the transistor? (Discussion: The only meaningful conductance for a basically nonlinear device such as a junction diode is an incremental one: $g = dI/dV$. Accordingly, the output conductance of the transistor in the common-base configuration is $g_{ob} = \partial I_c/\partial V_{cb}|_{I_e}$.) In the forward mode, the output conductance (the slope of the characteristic curves) is very small, something of the order of $0.2 \times \times 10^{-6}$ mhos for $I_c \simeq 1$ mA. In the reverse-active mode, the slope is clearly very much a function of both current and voltage. Near $V_{eb} = 0$, $g_0$ appears to be about 15 $\mu$mhos for $I_e = 1$ mA. As the emitter-base voltage increases, the output conductance increases more and more rapidly until a breakdown of some sort is evident.

In order that you not get the idea that these changes in slope are peculiar to the reverse mode, look for a moment at Figure 7.9. This is the same display as Figure 7.8*a*—the collector characteristics for the common base configuration—but the horizontal gain has been reduced by a factor of 20. In the

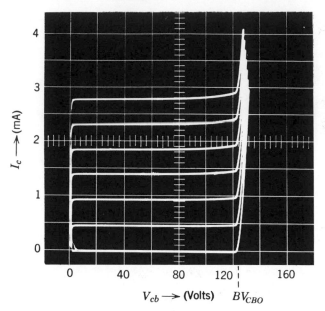

**Figure 7.9** The common-base collector characteristics for the 2N707A transistor. This is essentially the same display as Figure 7. 8a with a change in the horizontal scale to permit the demonstration of breakdown in the forward-active mode.

forward mode, breakdown occurs at about 120 V. At collector currents of the order of 1 mA and collector voltages of the order of 90 V, a noticeable amount of curvature is visible. The rather startling difference between Figures 7.8b and 7.9 shows that getting better performance in the forward direction is generally done at the expense of performance in the reverse mode. Such a conclusion is hardly unique to transistors and it seldom if ever bothers anyone. (When was the last time you drove your car in reverse for any great distance?) However, it is important to note that the difference between Figure 7.8a and 7.8b is merely a matter of degree; the basic bilateral symmetry is really there.

Now back to the question of where the slope comes from. First let us dispose of the matter of the curvature near breakdown. As you must suspect, the breakdown is by avalanche in the reverse-biased junction. Avalanching represents the extreme form of generation within the depletion region. Since we ignored currents that arose from effects within the depletion region, it is not surprising that equations 7.12 and 7.13 contain no description of the transistor's behavior in the breakdown region. That still leaves us with one puzzle to solve: Why should the curvature near breakdown depend on the current? (Discussion: Avalanche multiplication is a geometric process and we are looking at it with a linear display. In other words, at the voltage for which the multiplication factor, $M$, is $2 \times$, if the input current is 1 mA, the output current is 2 mA. If the input current is 2 mA, the output would be 4 mA, and so forth. On a linear scale, that makes the effect look bigger and bigger as the input current increases. On a logarithmic plot, the curves would stay parallel.) Thus, there is no mystery at all. You are simply getting a chance to see the avanalnche multiplication rather clearly prior to the point at which it becomes self-sustaining (breakdown). Two simple but important points should be noted: As the multiplication factor gets larger and larger, $\alpha$ will eventually exceed unity and the base current will reverse. [In this context, $\alpha$ is the *measured* forward or reverse-current transfer ratio rather than the coefficient we defined in (7.8) and (7.9).] Second, since the input current is being held constant in the display of the collector characteristics, the excess current created by the avalanche process must be flowing in the base lead.

So much for avanche multiplication. Let us now take a look at what is leading to a finite slope even with no multiplication. This is a subject of vital interest, since it limits the performance of transistors operating well within their design range. Avalanche multiplication determines only the upper bound to the design range.

Take a moment to go back to Table 7.2 to see which of the coefficients of the currents might be voltage dependent in a way that would show up in Figure 7.8. You are being asked to rediscover the *Early effect* [J. M. Early, *Proc. IRE* **40**, 1401 (1952)]. A clue—but not the answer—is contained in the comment column of that table. [Discussion: The clue is contained in the statements that the depletion widths, $(a' - a)$ and $(b - b')$, are voltage dependent. Although that by itself would lead to some voltage dependence for the saturation currents, the saturation currents are too small to be seen in the figure. Thus, the clue is not the answer. However, the only way that the depletion widths can change, since the depletion regions are buried in the middle of the transistor, is to have the bulk regions shrink or expand. In the reverse-active mode, the emitter is reverse biased and the collector forward biased. As the emitter voltage is made more and more positive in Figure 7.8$b$, $(a' - a)$ gets larger and larger. Consequently, $a$ and $(b - a')$ are both getting smaller. $w_b = (b - a')$ enters directly into the principal term in the current we are observing—$I_{c5}$. If we may assume that $w_b/L_b \ll 1$, then, by using the Taylor series for $\sinh(x)$ to give $\operatorname{csch}(w_b/L_b) \Rightarrow L_b/w_b$, we obtain a simplified form showing that the output current is inversely proportional to the base width:

$$I_{c5} = -\frac{qA_e D_b n_i^2}{w_b N_{Ab}} \left[ e^{qV_{bc}/kT} - 1 \right] \tag{7.14}$$

Although (7.14) appears independent of $V_{eb}$, $w_b$ is a function of the reverse bias on the emitter junction. In a diffused transistor, where the base width is only a micron or two, relatively small motions of the depletion width can correspond to substantial variations in the output current.]

For the forward-active mode, we obtain an identical result. For that case we are observing principally $I_{e5}$ and under the same narrow-base-width assumption used above, we obtain

$$I_{e5} = -\frac{qA_\epsilon D_b n_i^2}{w_b N_{Ab}}\left[e^{qV_{be}/kT} - 1\right] \tag{7.15}$$

In this mode, with the emitter junction forward biased and the collector reverse biased, $w_b$ is a function of $V_{cb}$. As you can see from the characteristic curves, $w_b$ changes more rapidly with variations in the reverse-biased emitter voltage than with the reverse-biased collector voltage. This is as it should be, since once again, the transistor is optimized for operation in the forward mode.

A number of pages ago, I commented that in the double-diffused transistor, most of the important parameters were pushed in the proper direction by the process itself. That is, the requirement that each successive diffusion be heavier than the one that preceded it leads naturally to a better transistor. As an example, consider the Early effect that we have just discussed.

**Problem 7.1.**

For simplicity, let the transistor be made of three layers, each of which is uniformly doped. Let the doping be:

$$N_{De} = 10^{19}/\text{cm}^3$$
$$N_{Ab} = 10^{17}/\text{cm}^3$$
$$N_{Dc} = 10^{15}/\text{cm}^3$$

*Estimate* the difference in the base-width modulation that occurs for an equal voltage swing in the forward- and reverse-active modes.

*Solution.* For uniformly doped, abrupt junctions, the depletion widths are given by (3.51) and (3.52). In those equations, $V_D$ is replaced by $(V_D + V_{\text{applied}}) \simeq V_{\text{applied}}$, so the only parameter of interest is the doping factor. At the collector side, the depletion region in the base is proportional to

$$l_p \propto \left[\frac{N_{Dc}}{N_{Ab}(N_{Ab} + N_{Dc})}\right]^{1/2} \simeq 3 \times 10^{-10}\,\text{cm}^{3/2}$$

At the emitter junction, the depletion region is proportional to

$$l_p \propto \left[\frac{N_{De}}{N_{Ab}(N_{De} + N_{Ab})}\right]^{1/2} \simeq 3 \times 10^{-9}/\text{cm}^{3/2}$$

Thus, for the particularly simple example that we have picked, the base width modulation will be *ten times greater* in the reverse mode than in the forward mode.

In a moderately well-designed transistor, the base-width modulation should be a small fraction of the total base width. In that case, we may expand $(1/w_b)$ in a useful Taylor series:

$$(1/w_b) = (1/w_{b0})[1 + \Delta w_b/w_{b0} + \dots] \qquad (7.16)$$

Putting (7.16) back into (7.14) and (7.15) shows that the factor of ten increase in base-width modulation should show up as a factor of ten increase in the output conductance. It is rather satisfying to note that the difference in the slopes of Figure 7.8a and 7.8b is roughly a factor of eight. Thus, our order-of-magnitude estimate appears to be successful.

## WHAT MAKES A GOOD TRANSISTOR?

So far, things seem to have gone pretty well. By a direct application of junction-diode theory to the transistor structure, we appear to have acquired very effective physical and electrical models for predicting transistor behavior. We can (and will) go a step further with our models by explicitly accounting for charge storage throughout the transistor structure. Then we may obtain very useful frequency-dependent incremental models by differentiating the appropriate charge variables. All this model making, however, will avail us naught if we have no practical and specific concept of what constitutes a good transistor. Let us begin with a moderately long-winded account of a transistor in a primitive switching circuit. All of the elements that might be important in almost any circuit can be introduced in a practical way without a great deal of analysis to becloud the issue. Then we may return to model-making and some detailed analysis to see what can and cannot be done with the bipolar transistor structure.

The objective of the circuit we are going to look at, Figure 7.10, is to turn an

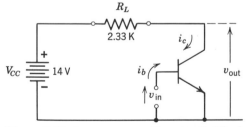

**Figure 7.10** A common-emitter circuit. The load resistor represents a 14 V, 84 mW automotive signal lamp.

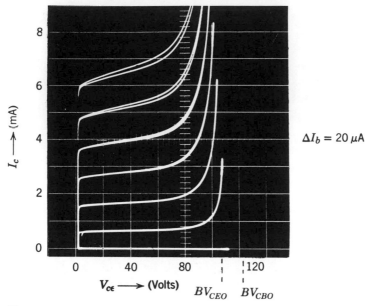

**Figure 7.11(a)** The common-emitter collector characteristics for the 2N707A transistor operating in the forward active mode.

automotive signal lamp on or off at the behest of a small (low-powered) signal. The role of the transistor in this as in almost all circuits—when one combines it with its sources of energy—is to provide power amplification. The power controlled should be much greater than the power used to operate the transistor. Sucess in the design and realization of a transistor can be measured in dB of gain in the particular regime in which the transistor is expected to work. For most circuits, the common-base configuration would not make optimum use of a transistor, since it provides no current gain. (In fact, since $\Delta I_c/\Delta I_e = = \alpha_F < 1$, the current gain is less than unity.) From the point of view of the transistor, the current flow is determined by the junction voltages. From the external point of view—that of the circuit—a current and a voltage are simultaneously required to drive the input terminals. If we use the base and emitter leads as the input terminals, the current that the drive (or input) circuit must provide (the remainder of the current being provided by the energy sources biasing the transistor) depends on which lead we pick as the common lead. If we choose the base lead as the common lead, as we have previously, then the drive circuit must deliver the full emitter current (see Figure 7.5). That's the largest of the three terminal currents. On the other hand, if we use the emitter as the common terminal, the drive circuit need only provide the base current, the smallest of the three terminal currents. The

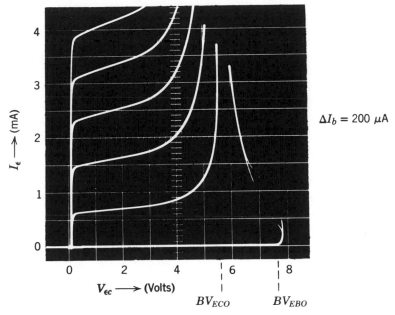

**Figure 7.11**(*b*) The same curves as Figure 7.11(*a*) with the emitter and collector leads interchanged. Note that ALL of the scales are changed in going from (*a*) to (*b*). The lowest value of base current in both figures is *slightly less than zero* to permit driving the collector voltage beyond the sustaining voltage, $BV_{CEO}$. In the reverse-active mode, the transistor is driven into the "avalanche mode," where it shows the negative incremental-output conductance typical of this mode of operation.

terminal voltage is the same for either of the choices. Thus, unless there are other compelling factors, the common emitter configuration is the one to use. This configuration is shown in Figure 7.10, which illustrates the circuit we are going to analyze.

Before charging into the analysis, we should rearrange our Ebers-Moll equations to fit the new configuration. Equations 7.13 and 7.3 provide us directly with the relationship between $I_c$ and $I_b$:

$$I_c = + \alpha_F(I_c + I_b) - I_{co}.\left[e^{qV_{bc}/kT} - 1\right]$$

$$= \frac{\alpha_F}{1 - \alpha_F} I_b - \frac{1}{1 - \alpha_F} I_{co} \left[e^{qV_{bc}/kT} - 1\right]$$

$$= \beta I_b - (1 + \beta) I_{co} \left[e^{qV_{bc}/kT} - 1\right] \tag{7.17}$$

$\beta = \alpha_F/(1 - \alpha_F)$ is the *common-emitter* forward-current transfer ratio. There is, of course, a $\beta_F$ and a $\beta_R$, but the latter is used so seldom that the subscripts hardly ever appear.

Although we obtained (7.17) from (7.13), the characteristic curves for (7.13) (Figure 7.8a) are totally unsuitable for (7.17). This is only a scale problem, but it is a rather important one. Since $\alpha_F \simeq 1$, the scale appropriate to a display of $I_e$ and $I_c$ is much too coarse to be used for $I_b$. Furthermore, if $\alpha \simeq 1$, very small variations in $\alpha$ will correspond to very large variations in $\beta$. A direct measurement of the common-emitter (CE) collector characteristics is called for. These are shown in Figure 7.11. You can see how much more spectacular the Early effect is when viewed in the common-emitter circuit. It is also evident that holding the base current constant emphasizes the avalanching in the collector junction. Why should that be so? [Discussion: A glance at Figure 7.4 and Table 7.1 shows that the avalanche current from the collector junction ($I_{b6}$ or $I_{c3}$) partially cancels the base current. If the base current is to be held constant, more voltage must be applied to the emitter junction. That additional voltage increases not only the base current but also the emitter current. In fact, the emitter current is increased by a factor of $(\beta + 1)$ times the increase in the base current, and the collector current is increased by a factor of $\beta$ times the increase in base current. Thus, the avalanche multiplication is fed back to the input of the transistor and amplified by a factor of $\beta$. It is very important to notice that the transistor is doing nothing different in this circuit from what it did in the common-base configuration. It is the external circuit that has changed. The transistor does whatever the terminal voltages demand of it regardless of its location within the circuit.]

Now let's return to the circuit of Figure 7.10. Our first task is to determine how much power the input generator must deliver to turn the lamp on. The lamp itself requires 84 mW at 14 V, but most of that will be provided by the 14 V bias battery, $V_{CC}$. What we want is the amount of power necessary to open or close the transistor switch. The first step—and the only problem— is to take the sum of the voltage drops around the outside loop in the figure. How do you find the drop across the transistor ($V_{ce}$)? $V_{ce}$ is conspicuously absent in (7.17), but $I_c$ in Figure 7.11 does show some dependence on $V_{ce}$. Since we have no *useful* analytical relationship between $I_c$ and $V_{ce}$, the only sensible approach is to use a graphical technique. It is easy enough to write $V_{ce}$ in terms of the voltage drops around the rest of the loop:

$$V_{ce} = V_{CC} - I_c R_L \qquad (7.18)$$

On a set of collector characteristics such as Figure 7.11, equation 7.18 plots as a straight line running from $V_{ce} = 14$ V, $I_c = 0$ to $V_{ce} = 0$, $I_c = 6$ mA. Such a line shows the combined action of the load and bias supply and is called a *load line*. (N.B. If the load were not linear, the load line would not be

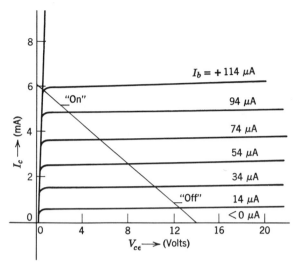

**Figure 7.12** The CE collector characteristics from Figure 7.11*a* with the load line from the circuit in Figure 7.10.

straight.) Since the collector characteristics give the *I-V* relationship across the same set of terminals, all possible values of $(I_c, V_{c\epsilon})$ that can be found in this circuit must lie at the intersection of the load line and the collector curves. This graphical solution is shown in Figure 7.12. Note that from the same figure we may obtain selected values of base current. Intermediate values can be obtained by a simple linear interpolation, per (7.17). (E.g., for a base current of 40 $\mu$A, Figure 7.12 gives us $I_c$ = 1.9 mA and $V_{ce}$ = 9.5 V.)

We now come to a question of some importance: At what base current is the transistor fully turned on? (Discussion: *In this circuit* the transistor is as on as it will get for base currents in excess of 115 $\mu$A.) Note carefully that for base currents in excess of 115 $\mu$A, the current through the lamp is limited by the lamp and the supply voltage rather than the transistor. Since the output current, $I_c$, is no longer dependent on the input current, $I_b$, the transistor is said to have *saturated*. (This is peculiar terminology indeed! The transistor could easily carry more current; it is the external circuit that is limiting the current here. Nonetheless, common parlance would have us blame the transistor. See the discussion on page 360.)

To answer the original question on how much signal power it takes to turn the lamp on, we must find the voltage that it takes to get a base current of 115 $\mu$A flowing. Although you might suspect that the answer would depend rather strongly on the type of transistor employed, it does not. For a silicon transistor at room temperature, it will take about 0.6 $\pm$ 0.05 V; for a germanium transistor it would require about 0.3 $\pm$ 0.05 V. How can it be independent of the transistor? Well, it isn't really independent, but since the voltage goes into the exponential while the transistor parameters make up the coefficient multiplying the exponential, very small variations in voltage make up for very large swings in the transistor's parameters. At room temperature, 57 mV corrects for an order-of-magnitude shift in the saturation current, and with one notable exception, only a few orders-of-magnitude shift are possible. The one exception is in $n_i^2$; it is subject to the same exponential dependence on band-gap as the current is to the applied voltage. Thus, for a given material and temperature, it is possible to assign an approximate $V_{be}$ for forward conduction, almost independent of the dimensions and doping of the junction involved.

The solution to the power gain in switching the lamp on is now at hand. It takes about $(0.55 \text{ V}) \times (0.115 \text{ mA}) = (0.063 \text{ mW})$ to hold the transistor (and thus the lamp) on. Since the lamp dissipates about 80 mW, the power gain for this switch is about 1270 (31 dB).

Now that we have the numbers, we should ask whether this particular transistor is doing a good job. The answer is just so-so. It is a transistor that has been "optimized" (Oh, that word!) for performance in a completely different regime, and optimization in one direction inevitably involves "pessimization" in most others. Let us see what is important in this low-frequency switching application. The output current and voltage swings (i.e., the power to be controlled) are completely specified by the external circuit. The output requirements on the transistor are that it be able to conduct the required current and stand off the required voltage. The input voltage is more or less fixed at about 0.6 V for silicon. The only parameter left, and therefore the one to be optimized, is $\beta$. *In this circuit*, the power gain available is proportional to $\beta$. Since the $\beta$ for the 2N707A is relatively low, it is not a particularly good choice for this job.

If you were designing a transistor for this circuit, you would find that simply maximizing $\beta$ would get you into trouble. As we discussed very briefly above, any avalanche multiplication that goes on in the collector junction is effectively multiplied by $\beta$. Thus, a transistor that breaks down in the common base circuit at a respectable 50 V or so may not stand off the 14 V required in the common emitter configuration. Just how bad things are for the CE circuit becomes clear when we look at equation 7.13 and note that the current that reaches the collector junction is multiplied by $M \geq 1$. Thus:

$$I_c = M[-\alpha_F I_e + I_{co}] \tag{7.19}$$

Substituting $(I_b + I_c) = -I_e$ above gives us the form equivalent to (7.17):

$$I_c = \left(\frac{\alpha_F M}{1 - \alpha_F M}\right) I_b + \left(\frac{M}{1 - \alpha_F M}\right) I_{co} \tag{7.20}$$

It is clear that things get pretty exciting when $\alpha_F M = 1$. When $\alpha_F M = 1$, the current will become self-sustaining. The voltage at which this occurs is called the *sustaining voltage*, $BV_{CEO}$. The notation indicates that the voltage is between *C*ollector and *E*mitter with the unnamed terminal (*B*ase) *O*pen. Thus, the *CB* collector breakdown is labeled $BV_{CBO}$. There is an important difference between these two limiting voltages. $BV_{CBO}$ is truly an upper limit; a transistor cannot be operated above this point because the collector is no longer serving its function. On the other hand, it is possible to operate above $BV_{CEO}$ if you are willing or want to use collector characteristics such as the one shown beyond $BV_{CEO}$ in Figure 7.11*b*.*

A final question that might arise in using a transistor as a low-frequency switch is the amount of power dissipated by the transistor in its on and off states. First, what is the power dissipation in the off state? [Answer: From (7.17), the collector current in the off state is $I_c = (1 + \beta) I_{co}$. If we could read this value in Figure 7.11, we could get the off-state power loss as $V_{CC} I_c$. Since we cannot read it, it must be rather small. However, if we were to raise the voltage until $M$ begins to exceed 1, or raise the temperature until $I_{co}$ becomes significant, this conclusion need not hold.]

The on-state power loss can be obtained as the product of the coordinates where the load line in Figure 7.12 crosses the almost vertical portion of the collector characteristics. Apart from possible ohmic drops, the saturated value of $V_{ce}$ should be somewhat less than $V_{be}$, corresponding to the almost vertical portion of the curves in Figure 7.7 or 7.8. (N.B. This means the collector junction would be slightly forward biased). For the silicon transistor, the saturated (or on-state) voltage drop for low current is of the order of 0.3 to 0.4 V. At higher currents, ohmic drops in the collector must be included. From Figure 7.12 the on-state demands that about $0.3 I_c$ watts be wasted in the transistor.

The situation that we have considered so far is classified as static switching. The transistor is either on or off; getting from one state to the other never entered the discussion. As soon as you ask just how fast something can be turned on or off, a whole new set of problems arise. For example, if we replace our lamp in Figure 7.10 by a resistor and use the voltage across the transistor to modulate the intensity of an electron beam in a cathode-ray tube, we would have a light source potentially capable of 10 to 100 MHz response. The question to ask is: How rapidly can the transistor turn the electron beam on and off? For purposes of the discussion, let us say that the beam is fully off if $V_{ce} \geq 12$ V and fully on if $V_{ce} \leq 2$ V. Let the drive signal be the perfectly square pulse shown in Figure 7.13*a*. If the signal current is perfectly square, what inhibits the collector current from being equally

---

* A rather complete analysis of the operation of transistors in this negative-resistance avalanche region can be found in S. L. Miller and J. J. Ebers, *Bell Sys. Tech. J.,* **34**, 883 (1955).

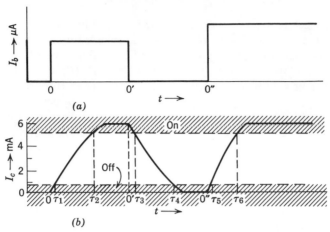

**Figure 7.13** The collector current as a function of time (*b*) when the base current is that of (*a*). The transistor is considered to be *on* in the upper hatched portion of (*b*) and *off* in the lower hatched portion. Note that the step in (*a*) is a third again as high as the pulse. Thus, $(\tau_6 - \tau_5)$ will be less than $(\tau_2 - \tau_1)$.

square? (Discussion: Briefly, the answer is *charge storage*. As we found in Chapter 6, associated with each junction voltage are distributions of fixed and free charge; the fixed charge is associated with the depletion region, while the free charge is stored in the bulk regions. These collections of charge are each responsive to the junction voltages through one or another very nonlinear relationship. However, no matter what the relationship, changing the voltage across a junction—and thus the current through it—can be accomplished only by supplying appropriate amounts of charge and *energy*. Since the signal source is a finite power supply, it takes it some time to charge or discharge the transistor.)

The message about switching speed should now be pretty clear. The initial effect of an abrupt change in the base current is an abrupt change in the *rate of change of the charge stored*. The junction voltage, and thus the collector current, changes as the charge changes. Initially the collector current follows the *integral* of the base current. It is only after the transistor "charges up" that the results in Table 7.2 or equation 7.17 apply.

You might despair of being able to solve circuit equations with charge storage being exponentially related to the junction voltage. We circumvented the same problem in the Ebers-Moll equations when we switched from the voltage to the *current-control* viewpoint [i.e., obtained (7.13) from (7.11) and (7.10)]. A similar advantage can be achieved here by viewing the operation of the transistor as being the result of *charge control*. The basis of this view is that the current, which is proportional to the derivative of the charge distribution in the bulk regions, and the bulk charge, which is proportional to the integral of the charge distribution over the bulk regions, will be functionally the same as long as the time scale is not so short that we must use the time-dependent form of the diffusion equation. Since the bulk charge and the diffusion currents are linearly proportional to each other, we may write time-dependent equations relating the input current, the stored charge, and the output current. If bulk charge storage dominates, the resulting equations are particularly simple. Just as in (7.13) and (7.17), the junction voltage will take care of itself (i.e., the *circuit* will supply the necessary voltage).

We are now in a position to determine qualitatively what the collector current would be for the base current shown in Figure 7.13*a*. Figure 7.13*b* shows the collector current for a pulse of base current that is barely able to switch the transistor in the time allotted and then for an input step of somewhat greater height. The transistor is initially in its *off* state. Upon application of the pulse of base current, the collector current begins to increase as the *integral* of the base current. Accordingly, it takes some time before it reaches the current of 0.8 mA where the cathode-ray tube begins to turn on. This delay is indicated as $\tau_1$ in 7.13*b*. As more and more of the base current is required to support injection into the emitter and recombination in the base, the proportion available for charging the base and emitter decreases. The collector current begins to rise less and less steeply; it would approach a steady state of $I_c = = \beta I_b$ if $I_b$ were small enough. In the case we are considering it stops rather abruptly as the collector bias reverses and the transistor saturates. Before it reaches the saturation current, the collector current exceeds the value at which the cathode-ray tube is fully on. This time is indicated as $\tau_2$ as on the graph. A very similar set of delays is observed for the shut-off cycle. $\tau_3$ is the time delay before the cathode-ray tube brightness begins to decrease; $\tau_4$ is the time delay before the tube goes dark.

An important and fundamental question should be asked and answered at this point: Could the switching times be shortened by increasing the amount of base drive? (*Answer:* Yes. The rate at which the transistor is charged is proportional to the base drive current. If the base current is well above the amount required to drive the transistor into saturation, most of the base current goes into charging the transistor. Under these conditions, the switching time is inversely proportional to the base current.) We obviously gain switching speed by pushing more current through the base. The large step in Figure 7.13 gives a shorter turn-on delay than the smaller pulse. Do we lose anything by increasing the base current? (*Answer:* Yes. Power gain. Without even considering what the transient base-emitter voltage looks like, it is evident that using more base current to get the same collector current means lower gain.) If we consider the steady-state power gain under conditions of high base drive, a rather interesting rule appears. The power switched is independent of the base drive, since it is determined solely by the external circuit. The base drive power is simply $V_{be}I_b$. Since $V_{be}$ is almost independent of the base current, being characteristic of the transistor material, the drive power is almost proportional to the drive current. The power gain is thus inversely proportional to the base current, while the switching speed $(1/\tau_2)$ is proportional to the base current. Accordingly, *the speed-gain product is essentially constant.* Even more precisely, it either stays constant or gets smaller as the base drive is increased. As you shall see, this rule that gain is purchased at the expense of speed haunts all amplifying devices. To get a better speed-gain product, you must improve the transistor itself. One could try either to get a higher $\beta$, thus improving the gain, or to reduce the charge storage, thus increasing the speed. Up to a point, it is not hard to improve both of these parameters together. However, this improvement is normally purchased at the expense of the maximum power that the transistor can switch. Transistor design and realization are far from trivial arts.

Let us review for a moment what this lengthy illustration has revealed. We have found that there are basically four parameters that must be considered in evaluating a transistor's effectiveness for particular switching applications: the current gain, $\beta$; the collector-junction breakdown voltage, $BV_{CBO}$; the power the transistor can dissipate; and the charge stored within the transistor in its on state. A completely equivalent set of parameters determine the transistor's performance in linear amplifier applications. Our next task is to see how the physical parameters of Table 7.2 may be manipulated to adjust these four parameters for a successful design.

## TRANSISTOR STRUCTURE AND TRANSISTOR PERFORMANCE

Let us begin by generalizing the Ebers-Moll definition of $\beta$ so that we may use it in situations where the Ebers-Moll coefficients are not constant. To do this, we need only define $\beta$ as the *incremental* CE forward current transfer ratio:

$$\beta = \Delta I_c / \Delta I_b \qquad (7.21)$$

This definition is not really at variance with (7.17), but it suggests that $\beta$ may be a function of the conditions under which it is measured. Going back Table 7.2 and Figure 7.4, we may write

$$\beta = \frac{-\Delta I_{e5}}{\Delta(I_{b1} + \ldots + I_{b4}) + \Delta(I_{b5} + \ldots + I_{b8})} \qquad (7.22)$$

Although we could proceed simply to put each of the terms from the table directly into (7.22), it is much more economical to take advantage of certain "truths" that are usually valid for practical transistors. First, the base width,

$w_b$, is always much less than $L_b$, the minority diffusion length in the base. Similarly, it is frequently the case that the emitter is short-based (i.e., $w_\epsilon/L_\epsilon \ll \ll 1$). Accordingly, it is profitable to expand the hyperbolic trig functions in a series, since we may drop all terms after the first:

$$\left.\begin{array}{l} \operatorname{csch}(x) \Rightarrow 1/x \\[2mm] \coth(x) \Rightarrow [1 + x^2/2]/x \end{array}\right\} \quad \text{ignoring terms in } x^3 \text{ and higher powers} \quad (7.23)$$

When avalanche multiplication is truly unipartant, the second bracket in the denominator of (7.22) as well as its increment should be negligible compared to the first term. With these approximations, we now evaluate (7.22) from Table 7.2. For normal operation the only strong voltage dependence for any of the terms is the exponential one on $V_{be}$. Accordingly, each of the ratios in (7.22) takes on the form:

$$\frac{\Delta I_a}{\Delta I_b} \Rightarrow \frac{\partial I_a/\partial V_{be}}{\partial I_b/\partial V_{be}}$$

Thus:

$$\frac{1}{\beta} = \frac{\Delta(I_{b1} + I_{b2})}{\Delta(-I_{\epsilon 5})} + \frac{\Delta(I_{b4})}{\Delta(-I_{\epsilon 5})} + \frac{\Delta(I_{b3})}{\Delta(-I_{\epsilon 5})}$$

$$= \frac{D_\epsilon w_b N_{Ab}}{D_b w_\epsilon N_{D\epsilon}} + \frac{w_b^2}{2L_b^2} + \frac{w_b N_{Ab}(kT/q\mathsf{E}_{max})/2}{D_b n_i (e^{qV_{be}/2kT}) \tau_\epsilon} \qquad (7.24)$$

It is helpful in obtaining the middle term to note that

$$\frac{I_{b4}}{-I_{\epsilon 5}} = \frac{(I_{\epsilon 4} + I_{\epsilon 5}) - I_{\epsilon 5}}{I_{\epsilon 5}}$$

Also, keep in mind that this entire evaluation is for the forward active mode (i.e., a moderate amount of forward bias on the emitter-base junction and reverse bias on the collector junction). The first two terms in (7.24) indicate the efficiency with which minority carriers are injected into the base and transported to the collector. The third term is the ratio of change in the depletion current to that in the collector current. Note that the first two are independent of emitter bias but that the third term is not. Since $\beta$ gets small as the terms in (7.24) get large, you can see that *as the forward bias gets small, $\beta$ must drop off drastically*. Thus, we find that there is a *lower* limit to the current that a transistor can carry in a useful way. As it happens, there are no variables in the third term of (7.24) that are really available to the designer. $N_{Ab}$ is constrained by other demands; so is $w_b$. Thus, the lower current limit for any silicon transistor is simply proportional to the area of of the emitter junction. We shall return to this shortly when we find that there is also an upper current density limit (e.g., a careful inspection of Figure 7.11 shows that although $\beta_F$ appears to increase monotonically with $I_c$, $\beta_R$ has a peak at about $I_e = 2$ mA). It is fairly easy to see how to maximize $\beta$: Make the base width as small as possible and dope the emitter heavily with respect to the base. The first maximizes the transport efficiency; the second minimizes reverse injection. In a modern double-diffused transistor, the base width is so narrow that the dominant term in (7.24) is usually the first one; for the al-loyed silicon structures of yore, the base width is so much larger and the emitter doping so much heavier that the second term is likely to dominate.

At this point you can see why GaAs, which makes excellent diodes, would be an awkward material to choose for a bipolar transistor. Diffusion lengths in GaAs are of the order of a fraction of a micron. This essentially rules out making the second term in (7.24) much less than 1. Note what the problem is: Minority carriers cannot diffuse across the base region before they re-combine. A diffusion length that is long compared to *practical* base widths is the fundamental material property required for a bipolar transistor. (With modern methods, a base can be created with widths from several tenths of a micron up.)

Equation 7.24 suggests that $\beta$ will be current dependent and small for very low currents. At higher currents, the equation indicates that $\beta$ should become constant. In the strongly current-dependent state, what is the func-tional relationship between the collector current and $\beta$? [Answer: In the portion of the $\beta$ versus $I_c$ curve that is dominated by the third term of (7.24), we have $\beta = -I_{e5}/I_{b3}$. Since $I_{e5}$ is proportional to $e^{qV_{be}/kT}$ while $I_{b3}$ goes as $e^{qV_{be}/2kT}$, $\beta$ increases as the square root of $I_c$. Notice that as long as $\beta = \beta(I_c)$, equation 7.17 is nonlinear.]

Equation 7.24 looks pretty good at the high-current end. Although high-injection effects will change the *I-V* relationship. We aren't considering $V_{be}$ in (7.17). Linearity is the primary advantage of adopting the "current-control" viewpoint. However, a caveat is in order. Look very carefully at the first two terms in (7.24). Are all of their components as independent of current as they seem? [Discussion: As long as the minority-carrier con-centration remains at the "low level," the first two terms of (7.24) are really quite constant. When high-injection-level effects begin to appear, however, several factors in these terms begin to change with the injection level. The factors that change are those associated with the base, since the base is much more lightly doped than the emitter. The first term in (7.24) gives the ratio of the forward to the reverse component of injection at the emitter junction. As it stands in (7.24), the first term expresses this ratio under the assumption that the majority-carrier concentrations are constant. If you will reread the paragraph following (6.48), you will note that, under conditions of high-level injection, the principal forward current ($I_{\epsilon 5}$) will increase as $e^{qV_{be}/2kT}$ while the reverse components ($I_{b1} + I_{b2}$) will continue to grow as $e^{qV_{he}/kT}$. Thus, the first term in (7.24) increases approximately proportionally with the collector current in the high-injection regime.

The second term in (7.24) increases because the lifetime of minority carriers decreases under conditions of high injection, as equation 5.17 shows. If you follow the arithmetic through the various squares and square roots, you will find that the second term also increases approximately as $I_c$. We must conclude, therefore, that under conditions of high injection, the $\beta$ of a transistor will fall off rather rapidly with increases in collector current.

It is also possible that the ohmic drops across the collector at high currents can partially or totally debias the collector junction. This will be discussed, but it too can cause $\beta$ to fall off at high current densities.]

Our "constant" of proportionality, $\beta$, turns out to be pretty nonconstant. The use of (7.17) is still effective, however, because most circuits tend to employ a transistor over a very limited portion of its dynamic range. On the other hand, if you wanted to use a transistor in a linear amplifier that was to work over five decades of current, you would certainly have an "interesting" problem.

## BASE SPREADING

When the base current begins to rise to substantial proportions, the one-dimensional analysis of the transistor becomes inadequate. We can no longer ignore the voltage drops associated with the transverse flow of current from the base ohmic contact into the active-transistor structure (Figure 7.4). The drop in the overlap diode is not critical, but any drops that may occur across the emitter-base junction enter directly into the exponential that determines the local injected-current density. In Figure 7.4, this drop would have the effect of decreasing the forward bias at the top of the figure; more of the current would flow near the bottom. (In the usual transistor structure, Figure 7.2, the base surrounds the emitter. Thus, the current is pushed out from the center to the periphery of the emitter. The effect is called, accordingly, *base spreading*.) As the current begins to flow more and more at the edge of the emitter junction, the effective junction area falls off. High-level injection effects become apparent at much lower currents than might be expected from the extension of the low-level analysis. The only way to avoid base spreading is to give the emitter a shape that is "all edge." All medium- and high-power transistors have highly convoluted emitter shapes—star-shaped figures, comb-shaped figures, and the like—which greatly increase the circumference-to-surface ratio. A typical comb-shaped (interdigitated) pattern from a very-

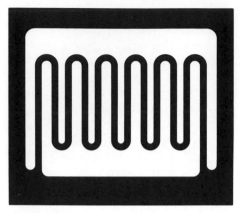

**Figure 7.14.** The metallization pattern of a high-power transistor. The comb-shaped pattern with seven base fingers and five emitter fingers is designed to maximize the periphery for the given emitter area. Such designs substantially reduce the base-spreading resistance.

high-power transistor is shown in Figure 7.14. It should be noted that base-spreading will not change the reasoning which suggested that $\beta$ will fall off quite rapidly at high current densities. It merely indicates that unless special care is taken, the effect will occur at lower currents and be more severe than we otherwise would have supposed. (For a detailed analysis of base spreading, see R. D. Thornton et al., *Characteristics and Limitations of Transistors*, Wiley, 1966.) Since the base doping is constrained to lie between the collector and emitter doping levels, and since most of the other constants in the expression for $I_{e5}$ do not vary that much from one silicon transistor to another, we can estimate the current density at which high-level injection will begin to be apparent (assuming that base-spreading is *not* a problem). We should expect that these effects would begin to manifest themselves when $n_p(a') \cong$ $\cong p_{p0} = N_{Ab}$. At that point

$$J_{e5} = qD_b N_{Ab}/w_b \simeq (1.6 \times 10^{-19})(20)(10^{17})/(10^{-4}) = 3 \times 10^3 \text{ A/cm}^2$$

This number may look like a lot of current to you, but go back and look at Figure 7.1. The typical not-too-tiny transistor has an emitter area of only about $10^{-4}$ cm$^2$. (Much smaller devices are getting to be very common, especially in integrated circuits, where the area may be smaller than $10^{-6}$ cm$^2$. That's small!) Thus, you can expect to see some substantial $\beta$ falloff for currents in the tenths of amperes. Bigger devices will give proportionately higher current values; smaller ones will give lower values. It is also possible (and frequently the case) to have $\beta$ fall off at earlier values for other reasons, some of which we shall examine in detail.

Given a plausible upper limit, it's interesting to see what we can get for a lower limit. The third term in (7.24) is dominant in the low-current, low-$\beta$ range. It will cease being important, let us say, when it is less than 1/10 of the sum of the first two (nominally current-independent) terms. Choosing a $\beta$, we may calculate an approximate current for this 1/10th point. Let us say that we choose $\beta = 100$. Then we want the third term of (7.24) to be equal to $10^{-3}$, from which

$$\frac{w_b N_{Ab}(kT/q E_{max})\, e^{-qV_{bc}/kT}}{D_b n_i \tau_\epsilon} = 10^{-3}$$

Putting in values from the example above and choosing $E_{max} = 10^5$ V/cm and $\tau_\epsilon = 10^{-7}$ sec leads to the result that $e^{qV_{b\epsilon}/kT} \approx 2 \times 10^{10}$. Putting that number back into the expression for $J_{\epsilon 5}$ gives a current density of about 1 A/cm$^2$. Our order-of-magnitude estimate suggests that $\beta$ can be reasonably constant over a span of only about three orders of magnitude of collector current—perhaps 0.1 to 100 mA in the transistor of Figure 7.1. If base spreading enters the analysis, the flat $\beta$ range will be even smaller.

An interesting example of the importance of base spreading shows up in Figure 7.11$b$. Note that $\beta_R$ has its maximum at a measly 2 mA. If you go back to (7.22) to see why this $\beta$ falloff should occur, you should be struck by the fact that since the collector is *more lightly* doped than the base, high injection should *enhance* $\beta_R$ by improving the forward-to-back injection ratio. However, $\beta_R$ is as low as it is primarily because of the shunting effect of the overlap diode. Anything that further forced the current to favor the overlap diode would depress $\beta_R$. Transverse currents in the base—especially in the lightly doped material that lies near the base-collector interface in a double-diffused transistor—would have just such an effect. They would spread the injected current right out into the overlap diode. Since the base currents in the active portion of the transistor are relatively high for the reverse-active mode, the effect is quite noticeable even at moderate emitter currents.

## MINIMIZING COLLECTOR VOLTAGE EFFECTS

In Figure 7.11, it is very evident that the collector current is not as independent of collector voltage as equation 7.17 would have us believe. We attributed the finite output conductance to two effects—base-width modulation and amplified avalanche multiplication. Neither of these effects is very desirable—although as usual, engineers can make a virtue of necessity: By operating a transistor at collector voltages above the sustaining voltage, $V_{EBO}$, it is possible to achieve some very fast switching times. This adaptation of breakdown is called *avalanche-mode* operation. [S. L. Miller and J. J.

Ebers, *Bell Sys. Tech. J.* **33**, 883 (1955).] In most applications, one wants a small output conductance. How does one design a transistor that gives the flattest possible collector curves? (Discussion: The answer on base-width modulation comes directly from problem 7.1. The extension of the collector-junction depletion width into the base is proportional to

$$\left[ \frac{N_{Dc}}{N_{Ab}(N_{Ab} + N_{Dc})} \right]^{1/2}$$

Thus, to minimize the dependence of the base width on collector voltage, we want the base heavily doped with respect to the collector.) How well does that conclusion fit with our desire to have little avalanche multiplication in the depletion region? (Discussion: For once we seem to win one thing without losing another. Avalanche multiplication is extraordinarily field dependent. The maximum field in the junction is proportional to [from (6.65)]:

$$\left[ \frac{N_{Ab}N_{Dc}}{N_{Ab} + N_{Dc}} \right]^{1/2}$$

Since we have already established that we want $N_{Ab} \gg N_{Dc}$, we must minimize avalanching by getting as low a collector doping as possible.) There is one liability to this solution; low collector doping implies high collector resistance. Putting a resistor in series with the load lowers the fraction of useful power delivered to the load. It also deleteriously affects the high-current and high-frequency performance of the transistor. All is not lost, however. Using a gaseous-epitaxy scheme, one may grow a thin, very lightly doped $n$-type layer on a heavily doped $n$-type substrate. The subsequent diffusions penetrate only part of the way into the epitaxial layer. The resulting structure would be labeled $n^{++}p^{+}\nu n^{++}$. One $(+)$ indicates moderate doping, two heavy doping, and the substitution of the Greek letter for the Roman indicates an exceptionally light doping (i.e., $\nu$ for $n$ and $\pi$ for $p$). As we shall see in a moment, doping the collector junction in this way, to optimize its DC performance, also minimizes the charge storage. (It probably was employed in the 2N707A we are using for our example, since that transistor is intended for high-frequency service.) At the collector junction in a diffused bipolar transistor, regardless of the transistor's intended application, the designer has many things going for him.

An interesting and important effect can be observed in lightly doped collector junctions under the influence of large current densities. If the flux of carriers from the emitter is large enough, those that are passing through the collector junction at any one time constitute a significant amount of free charge in the collector depletion region. The sign of these charges is such that they effectively increase the base doping and decrease or even invert the collector doping. The result is to shift the collector junction away from the emitter junction—an effect called *base stretching*. (It is sort of the opposite of the Early effect.) This will obviously cause further and sometimes very substantial deterioration of $\beta$ at high current densities. Things may be going well at the diffused collector junction, but you could hardly expect everything to work out perfectly. It would practically violate the Second Law of Thermodynamics to have nature entirely on your side.

## A RECAPITULATION

We have found the following rules to hold:

**1.** One maximizes $\beta$ by doping the emitter heavily with respect to the base and by making the base thin with respect to the minority-carrier diffusion length.

**2.** $\beta$ will be reasonably constant over a range limited at best to about three orders-of-magnitude change in collector current. On the low-current side $\beta$ falls off because of recombination in the emitter-base depletion region; on the high-current side it falls off because of high-injection-level effects, which may be enhanced or even dominated by base spreading and base stretching.

**3.** One minimizes the Early effect and avalanche multiplication effects by keeping the collector doping as low as possible.

**Problem 7.2.**

This problem should serve two functions; helping you to familiarize yourself with the labyrinth of currents that we have examined and giving you some insight into which of these currents are likely to be important at any one time. You are going to be asked to compare the performance of a germanium and silicon transistor each constructed as shown in Figure 7.1. Data for the transistor structure include

$A_\epsilon = 10^{-4} \text{ cm}^2$  $\qquad$ $A_{od} = 4A_\epsilon$

$w_\epsilon = w_b = 2 \ \mu\text{m}$  $\qquad$ $w_c = 150 \ \mu\text{m}$  $\qquad$ $w_{od} = 4 \ \mu\text{m}$

$N_{De} = 10^{19}/\text{cm}^3$  $\qquad$ $N_{Ab} = 10^{17}/\text{cm}^3$  $\qquad$ $N_{Dc} = 5 \times 10^{15}/\text{cm}^3$

(The length $w_{od}$ is the distance from the collector junction to the base contact in Figure 7.1. The simplification of Figure 7.3 makes $w_b = w_{od}$, but this is not really the case.)

Data for the transistors that differ because of differences in the materials:

|                    | Germanium                | Silicon                  |                        |
|--------------------|--------------------------|--------------------------|------------------------|
| $n_i^2$            | $6.25 \times 10^{26}$    | $1.21 \times 10^{20}$    | $/cm^6$                |
| $D_\epsilon$       | 25                       | 5                        | $cm^2/sec$             |
| $D_b$              | 80                       | 30                       | $cm^2/sec$             |
| $D_c$              | 40                       | 10                       | $cm^2/sec$             |
| $\tau_\epsilon$    | 0.1                      | 0.04                     | $\mu sec$              |
| $\tau_b$           | 1                        | 0.1                      | $\mu sec$              |
| $\tau_c$           | 100                      | 1                        | $\mu sec$              |

The diffusion lengths calculated from the above data come out:

|          | Germanium | Silicon |         |
|----------|-----------|---------|---------|
| $L_\epsilon$ | 16    | 4.4     | $\mu m$ |
| $L_b$    | 89        | 17      | $\mu m$ |
| $L_c$    | 634       | 32      | $\mu m$ |

Compute the peak $\beta$ for the two transistors. Then calculate $I_{co}$ for the germanium transistor.

[*Answer:* $\beta$ is obtained directly from the first two terms in (7.24). The only caution is to note that the emitter is not quite short-based in the silicon transistor. Accordingly, we must use the full $(1/L_e)$ coth $(w_e/L_e)$ in place of the simple $(1/w_e)$. The correction amounts to about 7% in the first term of $1/\beta$. Evaluating the two terms for each transistor leads to

$$\text{Ge:} \quad 1/\beta = 0.0031 + 0.0003 \quad \text{or} \quad \beta = 290$$

$$\text{Si:} \quad 1/\beta = 0.0018 + 0.0069 \quad \text{or} \quad \beta = 115$$

Note that the reverse injection term is better (smaller) in the Si mostly as the result of the better ratio of diffusion coefficients. (This advantage would go the other way in a *p-n-p* transistor.) On the other hand, recombination in the base is much worse in the silicon transistor. This is the result of both a shorter excess-carrier lifetime and a poorer carrier mobility. Both of these are inherent in the material, although for a given material, one batch may be substantially better than another. This batch-to-batch variation is likely to be especially noticeable for the silicon transistor, since $\beta$ depends more strongly on the minority-carrier lifetime, which is quite sensitive (although not very predictably) to minor variations in impurity content and processing.

A comment on the numbers we have just obtained: We have ignored for the moment the fact that all of the diffused-impurity concentrations will be rather strongly graded across the transistor structure. The electric fields that are caused by these gradients in the impurity concentrations accelerate minority carriers *toward regions of lower impurity concentration*. Thus, the base is effectively shortened and the emitter lengthened. (This is equivalent to saying that the minority carriers spend less or more time in traversing the region, which is really what happens.) A quick glance at (7.24) shows that this gradient in the concentrations will tend to improve both terms in $1/\beta$. That sounds pretty good, and it is. However, if you go back to Figure 6.10, you will see that there is an unhappy surprise awaiting the transistor designer. If you look at the diffusion profile for which $\sqrt{4Dt} = 2\ \mu$m, you will note that if $N_{Ab} = N_{Dc}$ at 4 $\mu$m, the basing doping at 2 $\mu$m will be about $10^{18}/$cm$^3$. That's an order of magnitude above that we had hoped for. Furthermore, to get the emitter junction at a depth of 2 $\mu$m will require a rather graded profile such as that for $\sqrt{4Dt} = 1\ \mu$m. This means that the effective emitter doping at the junction will be rather close to that of the base. That sounds pretty bad and it is. Our estimate of the reverse-injection ratio is quite optimistic compared to the realities that can actually be achieved by simple double diffusion. To get the injection efficiency that yields $\beta$'s much above 100 requires both a modicum of cleverness by the designer as well as great skill in controlling the fabrication process. Room temperature $\beta$'s of 50 to 100 in a silicon small-signal transistor and somewhat less for a power transistor are evidence of reasonably good design and fabrication. Higher $\beta$'s usually entail some compromises on other desirable features.

The second part of the problem was to compute $I_{co}$ for the germanium transistor. We must proceed from the definition $I_{co} = (1 - \alpha_F \alpha_R) I_{cs}$. Considering the $\beta$'s that we have just obtained, it is safe to say that $\alpha_F \simeq 1$. That leaves $\alpha_R$ and $I_{cs}$. In computing both of these terms, we must include the overlap diode. Our work is much easier if we may ignore the depletion currents, $I_{c3}$ and $I_{od2}$. (In fact, without specifying the applied voltage, we couldn't calculate them anyway.) There is a reason to focus on the diffusion currents, in any case. The reverse current in the collector junction is normally dominated by emitter injection. It is only when the value of $I_{co}$ becomes relatively large that it is of much interest. When $I_{co}$ is large, the bulk-region diffusion currents are dominant; when the depletion terms dominate, we may generally ignore $I_{co}$. This is why you were asked to calculate only the germanium case.

First, let us obtain $\beta_R = \alpha_R/(1 - \alpha_R)$. We may obtain $\beta_R$ from (7.24) by changing $\epsilon$ to $c$ and adding the term for the overlap diode: $I_{od}/I_{c5}$. From Table 7.2 (noting that $w_{od} \neq w_b$) we have

$$I_{od} = A_{od} n_i^2 q \left[ \frac{D_c}{w_c N_{dc}} + \frac{D_b}{w_{od} N_{Ab}} \right] \tag{7.25}$$

In the reverse-active mode, transport across the base is the same (assuming uniform doping) but the injection efficiency is terrible. Accordingly, we need only the first term from (7.24) plus the overlap-diode term:

$$1/\beta_R = \frac{D_c w_b N_{Ab}}{D_b w_c N_{Dc}} + \frac{A_{od}}{A_\epsilon}\left[\frac{D_c w_b N_{Ab}}{D_b w_c N_{Dc}} + \frac{w_b}{w_{od}}\right]$$

$$= 0.13 + 2.53 \Rightarrow \beta_R = 0.376 \quad \text{or} \quad \alpha_R = 0.273$$

Notice that the major reason for the very low $\beta_R$ is the overlap-diode current. Its dominance is primarily the result of its much larger area.

Our last task is to obtain $I_{cs}$. We may write down the diffusion components from Tables 7.1 and 7.2:

$$I_{cs} = (I_{c1} + I_{c2}) + (I_{c4} + I_{c5}) + I_{od1} \quad \text{(collector reverse biased)}$$

$$= qn_i^2\left\{\left[\frac{D_c}{w_c N_{Dc}} + \frac{D_b}{w_b N_{Ab}}\right]A_\epsilon + \left[\frac{D_c}{w_c N_{Dc}} + \frac{D_b}{w_{od} N_{Ab}}\right]A_{od}\right\}$$

$$= 6.7 \times 10^{-8} \text{ amperes}$$

About 32% of $I_{cs}$ is flowing in the overlap diode.

Finally, we can write down our answer:

$$I_{co} \cong (1 - \alpha_R)I_{cs} = 4.8 \times 10^{-8} \text{ amperes}$$

Had we done the same problem for silicon—that is, calculate $I_{co}$ ignoring the depletion current—we would have obtained an answer of the order of $10^{-12}$ to $10^{-14}$ amperes. However, its measured value, and the value calculated from the depletion current terms would be about three orders of magnitude higher.]

### Problem 7.3.

Estimate the junction breakdown voltages for the common base configuration for the silicon transistor in problem 7.2.

(*Answer:* From Figure 6.19, we may read the answers directly. For the dopings given in problem 7.2, we have

$$BV_{CBO} = 100 \text{ V} \qquad BV_{EBO} = 14 \text{ V}$$

Note that these are not very different from those of the transistor shown in Figure 7.8. In fact, they are "typical" for silicon transistors. The effort to obtain an effective collector junction—one with low capacity and little evidence of the Early effect—leads to rather high breakdown voltages. On the other hand, the effort to achieve an effective emitter junction—one with very high injection efficiency operating into a moderately heavily doped base—inevitably leads to a low breakdown voltage. In switching applications where the emitter-base voltage may swing into the reverse-bias region, low emitter-base breakdown voltage may present quite a problem.)

## CHARGE STORAGE IN THE TRANSISTOR

Along with good static characteristics, a successful transistor design must keep charge storage to a minimum. From Figure 7.4, it is easy to see that there are five regions in which charge will be stored in the active transistor plus three in the overlap diode. These comprise charge storage in each of the bulk regions plus charge storage in the injection-depletion regions.

What I am going to do here is a DC charge-storage analysis. Since this type of analysis has already been shown to lead to errors [see (6.35) and the subsequent discussion], you might wonder why I bring it up again. If you recall that discussion of why the DC analysis gives an incorrect value for the incremental capacity, you will remember that the problem was that the DC analysis fails to account for the phase shift inherent in the distributed charge distribution. In other words, one is really charging an RC line, not just a capacitor. Although the same problem besets a capacity calculation for the bulk-charge storage in a transistor, it is dwarfed by the very similar problem of the distributed character of the base charging line and by the relatively large percentage of uncertainty in the charge distribution in the base. We will deal with some of the ramifications of both the nonuniform charge distribution and the base charging line at a later time. For the moment, the very simple DC analysis will establish a usefully accurate set of relationships between the structure of the transistor and the capacitive behavior observed at its terminals.

We may once again segregate the components of charge according to the voltage that they respond to. However, there is little reason to keep separate all of the individual components as we did with the current. Instead, we divide the equations into diffusion and depletion terms, thus:

$$\Delta Q_\epsilon = \Delta Q_{\epsilon\,\text{diff}} + \Delta Q_{\epsilon\,\text{dep}} \qquad \text{(responds to } V_{b\epsilon})$$

$$\Delta Q_c = \Delta Q_{c\,\text{diff}} + \Delta Q_{c\,\text{dep}}$$

$$\Delta Q_{od} = \Delta Q_{od\,\text{diff}} + \Delta Q_{od\,\text{dep}} \qquad \text{(responds to } V_{cb})$$

$$\tag{7.26}$$

(Notice that each of the terms on the right includes charge stored on *both* sides of the metallurgical junction.) We may write down explicit analytic expressions for each of the terms above by direct application of our diode solutions, (6.26) and (3.51). However, only two really different forms arise, so let us evaluate $Q_{\epsilon\,\text{diff}}$ and $Q_{c\,\text{dep}}$ as samples. The other terms can be obtained from them by a rather simple exchange of subscripts. Accordingly:

$$Q_{\epsilon\,\text{diff}} = A_\epsilon q n_i^2 \left\{ \frac{L_b\left[\cosh\left(\dfrac{w_b}{L_b}\right) - 1\right]}{N_{Ab}\,\sinh\left(\dfrac{w_b}{L_b}\right)} + \right.$$

$$\left. + \frac{L_\epsilon\left[\cosh\left(\dfrac{w_\epsilon}{L_\epsilon}\right) - 1\right]}{N_{D\epsilon}\,\sinh\left(\dfrac{w_\epsilon}{L_\epsilon}\right)} \right\} [e^{qV_{be}/kT} - 1] \qquad (7.27)$$

$$\Delta Q_{c\text{dep}} = A_c q N_{Dc} \left[ l_n(V_{cb}) - l_n(0) \right] =$$

$$= A_c \left[ \frac{2\varepsilon q N_{Ab} N_{Dc}}{N_{Dc} + N_{Ab}} \right]^{1/2} [(V_{Dc} + V_{cb})^{1/2} - (V_{Dc})^{1/2}] \qquad (7.28)$$

where $A_c = A_\epsilon + A_{od}$. Frequently the short-based approximation is appropriate to both terms in (7.27), which simplifies it rather nicely to

$$\Delta Q_{\epsilon\,\text{diff}} = \frac{A_\epsilon q n_i^2}{2} \left\{ \frac{w_b}{N_{Ab}} + \frac{w_\epsilon}{N_{D\epsilon}} \right\} [e^{qV_{be}/kT} - 1] \qquad (7.27')$$

{N.B. $[\cosh(x) - 1] \Rightarrow x^2/2$.} Note that $I_{\epsilon b}$ and $\Delta Q_{\epsilon\,\text{diff}}$ have the same voltage dependence. Thus the collector current and the charge stored in the bulk portions of the transistor increase at the same rate. On the other hand, the depletion capacity varies in a very different way. It will approach a minimum for large reverse bias, but its dynamic range is likely to be less than an order of magnitude, especially when the junctions are diffused.

It is important to observe that some of the things that contribute to commendable DC performance also contribute to minimizing charge storage. Others, however, tend to go the wrong way. On the negative side, heavily doping the emitter certainly tends to maximize $Q_{\epsilon\text{dep}}$ at the same time as it maximizes $\beta$. On the other hand, the lightly doped collector junction, which gives the largest $BV_{CBO}$ and smallest amount of base-width modulation, also contributes to minimizing $Q_{c\text{dep}}$.

Since $Q_{\epsilon\,\text{diff}}$ and $I_{\epsilon 5}$ are proportional to each other, minimizing the charge stored in the bulk regions means minimizing $Q_{\epsilon\,\text{diff}}/I_{\epsilon 5}$. That ratio is given by

$$Q_{\epsilon\,\text{diff}}/I_{\epsilon 5} = (w_b^2/2D_b) + \frac{w_\epsilon w_b N_{Ab}}{2D_b N_{D\epsilon}} \qquad (7.29)$$

The first term in (7.29) gives the coulombs of charge stored in the neutral portion of the base per ampere of collector current. It becomes small very rapidly with decreasing base width. The second term gives the coulombs of charge stored in the neutral portion of the emitter per ampere of collector current. This term becomes small if the emitter is heavily doped with respect to the base and if the base and emitter regions are made as thin as possible. Making the base thin and doping the emitter heavily with respect to the base are the primary steps needed for a high $\beta$. Thus, a transistor with a high $\beta$ will have a minimum of charge storage in the neutral regions.

Probably the best way to get some feeling for the amount of charge stored is to calculate these values for a particular transistor. Let us use the transistor of problem 7.2 and the switching problem described in the section on "What makes a good transistor?" (page 379):

**Problem 7.4.**

Using the silicon transistor of problem 7.2, calculate the change in charge storage in switching the transistor from the state $[V_{ce} = + 14 \text{ V}, I_c = 0]$ to the state $[V_{ce} = + 0.6 \text{ V}, I_c = + 5.8 \text{ mA}]$. If the transition has to take place in $10^{-8}$ seconds, how much base current would be required simply to charge the base? (I.e., ignore the current necessary to support the flow of collector current.

[*Solution:* First a comment. This is one of those problems in which you have to calculate a lot of different numbers, and you might well say: "Why bother?" The answer is that only through using these expressions can you get familiar enough with the material in these "practical" chapters to have any functional use of it. If you have set yourself the goal of knowing something about solid-state device theory, this, sad to say, is where the learning really occurs. Now on with the grubby work.

We may quickly dismiss $\Delta Q_{c\,\text{diff}}$ and $\Delta Q_{od\,\text{diff}}$ as being negligible in a reverse-biased diode. That leaves us three depletion regions and one pair of bulk regions to analyze. For the depletion-region analysis we will need the diffusion voltage. Taking $kT = 1/40 \text{ eV}$ and the data from problem 7.2, equation 3.49 gives us

$$V_{Dc} = 0.725 \text{ V}$$

$$V_{De} = 0.915 \text{ V}$$

We will also need the forward bias on the emitter when $I_{e5} \cong - I_c = = - 5.8 \text{ mA}$. Using the expression in Table 7.2, we obtain

$$V_{be} = 0.69 \text{ V} \qquad \text{for} \qquad I_{e5} = - 5.8 \text{ mA}$$

We now have all the data to put into (7.28) to obtain the depletion-region charges:

$$\Delta Q_{\epsilon \, dep} = 5.9 \times 10^{-12} \text{ coulombs}$$

$$\Delta Q_{c \, dep} = 9.3 \times 10^{-12} \text{ coulombs}$$

$$\Delta Q_{od \, dep} = 4 \Delta Q_{c \, diff} = 37 \times 10^{-12} \text{ coulombs}$$

The diffusion charge is all that remains. From equation (7.28) that is

$$\Delta Q_{\epsilon \, diff} = 3.9 \times 10^{-12} \text{ coulombs}$$

Note well that the dominant-charge storage term comes from the collector-base depletion region, and, of that charge, the principal part is in the overlap diode. This is the usual case and it points up the reason why a very large effort is expended in minimizing charge storage in the collector junction.

Totalling up the charge that must be supplied to switch the transistor, we find $\Delta Q_{total} = 75$ pC. To switch that transistor in 10 $nS$ would require an average base charging current of 7.5 mA. The steady-state current (supporting 5.8 mA) would be only $5.8/\beta = 0.055$ mA. Thus, switching at this relatively fast speed would require base drive 137 times that required at low speed! In fact, the principal part of the collector current during the switching cycle would be flowing through the base, and the base current would exceed the collector current. As you can see, during the switching cycle, the speed we have demanded has turned our transistor from an active device with substantial gain to an almost passive device strongly reminiscent of a simple capacitor. Speed costs gain and vice versa.]

Up to this point, we have focused our attention on activity within the transistor itself. Our next task is to examine the terminal characteristics, representing the transistor by somewhat simpler but less accurate electrical analogs. Before beginning that task, it is appropriate to pause and summarize what we have seen so far.

## CHAPTER SUMMARY

This chapter's purpose was to develop insight into the internal behavior of the bipolar transistor. The physical model that was constructed is a relatively direct development of junction-diode theory, although the results of the analysis range far from the behavior of simple diodes.

The chapter began with a description of the structure of the bipolar transistor. Its high geometrical-aspect ratio suggested that a quasi-one-dimensional model be adopted for the analysis of the current flows. Such a model divides the device into an "active" transistor—the portion directly under the emitter—and an "overlap" diode that shunts the base-collector junction of the active transistor. Only base majority carriers flow between these two regions, traveling laterally in the plane of the junctions. Apart from a very small amount of "fringing" at the junction edges, the remainder of the current flows are all normal to the plane of the junctions. This permits a one-dimensional analysis for almost all of the terms of importance in transistor operation. Only the transverse base current and such fringing-dependent terms as avalanche multiplication are resistant to this simplification.

Once the model of Figure 7.3 was adopted, we were able to analyze the terminal currents as functions of the terminal voltages by a direct application of junction-diode theory to the structure of the transistor. By dividing the carrier fluxes into those that are injected at the emitter and those that are injected at the collector, it is possible to represent the junctions as essentially ohmic contacts (perfect sinks) for minority carriers incident upon them. For majority carriers, the junctions retain their normal barrier properties. Thus, each junction becomes an independent source of injected current and a dependent source of collected current. This bilateral symmetry, with each junction being both an independent junction diode and a dependent current generator, is the basis of the Ebers-Moll model of the transistor. To obtain the functional dependence of the Ebers-Moll currents on the terminal voltages, the currents were divided into components on the basis of the source and destination of the particular carriers in question. It is easy to write the terminal currents in terms of these carrier fluxes, while at the same time, by dividing the fluxes according to their place of recombination, each individual flux can be obtained directly from equivalent results for the junction diode. Apart from some not-very-restrictive simplifications of the depletion region recombination-generation rate, only two assumptions were made in these derivations. The doping profiles were assumed to be flat within any one region, and the voltage drops across the bulk regions were assumed to be insignificant. Both of these restrictions will be removed in subsequent analysis (next chapter).

Once we had the components of the terminal currents in hand, it was a direct step to the Ebers-Moll equations. The short-circuited diode saturation currents, $I_{es}$ and $I_{cs}$, were defined and obtained, and then the current transfer ratios for the dependent generators, $\alpha_F$ and $\alpha_R$, were derived. A second set of current coefficients, $I_{eo}$ and $I_{co}$, were obtained algebraically from the first. The first set is most useful in "voltage-control" analysis, while the second is best used in "current-control" analysis. Although these coefficients would be independent of bias in the first-order Ebers-Moll model, both the measured CB characteristics and the analytical expressions for the coefficients show that this first-order model is a good but not perfect representation. The first-order model predicts that $\partial I_c/\partial V_{cb} = 0$ as long as the collector junction is reverse biased. At $I_c = 1$ mA in the sample transistor, however, we found the small but nonzero value of $\partial I_c/\partial_{CD} \simeq 0.2 \times 10^{-6}$ mhos, while in the reverse mode of operation, the value of $\partial I_\epsilon/\partial V_{eb}$ at $I_\epsilon = 1$ mA was roughly two orders of magnitude higher and quite voltage dependent to boot. An inspection of the analytical expressions for the Ebers-Moll parameters revealed that the slope is the result of base-width modulation (the Early effect), while the curvature reveals the presence of multiplication in the reverse-biased junction. The physical model showed that both of these undesirable effects can be minimized by having the far side of the collecting junction very lightly doped. However, this improvement in the output characteristics is purchased at the expense of injection efficiency (i.e., lower $\alpha$). Thus, if the junction doping is arranged to improve the forward characteristics, the reverse characteristics suffer rather badly. Since transistors are normally employed in unidirectional amplifiers, optimizing the forward characteristics at the expense of the reverse makes excellent sense.

To develop some insight into the properties that characterize a "good" transistor, we next indulged in a rather lengthy analysis of a primitive CE switching circuit. The first step was to convert the Ebers-Moll equations from the CB form to the CE form, switching the input current from $I_e$ to $I_b$ and the current-transfer ratio from $\alpha$ to $\beta$. A set of characteristic curves taken for the CE mode showed that controlling the base current rather than the emitter current made the variation in the Ebers-Moll parameters much more evident, but the output compensated for this failing by providing substantial current gain.

We first examined the transistor's performance in a static-switching example in which the transistor was driven from cutoff ($I_c = 0$) to saturation ($V_{ce} = 0$). It was found that the power gain for the switch was determined principally by $\beta$ and the maximum collector voltage. For the sample transistor in the simple switching circuit, we obtained a power gain of 31 dB, which was not particularly spectacular. Selection of a transistor with a higher $\beta$ could have easily added another 6 dB of gain. For static-switching applications, the higher $\beta$ would have only one potential liability: it would probably be accompanied by a lower CE breakdown voltage, $BV_{CEO}$. The physical model showed that this lower breakdown voltage is the result of avalanche multiplication in the collector junction. For the open-base case, the collector current becomes indeterminate (self-sustaining) when $\alpha_F M \Rightarrow 1$. Since this amount of multiplication is finite, it requires substantially less collector voltage than the CB breakdown, for which $M \Rightarrow \infty$. Thus $BV_{CEO} < BV_{CBO}$. As $\beta$ gets very large, the amount of multiplication required to reach the sustaining voltage falls off, so the difference between $BV_{CEO}$ and $BV_{CBO}$ gets larger.

We turned briefly to the question of dynamic switching behavior. Charge storage in the bulk and depletion regions of the transistor were found to limit the speed with which the transistor could be switched between its "on" and "off" states. Although the details of the analysis were saved until later, we arrived at the rather important conclusion that by increasing the base drive, the transistor's switching speed could be increased. Unfortunately, the speed is purchased by giving up power gain. For sufficiently high base drive, it was found that the product of the switching speed and the gain is a constant. This represents the best that one can do. In most situations, one gives up more than one gets back.

We then returned to the physical model of the transistor to learn what things could be done to improve its DC performance. Most of our attention was focused on how the transistor's structure and bias point established $\beta$. We found that $\beta$ was determined by three factors: injection efficiency, recombination in the emitter-depletion region, and recombination in the base. From these three factors alone, we were able to conclude that $\beta$ is very low at both low and high emitter-current densities. The reasonably flat middle section can encompass no more than three orders of magnitude swing in collector current. These limitations are imposed at the low end by recombination in the emitter junction and at the high end by a loss in injection efficiency resulting from a high injection level in the base. Later in the chapter we saw that the realized high current limit was frequently substantially below the onset of high-injection-level limitations. This reduction in current-carrying capacity was the result of a combination of base spreading, base stretching, and collector debiasing. All of these effects came about because of voltage drops of ohmic or space-charge origin that had been ignored in the initial physical model.

The current-carrying capabilities of a transistor at the low current limit are only remotely adjustable by the designer. The only readily available parameters are generally well constrained by other demands in the design. Thus, the low current limit is dictated by the area of the emitter junction. To some extent, one may extend the upper current limit by careful design that prevents any of the four factors that limit the current density from setting in too soon. Unfortunately, one is also limited at the upper end by the other requirements of a good transistor. For example, a high collector breakdown voltage and a minimum amount of base-width modulation require a very lightly doped collector; on the other hand, a very lightly doped collector emphasizes collector debiasing and base stretching.

Base spreading provides some special analytical difficulties. Although it is clear enough that the problem arises because the small voltage drops associated with the transverse flow of base current enter into the equation for the injected-current density, the exponential character of that relationship makes the effect reasonably severe and at the same time renders it hard to handle analytically. We concluded that to minimize the effect, one should design emitter shapes in which no part of the emitter is very far from an edge.

Long narrow stripes, star-shaped patterns, and interdigitated designs are most frequently employed. Although the palliative to the condition is clearly visible, the lack of a good method of analysis makes it very difficult to predict the extent to which base spreading limits a particular transistor's current-carrying capacity. In fact, efforts to model the effects of the voltage drops across the base are undoubtedly the weak point of the whole analysis.

The maximum value that $\beta$ achieved was determined by two factors that, within limits, were well within the control of the designer. These were the ratio of the emitter-to-base doping level, which determined the injection efficiency, and the ratio of the base width to the base minority diffusion length, which determined the transport efficiency. For diffused germanium transistors, the injection efficiency is the dominant term; for silicon transistors, both terms are important. This has the important result of making $\beta$ in a silicon transistor quite temperature dependent as a direct consequence of the minority-carrier lifetime increasing with temperature.

The structure analysis concluded with an examination of the charge storage in the transistor. There is a charge associated with each of the current terms that we had considered before, but we found it convenient to collect terms on the basis of whether they were associated with diffusion through the bulk regions or with the motion of the edges of the depletion regions. We found that the principal diffusion term is proportional to the collector current, while the depletion terms followed the junction voltage according to the power law for the doping profile of the junction. Using the switching example developed in earlier sections, each of the charge storage terms was determined. We found that because of the large voltage swing across the collector junction, the largest amount of charge storage occurred there, with roughly 80% of that charge being stored in the junction of the overlap diode. This result is directly associated with the Miller effect (i.e., for the small-signal model, the relatively small incremental capacity associated with the base-collector junction has a very large loading effect on a high-gain amplifier because of the large voltage swing across that capacitor). Efforts to minimize charge storage in the transistor are generally compatible with efforts to obtain good DC characteristics. For example, the narrow base width that is needed for a high $\beta$ also minimizes the diffusion charge storage; the lightly doped collector that yields a very high $BV_{CBO}$ also minimizes the depletion-layer charge storage at the collector. The only conflicts are in the heavy emitter doping that gives good injection efficiency but high junction capacity and the possibility of incompatibility between the light doping of the collector and high current operation.

**Problems**

**1.** In an integrated circuit, all of the diffusions of a given type must be done together. Devices are differentiated primarily by their areas. Since the signal grows in magnitude from input to output, the devices to handle the signal may do so too. With the data given below, select the emitter area to put the center of the $\beta$ flat zone at $I_c = 0.1$ mA. Compute the $\beta$ at that current for $V_{ce} = 10$ V. How much would $\beta$ change if $V_{ce} = 20$ V?

Data:  $A_{od} = 4A_\epsilon$

| $w_\epsilon = 2\ \mu m$ | $w_b = 1.5\ \mu m$ | $w_{od} = 3.5\ \mu m$ |
|---|---|---|
| $N_{D\epsilon} = 10^{19}/cm^3$ | $N_{Ab} = 10^{17}/cm^3$ | $N_{Dc} = 10^{16}/cm^3$ |
| $D_\epsilon = 5\ cm^2/sec$ | $D_b = 30\ cm^2/sec$ | $D_c = 10\ cm^2/sec$ |
| $\tau_\epsilon = 0.04\ \mu sec$ | $\tau_b = 0.1\ \mu sec$ | $\tau_c = 1.0\ \mu sec$ |

2. Our static lamp-switching problem suggests that the Early effect (base-width modulation) is not important in switching problems. As you might suspect, however, it does appear in the dynamic switching problem.

a) You are asked to set up the differential equations for the active region of switching (between $\tau_2$ and $\tau_3$ in Figure 7.13b) ignoring, for the moment, the fact that $w_b$ is, indeed, a function of the collector voltage. Let the base-current drive be barely adequate for switching. Allow yourself the following simplifications: Let $V_b$ and $I_b$ be constants during the switching period and assume that the charge storage may be adequately represented by only two terms: $Q_{ediff}$ and $Q_{cdep}$. The schematic below may be of some help, but keep in mind that the capacitors are not of the friendly, linear variety.

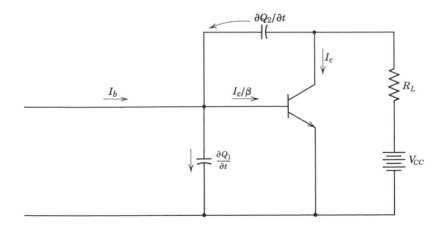

b) Now go over the derivation and determine which terms will be modified by the variation in the base width. Note carefully that the Early effect lengthens the switching time.

*Answer:* From Kirchhoff's laws and equations 7.27 and 7.28, one obtains two unfortunate coupled equations:

$$\frac{\partial Q_2}{\partial t} = -K_2 R_L \left[ V_{Dc} + V_{CC} - \left( I_c + \frac{\partial Q_2}{\partial t} \right) \right]^{-1/2} \left( \frac{\partial I_c}{\partial t} + \frac{\partial^2 Q_2}{\partial t^2} \right)$$

$$K_1 \frac{\partial I_c}{\partial t} = I_b - \frac{I_c}{\beta} + \frac{\partial Q_2}{\partial t}$$

where:

$$K_1 = \frac{w_b^2}{2D_b} + \frac{w_e w_b N_{Ab}}{2D_b N_{De}}$$

and

$$K_2 = A_c \left[ \frac{2\varepsilon q N_{Ab} N_{Dc}}{N_{Dc} + N_{Ab}} \right]$$

It is clear that both $K_1$ and $\beta$ depend on the basewidth. $K_1$ grows as $V_c \to 0$ while $\beta$ shrinks. These terms are in the second equation in such a way that it is clear that $\partial I_c / \partial t$ is decreased by the base width modulation.

3. The switching mode that is most economical of power in the idle state is one in which the switching element is either fully on (no voltage drop) or fully off (no current). For computation, such switching is called *saturated logic*. The problem with saturated logic is that it is not as fast as logic that does not require such wide swings of the charge state of a transistor. In particular, in driving the transistor well into the "on" state, the collector junction is somewhat forward biased. This leads to an accumulation of minority charge, particularly in the lightly doped collector regions. Getting into the "off" state involves removing that charge before the transistor can get into the active region where the collector current begins to respond to the "turn-off" signal at the base. The partial solution to this switching delay is to dope the whole transistor structure with gold. (The enormous diffusivity of gold precludes doping less than all of the transistor.) This lowers the lifetime throughout the transistor, but the change is greatest in the lightly doped collector regions. The net result is to make the collector region short-based, drastically reducing the amount of charge stored in the collector in the saturated "on" state. (See problem 5 at the end of Chapter 6 for a similar approach to switching diodes.) the gold is not without its effect on the rest of the transistor. Picking numbers appropriate to a modern n-p-n silicon transistor, discuss how the shortened lifetime (to about $2 \times 10^{-9}$ sec) will alter the current gain and charge storage

throughout the transistor, and make a judgment on whether such treatment would improve the speed of a transistor that operated only in the forward active mode (i.e., one that is never driven fully on or off).

4. Would SiC make a suitable material for bipolar transistors? Data on SiC can be found in the sixth problem at the end of Chapter 3. (Hint: A good transistor must have an adequate value of $\beta$, breakdown voltage, and current handling capability. You may assume that any nonfundamental technological problems associated with actual fabrication could be solved.)

*Answer:* Not at room temperature. At 300°K, the emitter current is dominated by the depletion current, which contributes nothing to the collector current. At elevated temperatures (diodes have been operated in excess of 800°C!), SiC bipolar transistors may be useful.

5. The lateral *p-n-p* transistor shown in the figure is a useful isolation device in an *n-p-n* monolithic integrated circuit. A normal *n-p-n* double-diffused transistor is shown as well as the lateral *p-n-p* so that you can see how the diffusion steps would proceed and how the diode isolation of the two devices is obtained. Note that the vertical and horizontal scales are very different, as they must be to do a sensible drawing of a planar device. Using the configurational and doping data in the figure and sensible values for any parameters you might need, make a good engineering guess as to the maximum $\beta$ for this device.

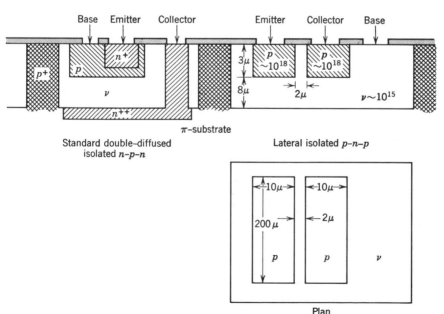

## CHAPTER 8

# DEVICES THAT WORK PRINCIPALLY BY DIFFUSION: (C) ELECTRICAL MODELS AND CIRCUIT PERFORMANCE OF THE BIPOLAR TRANSISTOR

The previous chapter dealt with the bipolar transistor in terms that were directly related to the activities internal to the transistor. Most users of transistors need representations that are more amenable to circuit analysis. In fact, it is difficult to assess the quality of a transistor except in terms of its performance in a particular circuit. Accordingly, this chapter is devoted to circuit models of the transistor. We will not abandon the physical models of the previous chapter. It is through them that we will not only construct the circuit models but also set realistic limits to circuit performance.

## SMALL-SIGNAL ANALYSIS AND THE LINEAR AMPLIFIER

It is frequently very desirable to be able to describe or model all of the elements of an amplifying circuit in terms of linear equations. This is certainly the case if you want to be able to use any of the powerful and efficient techniques of linear-circuit analysis. Furthermore, if the circuit does not lend itself to linear representation, you will find it most difficult to manufacture an amplifier that does not seriously distort the signal it is processing. (However, keep in mind that serious distortion of the signal is precisely what many circuits are intended to create. Limiters, modulators, digitizers and demodulators—to name a few—all put out something quite different from what was put in. Not only is nonlinearity not bad, we would be rather helpless without it. But there are many circuits that require faithful linear magnification of the signal, and these are the ones that are discussed in the following pages.)

The method we will employ to get a linear representation of the transistor is the same one we used on the junction diode. We will take advantage of the fact that, over a limited but useful range, any curve can be represented reasonably accurately by its tangent. Then, if we have a relationship between $I$ and $V$, the conductance for that pair is simply $\partial I/\partial V$. Similarly, the capacity associated with $Q(V)$ is $\partial Q/\partial V$. (We do not normally run into the problem of having to use the time-dependent diffusion equation with the transistor because the base width and emitter width are so much shorter than the diffusion length that, even with the foreshortened AC diffusion length, the short-based approximation is quite valid. By the time this approximation has begun to fail, the usual transistor has ceased functioning in any useful fashion.)

From a circuit point of view, the transistor is a two-port. Regardless of which terminal is common, there are only two pairs of terminal variables, and if they are coupled, it takes only two equations to specify one pair (your choice) in terms of the other. The three choices may be written in matrix form as

$$\begin{bmatrix} v_1 \\ v_2 \end{bmatrix} = \begin{bmatrix} z_{11} & z_{12} \\ z_{21} & z_{22} \end{bmatrix} \begin{bmatrix} i_1 \\ i_2 \end{bmatrix} \qquad \text{impedance parameters}$$

$$\begin{bmatrix} i_1 \\ i_2 \end{bmatrix} = \begin{bmatrix} y_{11} & y_{12} \\ y_{21} & y_{22} \end{bmatrix} \begin{bmatrix} v_1 \\ v_2 \end{bmatrix} \qquad \text{admittance parameters}$$

$$\begin{bmatrix} v_1 \\ i_2 \end{bmatrix} = \begin{bmatrix} h_{11} & h_{12} \\ h_{21} & h_{22} \end{bmatrix} \begin{bmatrix} i_1 \\ v_2 \end{bmatrix} \qquad \text{hybrid parameters}$$

Note that all of the parameters and terminal variables are written in lower-case letters, indicating that they are *incremental quantities* that are probably functions of the bias point selected. In practice, alphabetic subscripts are usually employed to give the reader information on what the parameter or variable is. For example, in the common-emitter circuit of Figure 7.10, the hybrid parameters would be written

$$\begin{bmatrix} v_i \\ i_0 \end{bmatrix} = \begin{bmatrix} h_{ie} & h_{re} \\ h_{fe} & h_{oe} \end{bmatrix} \begin{bmatrix} i_i \\ v_o \end{bmatrix}$$

The subscript $e$ indicates that performance in a common-emitter circuit is being described. $i$ and $o$ stand for input and output respectively. (Once the circuit is specified, there is no confusion between input and output terminals.) Finally, $f$ and $r$ stand for forward and reverse respectively. Thus, $h_{ie} = \Delta v_i/\Delta i_i|_{v_o=0}$ is the input impedance with the collector shorted; $h_{re} = \Delta v_i/\Delta v_o|_{i_i=0}$ is the reverse voltage-transfer-ratio with the base open; $h_{fe} = \Delta i_o/\Delta i_i|_{v_o=0}$ is the forward current-transfer-ratio (i.e., $\beta$) with the collector shorted; and $h_{oe} = \Delta i_o/\Delta v_o|_{i_i=0}$ is the output admittance with the base open. Note that each of these defines a very specific electrical measurement. For example, $h_{ie}$ is found by measuring the ratio of the increments of input current and voltage with the collector voltage held constant (frequently called *incrementally short-circuiting* the output).

Although one could measure the *h* parameters as functions of the bias point, temperature and frequency, for most human computations, a table of values or even an algebraic representation of the table is unsuitable. People prefer to think in terms of familiar linear circuit elements. Furthermore, if the model is to be suitable for human consumption, it should be relatively simple, and its elements should relate directly to the physical activities going on in the device. If our physical analysis is complete enough, the electrical model should apply at most frequencies and scale in a known way as the bias or temperature is varied.

It is important to recognize right from the start that there is no unique model. For example, a perfectly acceptable "T" model can be converted into an equally good "$\pi$" model by electrical circuit manipulations that have nothing to do with the intrinsic device. Which model you choose is determined by your own personal prejudices. Most people prefer the T circuit for the common-base configuration and the $\pi$ circuit for the common-emitter or common-collector (emitter-follower) circuits. Let us begin with the T.

We wish to get a linear circuit to represent the incremental or *small-signal* behavior of the CB circuit such as that shown in Figure 8.1*a*. Note that no batteries of any sort appear in Figure 8.1*a*. This is the conventional way of drawing circuit diagrams, and it corresponds to the usual way of distributing power in a complex circuit. The various voltages that are required are distributed throughout the circuit on bus lines. In the diagram, the bus lines' potentials with respect to the common or ground line are marked, but the actual source of the voltage is seldom shown. However, it is generally presumed that the source is equivalent to an ideal battery—a DC voltage source that is truly an *AC short circuit*. This means that in the incremental analysis, the DC bus lines such as the 20 V line in Figure 8.1*a* are all equivalent to and thus map into the common line.

A second convention is also to be observed in the figure. Those circuit elements that are really resistors or capacitors are indicated by capital letters with appropriate upper-case subscripts that indicate their location and/or function. For example, $R_L$ is the load resistor while $R_E$ is a resistor in series with the emitter lead. In Figure 8.1*b*, where the transistor is replaced by its incremental equivalent circuit, the components representing the ratios of the incremental variables are indicated by lower-case letters with lower-case subscripts. For the T circuit, with its common node, a single subscript identifies the arm in which the particular component is found. For example, $g_e$ is a conductance running from the point *e* to the point *b'*. (Note, by the way, that the standard *e* is usually employed to indicate the emitter rather than the $\epsilon$ we have used up to this point.)

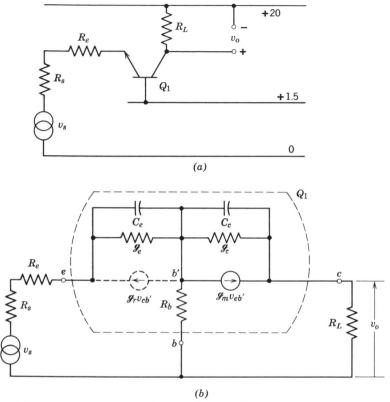

(a)

(b)

**Figure 8.1** A common-base circuit (a) and its small-signal T equivalent (b). Actual circuit components are indicated by capital letters. Incremental components are specified by a lower-case letter (e.g., r for resistance or g for conductance), with a subscript indicating the external terminal (e.g., $r_e$ is a resistor running from the point labeled e to the common point of the T labeled b').

To obtain the incremental representation of the transistor, we need only write the stored charges and the terminal currents in terms of the terminal voltages. Then, by differentiation, the incremental circuit is at hand. Let's begin with the simplest large-signal representation that we have—the Ebers-Moll model and equation 7.11. That equation related the collector current to the terminal voltages for the CB circuit. However, we need not be as naive as we were when we first wrote (7.11). We know that $I_{es}$ and $\alpha_F$ will be slowly varying functions of the collector voltage as a result of base-width modulation. Furthermore, there can be a noticeable voltage drop between the base ohmic contact and the base side of the emitter depletion region. Accordingly, to be complete, we should write (7.11) for a properly biased transistor in the CB circuit as

$$I_c = I_{cs}(V_{cb'}) + \alpha_F(V_{cb'})I_{es}(V_{cb'})e^{qV_{b'e}/kT} \tag{8.1}$$

Equation 8.1 allows for the voltage drop across the base by introducing the fictitious point, $b'$, lying presumably on the base side of the emitter depletion region. Thus, $V_{b'e}$ is the voltage across this depletion region, and $V_{b'e} \leqq V_{be}$.

Unless avalanche multiplication is running rampant, the $I_{cs}$ term from (8.1) should be either small or not particularly voltage dependent. Thus, in differentiating (8.1), the $I_{cs}$ term does not enter, and we have

$$dI_c = i_c = \frac{\partial I_c}{\partial V_{eb'}} v_{eb'} + \frac{\partial I_c}{\partial V_{cb'}} v_{cb'}$$

$$= g_m v_{eb'} + g_c v_{cb'} \tag{8.2}$$

From (8.1) we obtain explicit expressions for $g_m$ and $g_c$:

$$g_m = (q/kT)(I_c - I_{cs}) \tag{8.3}$$

$$g_c = \left( \frac{1}{\alpha_F} \frac{\partial \alpha_F}{\partial V_{cb'}} + \frac{1}{I_{es}} \frac{\partial I_{es}}{\partial V_{cb'}} \right)(I_c - I_{cs}) \simeq I_c \frac{\partial \log(\alpha_F I_{es})}{\partial V_{cb}} \tag{8.4}$$

In most practical applications, $I_c \gg I_{cs}$ so that both of the parameters we have just defined *scale proportionally with* $I_c$. It is important that you note this scaling rule, since it is one of our principal objectives. The computed small-signal gain and bandwidth of the transistor will obviously depend on the values of the components in the linear representation of the transistor; thus, this scaling relates the performance of the transistor to its DC bias point.

Let us now begin to choose some discrete components that would behave as (8.2) says the transistor does. $g_m$ is the ratio of the incremental output current to the incremental change in voltage across the emitter junction. Dimensionally it is a conductance, but functionally it indicates a dependent source of current. Accordingly, it is called the *transconductance* and it is represented in Figure 8.1b by the dependent generator. The second coefficient in (8.2) is readily identified with the output conductance of the generator. These two components form one arm of the "T." From equation 7.10 we may extract the other arm of the T. Just as in (8.2), we have

$$dI_e = i_e = \frac{\partial I_e}{\partial V_{eb'}} v_{eb'} + \frac{\partial I_e}{\partial V_{cb'}} v_{cb'}$$

$$= g_e v_{eb'} + g_r v_{cb'} \tag{8.5}$$

Then, from (7.10), with the same assumption about the $I_{cs}$ term being either very small *or* constant, we obtain the two conductances in (8.5):

$$g_e = (q/kT)(I_e - \alpha_R I_{cs}) \simeq g_m(1/\alpha_F) \tag{8.6}$$

$$g_r = \left( \frac{1}{I_{es}} \frac{\partial I_{es}}{\partial V_{cb'}} + \frac{1}{\alpha_R} \frac{\partial \alpha_R}{\partial V_{cb'}} \right)(I_e - \alpha_R I_{cs}) \simeq \frac{I_c}{\alpha_F} \frac{\partial \log \alpha_R I_{es}}{\partial V_{cb'}} \tag{8.7}$$

Once again, we may identify the transconductance, $g_r$, with a generator and the conductance, $g_e$, with the conductance shunting that generator. In normal operation, the shunt conductance is so large that the relatively tiny current source, $g_r v_{cb'}$, is effectively shorted out. Accordingly, in Figure 8.1b, the generator is shown as a dotted line in the emitter leg of the T. Only when it becomes important to know exactly what $i_b$ is, will we have to put $g_r$ back in.

Finally, we must examine the voltage drop, $V_{bb'}$. Here life is not too kind to the would-be circuit analyzer. We may divide the drop across the base into the component across the overlap diode and the component across the active junction itself. It is this second component that gives us the most grief. Since the local emitter current-density depends exponentially on the local emitter-base voltage, the $IR$ drop across the base goes directly into the exponential function that must be integrated across the surface of the emitter junction to obtain the current. The relationship so obtained[*] is sufficiently complicated that it cannot be modeled by a readily scaled resistor in the base lead except in the low-current limit where base spreading is insignificant anyway. This does not complete our tale of woe about $r_b$; there is more to come. It certainly is the weakest part of the model. And yet, we can use the model of 8.1b with some confidence in most circuits because $r_b$ is *usually* not very important. Then why include it at all? The answer is that optimizing the performance of many circuits requires minimizing the resistance in series with the base lead. $r_b$ obviously represents the minimum value possible, and thus it is often the limiting element in maximizing the gain or bandwidth in a particular circuit application. We need something in the base lead to keep us honest, but when it becomes really important, the value of $r_b$ is likely to be the least precise quantity in the analysis.

Before we add the capacitors shown in Figure 8.1b, calculate the $h$ parameters, the voltage gain, and the input and output conductances for the circuit in Figure 8.1b. For numerical evaluation, let $\alpha_F = 0.98$, $I_c = 1$ mA, $r_b = = 100$ ohms, $g_c = 0.0001$ mho, $R_S + R_E = 100$ ohms, and $R_L = 1000$ ohms. With the exception of $g_c$, which is about one hundred times too large, these values are more or less typical for a modern small-signal silicon transistor. [*Solutions*: The hybrid parameters are independent of the external circuit,

---

[*] See R. D. Thornton et al., *Characteristics and Limitations of Transistors*, pp. 23–35, Wiley, 1966 (i.e., SEEC series, Vol. 4).

so Figure 8.2$a$ shows the portion of the circuit that we need. Figure 8.2$b$ shows the equivalent circuit for the hybrid parameters. For future convenience let us do the problem using admittances or impedances rather than the simple conductances. From (8.3) and (8.6), we have $g_m = 0.04$ mhos and $g_e = = 0.041$ mhos. To get the $h$ parameters, we must solve

$$\begin{bmatrix} v_i \\ i_o \end{bmatrix} = \begin{bmatrix} h_{ib} & h_{rb} \\ h_{fb} & h_{ob} \end{bmatrix} \begin{bmatrix} i_i \\ v_o \end{bmatrix}$$

With simple circuit analysis, first shorting the output and then opening the input, we obtain

$$\left.\begin{aligned}
h_{ib} &= z_e + (r_b \parallel z_c)(1 - g_m z_e) & &= 26.4 \text{ ohms} \\
h_{fb} &= -g_m z_e - \frac{y_c}{y_c + y_b}(1 - g_m z_e) & &= -0.98 \\
h_{rb} &= \frac{r_b}{z_c + r_b} & &= 0.01 \\
h_{ob} &= [z_c + r_b]^{-1} & &= 0.99 \times 10^{-4} \text{ mhos}
\end{aligned}\right\} \quad (8.8)$$

The last two terms are particularly easy, since opening the emitter (incrementally) quite effectively turns the (incremental) generator off. Note also how large the short-circuited input conductance is. Matching into such a low resistance is very difficult in most circuits, so the common-base mode of operation will often make inefficient use of the available signal power.

In the limit of low frequency where the capacitive elements may be ignored, the fact that $r_e = \alpha/g_m$ gives us

$$h_{ib} = r_e + \frac{1 - \alpha}{g_b + g_c}$$

$$h_{fb} = \alpha + \frac{g_c}{g_c + g_b}(1 - \alpha)$$

Thus, for large $\alpha$, the input conductance is $qI_c/kT$, and the forward-current transfer ratio is $\alpha$.

The input impedance of the loaded amplifier is obtained by setting $v_0 = = -i_o R_L$ and solving for $v_i/i_i = z_i$. (N.B. This is the impedance looking into the transistor terminals; it does not include $R_E$.) The result is

$$z_i = h_{ib} + h_{rb}\left[\frac{-h_{fb}}{1/R_L + h_{ob}}\right] \quad (8.9)$$

Evaluating for this circuit, we get $r_i = 35.3$ ohms.

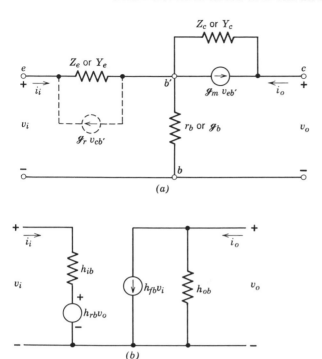

**Figure 8.2** (a) The T model of the transistor and (b) the "h" parameter equivalent circuit. The terminal variables are the same in both representations.

The output admittance is obtained in a similar fashion by setting $v_i = -i_i(R_E + R_S)$ and solving for $i_o/v_o$. The result is

$$g_o = h_{ob} - h_{fb}h_{rb}/[R_E + R_S + h_{ib}] \tag{8.10}$$

For this circuit:

$$g_o = 1.78 \times 10^{-4} \text{ mhos} \qquad \text{or} \qquad r_o = 6.2 \text{ Kohms}$$

The effect of loading the input and the output of the amplifier is to increase the input impedance and decrease the output impedance.

The final item to be obtained is the voltage gain. Two reasonably useful voltage gains can be defined. These are the intrinsic voltage gain, $A_v = v_o/v_i$, and the realized voltage gain, $A'_v = v_o/v_s$. The second takes into account the mismatch between the signal source and the amplifier, and is related to the former by

$$A'_v = A_v[z_i/(z_i + R_E + R_S)] \tag{8.11}$$

Solving for the intrinsic voltage gain gives

$$A_v = \frac{- h_{fb} R_L h_{ib}}{1 - R_L \left( \dfrac{h_{fb} h_{rb}}{h_{ib}} - h_{ob} \right)} \xrightarrow{\omega \to 0} - \frac{g_m R_L}{1 + (g_m R_L h_{rb} + h_{ob} R_L)} =$$

$$= \frac{A_o}{1 + A_o (h_{rb} + h_{ob}/g_m)} \tag{8.12}$$

The quantity $A_0 = - g_m R_L$ is the gain that the amplifier would have had if the simple Ebers-Moll model (i.e., no base-width modulation) held. Note that as we increase $R_L$ to get more and more gain, we hit a maximum gain of

$$A_{\max} = - g_m / (h_{rb} g_m + h_{ob}) \tag{8.13}$$

The gain is limited by the shunt conductance across the generator and the feedback into the input circuit resulting from $r_b$. To achieve higher gain, we need a better transistor—that is, one with smaller values of $r_b$ and $y_c$. With the parameters that we have, the various gains are $A_{\max} = - 80$, $A_0 = - 40$, $A_v = - 27$, and finally, the realized gain is $A_v' = -7$. Since the current gain is approximately unity, the power gain is also 7, or, in the usual logarithmic scale, 8.4 dB.

As you can see, we did not get much gain out of this transistor. Three things conspired to minimize the gain. For two of them, changing to a common emitter circuit would help substantially. These are the losses in gain due to the low current gain and the awkwardly low input impedance. The third cause is in the transistor itself—only a trip to the local transistor dispensary can correct this one. However, it is reasonable to ask whether we have gotten all we could have out of this transistor. That depends on how much you are able to modify the given circuit. A great deal of improvement could be achieved if the source impedance could be lowered, and there is another factor of 3 to be gleaned by raising $R_L$. Unfortunately, these impedances are frequently specified by other requirements in the circuit. The only other hope is to in-increase $g_m$ by raising $I_c$. This is where we must apply the scaling rules that we have developed for each of the small-signal parameters. $A_0$ will increase linearly with $I_c$ if you are willing to hold $R_L$ constant. From (8.3), (8.4), and (8.8) we can conclude that $A_{\max}$ should decrease at about the same rate, since $g_c$ is proportional to $I_c$. This means that we can gain something by increasing the current but only if $A_0$ is quite a bit less than $A_{\max}$. Finally, if $I_c$ is approaching the values at which high-level injection is important, $g_m$ goes as $qI_c/2kT$. As the collector current continues to rise, the series resistance of the collector (which we've ignored) becomes important and the net result of the increase in current could well be a net reduction in gain.

Note also that the input impedance is falling, going approximately inversely as $I_c$. In a circuit as badly matched as this one is, this loss of signal due to mismatch is likely to be the dominant effect, canceling out any gains achieved by the higher transconductance.]

You might conclude from this exercise that the common-base circuit has no redeeming features whatever. This is not the case. It is certainly not as generally useful as the CE circuit, but there are places where it is the better choice. These comprise circuits in which the impedance levels are intrinsically very low. For example, in a current regulator circuit or VHF circuitry the input impedances are usually quite appropriate to the CB arrangement. But since the CE circuit is the most widely used, let us turn our attention to it as we develop the frequency-dependent terms in the model.

## CHARGE STORAGE AND SMALL-SIGNAL CAPACITY

In (7.26) are listed six quantities of stored charge and the principal voltage to which the charges respond. If nature were on our side, all we would have to do is differentiate each of these charges with respect to the appropriate voltage and we would have a perfectly reasonable small-signal capacity to put in our small-signal representation. Naturally, things don't work out quite that well. For those charges that respond to $V_{eb'}$, simple differentiation does well enough and we get

$$c_e = \frac{\partial(\Delta Q_e)}{\partial V_{eb}} = \frac{\partial Q_{e\,\text{diff}}}{\partial V_{eb}} + \frac{\partial(\Delta Q_{e\text{DEP}})}{\partial V_{eb}}$$

For most transistors, the *rate of change* of the diffusion charge (although not necessarily the charge itself) is much larger than the rate of change of the depletion charge. (N.B. This would probably not be true for the 2N707A that we have examined, since it is designed for service in the 100 MHz range, but in the run-of-the-mill transistor, the diffusion capacity dominates by almost an order of magnitude.) Since the diffusion charge increases exponentially with voltage, it is certainly a likely candidate for having a large slope. If it does dominate the incremental charge storage, we have from (7.27):

$$c_e = q Q_{e\,\text{diff}}/kT = (Q_{e\text{diff}}/I_{e5})g_m \qquad (8.14)$$

The last form may be evaluated from (7.29), the ratio in the bracket being typically $10^{-9}$ to $10^{-10}$ seconds. Note that if the diffusion charge dominates, the capacity increases linearly with the current. For depletion terms, this will *not* be the case.

Now let us turn to the other two charges: $\Delta Q_c$ and $\Delta Q_{od}$. There is no doubt that under reverse bias the charge responding to the collector-base voltage is dominated by the depletion terms. Although these are given clearly enough by an expression such as (7.28), a problem arises as soon as you try to put circuit elements into Figure 8.1b to represent the charge derivatives. Although $V_{b'c}$ is not *very* different from $V_{bc}$, there is a difference and it can be a very important one. In fact, if we would be extremely accurate, we must observe that the collector depletion layer is distributed all along the path that the transverse base currents must follow. The "equivalent circuit element" is thus a lossy or RC transmission line. Fortunately, over the bandwidth of most transistor circuits, we may approximate the transmission line by its simplest lumped equivalent—a single $\pi$ section. The need for doing this arises precisely because $r_b$ represents the ultimate limit on the rate at which charge can be supplied to the collector junction. Thus, if $r_b$ is an important element in a circuit, the $\pi$ section is required; if not, then it doesn't really matter which fraction of the total junction capacity is called $c_c$ and which $c_{od}$.

In the model of the transistor shown in Figure 8.1b, only $c_e$ and $c_c$ are shown. The junction capacities between terminals $e$, $c$, and $b$ are lumped with the unavoidable lead and packaging capacities, and are thus not part of the "internal" model. A fairy tale this may be, but only if you want to push transistors to their ultimate limits will you need to use a much better model. No one ever tries to use a model sufficiently complex to be equally applicable in all frequency ranges. But unless microwave circuits are your special pleasure, the T model of 8.1b and the $\pi$ model we are about to derive from it will serve you well.

We cannot be quite so general about scaling the depletion capacities as we were about the diffusion terms. The way that the depletion region grows with increasing reverse bias depends on the junction doping profile. For the step profile, equation 7.28 applies and we get

$$c_c \propto (V_{cb})^{-1/2}$$

For other profiles, other exponents will apply. The only general rule is that the greater the reverse bias on the collector junction, the smaller the collector junction capacity will be. As we found in the primitive switching analysis that we did before, $c_c$ turns out to be the critical capacity in most circuits. This is true in spite of the fact that $c_c$ is typically an order of magnitude less than $c_e$, because only the input signal appears across $c_e$ while the amplified signal charges $c_c$. Since one usually gets more than an order of magnitude of amplification per stage, $c_c$ has at least as big an influence as the much larger $c_e$. This is called the *Miller effect* and we shall look at it analytically in a moment. First, let us get some order-of-magnitude estimates for these capacities by using the data developed in problem 7.5.

**Problem 8.1**

Using the data of problem 7.4 on page 401, calculate $c_e$ and $c_c$ under the bias conditions: $V_{cb} = 14$ V, $I_c = 2$ mA. Include the depletion capacity for the emitter junction.

*Solution.* Since all of the $Q$'s for problem 7.4 have already been calculated, problem 8.1 can be solved by appropriately scaling those answers. First, we will need the relationship between the charge stored in the depletion layer and the incremental capacity. If (7.28) holds (step junction), then, lumping all of the constants into a single coefficient, we have:

$$Q = A\left[(V + V_D)^{1/2} - (V_D)^{1/2}\right] \tag{8.15}$$

from which:

$$c = dQ/dV = \frac{A}{2(V + V_D)^{1/2}} \tag{8.16}$$

and in the limit as $V$ gets much larger than $V_D$:

$$c = Q/2V \tag{8.16'}$$

The second form will work reasonably well for the collector junction, but the first must be used for the emitter junction. Since the charges stored in the collector junction at 14 V reverse bias are already computed in problem 7.4, we have directly

$$c_c = [\Delta Q_{c\,\text{dep}} + \Delta Q_{od\,\text{dep}}]/2V = 1.44\,\text{pF}$$

The diffusion charge scales with the collector current (equation 8.14), so we have

$$c_{e\,\text{diff}} = qQ_{e\,\text{diff}}/kT = 40 \times 3.9 \times (2/5.8)\,\text{pF} = 54\,\text{pF}$$

Note that $c_{e\,\text{diff}}$ is substantially larger than $c_c$ in spite of the fact that $\Delta Q_c$ is much larger than $Q_{e\,\text{diff}}$. It's the rate of change, not the total, that counts in the calculation.

The last term to be computed is $c_{e\,\text{dep}}$. First we must find the voltage across the emitter junction. From problem 7.4 we know that at $I_c = 5.8$ mA, the emitter bias was 0.69 V and the depletion charge stored was 5.9 pC. The new bias may be obtained as

$$\Delta V_{be} = (kT/q)\log(I_{c\,\text{new}}/I_{c\,\text{old}}) = 26.6\,\text{mV}$$

Thus, the new forward bias point is $V_{be} = 0.67$ V. Using (8.15) and the data from problem 7.4, we may obtain $A = 19.8\,\text{pF}/\text{V}^{1/2}$. N.B. The $V$ in (8.15) is the amount of applied *reverse* bias, so the forward bias goes in as a *negative* number. Then, directly from (8.16) using the value of $V_D$ computed in problem 7.4, we get

$$c_{e\,\text{dep}} = 19.8/(2 \times 0.498) = 20\,\text{pF}$$

As you can see, although $c_{e\,\text{dep}} < c_{e\,\text{diff}}$, it is not really insignificant. On the other hand, at 5.8 mA, $c_{e\,\text{diff}} = 156$ pF but $c_{e\,\text{dep}}$ is only 21 pF. Thus, whether $c_{e\,\text{dep}}$ is important or not depends on the bias point chosen.

Our results can be summed up by adding a 74 pF capacitor in parallel with $r_e$ and a 1.7 pF capacitor in parallel with $r_c$ as shown in Figure 8.1$b$.

## THE $\pi$ MODEL AND BANDWIDTH CALCULATIONS IN THE CE CONFIGURATION

Getting the most gain from a transistor usually requires the CE circuit. To do a small-signal analysis of such a circuit, we could continue with the T model, simply interchange the emitter and base leads and crank away. What you get is shown in Figure 8.3$a$. Note one vital difference between 8.3$a$ and 8.2$a$. The generator that is hinted at in Figure 8.2$a$ is bolted solidly into the circuit of 8.3$a$. Although that generator makes little difference in the emitter current, it can put out a substantial portion of the much smaller base current. Thus, for a realistic representation of the CE circuit, the second generator must be there. That's a bit of a nuisance, to say the least. Most designers would much rather use the $\pi$ model of Figure 8.3$b$, in which there is only one generator. The two circuits are NOT equivalent, but by properly choosing the three impedances, most real transistors can be as well represented by the single-generator $\pi$ circuit as the two-generator T. Another way of saying the same thing is that the *important* effects of the emitter generator can be modeled for a normal transistor by adding the resistor $r_{ce}$; this resistor supplies a path from $e$ to $c$ that does not pass through $y_e$.

Our procedure is to set equal the $h$ parameters for the two circuits contained within the dashed lines. This will take one reasonable approximation. Then, setting $r_{bb'} = r_b$ will make the two circuits effectively equivalent. Finding the $h$ parameters for the two boxes yields

$$
\begin{array}{ccc}
\text{T} & & \pi \\
\end{array}
$$

$$
h_{ie} = \left[-g_m - g_r + y_e + y_c\right]^{-1} \iff (y_{b'e} + y_{b'c})^{-1}
$$

$$
h_{re} = \frac{(y_c + g_r)}{v_e + g_r - g_m + y_c} \iff y_{b'c}h_{ie} \tag{8.17}
$$

$$
h_{fe} = (-g_m + y_c)h_{ie} \iff (-g_m + y_{b'c})h_{ie}
$$

$$
h_{oe} = \frac{y_c(y_e + g_r) + g_r(g_m - y_c)}{y_e + g_r - g_m + y_c} \iff y_{ce} + (g_m + y_{b'e})y_{b'c}h_{ie}
$$

Any transistor worthy of the name has $g_m \gg g_r$. If we may ignore $g_r$ in the expressions above whenever it is added or subtracted from $g_m$, the four

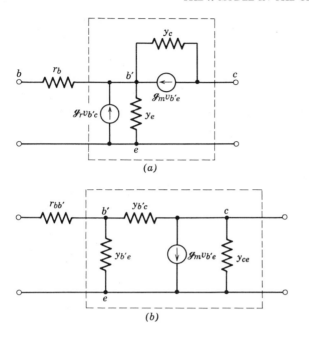

**Figure 8.3.** (*a*) T model of the CE circuit; (*b*) π model for the same circuit. The dashed box encloses the group of elements that must be properly chosen to make the two circuits identical in function.

equations become equal if

$$y_e - g_m = y_{b'e}$$

$$y_c = y_{b'c} \tag{8.18}$$

$$g_r(g_m - y_c)h_{ie} = y_{ce}$$

Ignoring $g_r$ compared to $g_m$ may seem clear enough, but we could run into a little trouble if $\alpha$ is exceptionally high. For example, if you look at the real part of $h_{ie}^{-1}$ above, you will see that it is

$$Re(h_{ie}^{-1}) = g_e - g_m + g_c - g_r = g_m\left(\frac{1}{\alpha} - 1\right) + g_c - g_r$$

Under certain operating conditions the $g_m(1/\alpha - 1)$ term can get small enough so that $g_r$ would not be really insignificant. For these rather special cases, our π model will be somewhat inaccurate; in general, however, it will serve us remarkably well.

Take a look at the real and imaginary parts of (8.18). From the first equation we get

$$g_{b'e} = g_e(1 - \alpha) = \frac{g_e}{1 + \beta} = \frac{g_m}{\beta} \quad \text{and} \quad c_e = c_{b'e}$$

This makes excellent sense: By switching from the emitter to the base lead, we have reduced the current by $1/(\beta + 1)$, but the charge stored has remained the same. From the second equation we get

$$g_c = g_{b'c} \quad \text{and} \quad c_c = c_{b'c}$$

The third equation looks somewhat more formidable, but if we take advantage of the relative size of the different terms, it quickly reduces for all practical purposes to a simple resistor. First, over the entire useful range of operation, $g_m \gg |y_c|$. The third equation thus becomes

$$y_{ce} = \frac{g_r g_m}{g_{b'e} + s(c_{b'e} + c_{b'c})} = \frac{\beta g_r}{1 + s/\omega_\beta} \quad \text{where} \quad \omega_\beta = \frac{g_{b'e}}{c_{b'e} + c_{b'c}}$$

The angular frequency, $\omega_\beta$, turns out to be less than but not too far from the maximum useful frequency for the transistor. In general, we will want to operate the transistor with loads that are substantially less than $g_{ce} = \beta g_r$; at the higher operating frequencies it is absolutely mandatory. Why we want to shunt $y_{ce}$ so strongly will be examined in detail in a short while, but the upshot of it is that we need not concern ourselves with the reactive part of $y_{ce}$. It will never come back to haunt us.

At this point, we may also estimate the size of $g_{ce}$. If you compare (8.4) with (8.7) and reexamine the physical derivation of the Ebers-Moll parameters, you will conclude that $g_r$ is less than but of the same order of magnitude as $g_c$. For the sample transistor, we saw that $g_c$ was a fraction of a $\mu$mho. Thus, we should expect $g_{ce} = \beta g_r$ to be of the order of $10^{-5}$ mhos or so. This turns out to be a reasonable guess for most transistors. Values typically range from 10 to 40 $\mu$mhos at collector currents of 1 mA.

Our $\pi$ model is now complete and connected rather solidly to the physical analysis of transistor action that preceded it. It is now time for an exercise — one that will be used to make the model easier to use.

As an exercise whose result we shall use immediately, calculate the $y$ parameters for the *circuit within the box* in Figure 8.3 and draw the circuit appropriate to the $y$ parameter representation.

*Solution.* In the admittance representation, we have

$$\begin{bmatrix} i_i \\ i_o \end{bmatrix} = \begin{bmatrix} y_{ie} & y_{re} \\ y_{fe} & y_{oe} \end{bmatrix} \begin{bmatrix} v_i \\ v_o \end{bmatrix}$$

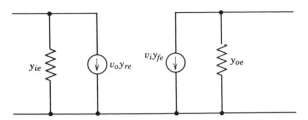

**Figure 8.4** $y$-parameter model for a two-port.

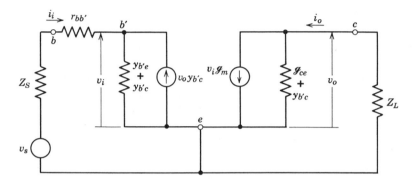

**Figure 8.5** The $y$-parameter representation of a single-stage bipolar transistor amplifier with the $y$ parameters given in terms of the $\pi$ parameters.

With the output terminals shorted, circuit analysis quickly gives the first column of $y$ parameters; shorting the input then yields the second column. The results are

$$y_{ie} = y_{b'e} + y_{b'c}$$

$$y_{re} = - y_{b'c}$$

$$y_{fe} = g_m - y_{b'c} \simeq g_m \tag{8.19}$$

$$y_{oe} = y_{b'c} + g_{ce}$$

In general, the admittance representation would have a circuit diagram such as that shown in Figure 8.4. Putting in the values from (8.19) and including $r_{bb'}$, the source, and the load yields the circuit of Figure 8.5.

Figure 8.5 is complete enough and certainly could be used for circuit calculations, but it would be very nice if we could separate the input and the output circuits. Then we could analyze each of them without considering the other. Of course, the two halves are coupled, as the two dependent generators show. The best we can achieve is a pair of partial circuits that show how the input circuit behaves if the output load is fixed and to some extent

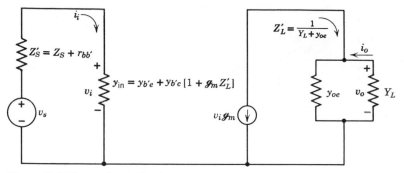

**Figure 8.6** The unilateralized version of the circuit in Figure 8.5.

vice versa. Such a step is called *unilateralization,* and it frequently provides some valuable insights into the electrical behavior of the amplifying element. To unilateralize the input circuit, we need only replace $v_o$ by its equivalent: $v_o = v_i g_m Z'_L$. Making this exchange in the dependent generator of the input circuit converts that generator into a passive admittance. Note that this is possible only because the load is fixed.

The unilateralized circuit is shown in Figure 8.6. It clearly shows the Miller effect that was mentioned before. That is, from the viewpoint of the input terminals the admittance, $y_{b'c}$, appears to be multiplied by the intrinsic gain of the amplifier, $g_m Z'_L$. This is, of course, just another way of saying that the voltage across $y_{b'c}$ is $v_i - v_o$, but either point of view is equally accurate. For example, if you wanted a 100 $\mu$F capacitor but had only a 1 $\mu$F capacitor, you could "make" your 100 $\mu$F capacitor by putting the 1 $\mu$F capacitor across the base-collector terminals of a transistor whose intrinsic gain was 99 and then using the base emitter pair of terminals as the "capacitor." From an energy-storage point of view, you've done nothing more than use the transistor valve and its power supply to charge the capacitor a great deal more than the input source could. However, since this extra charging current runs through the base lead and is proportional to the signal amplitude, as far as the signal source is concerned, there is a much larger capacitor across the line.

Figure 8.6 has a very peculiar property: It is correct only in going from left to right. The principal fault with the circuit—and also its principal advantage—is that it shows the output-circuit elements as if they were independent of the input circuit. Thus, if we ask what the output impedance of the amplifier is, Figure 8.6 would suggest that the answer is $y_{oe}$. However, asking for $i_o/v_o$, from a circuit derived on the basis of making $i_o/v_o = - Z'_L$, is not apt to be very productive. What the unilateralized circuit of Figure 8.6 does—and very nicely at that—is to predict what the output of the amplifier will be for a given $Z'_L$. This is the situation that applies in many practical problems, such as the calculation of an amplifier's bandwidth. Figure 8.6 is useful, but it's not the whole story.

Should you want to know what the output impedance is—for example, if you wanted to match the amplifier to the load to achieve the maximum power gain—you must go back to the circuit of Figure 8.5 and unilateralize it going the other way. If we ask what $i_o/v_o$ is—an admittance that is measured by putting a voltage generator across the output terminals and measuring the current—the answer is obtained quite readily from Figure 8.5 by analyzing the circuit with the *independent* voltage generator, $v_s$, appropriately shorted out. The result is

$$y_{\text{out}} = \frac{g_m y_{b'c}}{Y_s' + y_{b'e} + y_{b'c}} + y_{oe} \tag{8.20}$$

Thus, unless the effective source admittance is enormous, $y_{\text{out}}$ will be greater than $y_{oe}$. For typical low-frequency circuits, $g_{\text{out}}$ will be 2 to 4 times $g_{ce}$. Now that we have our gaggle of models, it is time to proceed with some useful and possibly enlightening circuit analysis. We will have two purposes in mind. Our first objective is to obtain analytical expressions for the voltage, current, and power gain as functions of frequency. Our second and related objective is to establish techniques for measuring small-signal parameters and determining what information along these lines can be gleaned from the typical transistor specification sheet.

Our first step should be to eliminate from Figures 8.5 and 8.6 any terms that are always or at least usually completely insignificant. For example, unless one were to heavily load the base-collector terminal pair, the term $|y_{b'c}|$ would always be much less than $g_m$. Thus, $y_{b'c}$ can be dropped from terms containing $(g_m - y_{b'c})$, as had already been done in the figure. Similarly, if the base-collector pair is not heavily loaded or the load impedance very large or inductive, $g_{b'e}$ will be much larger than $Re[y_{b'c}(1 + g_m Z_L)]$. Finally, for most circuits and most transistors, $|y_{oe}| \ll |1/Z_L|$. Given all of these "if's" and "usually's," the circuit of Figure 8.6 reduces to the simpler one of Figure 8.7. The chief convenience in Figure 8.7 over Figure 8.6 is that Figure 8.7 makes the effects of frequency much more obvious. Since our next task—and one that is of great use in practical circuit design—is to calculate the various gains as a function of frequency, the circuit of Figure 8.7 is our starting point.

For some practice in using the small-signal model of Figure 8.7, consider the following problem:

**Problem 8.2**

Using the results of problems 8.1 on page 420 and 7.4 on page 401, calculate the small-signal voltage gain $(A_v)$ and current gain $(A_i)$ under the following

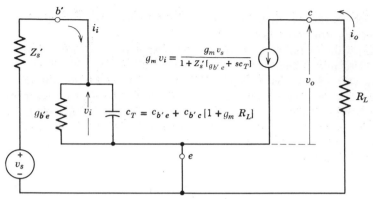

**Figure 8.7** A simplified version of the unilateralized model of Figure 8.6. This model is adequate for many circuit applications.

conditions: $I_C = 2$ mA, $V_{CB} = 14$ V, $\beta = 40$, $r_{bb'} = 100$ ohms, $R_s = 400$ ohms, and $R_L = 2$ K ohms.

*Discussion.* Circuit analysis in Figure 8.7 very quickly yields

$$A_i = i_o/i_i = \frac{g_m}{g_{b'e} + sc_T} = \frac{\beta}{1 + sc_T r_{b'e}} = \frac{\beta}{1 + s/\omega_1} \tag{8.21}$$

$$A_v = v_o/v_i = -\frac{g_m R_L}{1 + R_s'[g_{b'e} + sc_T]} = -\left(\frac{\beta R_L}{r_{b'e} + R_s'}\right)\left(\frac{1}{1 + \dfrac{sc_T}{g_{b'e} + G_s'}}\right)$$

$$\equiv A_{\text{mid}} \frac{1}{1 + s/\omega_2} \tag{8.22}$$

Note that both $A_i$ and $A_v$ have the same form—a simple pole—but that $\omega_1 = 1/c_T r_{b'e}$ and $\omega_2 = (g_{b'e} + G_s')/c_T$ will be different unless the signal source is a current source (i.e., $G_s' \ll g_{b'e}$). To evaluate the two gains, we need $g_m = qI_C/kT = 0.08$ mhos, $r_{b'e} = \beta/g_m = 500$ ohms, and $c_T = c_{b'e} + c_{b'c}(1 + g_m R_L)$. From problem 8.1, we can get the several capacities to give $c_T = = 74 + 2.2(1 + 160) = 428$ pF. Look at how the Miller capacity dominates $c_T$! A measly 3% of the intrinsic capacity is multiplied into 83% of the effective capacity.

We may now evaluate our gains. $\beta$, of course, was given. Thus, the midband current gain is 40 and $\omega_1 = 4.67 \times 10^6$ rad/sec ($f_1 = 0.744$ MHz). The midband voltage gain, $A_{\text{mid}}$, is 80 and $\omega_2 = 2\omega_1$.

By the way, the reason that these gains are called "midband" rather than "low frequency" is that one frequently ties circuits together with relatively large capacitors that easily pass the AC signal but isolate the DC voltages in one stage from those in the next. As the frequency gets low enough, these coupling capacitors begin to limit the gain. Thus, a complete circuit analysis would show that the gain first increased with increasing frequency, then remained relatively constant for a substantial increase in frequency, and finally decreased with increasing frequency at higher frequencies. The flat section in the middle is the *midband*. Of course, it is possible to design amplifiers that are flat down to 0 frequency; in the design of integrated monolithic circuits, where capacitors are very expensive but transistors and diodes very cheap, this is almost always done. Still, the name *midband* is applied to the flat portion no matter how close to DC the low end is.

We now come to a question that looks easy but isn't. What is the power gain? You can dash off with a flurry of proper definitions and come up with the perfectly correct answer:

$$A_p = \frac{Re(v_o i_o^*)}{Re(v_s i_s^*)} = \frac{\beta^2 R_L \left[ 1 + \dfrac{R_s'}{R_{in}} \left( \dfrac{\omega}{\omega_1} \right)^2 \right]}{R_{in} \left[ 1 + \left( \dfrac{R_s'}{R_{in}} \dfrac{\omega}{\omega_1} \right)^2 \right] \left[ 1 + \left( \dfrac{\omega}{\omega_2} \right)^2 \right]} \quad (8.23)$$

where $R_{in} = r_{b'e} + R_s'$. This answer, although impeccably correct, is probably not very useful since the signal, as far as the user is concerned at least, is either a current or a voltage. Let us say that it is a voltage. Then, as long as the voltage gain is maintained, the user probably couldn't care less about the extra power drawn from the signal source at high frequencies. As far as the user is concerned, the power gain goes as the square of the voltage gain:

$$A_p = \left( \frac{R_L}{r_{b'e} + R_s'} \right) \left| \frac{v_o}{v_s} \right|^2 = \beta A_{mid} \frac{1}{1 + (f/f_2)^2} \quad (8.24)$$

The difference between (8.23) and (8.24) is really not very great, but when someone gives a gain for an amplifier, he is almost always talking about (8.24).

We now come to the question of *bandwidth*. By definition, the bandwidth is the frequency difference between the upper and lower *half-power points*. That is, you find the frequencies at which the power gain (8.24) is one half of the midband gain. The difference between these two frequencies is the bandwidth. The bandwidth is usually large enough so that the upper half-power point is approximately the bandwidth.

There is a technically important presentation of the information we have just developed. It is called a *Bode plot* or a *gain-phase* plot. Actually, it is two plots: the log of the power gain versus the log of the frequency and the phase shift in the voltage ratio as a function of the log of the frequency. It

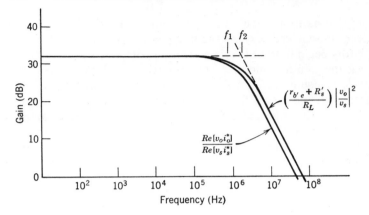

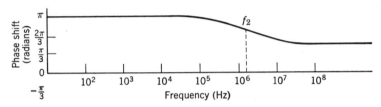

**Figure 8.8** The Bode plot [log (gain) in dB and phase in radians versus log (frequency) in Hz]. Both the true power gain and the squared voltage gain are shown on the same scale (results from problem 8.2). The phase shift is for the voltage.

is customary to use dB on the gain scale. (A *decibel* is 10 times the log of the ratio of two powers; since the power ratio is proportional to the square of the voltage ratio, it is equally common to see a dB defined as 20 times the log of a voltage ratio.) Thus, the half-power points are also referred to as $-3$ dB points [i.e., $\log(1/2) = -0.3$]. Figure 8.8 shows the Bode diagram for the simple amplifier that we have just analyzed. Both (8.23) and (8.24) are shown in the gain plot. At the upper end, both fall at the same rate [20 dB per decade of frequency, which is the same thing as 6 dB per octave (factor of 2) of frequency], but the true power-gain at these high frequencies is 3 dB less than the signal power-gain. As you can see by a quick glance at (8.24), the $-3$ dB point occurs at $f = f_2$. This is where the asymptotes of the midband gain and the 20 dB per decade segments cross. It is also the point at which the phase shift has decreased by 45°. [N.B. Although books such as this one, which barely touch on the circuit applications of transistors, tend to emphasize the gain characteristics, there are many applications in which the phase shift is the critical parameter. An obvious one is in preventing an amplifier from breaking into spontaneous oscillation, but there are others equally important. Consider, for example, using our one-stage amplifier to amplify a square wave. At 1 KHz, the wave form would be faithfully magnified by 32 dB. A 1 MHz square wave, on the other hand, although amplified almost as much ( $\sim$ 30 dB), would be badly distorted by the phase shift between its first and third harmonics ( $\sim$ 12°). Doing a meaningful analysis of an amplifier's performance requires *both* graphs in the Bode plot.]

In the primitive switching analysis that we did earlier in the previous chapter, we found that switching speed was purchased at the expense of gain. A similar result is obtained in the small-signal regime. The midband gain is proportional to $R_L$ as long as $R_L \ll |z_{oe}|$:

$$A_{\text{mid}} = -\frac{\beta R_L}{r_{b'e} + R_s'}$$

If $g_m R_L$ is sufficiently large, then $f_2$, the upper 3 dB point, is inversely proportional to $R_L$:

$$f_2 = \frac{G_s' + g_m/\beta}{2\pi\left[c_{b'e} + (1 + g_m R_L)c_{b'c}\right]} \Rightarrow \frac{G_s' + g_m/\beta}{2\pi g_m c_{b'c} R_L} \tag{8.25}$$

Thus, for sufficiently high gains, the *gain-bandwidth product* is a constant. That is the best you can do, since when $g_m R_L$ gets lower, you begin to lose gain without getting a commensurate increase in bandwidth.

## MEASUREMENT OF THE SMALL-SIGNAL PARAMETERS

In the $\pi$ model of Figure 8.3b, there are five parameters to be determined. Since two of them are complex, we have seven numbers that we need to make the model represent a particular transistor in a *particular regime of bias and temperature*. Keep in mind that it is necessary to scale the small-signal parameters when changing the operating conditions. Some of the parameters, such as $g_m$, are readily converted; others, especially $r_{bb'}$, change in more obscure ways. Thus, it is necessary to be able to measure these coefficients under conditions that roughly resemble the conditions under which the transistor will operate.

The tools used to make these measurements usually include relatively sophisticated bridges, generators, and meters, and the measurements are frequently performed in carefully controlled environments. Speed, accuracy, and reproducibility demand this sophistication. The principles that underlie the measurements, on the other hand, are really quite simple. We shall deal only with the principles. Should you need information on specific techniques, you might begin by consulting R. D. Thornton et al., *Handbook of Basic Transistor Circuits and Measurements*, Vol. 7 of the SEEC series, Wiley, 1966.

**Table 8.1.** Measurements Necessary to Determine the $\pi$-Model Circuit Parameters[a]

| Quantity Measured | The Circuit Parameter in Terms of the Parameters Important at the Measurement Frequency | Useful Approximation | Comments |
|---|---|---|---|
| | **LOW-FREQUENCY MEASUREMENTS** | | |
| (1) $I_C, T$ | $g_m = qI_C/kT$ | — | Goes to roughly half this value at high current densities |
| (2) $h_{fe} = i_c/i_b$ | $h_{fe} = \dfrac{g_m - g_{b'c}}{g_m/\beta}$ | $h_{fe} = \beta$ | Depends on both $I_C$ and $T$ |
| (3) — | $g_{b'e} = g_m/\beta$ | | |
| (4) $h_{ie} + r_{bb'} = \dfrac{v_{be}}{i_b}$ | $h_{ie} = (g_{b'e} + g_{b'c})^{-1}$ | $h_{ie} = r_{b'e}$ | This measurement determines $r_{bb'}$ but see comment 1 in the text and read the table footnote |
| (5) $h_{re} = v_{be}/v_{ce}$ | $h_{re} = \dfrac{g_{b'c}}{g_{b'c} + g_{b'e}}$ | $h_{re} = g_{b'c}/g_{b'e}$ | $g_{b'c}$ depends on $I_C$ directly and on $T$ through $I_{es}$ |
| (6) $h_{oe} = i_c/v_{ce}$ | $h_{oe} = g_{ce} + \dfrac{g_m + g_{b'e}}{1 + \dfrac{g_{b'e}}{g_{b'c}}}$ | $h_{oe} = g_{ce} + g_{b'c}$ | $g_{ce}$ is directly proportional to $I_C$ and depends on $T$ through $I_{cs}$ |

| Quantity Measured | The Circuit Parameter in Terms of the Parameters Important at the Measurement Frequency | Useful Approximation | Comments |
|---|---|---|---|
| | MODERATE FREQUENCY ($g_{bb'} \gg \left| sc_{b'c} \right| \gg g_{b'c}$) | | |
| (7) $h_{ob} = i_c/v_{cb}$ | $h_{ob} = \dfrac{y_{b'c}}{1 + r_{bb'} y_{b'c}}$ | $h_{ob} = sc_{b'c}$ | See comment 2 in the text |
| | HIGH FRQUENCY (but $\left| y_{b'c} \right| \ll g_m$) | | |
| (8) $h_{fe} = i_c/i_b$ | $h_{fe} = \dfrac{g_m - y_{b'c}}{y_{b'e} + y_{b'c}}$ | $h_{fe} = \dfrac{\beta}{1 + \dfrac{s(c_{b'c} + c_{b'e})}{g_{b'e}}}$ | See comment 3 in the text |

[a] The $h$ parameters are chosen because they are measured with the base incrementally open circuited. Thus, our definition of the $h$ parameters as applying only to the circuit parameters within the dashed lines in Figure 8.3 does not lead us into any difficulty except for the notational oddity of line (4) above. Since the point $b'$ is inaccessible, the measured $y$ parameters would not agree with (8.19). Shorting $b$ to $e$ is not the same as shorting $b'$ to $e$.

To separate the capacitive from the resistive components, we may choose a frequency that is sufficiently low so that the capacitive susceptance is negligible compared to the conductance for all of the complex terms. A frequency of about 1000 Hz is usually low enough and very convenient. Once these low-frequency elements are determined, we may obtain the capacitive terms by measurements at intermediate and high frequencies, choosing the frequency to put the capacitive susceptance in the most convenient relationship to the conductance terms.

A "complete" set of such measurements is given in Table 8.1. With the exception of $I_C$, all of the measurements involve the determination of one of the hybrid parameters. Since the $h$ parameters are terminal variables, they can be measured directly. The first column in the table gives the quantity measured. The second column relates that measurement to the $\pi$-model parameters through (8.17). The third column shows the approximation suitable for all or almost all transistors. Of the seven measurements listed, three require some detailed comment because the results are not quite what they seem to be.

**Comment 1.** In the determination of $r_{bb'}$ (measurement 4), we once again must admit to the inadequacy of this part of our physical model. At the low frequency at which measurement 4 is performed, the distributed nature of the base voltage drop and its accompanying charge storage does not reveal itself. At the higher frequencies, where $r_{bb'}$ really becomes important, we are dealing with at least a $\pi$ network rather than a simple $r$ terminated at its far end by a capacitor. It is also true, unfortunately, that $r_{bb'}$ does not scale in a very predictable fashion with collector current. What saves the model and makes it useful, at least for preliminary design work, is the fact that $r_{bb'}$ can usually be lumped in with $R_s$. The error in the sum of the two resistances is normally acceptably small. The $\pi$-section model of $r_{bb'}$ and the various capacities can be used for further refinement, if necessary, as discussed in comment 2 below.

**Comment 2.** The measurement of $h_{cb}$ at intermediate frequency—one that makes $|r_{bb'}y_{b'c}| \ll 1$—would give $c_{b'c}$ directly if that were the only capacity connecting the collector to the base terminal. Actually, of course, the overlap diode capacity that makes up part of both capacitors in the $\pi$ section connecting $b$ to $b'$ is included in this measurement, as is the small but significant capacity associated with the package (i.e., *header*) that contains the transistor. This total capacity is called $c_{ob}$. The measurement of $h_{ob}$ gives us no clue as to how to separate $c_{ob}$ into its component parts. As a matter of fact, if we draw out the full $\pi$ model (Figure 8.9), including all of the "external" capacitors, it is fairly obvious that a measurement at a

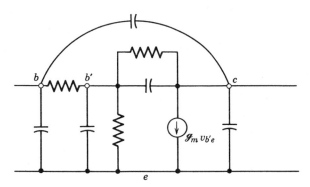

**Figure 8.9** The complete $\pi$ model including the "external" capacities. Of the three external capacities, $c_{bc}$ is the largest, since it includes a fraction of the overlap diode's capacity. It is also more important, since it is multiplied by the Miller effect when the transistor is used in a CE amplifier.

single frequency could not possibly unsnarl the resulting mess. If it must be accurately determined, one is obliged to measure one or several of the $h$ parameters as functions of frequency. Then, a "best fit" procedure can be used to obtain a pretty good picture of how to divide the pieces that make up $c_{ob}$. Another less tedious but less accurate approach is simply to divide $c_{ob}$ in half. From what we saw in a rather typical example (problem 7.4), the overlap diode contributes the largest single piece of $c_{ob}$. Part of this over-lap-diode capacity must be put into $c_{b'c}$ and part into $c_{bc}$. The remainder of $c_{bc}$ is header capacity; the rest of $c_{b'c}$ is the active collector junction's capacity. A 50–50 split does not look too unreasonable and usually turns out to be roughly correct. Since most transistor amplifiers have their 3 dB point well below $f_\beta$, the principal reason for trying to unravel $c_{ob}$ is to get an estimate of $c_{b'e}$. Unless $c_{b'e}$ is nearly as small as $c_{b'c}$, small errors in estimating $c_{b'c}$ will not be very important. For the amplifier itself, with $R_s \simeq r_{b'e} \gg r_{bb'}$, $c_{ob}$ is the capacity to use for the analysis.

**Comment 3.** The measurement of the forward current transfer ratio at high frequency is used to obtain the last remaining component, $c_{b'e}$. Although measurement 8 suggests that a single point is all that is needed, one would normally take measurements at several frequencies out to the point where $|h_{fe}|$ had fallen by a factor of five or more. One may then locate rather accurately the frequency at which $|h_{fe}|$ had decreased to $1/\sqrt{2}$ of $\beta$. This frequency is defined as $f_\beta$ and from line 7 in the table it is given by

$$f_\beta = 1/2\pi(c_{b'c} + c_{b'e})r_{b'e} \tag{8.26}$$

A more frequently quoted number is obtained by plotting $|h_{fe}|$ versus $f$

on a log-log plot. For $f > f_\beta$, the slope of the line should rapidly approach $-1$. By extrapolating this asymptote to the point at which $|h_{fe}| = 1$, one obtains the frequency, $f_T$, which is given by line 8 in the table as

$$f_T = g_m/2\pi(c_{b'c} + c_{b'e}) = \beta f_\beta \tag{8.27}$$

Given either $f_T$ or $f_\beta$ and a plausible value for $c_{b'c}$, the value of $c_{b'e}$ is determined. $c_{b'e}$ will be a function of collector current, while $c_{b'c}$ will depend on the reverse bias on the collector junction. Keep in mind that $f_T$ is obtained by extrapolation, not by direct measurement. By the time that $|h_{fe}| \Rightarrow 1$, not only are we obliged to put all of the capacitors from Figure 8.9 into our working model, but also we may no longer neglect $y_{b'c}$ in comparison to $g_m$. Thus, an actual measurement of the frequency at which $|h_{fe}|$ went to unity would yield an answer grossly different from what is meant by $f_T$.

## SCALING THE SMALL-SIGNAL PARAMETERS WITH VARIATIONS IN TEMPERATURE

The parameters of both the large- and small-signal models show considerable variation with temperature. This means that unless a transistor is going to function at all times in a rigorously controlled thermal environment, the designer is obliged to allow for large parametric variations, even if the transistors installed have been carefully and individually selected. (Again making a virtue of necessity, once the circuit has been designed with sufficient tolerance for parameter variation, it is unlikely that careful selection of the individual transistors will be necessary. The variation of parameters within a given transistor type will probably not be any worse than the variations induced in an individual transistor by temperature changes.)

Our physical model of the transistor stands up very well in a comparison of its predictions with the measured temperature variation of both the large- and small-signal parameters. The easiest approach to getting these predictions is to ask first which of the material parameters will change with temperature. Those likely to be important include the carrier concentrations, the mobility and diffusion coefficients, and the excess-carrier lifetime. Since the materials used for transistor fabrication are usually doped well into the extrinsic region by impurities that are totally ionized at room temperature (and well below), the majority-carrier concentrations are independent of temperature. Then, from (3.28) and (3.23), we find that the minority-carrier concentration should be proportional to $n_i^2$, which goes as $T^3 e^{-E_g/kT}$. This extremely strong temperature dependence is characteristic of all of the Ebers-Moll current coefficients such as $I_{es}$, $I_{cs}$, and $I_{co}$.

The temperature dependence of the mobility is found to be proportional to $T^{-n}$ where $n$ ranges from 1.5 to 2.5. Since the mobility and diffusion coefficients are related by the Einstein relationship, we get $D \propto T^{1-n}$.

The temperature dependence of the excess-carrier lifetime is not so obvious. We have the Shockley-Read theory of Chapter 5 (equation 5.17), but that contains an obscure quantity called the capture cross-section. With what we have developed in this book, we have no way of predicting whether this quantity should be strongly temperature dependent or not. If we take equation (5.17′) and select a particular example—such as a $p$-type sample with the trap level lying in the lower half of the forbidden band—we may reduce the lifetime equation to fairly manageable proportions. For the example selected, $n' = p' \ll p_0$ and $E_t < E_i$. Put into (5.17′) with negligible terms thrown out, we are left with

$$\frac{1}{n'} \frac{\partial n'}{\partial t} = \frac{1}{\tau} = \frac{(1/\tau_{eo})}{1 + (p_t/p_0)} \tag{8.28}$$

In Chapter 5 we considered what happened if the trap level were very deep; $p_t$ would be so small that the lifetime would be simply $\tau_{eo}$. More quantitatively, since $p_t = N_v e^{-(E_t - E_v)/kT}$, (8.28) may be written

$$\tau = \tau_{eo} \left[ 1 + \left( \frac{N_v}{N_A} \right) e^{-(E_t - E_v)/kT} \right] \tag{8.28′}$$

Considering that $N_v \gg N_A$, it is obvious that for sufficiently large $T$, the lifetime must decrease exponentially with $(1/T)$. How high the temperature must be is determined by the doping $(N_A)$ and the trap level $(E_t)$. As it turns out, most samples of Ge and Si that are not deliberately doped with a deep, lifetime-killing trap (e.g., Cu or Au) show the simple exponential dependence on $(1/T)$ for at least 100°C on either side of room temperature. Figure 8.10 shows typical data for lightly doped samples of Ge. It also includes an example of the behavior of a Cu-doped Ge sample. The slope for that sample is positive. The positive slope is attributed to the temperature dependence of copper's capture cross-section.

What we may conclude from Figure 8.10 is that, except for a few anomalous cases, excess-carrier lifetime will increase relatively rapidly with increasing temperature. An important result of this strong temperature dependence is that it gives the diffusion length a *net positive temperature dependence*.

Our next step is to find the partial derivative of each of the $\pi$-model parameters with respect to $T$. This requires that we hold as many of the bias parameters as constant as possible while varying the temperature. The two most critical bias parameters are $I_C$, which affects almost everything, and $V_{CE}$, which interacts with the collector depletion capacity and the base-width modulation terms. If we wish to hold the collector current constant while $\beta$ and $I_{es}$ vary, we are obliged to let $V_{BE}$ and $I_B$ adjust themselves accordingly.

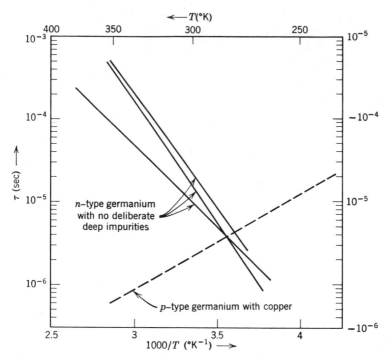

**Figure 8.10** Excess-carrier lifetime in germanium versus temperature. The three *n*-type samples are typical of material with no special "lifetime killers" (e.g., Cu, Au, or Ni) added. The *p*-type sample shows the effect of the addition of Cu. The data are from R. M. Baum and J. F. Battey, *Phys. Rev.* **98**, 923 (1955).

As it turns out, holding $I_C$ constant is about the best simple thing that a circuit designer can do to help stabilize his circuit against thermal drifting, so we will be examining the sort of shifts that must be expected in actual operation. This is not to say that more sophisticated schemes cannot do better; they can. In fact, sophisticated bias stabilization is probably the rule rather than the exception. However, holding $I_C$ relatively constant is a good place to begin, and so we shall.

To hold the collector current constant with increasing $T$, $I_{es}e^{qV_{BE}/kT}$ must remain constant or decrease. If $I_{cs}$ has increased to the point where it makes up a substantial fraction of $I_C$, we would have to *decrease* the injected current to hold the collector current constant. Let us assume for the moment that $I_{cs}$ is not important. Thus, as $I_{es}$ has increased, we have decreased $V_{BE}$ to hold $I_{es}e^{qV_{BE}/kT}$ constant. Let us see what $I_c$ stabilization implies for the variation in $\beta$.

$\beta$ is determined by the three terms of equation 7.24. The first term—the ratio of the reverse-to-forward injection rates—is almost independent of temperature (the two $D$'s do not drift at quite the same rate, but this gives less than a $T^{1/2}$ dependence) and completely independent of forward bias as long as the injection level is not high. The second term—the fraction of the carriers injected into the base that recombine in the base—is independent of emitter bias but quite dependent on temperature through the variation of the diffusion length. The third term—the ratio of the depletion-region recombination current to the collector current—varies with temperature through the $T$ in the numerator, through $D_b$, and possibly through $\tau_\epsilon$. The term $n_i e^{qV_{BE}/2kT}$ in the denominator is essentially the square root of term we held constant above, so it does not enter into the temperature dependence. Given the range of temperature dependence exhibited by $D$, it is unlikely that the third term in (7.29) will vary much more rapidly than $T^{+1/2}$.

Since transistors function well only when the first two terms in (7.24) dominate, we may consider two cases. The first, generally applicable to modern germanium transistors, has $1/\beta$ dominated by the reverse-injection term. The base-recombination term is so small that it does not contribute. For this case we would expect that $\beta$ will be rather independent of temperature. The other case, where $1/\beta$ is partially or totally determined by recombination in the base region, is applicable to silicon transistors. For this case, the carrier-lifetime dependence of Figure 8.10 shows that $\beta$ should increase relatively rapidly with temperature. An obvious but important corollary is that for silicon transistors, $\beta$ will rapidly fall to a useless value as the temperature decreases. Experiment bears out these conclusions. For a silicon transistor, $\beta$ can easily double in going from 25°C to 100°C and be cut in half going from 25° to $-$ 50°C.

Germanium transistors would show much, much less variation in $\beta$ over that same range. However, although this result does commend Ge transistors for use at low temperatures, the advantage of a relatively constant $\beta$ at high temperature is lost because $I_{co}$ grows to the point where it dominates the collector current. This, of course, is just a result of having such a large $n_i^2$ (or a smaller bandgap; either way you want to look at it, the result is the same), but it effectively limits Ge transistors to operation below about 65°C. Silicon, on the other hand, can operate well up to temperatures of the order of 150°C and above. This high-temperature capability and the high junction break-down voltages inherent in Si have all but pushed once-dominant Ge out of the transistor market.

Once we have established the temperature dependence of $\beta$, it is a relatively easy matter to see how the small-signal parameters would vary at constant $I_C$. This information is presented in Table 8.2.

**Table 8.2.** Variation of the Small-Signal $\pi$-model Parameters with Temperature Under Conditions Where the Collector-Emitter Bias and the Collector Currents Are Held Constant

| Term | Equation | $T$ Dependence ($I_C$ Constant) | Comments |
|---|---|---|---|
| $g_m$ | $qI_C/kT$ | $\propto 1/T$ | For Ge transistors, $g_m = q(I_C - I_{co})/kT$, which means that $g_m$ will fall more rapidly than $1/T$ as $I_{co} \Rightarrow I_C$ |
| $r_{b'e}$ | $\beta/g_m$ | $> T$ | Much more $T$ dependent in Si than Ge |
| $g_{b'c}$ | | Depends on growth of the collector depletion region, which is essentially independent of $T$ | |
| $g_{ce}$ | $\beta g_r$ (7.36) | Increases as $\beta$ increases | Much more $T$ dependent in Si than Ge |
| $r_{bb'}$ | $1/\mu$ | $\propto T^n$ ($1.5 < n < 2.5$) | The lateral drop across the base is mostly of ohmic origin, so the principal determining factor is the majority-carrier mobility |
| $\left. \begin{array}{c} c_{b'c} \\ c_{ob} \end{array} \right\}$ | | Predominantly depletion region capacity, so these two terms are generally temperature independent | |
| $c_{b'e}$ | (7.27) and (7.29) | $c_{b'e}$ comprises a depletion term, which is essentially independent of temperature apart from the variation of $V_{BE}$ needed to hold $I_C$ constant, and a diffusion term that may be written (7.27): $$c_{e\,\text{diff}} = (qQ_{e\text{diff}}/kT)$$ $$= (qI_C/kT)(Q_{e\text{diff}}/I_C)$$ From (7.29), the temperature-dependence is given by $$c_{e\text{diff}} \propto 1/D_b T \propto T^m \, (-0.5 < m < 0.5)$$ | How this term varies obviously depends on the relative magnitudes of the two components; the $m$ in the result to the left is $+$ for all silicon and for germanium $p$-$n$-$p$ transistors. The germanium $n$-$p$-$n$ types show the negative temperature coefficient |

## MANUFACTURER'S SPECIFICATIONS

With the rules we have for scaling small-signal parameters with changes in $I_C$ and $T$, it is not always necessary to go through a set of measurements to determine the characteristics of transistors of a given type. If great accuracy is not required, the spec sheets put out by the transistor's manufacturer usually contain enough information, at least for a preliminary design. You should always recognize that these spec sheets represent typical rather than actual values. Usually the manufacturer includes *worst-case* values, maximum ratings for safe operation, and typical values. Some manufacturers further support their products by publishing the results of some fairly complete testing of a statistically significant sample of their output. Such a set of data sheets for a Fairchild 2N708 *n-p-n* diffused silicon transistor are shown in Figure 8.11. Since the 2N708 transistor is intended for both switching and linear-amplifier service, data applicable to both are shown. Most of the effects that we have discussed are clearly demonstrated—the large variation of $\beta$ with $T$ and from unit to unit, the variation of $\beta$ with $I_C$, the strong temperature dependence and the voltage dependence of $I_{CO}$ $(I_{CBO})$, and even such things as the improvement in $h_{fe}$ that can be obtained at high frequency by increasing $V_{CE}$ (and thus decreasing $c_{ob}$) (see Figure 8.11$d$).

As an exercise in scaling and also to familiarize yourself with the technique in going from typical data to small-signal-model parameters, consider the following problem.

### Problem 8.3

Using data gleaned from Figure 8.11, find the value of $c_{b'e}$ for the bias conditions, $V_{CE} = 10$ V, $I_C = 2$ mA. Then find the power law that relates $c_{b'c}$ to $V_{CB}$.

*Solution.* The data available from the spec sheets is $c_{ob}$ at $V_{CB} = 10$ V; a graph of $h_{FE}$ versus $I_C$ (Figure 8.11$b$ in the third column); and a graph of $|h_{fe}|$ at 100 MHz versus $I_C$ for $V_{CE} = 5$ V and again for $V_{CE} = 10$ V (Figure 8.11$d$, first column). The first task is to use row 8 in Table 8.1 to get $c_T = c_{b'c} + c_{b'e}$. To use that expression for $h_{fe}$ versus $\omega$, it is necessary to have a value for $h_{fe}$ at low frequency. The spec sheets don't give us exactly what we want, but they do give us $h_{FE}$, which is the low-frequency *large-signal* current gain. Although these two numbers do differ somewhat ($i_c/i_b \neq I_C/I_B$), $h_{FE}$ is what we have so we must "make do" with it. Since $h_{FE}$ increases monotonically with $I_C$ up to about 10 mA, and since $\partial I_C/\partial I_B$ must decrease before $I_C/I_B$ can do so, the difference between $h_{FE}$ and $h_{fe}$ should be very small in the vicinity of 2 mA collector current. Taking the *typical* value for $h_{FE}$ at 2 mA gives us $h_{fe} = 44$. At $V_{CE} = 10$ V, $I_C = 2$ mA and $f = 100$ MHz, $|h_{fe}|$ from

# 2N708

## NPN HIGH-FREQUENCY AND LOW-STORAGE

### DIFFUSED SILICON PLANAR* TRANSISTOR

PHYSICAL DIMENSIONS
in accordance with
JEDEC (TO-18) outline

The 2N708 is an NPN silicon transistor designed specifically as a high-speed saturated logic switch to replace the 2N706 (A, B), 2N753 mesa series. In addition the 2N708 is oriented toward Satellite and Conventional small-signal, RF, and all digital type circuits.

The Fairchild PLANAR structure extends the range of useful current gain down to the microampere region. Other features are lower leakage current, increased maximum ratings, reduced storage time, higher beta, and lower saturation voltage relative to its predecessors.

Typical three sigma limits for B E T A and S A T U R A T E D $V_{CE}$ are included to completely characterize the 2N708 as a switch, thus promoting design over a wide family of operating conditions. These transistors are de-signed to meet the environmental requirements of MIL-S-19500.

## ABSOLUTE MAXIMUM RATINGS [Note 1]

Maximum Temperatures

| | | | |
|---|---|---|---|
| Storage Temperature | | | $-65°C$ to $+300°C$ |
| Operating Junction Temperature | | | 200°C Maximum |
| Lead Temperature (Soldering, No Time Limit) | | | 300°C Maximum |

Maximum Power Dissipation

| | | | |
|---|---|---|---|
| Total Dissipation at Case Temperature | 25°C | [Note 2 & 3] | 1.2 Watts |
| at Case Temperature | 100°C | [Note 2 & 3] | .68 Watt |
| at Ambient Temperature | 25°C | [Note 2 & 3] | .36 Watt |

Maximum Voltages

| | | |
|---|---|---|
| $V_{CBO}$ | Collector to Base Voltage | 40 Volts |
| $V_{CER}$ | Collector to Emitter Voltage ($R_{BE} \leq 10 \Omega$) [Note 4] | 20 Volts |
| $V_{CEO}$ | Collector to Emitter Voltage [Note 4] | 15 Volts |
| $V_{EBO}$ | Emitter to Base Voltage | 5.0 Volts |

**ELECTRICAL CHARACTERISTICS** (25°C Free Air Temperature unless otherwise noted)

| SYMBOL | CHARACTERISTICS | MIN. | MAX. | UNITS | TEST CONDITIONS |
|---|---|---|---|---|---|
| $h_{FE}$ | DC Pulse Current Gain [ Note 5 ] | 30 | 120 | | $I_C = 10$ mA, $V_{CE} = 1.0$ V |
| $h_{FE}$ ($-55°C$) | DC Pulse Current Gain [ Note 5 ] | 15 | | | $I_C = 10$ mA, $V_{CE} = 1.0$ V |
| $h_{fe}$ | DC Current Gain | 15 | | | $I_C = 0.5$ mA, $V_{CE} = 1.0$ V |
| $V_{BE}$ (sat) | Base Saturation Voltage | .72 | .80 | Volts | $I_C = 10$ mA, $I_B = 1.0$ mA |
| $V_{CE}$ (sat) | Collector Saturation Voltage | | .40 | Volts | $I_C = 10$ mA, $I_B = 1.0$ mA |
| $V_{BE}$ (sat) | Base Saturation Voltage ($-55°C$) | | .90 | Volts | $I_C = 7.0$ mA, $I_B = 0.7$ mA |
| $V_{CE}$ (sat) | Collector Saturation Voltage ($-55°C$ to $+125°C$) | | .40 | Volts | $I_C = 7.0$ mA, $I_B = 0.7$ mA |
| $h_{fe}$ | High Frequency Current Gain (f = 100 MHz) | 3.0 | | | $I_C = 10$ mA, $V_{CB} = 10$ V |
| $C_{ob}$ | Output Capacitance | | 6.0 | pF | $I_F = 0$, $V_{CB} = 10$ V |
| $r_b'$ | Base Spreading Resistance [Note 6](f = 300 MHz) | | 50 | ohms | $I_C = 10$ mA, $V_{CB} = 10$ V |
| $I_{CBO}$ | Collector Cutoff Current | | 25 | $m\mu A$ | $I_E = 0$, $V_{CB} = 20$ V |
| $I_{CBO}$ (150°C) | Collector Cutoff Current | | 15 | $\mu A$ | $I_E = 0$, $V_{CB} = 20$ V |
| $BV_{CBO}$ | Collector to Base Breakdown Voltage | 40 | | Volts | $I_C = 1.0$ $\mu A$, $I_E = 0$ |
| $V_{CER}$ (sust) | Collector to Emitter Sustaining Voltage [Note 4 & 5] | 20 | | Volts | $I_C = 30$ mA, $R_{BE} \le 10$ $\Omega$ (pulsed) |
| $V_{CEO}$ (sust) | Collector to Emitter Sustaining Voltage [Note 4 & 5] | 15 | | Volts | $I_C = 30$ mA, $I_B = 0$ (pulsed) |
| $BV_{EBO}$ | Emitter to Base Breakdown Voltage | 5.0 | | Volts | $I_C = 0$, $I_E = 10$ $\mu A$ |
| $I_{EBO}$ | Emitter Cutoff Current | | 0.1 | $\mu A$ | $I_C = 0$, $V_{EB} = 4.0$ V |
| $I_{CEX}$ (125°C) | Collector-Emitter Cutoff Current | | 10 | $\mu A$ | $V_{CE} = 20$V, $V_{BE} = .25$ V |
| $\tau_1$ | Charge Storage Time Constant [Note 7] [See circuit of page 3] | | 25 | nsec | $I_C = I_{B1} \approx 10$ mA, $I_{B2} \approx -10$ mA |
| $t_{on}$ | Turn On Time [See circuit on page 4] | | 40 | nsec | $I_C \approx 10$ mA, $I_{B1} \approx 3.0$ mA, $V_{BE} = -2.0$ V |
| $t_{off}$ | Turn Off Time [See circuit on page 4] | | 75 | nsec | $I_C \approx 10$ mA, $I_{B1} \approx 3.0$ mA, $I_{B2} \approx -1.0$ mA |

**FAIRCHILD**

SEMICONDUCTOR

A DIVISION OF FAIRCHILD CAMERA AND INSTRUMENT CORPORATION

\* Planar is a patented Fairchild process.

313 FAIRCHILD DRIVE, MOUNTAIN VIEW, CALIFORNIA, (415) 962-5011, TWX: 910-379-6435

COPYRIGHT FAIRCHILD SEMICONDUCTOR 1966 • PRINTED IN U.S.A. 2320-493-126 10M

MANUFACTURED UNDER ONE OR MORE OF THE FOLLOWING U.S. PATENTS: 291877, 3015048, 3025589, 3064167, 3108359, 3117260; OTHER PATENTS PENDING.

**Figure 8.11a**

## FAIRCHILD TRANSISTOR 2N708

## TYPICAL ELECTRICAL CHARACTERISTICS

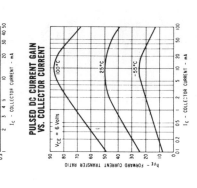

**BASE SATURATION VOLTAGE VS. COLLECTOR CURRENT**

$I_C = 10 I_B$

**PULSED DC CURRENT GAIN VS. COLLECTOR CURRENT**

$V_{CE} = 6$ Volts

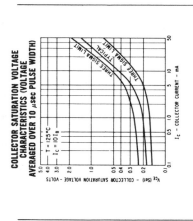

**COLLECTOR SATURATION VOLTAGE CHARACTERISTICS (VOLTAGE AVERAGED OVER 10 $\mu$sec PULSE WIDTH)**

$T = 125°C$
$I_C = 10 I_B$

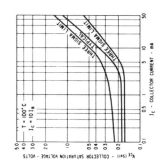

$T = 100°C$
$I_C = 10 I_B$

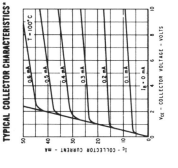

**TYPICAL COLLECTOR CHARACTERISTICS***

$T = 100°C$

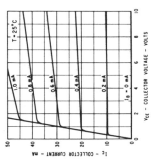

$T = 25°C$

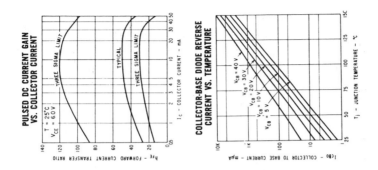

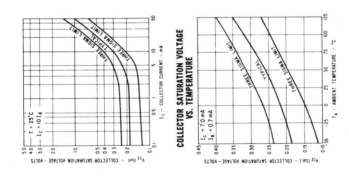

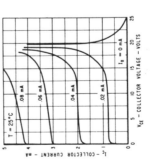

* Single family characteristics on Transistor Curve Tracer.

**Figure 8.11b**

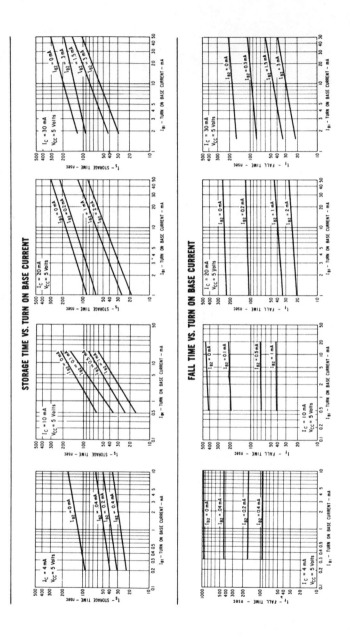

FAIRCHILD TRANSISTOR 2N708

SATURATED SWITCHING CHARACTERISTICS-TYPICAL DELAY, RISE, STORAGE AND FALL TIMES

STORAGE TIME VS. TURN ON BASE CURRENT

FALL TIME VS. TURN ON BASE CURRENT

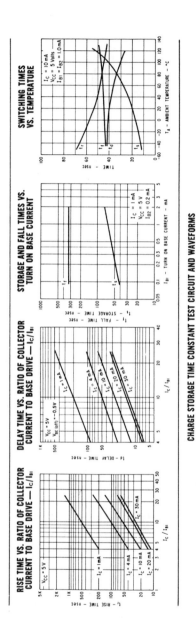

**Figure 8.11c**

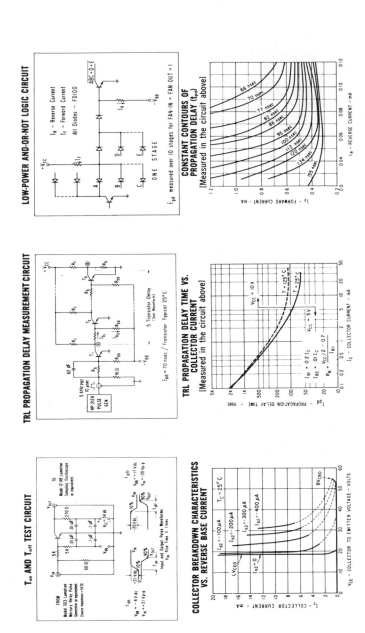

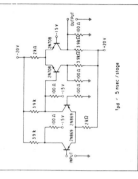
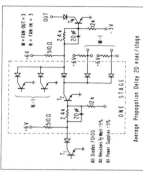
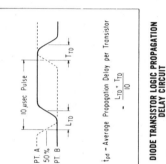
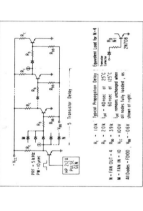
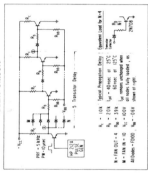

**COMPLEMENTARY HIGH-SPEED NON-SATURATED CURRENT STEERING LOGIC CIRCUIT**

$t_{pd} < 5$ nsec/stage

**HIGH-SPEED LOW-LEVEL NAND CIRCUIT**

M = FAN OUT = 3
N = FAN IN = 3
ONE STAGE
Average Propagation Delay 20 nsec/stage

All Diodes FD100
All Resistors ½ Watt ±5%
All Power Supplies ±5%

**DEFINITION OF PROPAGATION DELAY**

10 μsec Pulse

$t_{pd}$ = Average Propagation Delay per Transistor

$$= \frac{L_{TD} + T_{TD}}{10}$$

**DIODE TRANSISTOR LOGIC PROPAGATION DELAY CIRCUIT**

5 Transistor Delay

$R_1 = 10k$  Typical Propagation Delay : at 25°C
$R_2 = 20k$  $t_{pd} = 40$ nsec at 25°C
$R_{BB} = 39k$  $60$ nsec at 125°C
$V_{CC} = 100V$  $t_{pd}$ remains unchanged when
$V_{BB} = -06V$  all nodes fully loaded , as shown at right.

N = FAN OUT = 4
M = FAN IN = 10
All Diodes FD100

**SMALL SIGNAL FORWARD CURRENT GAIN VS. COLLECTOR CURRENT**

f = 100 MHz

$V_{CE} = 10$ Volts
$V_{CE} = 5$ Volts

$I_C$ - COLLECTOR CURRENT - mA

$h_{fe}$ - SMALL SIGNAL FORWARD CURRENT TRANSFER RATIO

$T_C = 125°C$

$I_{b2} = -100μA$
$I_{b2} = -200μA$
$I_{b2} = -300μA$
$I_{b2} = -400μA$
$LV_{CEO}$
$I_{b2} = 0$
$BV_{CBO}$

$V_{CE}$ - COLLECTOR TO EMITTER VOLTAGE - VOLTS

$I_C$ - COLLECTOR CURRENT - mA

**NOTES:**

(1) These ratings are limiting values above which the serviceability of any semiconductor device may be impaired.
(2) These are steady state limits. The factory should be consulted on applications involving pulsed or low duty cycle operation.
(3) These ratings give a maximum junction temperature of 200°C and junction-to-case thermal resistance of 146°C/watt (derating factor of 6.85 mW/°C); junction-to-ambient thermal resistance of 486°C/watt (derating factor of 2.06 mW/°C).
(4) Rating refers to a high current point where collector-to-emitter voltage is lowest. For more information send for Fairchild Publication APP-4.
(5) Pulse Conditions: length = 300 μsec; duty cycle = 1%.
(6) $r'_b = h_{ie}$ (Real Part) — Measured with GR #1607-A Bridge.
(7) Measured on Sampling Scope. PW ≤ 400 nsec.

Fairchild cannot assume responsibility for use of any circuitry described. No circuit patent licenses are implied.

**Figure 8.11 d**

the graph is about 2.5. Using the expression from Table 8.1, we then get

$$2.5/44 = 1/[1 + (\omega/\omega_\beta)^2] \quad \text{or} \quad \omega_\beta = 1.54 \times 10^8/\text{sec} \quad \text{and}$$

$$\omega_T = \beta\omega_\beta = 6.77 \times 10^9/\text{sec}$$

Then, by applying (8.27), we obtain

$$c_T = c_{b'e} + c_{b'e} = 11.8 \text{ pF}$$

Now we come to the second problem with the spec sheets. We are given $c_{ob} = 6$ pF at $V_{CB} = 10$ V. Unfortunately, for the problem we are now doing, $V_{CB} = V_{CE} - V_{BE} = 9.4$ V. How much difference would that make? We can easily estimate it. If we assume that the junction is approximated by a linear-doping profile, we get $c \propto (V + V_D)^{-1/3}$; if we assume an abrupt junction, $c \propto (V + V_D)^{-1/2}$. Since the junction is probably in very lightly doped material, the diffusion potential is of the order of 0.7 V. Using the 1/3 power law to extrapolate, $c_{ob} \simeq 6.1$ pF; for the 1/2 power law, $c_{ob} \simeq 6.2$ pF. These differences would be easily measured, but their significance is open to question when you consider the problem of making appropriate assignments for the part of $c_{ob}$ that is "outside" and the part that is supposed to be "inside" the transistor. If we guess 50%, then $c_{b'c} \simeq 3$ pF and $c_{b'e} \simeq 8.8$ pF. A guess of 75% gives $c_{b'c} \simeq 4.5$ pF and $c_{b'e} \simeq 7.3$ pF. Which of these provides the better "working" number? The last part of the problem provides at least a clue.

Reworking the calculation of $c_T$ using the curve for $|h_{fe}|$ versus $I_c$ taken at $V_{CE} = 5$ V yields $c_T = 12.6$ pF. Thus, dropping the collector-base voltage by 5 V has increased the value of $c_{b'c}$ by 0.8 pF. The change in total potential across the junction was 5 V, from about 10.1 to 5.1 V. We may check the power law relating the potential to the capacity to see what values of $c$ give reasonable values of $n$ in

$$c_1/c_2 = (V_2/V_1)^n$$

Taking the 50% value leads to

$$3/3.8 = (5.1/10.1)^n \quad \text{giving} \quad n = 1/2.9$$

Taking the 75% value leads to

$$4.5/5.3 = (5.1/10.1)^n \quad \text{giving} \quad n = 1/4.2$$

The 50% value looks better, agreeing rather well with the $n = 1/3$ that we would obtain for the linearly graded junction. With no other data at hand, $c_{b'c} \simeq 3$ pF at $V_{CE} = 10$ V would be a reasonable number to try for a preliminary design.

## Problem 8.4

Using the data from problem 8.3 and Figure 8.11, calculate the gain and bandwidth for an amplifier using a 2N708 as the active element with a bias point, $I_C = 2$ mA, $V_{CE} = 10$ V; an output load of 600 ohms; and a signal source resistance of 500 ohms.

*Solution:* This problem is a straightforward application of equation 8.22. According to (8.22), the midband gain is

$$A_{\text{mid}} = \beta R_L / R_{\text{in}} = 44 \times 600 / (500 + r_{bb'} + r_{b'e})$$

For $r_{b'e}$ we have $\beta/g_m = \beta k T / q I_C = 550$ ohms. $r_{bb'}$ is a little bit more of a problem. We have the manufacturer's statement that $r_{bb'} = 50$ ohms at 300 MHz; that is undoubtedly too low a value for $r_{bb'}$ at low frequency, but we will use it and return to this question in a moment. Setting $r_{bb'} = 50$ ohms gives

$$A_{\text{mid}} = 24$$

The 3 dB frequency is also obtained from equation 8.22. Its value is

$$\omega_{3\text{dB}} = \frac{g_{b'e} + G'_s}{c_T} = \frac{0.00182 + 0.00182}{c_{b'e} + (1 + g_m R_L)c_{b'c}} = 2.33 \times 10^7 / \text{sec}$$

or

$$f_{3\text{dB}} = 3.71 \text{ MHz}$$

This is a moderately broadband amplifier; to achieve it we have had to use a pretty good transistor AND work at relatively low impedance (and gain) levels.

Now let us return to the subject of $r_{bb'}$ (what the manufacturer calls $r_{b'}$ in Figure 8.11a). The spec sheets inform us (note 6, Figure 8.11d) that the value of $r_{bb'}$ given is the real part of $h_{ie}$ measured at 300 MHz. Our complete model (Figure 8.9) shows that at sufficiently high frequency, $c_{b'e} + c_{b'c}$ would "short out" $r_{b'e}$, leaving an input circuit comprised essentially of $r_{bb'}$ shunted by $c_{be} + c_{bc}$. Thus, the real part of the input *admittance*, $y_{ie}$, should become a constant. [Why Fairchild chose to measure $h_{ie} = 1/y_{ie}$ is certainly not obvious. The usual measurement is $y_{ie}$. Any bridge that measures $h_{ie}$ also determines $y_{ie}$, but $g_{bb'} = \text{Re}(y_{ie}) \neq 1/\text{Re}(h_{ie})$.] The problem of determining a satisfactory value for $r_{bb'}$ is difficult for two very practical reasons. These are most clearly seen by writing down what our model of the CE input says $y_{ie}$ should be. From Figure 8.9 with the output short circuited we have

$$y_{ie} = sc_1 + \left( \frac{g_{bb'} g_{b'e}}{g_{bb'} + g_{b'e}} \right) \left[ \frac{1 + sc_T / g_{b'e}}{1 + sc_T / (g_{bb'} + g_{b'e})} \right] \tag{8.29}$$

where $c_1 = c_{be} + c_{bc}$ and $c_T = c_{b'e} + c_{b'c}$. The circuit and two plots of $|y_{ie}|$ are shown in Figure 8.12. The upper curve, drawn with the pole and two zeros nicely separated, is what you would like to meet in practice. In principle, all you would have to do is find a frequency out beyond $\omega_\beta$ at which $y_{ie}$ is real. Its value at that point is $g_{bb'}$.

Now look at Figure 8.12.$b$. This is a plot of $y_{ie}$ from (8.29) using the values appropriate to the 2N708. It is not nearly so pleasant to behold. At low frequencies, $r_{bb'}$ is masked by $r_{b'e}$. At high frequencies, it is thoroughly enmeshed in the interaction between the closely spaced pole and zero. Looking at (8.29), you might hope to be able to go to a high enough frequency so that the real part of $y_{ie}$ was $g_{bb'}$ and yet not so high that the imaginary part, $sc_1$, would be masking what you wanted to measure. Unfortunately, as you push on beyond the second zero, you would find that the measured data failed to agree with what our model predicted. It is simply that the distributed character of the base resistance and capacity would make itself evident. Thus, our problems are very fundamental: at frequencies for which our model is adequate, $r_{bb'}$ is hidden by the other components. At frequencies for which the model might give reasonably clear data, the model itself is inadequate.

The next obvious question is: Why should Fairchild or anyone else publish a piece of data that is so hard to interpret? As you can see by looking at the gain and bandwidth that we have just calculated, for a run-of-the-mill amplifier, the performance is not very sensitive to the value of $r_{bb'}$. The gain and bandwidth are limited at the input by $R_s$ and $r_{b'e}$. On the other hand, we are not obliged to use this transistor at frequencies well below $f_\beta$. Let us say that we want to use the 2N708 to amplify signals out to 100 MHz. If we were willing to lower our load resistance to 100 ohms, how large would $g_{b'e} + G'_s$ have to be to put the 3 dB point at 100 MHz? Running these numbers through the mill again gives $g_{b'e} + G'_s = 0.023$ mhos. This is much larger than $g_{b'e}$, so we have found that $R_s + r_{bb'} \simeq 1/0.023 \simeq 43$ ohms. What the manufacturer tells us is that this cannot be done, since he reports a greater value than 43 ohms for $r_{bb'}$ alone. Thus, Fairchild has done a useful thing in publishing a value for $Re(h_{ie}) \simeq r_{bb'}$ at a frequency well in excess of $f_\beta$. However, to then assume that $r_{bb'}$ had really been measured would be a mistake. If you did take the trouble to measure $r_{bb'}$ at low frequencies, you would probably find a value closer to 100 ohms.

It is all very well and good to know how to scale all of the large- and small-signal parameters, but given the fact that these parameters are dependent on both bias and temperature, how can a circuit be designed that doesn't take off on its own every time the wind blows? Two partially independent problems confront the designer here. He must be sure that the transistor always stays within its power-dissipation limits, and he probably wants his amplifier to perform in much the same fashion at 100°C as it does at room

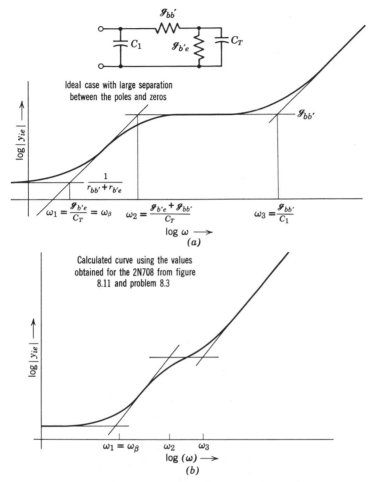

**Figure 8.12** Logarithmic plots of $|y_{ie}|$ versus angular frequency. (a) shows an ideal, clear-cut diagram with wide separation between the roots. $g_{bb'}$ is the value of $y_{ie}$ in the flat portion of the curve. (b) shows what the curves usually look like, making the determination of $g_{bb'}$ very ambiguous. Both curves are derived for the input circuit shown above, which is obtained from Figure 8.9.

temperature. The subject of making semiconductor circuits reasonably temperature insensitive—usually called *bias stabilization*—can be made arbitrarily complicated. A thorough discussion of bias stability is a subject most suitably handled in terms of multistage circuit analysis using the concepts of feedback (see P. E. Gray and C. L. Searle, *Electronic Principles*, Wiley, 1969; R. D. Thornton et al., *Multistage Transistor Circuits*, SEEC

Vol. 5, Wiley, 1965; or Basil Cochrun, *Transistor Circuit Engineering*, Macmillan, 1967). For our purposes, however, the following two fairly simple problems should illustrate what must be done and at least one reasonably practical way to do it.

## Problem 8.5

Consider the very simple amplifier circuit of Figure 8.13. If the transistor has $\beta = 100$ and $I_{co} = 10^{-10}$ A at room temperature ($q/kT = 40$ V$^{-1}$), find $R_b$ so that $I_C = 2$ mA and $V_{CE} = 5$ V. Estimate the intrinsic midband gain of the amplifier. Then find $I_C$, $V_{CE}$, and $A_{mid}$ at 100°C, assuming that at that temperature, $\beta = 200$ and $I_{co} = 5 \times 10^{-7}$ A.

*Answer:* $I_{co}$ is quite obviously trivial. Accordingly, $I_C = 2$ mA requires that $I_B = I_C/\beta = 0.02$ mA. The voltage drop across the emitter base junction is about 0.6 V, leaving about 9.4 V across the resistor, $R_b$. Thus, $R_b$ is given by

$$R_b = 9.4/0.02 = 470 \, \text{Kohm}$$

Note that the answer is only weakly dependent on our assumption about $V_{BE}$. Estimating $r_{bb'}$ to be about 100 ohms, the intrinsic gain of the amplifying stage is

$$A_{mid} = \frac{-\beta R_L}{r_{bb'} + r_{b'e}} = \frac{-100 \times 2500}{100 + 1250} = -185$$

The second half of the problem has some surprises. The base current is still about 0.02 mA; it is pretty independent of the transistor's state. Thus, $I_C = \beta I_B = 4$ mA and $V_{CE} = 10 - R_L I_C = 0$. The amplifier has obviously saturated since $V_{CE} < V_{BE}$. Thus, the gain has gone to 0; the amplifier has turned itself fully on and has nothing else it can do for us.

Although a good many of the transistor's parameters change with temperature, it should be evident in the example above that the change in $\beta$ was what clobbered the amplifier. It would be nice to make the circuit self-correcting against changes in $\beta$, so that, as $\beta$ increased, $I_B$ was reduced appropriately. This would hold $I_C$ approximately constant and with it all of those parameters that are proportional to $I_C$. Not only would this make the amplifier much less temperature sensitive; it would also mean that variations in $\beta$ from one transistor to another of the same type would not seriously affect the amplifier's performance. Are there any liabilities? Let's take an example to see.

## Problem 8.6

To minimize the changes in $I_C$ from any cause, it is necessary to sense the collector current and adjust the base drive accordingly. One reasonably simple and effective way of doing this is to put a resistor in the emitter lead

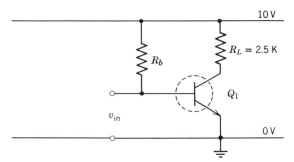

**Figure 8.13** The simplest form of the CE circuit. This method of applying bias to the transistor is particularly sensitive to variations in $\beta$.

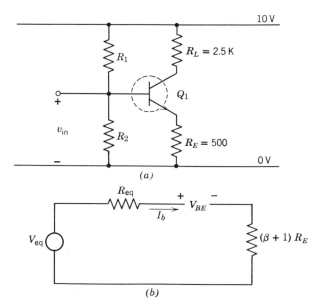

**Figure 8.14** (a) A CE circuit that uses emitter degeneration to stabilize the collector current. (b) An equivalent circuit for the base-emitter loop. The bias bus and the voltage divider are replaced by their Thevenin equivalent and $I_E R_E$ by its algebraic equivalent, $(\beta + 1) R_E I_B$.

as shown in Figure 8.14a. Making a Thevenin equivalent for the base drive circuit and noting that the voltage across $R_E$ is $(I_B + I_C)R_E = (\beta + 1)I_B R_E$, the base loop can be drawn as shown in Figure 8.14b. The active element, $Q1$, essentially multiplies the resistance of $R_E$ as seen by the base-drive circuit.

From 8.14$b$ the base current is given by

$$I_B = \frac{V_{eq} - V_{BE}}{R_{eq} + R_E(\beta + 1)}$$

It is immediately obvious that the larger $R_E$ is, the less dependent $I_C$ will be on $\beta$. It is also evident that the smaller we can make $R_{eq}$, the more stable the circuit will be. However, $R_{eq} = (G_1 + G_2)^{-1}$ also shunts the input, so we may not make it too small. Let us say that we arbitrarily set $G_1 + G_2 = = 5 \times 10^{-5}$ mhos $(R_{eg} = 20$ Kohms$)$. Then, using the same transistor as above in problem 8.5, choose $R_1$ and $R_2$ so that you obtain the bias point: $I_C = = 2$ mA, $V_{CE} = 4$ V. Then calculate $A_{mid}$ at room temperature and at $100°$C. *Answer*: Having decided on the value of $R_{eq}$, $V_{eq}$ is obtained directly from Figure 8.14$b$ as

$$V_{eq} = V_{BE} + [R_{eq} + (\beta + 1) R_e] I_C/\beta = 2 \text{ V}$$

The elements of the voltage divider must then be $R_1 = 100$ Kohms, $R_2 = = 25$ Kohms. At $100°$C, the transistor's $\beta$ has risen to 200, and the forward bias required to keep 2 mA flowing has decreased to roughly 0.4 V. Solving again in the above equation for the collector current gives 2.6 mA. Thus, for a 100 percent increase in $\beta$ plus a 33 percent decrease in $V_{BE}$, the collector current has changed by only 33 percent. Completing the bias analysis, $V_{CE}$ has decreased from its room-temperature value of 4 V to 2.2 V. Thus, our emitter resistor has indeed done its job in keeping the amplifier working at these elevated temperatures. But now to the second question: What's happened to the gain?

The small-signal circuit that we get by putting an emitter resistor into Figure 8.3$b$ is not particularly convenient. It would be much nicer to have a unilateralized circuit like Figure 8.6. A bit of algebra will always allow us to do this conversion, so why not? (The same algebra would get you the answer directly, it is true, but this circuit is so widely used that it pays to develop a more convenient representation.) What we are starting with and what we want to end up with are shown in Figure 8.15. Compare Figure 8.15$b$ with Figure 8.6. They look the same, and, in fact, they are the same with but two exceptions. These changes are

$$g_m \Rightarrow g'_m = g_m/U \qquad \text{and} \qquad y_{b'e} \Rightarrow y'_{b'e} = y_{b'e}/U \qquad (8.30)$$

where

$$U = 1 + (y_{b'e} + g_m) R_e$$

which reduces at low frequency to $U = 1 + (1/\beta + 1) g_m R_e \approx g_m R_e$. Since the Miller capacity is also reduced by the factor $(1/U)$ if $g'_m R_L \gg 1$, the whole input admittance is reduced. If the source conductance is of the order of the input conductance or higher, it is evident that we will have gained bandwidth (see equation 8.22) while losing gain.

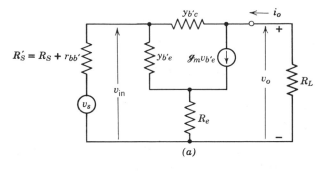

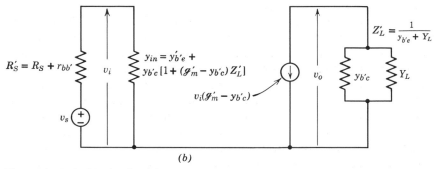

**Figure 8.15** A CE circuit with emitter degeneration. (*a*) gives the small-signal equivalent circuit using the π-model for the transistor. (*b*) shows the unilateralized version of (*a*). The circuit in (*b*) is identical with that in Figure 8.6, but two of the coefficients have changed in important ways. These are

$$g'_m = g_m/U \quad \text{and} \quad y'_{b'e} = y_{b'e}/U \quad \text{where} \quad U = 1 + (y_{b'e} + g_m)R_e$$

Now let us see what has happened to the gain. We may write the intrinsic gain as

$$A_{\text{mid}} = \frac{r'_{b'e}}{r'_{b'e} + r_{bb'}} \left[ -g'_m R'_L \right]$$

When $U$ is large, $A_{\text{mid}}$ goes to: $-g_m R_L = -g_m R'_L/g_m R_e = -R'_L/R_e$. Thus, the gain becomes *independent of the transistor*! $U$ is certainly large in our case, so $A_{\text{mid}} = -2K/0.5K = -4$. That's quite a comedown from the $-185$ we got out of the same transistor in the circuit of Figure 8.13.

You might well wonder whether we could not stabilize the amplifier without paying so fearful a price in gain. The answer is a guarded yes. If you are willing to give up DC gain, for instance, you could shunt $R_e$ with a large capacitor. Then, at frequencies for which $sC_e \gg G_e$, the amplifier would have its gain back. This is not always an acceptable solution, and there are many other ways of going about the problem. In all of them, however, you pay for stability by giving up potential gain of one sort or another. This should not seem any more ominous than any other conservation law. It merely puts a price on a combination of gain and stability.

## SOME METHODS AND TOOLS OF TRANSISTOR FABRICATION

Having developed some reasonably powerful methods for modeling transistor behavior, it would be very nice if we could go back to our physical model of transistor action to see just what can and cannot be accomplished in making the "ideal" transistor. Unfortunately, this is rather like trying to discuss the perfect game of chess. The number of different kinds of moves is very limited, but the combinations and interactions of the different steps is so complicated that the game defies closed-form analysis.

The fabrication of transistors in economically viable quantities is almost as much art as science. Some of the leaps ahead have been made on the basis of sound scientific reasoning and research, but many have been the outgrowth of some a priori outlandish ideas that have been made to work by a couple of very green thumbs in the laboratory. Thus, without a very subtle understanding of the processing limitations, it is difficult to be very omniscient about what can and will be done. Still, it is both useful and enjoyable to have a general idea of how transistors are made, so let us have a look at the steps necessary to fabricate a silicon and a germanium transistor that would be suitable for incorporation in a monolithic integrated circuit. I have two goals in mind here. First, I want to introduce some of the words that are used to describe both a process as well as the transistors that are made by the process. You are "in" if you know the words, but "out" if you don't. The second goal is to get back to the impurity profiles that you will find in real devices. This will lead naturally into a discussion of the *drift transistor* and what one can do to make a transistor that will give useful gain in the microwave range.

The two devices whose fabrication I am going to describe are rather similar small-signal transistors (i.e., intended for low power output but with relatively large gain) at moderately high frequency. Thus, they should have a moderately large $\beta$ and as little charge storage as possible. To be useful in a monolithic circuit, the transistors must have all three leads accessible from the top and be isolated from the substrate. Let us begin with the somewhat more familiar silicon processing steps.

With silicon it is so much easier to make an *n-p-n* than a *p-n-p* transistor that almost nobody who can avoid it makes silicon *p-n-p*'s. A quick review of this entire book won't help a bit; there is nothing fundamentally harder about *p-n-p*'s. In fact, in germanium they are easier to make. The problem with silicon arises mostly from minute amounts of pervasive surface contamination. This is a subject that we must deal with in chapter 10, so let us leave it until then. For the moment, what we need to know is that a $p^+n$ junction that comes to the surface is likely to have a very soft reverse characteristic, and without proper precautions it is apt to be so leaky at the surface that it will not make a useful emitter junction.

Our device, then, is to be pair of $n$-$p$-$n$ transistors on the same chip (hence *monolithic*) but electrically isolated from each other. Figure 8.16 shows or lists the various steps that could be used to create such a device. It is not THE way. There are many different approaches to the same goal. Nor is this a particularly choice way; the method selected here was to illustrate as many of the *different* types of steps as possible.

In the old days (e.g., the late 1950's), one started with a tank full of $SiCl_4$ or $SiH_4$, which is only a few steps away from starting with a shovel full of beach sand. Today, fortunately, one can buy excellent single-crystal silicon wafers properly oriented and doped to the user's specification from such chemical supply houses as Monsanto. Still other firms will polish these slices to a beautiful stress-free mirror finish on one side and a clean, etched surface on the other. At this point, you are completely ready to "do your own thing." For our purposes, we would order lightly doped (50–100 ohm-cm) $p$-type wafers. Our first task would be to oxidize the wafers to make a diffusion mask. We would then be ready to make some devices.

Our first problem is to isolate the two transistors from each other. Putting two $n$-type collectors into the $p$-type substrate accomplishes this very nicely. In essence, the collectors are then connected by a pair of back-to-back diodes. If the junctions are good, this will give us only capacitive coupling. (Bell Telephone's *beam-lead* technique eliminates even this coupling by etching apart the individual components from the back after they are supported by unetchable metal conductors [beam leads] on the front. This is an elegant but difficult solution to the isolation problem.)

The first problem that confronts us is that we want the collector side of the base-collector junction to be lightly doped. That will be difficult to accomplish by diffusing first the collector and then the base. Even if we could do it, the high-resistivity collector material would be very unsatisfactory for leading the collector current out through the top surface. What we would like to have is a *buried layer* of heavily doped $n$-type silicon covered by a lightly doped $n$-type layer. The buried layer would provide the low-resistance path to the outside, while the lightly doped layer would make a good collector junction. To accomplish this desirable end, we would first diffuse the buried layer into the $p$-type substrate as shown in the second frame of Figure 8.16. Then we can strip the oxide mask and grow a lightly doped $n$-type layer on top by gaseous epitaxy. This layer, of course, shorts all of the

A. Steps performed by the crystal manufacturer:

1. Polycrystalline Si of high purity produced by pyrolytic decomposition of $SiH_4$. The silane may be purified by distillation before decomposition.
2. Large single-crystal boules (as large as 2 1/2″ in diameter) grown by melting and refreezing the high-purity polycrystalline Si. In the process of freezing, the majority of the remaining impurities are segregated into the liquid. Impurities are deliberately added to produce the desired final doping.
3. The large boule is sliced like a salami into very thin (300 $\mu$m) wafers. These are shipped to the customer or a finishing lab.
4. The finishing house or customer electrolytically polishes one side of the wafer to a flat, mirror finish. The other side is etched to remove the surface damaged by sawing. This leaves the transistor manufacturer with the slice of lightly doped material shown below.

B. Steps performed by the device manufacturer:

1. Grow oxide layer by heating wafer in $O_2$. $SiO_2$ layer 0.8 to 1 $\mu$m thick.
2. Cover with photo resist.
3. Selectively expose and remove unwanted resist.
4. Buffered HF etch removes uncovered $SiO_2$.
5. Remaining resist removed chemically.
6. $n$-type diffusion for buried $n^{++}$ layer.

These four steps are called "step $A$" in subsequent processing.

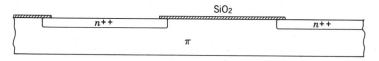

7. All the oxide stripped off.
8. Lightly doped $n$-layer grown by gaseous epitaxy. Some diffusion will occur from $n^{++}$ into $\nu$-layer.
9. Step $A$ to open base windows.
10. Short $p$-type deposition diffusion (boron).
11. Quick etch to remove boron glass.
12. Drive-in diffusion performed in $O_2$ to form $SiO_2$ layer over base window. This layer must be thick enough so that the subsequent emitter diffusion does not penetrate the glass and invert the base surface.

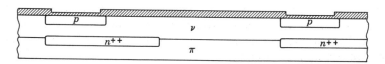

13. Step *A* performed to open emitter window.
14. *n*-type deposition and diffusion (phosphorus). Drive-in in $O_2$ atmosphere to passivate the surface.
15. Step *A* performed to open mesa window.
16. Nitric acid etch to cut mesa.
17. Exposed mesa surfaces passivated with $SiO_2$.
18. Step *A* performed to open ohmic contact windows.

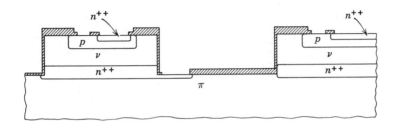

19. Aluminum evaporated over entire surface.
20. Mask and etch with phosphoric acid etch to remove unwanted Al.
21. Heat to 550°C in argon to microalloy Al to Si and make good ohmic contacts.
22. Wafer scribed and broken into individual devices.
23. Each chip bonded to its header and wires bonded from header to chip.
24. Chip and wires encapsulated and sealed.

**Figure 8.16.** Steps to produce diode-isolated *n-p-n* silicon transistors.

buried layers together, but that will not be the case after subsequent processing.

The next two steps are pretty much what you would expect. The base is diffused through windows in an oxide mask into the epitaxial layer, and then the emitter is diffused into the base region. Now comes the step that completes the isolation of the two transistors and provides access to the buried layer. By using a very tough photoresist or the $SiO_2$ itself as a mask, all of the unused epitaxial layer and a little bit of the original substrate are removed by a nitric acid etch. This leaves each of the transistors sitting up on a little *mesa* above the plain (plane if you prefer, but the Western-mesa analogy calls for plain) of the original substrate. Now there is no connection between the two transistors except through the isolation junctions. Then, to keep the surface free of the contamination that would lead to soft junctions and serious leakage, the exposed surfaces of the silicon are oxidized. This step is called *passivation* of the surface. Although surfaces can be passivated by depositing all sorts of adherent insulating films, the fact that silicon can be so well sealed by its own oxide is a considerable advantage not shared by most other semiconductors.

The devices are now completed by opening windows in the oxide where the ohmic contacts are to be made (the state of affairs in the last frame of Figure 8.16), metalization, separation, and packaging.

When you are looking at Figure 8.16, bear in mind that the aspect ratio in the picture is grossly distorted. This is the same drawing problem that we encountered in Figure 7.1. The real thing looks much more like Figure 7.2.

Let us now take a look at the processing of a germanium device with roughly the same objectives in mind. There are two very important differences between the processing that works with germanium and that applicable to silicon. The first concerns surface passivation. $GeO_2$ is, unfortunately, quite porous. Thus it is inapplicable as either a diffusion mask or a surface-passivation layer. It is necessary to deposit some other oxide or nitride by chemical or vacuum-deposition techniques. These are reasonably effective when done properly, but they are neither as easy nor as convenient as simple oxidations. The second problem concerns the diffusion rates and solubilities of the $n$- and $p$-type dopants. In silicon, P and B diffuse at roughly the same rate and have about the same solid solubility. This makes it relatively easy to obtain a double diffusion that puts both the emitter and collector junctions at desirable depths. In germanium, on the other hand, the $p$-type dopants have universally high solubilities and very low diffusivities, while the $n$-type dopants have high diffusivity but only moderate solid solubility. This makes a proper double-diffusion program very difficult to achieve. However, making a virtue of necessity, it does open up an equally good method of generating a profile like that in the silicon transistor. The method is called *postalloy diffusion* and works as shown in the second and third frames of Figure 8.17. A layer of arsenic-doped indium is deposited on the surface of the $p$-type germanium. The layer might contain as much as 1 % As (or InAs if you prefer). This layer is then alloyed with the germanium by heating well above the eutectic temperature and then cooling. The regrown germanium will be saturated with In (of the order of $10^{20}/cm^3$), but the As will be segregated mostly in the In, resulting in about $5 \times 10^{18}$ As/$cm^3$ in the regrown Ge (compared to $10^{20}/cm^3$ in the In). This is the situation shown in the second frame.

The remainder of the cycle is easy. First the remaining metal (In) on the surface is removed with a quick etch. Then a simple diffusion cycle is performed. What happens is quite automatic. The diffusion coefficient for arsenic is two and a half orders of magnitude greater than that for In. Thus, for all practical purposes, the In stays put while the arsenic diffuses rapidly into the lightly doped $p$ region. The net result is shown in the third frame of Figure 8.17. The $p^{++}$ on $\pi$ layer is converted to the $p^{++}n^{+}\pi$ structure that we want for our transistor.

1. Polished $n$-type Ge wafer purchased.
2. $SiO_2$ layer deposited by sputtering.
3. Step $A$ (see Figure 8.16) performed to open buried-layer window.
4. Buried layer diffused in ($p$-type).
5. Oxide stripped.
6. Lightly doped $p$-layer grown epitaxially.
7. Layer of indium doped with arsenic evaporated onto surface.
8. Mask and etch to remove unwanted metal.
9. Rapid heat to alloy temperature, then cool to grow $p^{++}$ layer counterdoped with As.

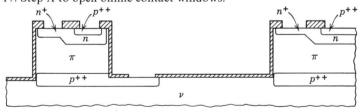

10. Postalloy diffusion.

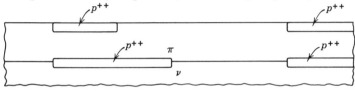

Impurity profile postalloy but prior to diffusion

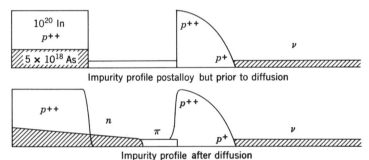

Impurity profile after diffusion

11. $SiO_2$ deposited by sputtering.
12. Step $A$ performed to open base (overlap diode) window.
13. $n$-type diffusion to "find" the base layer.
14. Mask made with "tough" resist, making window for mesa etch.
15. Mesa etch.
16. Deposit oxide.
17. Step $A$ to open ohmic contact windows.

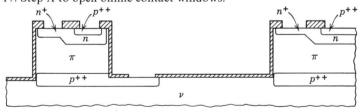

18–21. Metalize, bond to header, attach leads, encapsulate.

**Figure 8.17** Steps to produce diode-isolated, postalloy diffused germanium mesa $p$-$n$-$p$ transistors.

The only problem that remains is "finding" the base. At the end of the post-alloy diffusion cycle there is no obvious way of making a connection to the base. However, a masking step followed by an $n$-type diffusion makes contact with the base layer and provides a suitable place for an ohmic contact. This is the situation shown in the fourth frame of Figure 8.17. The rest of the steps are quite similar to those for the silicon transistor.

Let me stress once again that the processing just examined is only part of the story. Although the *essential* steps were discussed, the unessential ones, like getting the proper temperature cycles in the furnace, getting the etches to work in the same fashion today as they did yesterday, and keeping things clean when it counts, can frequently be the steps that really determine whether the devices will work. There are also many variants on the processes we have discussed as well as a great many other ways (some now discarded) for achieving a transistor that works.

Most of the transistors that are made today do utilize epitaxy and either double diffusion or postalloy diffusion to achieve their structures. Thus, it is not only reasonable but necessary that we ask whether the steep impurity gradients characteristic of the diffusion process and the $n^{++}vp$ structure of the collector junction will change our physical analysis in any important way.

## THE DRIFT TRANSISTOR

As far as excess carriers are concerned, the impurity-concentration gradients make themselves visible only through the electric fields they create. Our assumption has always been that drift was unimportant compared to diffusion in transporting minority carriers across the base. If you will go back to the place where this argument was first raised (in going from equation 5.25 to equation 5.26), you will see that it was based on the fact that the very few minority carriers available cannot contribute much of a drift current compared to what the majority carriers would provide. Another way of saying the same thing is to state that *if* the minority current is to be significant in a very extrinsic sample, the minority current must be flowing by diffusion. However, there is one very important exception to this mode of reasoning. If the electric field comes about in the achievement of thermal equilibrium, the field produces no net flow of the equilibrium carriers. Thus, we can have a large field (up to about 100 V/cm with ordinary diffusion profiles) with no current. There is no reason to assume that this large field will not strongly influence the flow of the excess carriers. We must include it in our calculation of the carrier flow across the base.

The first question to be asked is: Does the impurity gradient across the base (high concentration at the emitter junction and low at the collector) help or hinder the flow of minority carriers from emitter to collector? (Answer: The field is such that it prevents the diffusion of majority carriers from regions of high concentration to regions of lower concentration. Thus, it must push the minority carriers into the lightly doped regions. This helps the flow in the forward direction, but obviously hinders it in the reverse direction. Once again, the diffusion process seems to lead quite naturally to better forward characteristics.)

As long as the base is fairly narrow, so that recombination within the base is pretty small, the calculation of the effect of the majority carrier gradient on transistor performance is quite simple. All we must do is combine the electron-current equation (4.64) with the relationship between the electric potential and the majority-carrier concentration (3.42′):

$$J_e = q\mu_e n E + qD_e dn/dx \quad\quad \text{and} \quad\quad E = - d\psi/dx = (kT/q)\frac{1}{p}\frac{dp}{dx}$$

The result is

$$J_e = qD_e \left[ \frac{n}{p}\frac{dp}{dx} + \frac{dn}{dx} \right] = q\frac{D_e}{p}\frac{d(pn)}{dx} \tag{8.31}$$

Having assumed that the electron current traversing the base is essentially constant, we may integrate (8.31) to obtain the term $I_{\epsilon 5}$ in Table 7.1 (using the notation of Figure 7.3):

$$I_{\epsilon 5} = \frac{A_\epsilon qD_e p(a')n(a')}{\displaystyle\int_{a'}^{b} p(x)dx} = \frac{qD_e n(a') A_\epsilon}{w_b'} \tag{8.32}$$

where the effective base width, $w_b'$, is given by

$$w_b' = w_b \left( \frac{\bar{p}}{p(a')} \right)$$

and $\bar{p}$ is the average value of the majority-carrier concentration in the base. Since $p_0(a') > \bar{p}$, the field resulting from the concentration gradient effectively shortens the base. All of the quantities that depend on the base width—$\beta$, $c_{\text{ediff}}$, $I_{es}$, etc.—are affected beneficially. (On the other hand, in the reverse direction, things get worse rather than better because the integration in (8.32) goes from $b$ to $a'$ and $w_b' = w_b[\bar{p}/p(b)]$. Since $\bar{p} > p_0(b)$, the base is effectively lengthened for reverse operation.)

The same sort of reasoning applies to the gradient in the emitter. Since the majority-concentration gradient slopes toward the junction, the built-in field will *inhibit* the injection of holes from the base into the emitter. Thus, both the first and second terms of (7.24) are improved (made smaller) by the natural diffusion profiles. The only term that gets worse is the part of the diffusion charge that is stored in the emitter, and that term is generally so small that it doesn't count anyway.

We may conclude from this rather simple analysis that the diffused base will yield a higher $\beta$ and a lower $c_{b'e}$ than a uniformly doped base of the same dimensions. The only disadvantage of the strongly graded base is that the base doping at the emitter junction will be relatively high. Thus, the emitter junction breakdown voltage, $BV_{EBO}$, will be pretty low (cf. Figure 8.11a).

Although most transistors made today have a strongly graded base, in the earlier days of transistor manufacturing the achievement of such a base was no mean trick. Since they were rather special in those days, they became known as *drift transistors*. The term has since fallen into disuse because almost all transistors now have this feature.

## CURRENT AND FREQUENCY LIMITATIONS IMPOSED BY THE COLLECTOR JUNCTION

We have seen that there are some substantial advantages in having the collector side of the collector junction much more lightly doped than the base side. Such an arrangement gives the maximum collector breakdown voltage, minimum base-width modulation, and minimum collector and overlap-diode junction capacity. Thus, in the descriptive section on how transistors are made, the profiles sought after were either $n^{++}p\nu n^{++}$ or $p^{++}n\pi p^{++}$.

Although the $\nu$ and $\pi$ layers provide a number of very important advantages, they have an Achilles heel that limits both the maximum current and the upper frequency of operation. The very fact that the collector depletion region and usually a portion of the ohmic region of the collector lie in very lightly doped, high-resistivity material means that at moderate current densities the transistor will be subject to base stretching and to ohmic and space-charge voltage drops that may actually forward bias the collector. Although too many things start to happen at once to allow a completely quantitative analysis of this region of operation, a qualitative analysis yields both a good picture of what is happening as well as some quantitative results on the upper bounds of useful operation. Let us begin with a quick sketch of what happens at the collector junction as the current density begins to affect the transistor's performance.

To best separate the different phenomena that are transpiring at the base-collector junction, let us consider two relatively extreme examples. In the first, the lightly doped $\nu$ layer will be quite wide so that the depletion layer extends only a little way into the $\nu$ layer. The second case will have the $\nu$ layer thin enough so that the depletion layer stretches all the way to the very heavily doped $n^{++}$ portion of the collector. The first case is illustrated in Figure 8.18.

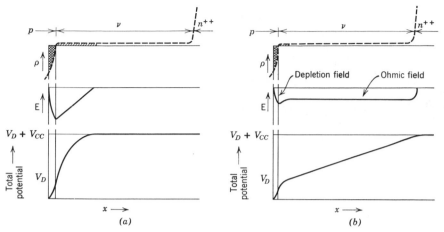

**Figure 8.18** The net-charge density, electric field, and total-potential distributions in the collector region of an $n^+pvn^{++}$ transistor. (a) gives these distributions for the transistor with the collector reverse biased by $V_{CC}$ and no current flowing. (b) gives the same distributions with $V_{CB} = V_{CC}$ and $I_cR_c = V_{CC}$.

In Figure 8.18a, the space-charge, field, and voltage distributions are shown for very low collector current. In Figure 8.18b, the same three parameters are drawn with the current high enough so that the ohmic drop in the lightly doped material has absorbed the entire collector-base bias. As long as the minority-current density is still well below the majority density in the ohmic portion of the collector, it is easy to find the current corresponding to the situation in Figure 8.18b. If we define the collector series resistance as

$$R_c = l/\sigma A_\epsilon \tag{8.33}$$

where $l$ is the length of the undepleted $v$ region for zero applied bias and $\sigma$ is its conductivity, then the case illustrated by Figure 8.18b occurs when

$$I_cR_c = V_{CB} \tag{8.34}$$

If one attempts to raise the current above this point without increasing $V_{CB}$, the collector junction becomes *forward biased*. Naturally, when this happens, everything goes to pot in a grand hurry—however, not quite as grand a hurry as you might guess from the simple exponential-diode equation. For the first few tenths of a volt after the junction becomes forward biased, the principal effect is that a large number of holes are injected from the base into the lightly doped $v$ layer. This immediately increases the conductivity of the resistive layer, so the rate at which the collector junction becomes forward biased as the current rises is markedly reduced. Thus, instead of the sharp turnover at a forward collector-base bias of about 0.6 V ($V_{CE} \Rightarrow 0$)

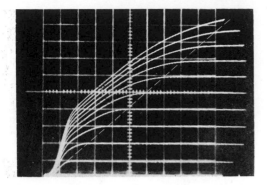

**Figure 8.19** CE collector characteristics at high current (10 mA/div) for a typical $n^+pvn^+$ double-diffused planar epitaxial transistor. The dashed line marks the locus of points where the ohmic drop in the lightly doped epitaxial layer just equals the applied collector bias. [From L. A. Hahn, *Trans. IEEE ED*-16, 654 (1969).]

as in the CE collector characteristics (Figure 7.11), we should expect the nice, flat, horizontal $I_C$ versus $V_{CE}$ curves to turn over slowly and collapse together as $V_{CE}$ drops below $I_C R_c$. This is precisely what is observed in most double-diffused epitaxial transistors. Figure 8.19, taken from a paper by L. A. Hahn, who first pointed out the extent to which the ohmic drop de-biased the collector junction, shows just this phenomenon. The dashed line is the locus of $I_C R_c = V_{CE}$, and as you can see, the fall off in $\beta$ and effective output impedance takes place completely to the left of that locus.

You might conclude from this analysis and Figure 8.19 that all you need to do is keep $V_{CE}$ large enough and you can extend $I_C$ as far as you want. It's not that such a conclusion is wrong, but you must also keep the transistor operating within its power-dissipation rating. Thus, there is a very definite upper bound to the collector current that a transistor can usefully handle.

Let us say that you wanted to increase the current that a particular transistor could handle. From (8.33) you might quite properly conclude that shortening the $v$ region would raise the current-carrying capability. Indeed it would, but as the region was shrunk to the thickness of the rather wide depletion layer, a new limitation would appear. This, of course, was the second case delineated above. By the time the entire $v$ region is depleted, the collector current is carried away from the junction through such heavily doped material that no noticeable ohmic drop occurs. Now, however, the space charge of the free carriers traversing the depletion region will begin to be important.

Figure 8.20 illustrates what happens. The figure on the left (Figure 8.20$a$)

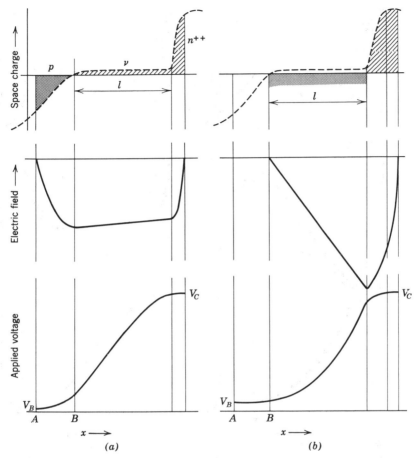

**Figure 8.20** The space charge, electric field, and voltage distribution at the collector junction of an $n^+pn^{++}$ transistor. (a) gives the distributions for the case of $I_C = 0$ with $V_{CB}$ large enough to deplete the entire $\nu$ region. (b) gives the same functions for the same collector voltage, but now with the current up to the point where the negatively charged depletion region is entirely in the inverted $\nu$ layer. The edge of depletion region in (b) has already shifted from A to B, stretching the base width and lowering $\beta$.

shows the situation with no current flowing. The applied bias is large enough to just deplete the entire $\nu$ region. (Any more bias would start driving the depletion region into the base, causing the output impedance to fall off. It was partly to avoid this Early effect that the lightly doped region was put in in the first place, so one would not want to operate much in excess of the voltage shown in the figure.) As the current increases, more and more carriers at any instant are in transit across the depletion region. Up until now, we have never considered this charge to be important. However, since the velocity of the carriers is quite finite regardless of the field, and the depletion layer in the very lightly doped material is apt to be longer than the base width itself, the space charge of the minority carriers must eventually become important. How big $I_C$ must be before space-charge effects are discernible we shall see in a moment.

What will happen as the free-carrier space charge becomes important is pretty clear. The electron's charge will first neutralize the fixed donor-charge and then proceed to make the region effectively p-type. This means that the base side of the depletion layer must move to the right as the current and its attendant space charge is increased. This is the phenomenon of base stretching that was mentioned before. Now the Early effect is turned topsy-turvy. As long as the depletion region is moving through the relatively heavily doped base, it will move very slowly. This means that the base-width modulation induced by the current will not be too severe. However, as soon as enough charge has been accumulated in the now-inverted v layer to move the edge of the depletion region to the metallurgical junction, we can expect the base stretching to become very rapid. Accordingly, we can specify the current at which the field goes to zero at the metallurgical junction (pv) as the effective upper limit to useful transistor performance. Figure 8.20b shows this limiting case for a situation in which the field in the depletion layer is everywhere large enough to saturate the drift velocity. Under this last assumption—which is only a little bit crude—we may write the space charge in the v layer as

$$\rho = qN_D - I_C/A_e v_{sat} \tag{8.35}$$

Since all the terms in (8.35) are constant, we may readily solve for the voltage across the v layer. At the critical collector current, this voltage must be almost the entire collector voltage drop, as you can see in the figure. Thus, we have

$$I_{C\max} = A_e v_{sat} \left[ qN_D + 2\varepsilon V_{CB}/l^2 \right] \tag{8.36}$$

To avoid the resistive limitations of (8.34), $V_{CB}$ must be large enough to deplete the whole v layer. For example, if $N_D = 10^{15}/\text{cm}^3$ and $l = 3$ $\mu$m, it would require about 5 V to deplete the lightly doped layer. Working through (8.36) for such an example gives a current density of about 1000 A/cm$^2$ or 0.1 A for a typical small-signal transistor's area.

At this point you may begin to wonder why there is all this worry about maximum current-density. After all, silicon is not so very expensive—it makes up almost 26% of the earth's crust—so buy yourself another gram. This, in fact, is the solution if current capacity is the only consideration. Transistors are available on the market that can handle 1000 amperes, and they do indeed have junctions with very large areas. The only practical limitation on the size of the junction is the presence of randomly distributed imperfections in the wafers. When such a blemish lies in the junction area, the device will not operate properly, if at all. Thus, to have much yield, the device junctions must be small enough so that the entire junction is reasonably likely to be blemish free.

The problem imposed by the current capacity of a given transistor lies not in a need to get more coulombs from here to there but in the desire to operate transistors at the highest possible frequencies. For high-frequency operation, increasing the junction area makes things worse, not better. The issue of frequency versus collector current can be clearly seen in equation 8.24. $g_m$ is proportional to $I_C$, the diffusion component of $c_{b'e}$ increases linearly with $I_C$ and the depletion component somewhat less rapidly, and finally $c_{b'c}$ is reasonably independent of $I_C$ for currents well below the maximum. Thus, $f_T$ should increase with increasing current—up to a point. As the collector current approaches its maximum useful value for the particular collector voltage, $c_{b'c}$ will begin to grow rapidly with $I_C$, either because of collector debiasing or space-charge accumulation in the depletion region. $f_T$ will then begin to fall with increasing current, deteriorating very rapidly as the maximum current is approached. A curve of $f_T$ versus $I_C$ for several values of $V_{CE}$ is shown in Figure 8.21. The transistor is a 2N699, a high-voltage n-p-n double-diffused unit developed in the early 1960's. The high collector breakdown voltage definitely indicates a lightly doped collector. Included in an insert in the figure is a typical set of collector characteristics for the 2N699. These curves clearly show the $R_c$ type of limitation with a value of $R_c \simeq 58$ ohms. Having seen the various constraints on current, voltage, and power, you are probably curious about how high a frequency can be obtained, given high-frequency operation as the sole goal. Briefly, the answer is something of the order of 10 GHz. To achieve useful gain at such a high frequency, the minimax philosophy is normally adopted. As we whittle away at the factors that limit the bandwidth, eventually all of the charge storage mechanisms will begin to share in the limit. To have the maximum delay a minimum (hence *minimax*), the delay is distributed equally among the various elements. Then all of them are minimized together to achieve the ultimate frequency of operation. (If, on the other hand, you were aiming for operation at some intermediate point—say, 0.5 GHz instead of 5 GHz—it would probably not pay to strive for the higher-frequency response. Inevitably, pushing the frequency to its upper limit will cost you power-handling ability at lower values.)

To see the interaction between the various components, consider that:

**1.** Minimizing charge storage in the base by shortening the base width increases the charging resistance, $r_{bb'}$.

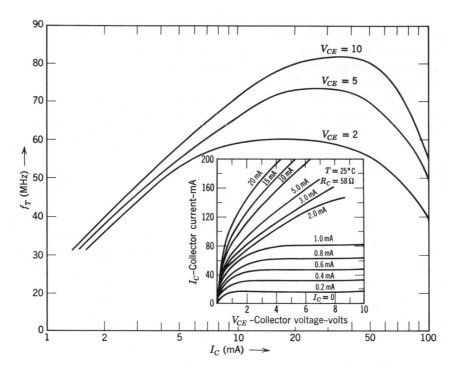

**Figure 8.21** $f_T$ versus $I_C$ for the 2N699 high voltage *n-p-n* silicon transistor. The insert shows the collector characteristics for the transistor. Collector debiasing is clearly evident.

**2**. Minimizing $r_{bb'}$ by doping the base relatively heavily decreases the injection efficiency, making charge storage in the emitter get larger. It also makes the emitter-depletion capacity get larger.

**3**. Minimizing the collector-depletion capacity by having a very long, very lightly doped collector layer maximizes either the series charging resistance, $R_c$, or the charge stored in transit across the collector-depletion layer.

Thus, all of the delay elements must be treated simultaneously. Push one down and another pops up. Obtaining an optimum solution is certainly far from trivial. Not only is the analysis difficult, but the result must also be realizable through a sequence of strongly interacting processing steps. The art and science of achieving such esoteric devices is obviously far beyond the scope of this book, so let us stop here.

## CHAPTER SUMMARY

Following Chapter 7's analysis of the structure factors in transistor design, we launched into a study of several electrical models of the small-signal
linear behavior of bipolar transistors. These models are based on a Taylor-
series approach in which any differentiable function may be represented over
a limited portion of the curve by the tangent to the curve. The "limited portion"
of the curve for transistors is generally big enough to allow the use of the small-
signal model for quite substantial swings about the bias point. Since the
models were to be linear, they could have been represented by standard
two-port $h$, $y$, or $z$ parameters. However, it is much easier to begin with a
model that represents each physical effect in a one-to-one fashion. The most
direct model of this kind is a differential representation of the Ebers-Moll
model. This is the so-called "T" model of the transistor. To make the model
reasonably accurate, it is necessary to include the variation of the Ebers-
Moll parameters with the applied voltage, but with a sophisticated physical
model at hand, this presented no obstacle. We found that we needed two
dependent current generators—$g_m$ and $g_r$—plus three resistors—$g_e$, $g_c$,
and $r_{bb'}$—to give an accurate picture of the transistor's low-frequency behavior. The addition of two capacitive elements—$c_e$ and $c_c$—permitted
the model to be used at reasonably high frequencies. At low frequencies, no
substantial difficulties in the modeling process were encountered. A simple
CB amplifier was analyzed with several interesting conclusions. We found
that the power gain in such a configuration was relatively low and that the
input impedance was too low to be useful in most applications. However,
for certain special applications such as series regulators or very-high-frequency amplifiers where the low input impedance and high collector breakdown voltages are substantial advantages, the CB circuit was the one to choose.

When we tried to extend the T model to high-frequency operation, we ran
into a difficulty that stems from a fundamental weakness in the physical
model. The fact that the base voltage drop is distributed along the overlap-
diode junction capacity means that our already inadequate model of the
base spreading resistance becomes even less effective as the frequency goes
up. We adopted the simplest compromise that had any hope of being adequate
and replaced the distributed RC line by a single CRC $\pi$-section. We were
then obliged to develop a heuristic rule for parceling up $c_{ob}$ between "internal"
and "external" capacities. A not overly convincing argument coupled with
an example from one particular transistor was used to establish a 50-50
split as a reasonable guess. The principal reason why a model based on such
weak analysis can serve our needs so adequately is that in most circuit applications, the exact value of $r_b$ is not very important. Most transistor amplifiers have bandwidths that are sufficiently below $f_\beta$ that lumping all of $c_{ob}$

together and throwing a hundred ohms into the base lead provides sufficient accuracy for circuit analysis.

We then turned to the CE circuit and the $\pi$ model. The $\pi$ model was obtained directly from the T model by requiring the $h$ parameters for both models in the CE configuration to be as equal as possible. An exact one-to-one correspondence was not possible, but the difference was extremely small in almost all cases.

Certain aspects of the CE circuit stand out very strongly in the $\pi$ representation. For example, the much higher power gain and input impedance (up a factor of $\beta$ compared with the CB circuit) as well as the substantially lower output impedance (down by roughly the same factor) are clearly evident. Several examples were used to develop methods for gain and bandwidth calculations. Unilateralization was accomplished by going over to a $y$-parameter representation and then eliminating the output voltage as an independent variable. The resulting circuit shows the Miller effect very clearly. The current, voltage, and power gains were obtained in analytical form from the unilateralized circuit. The power gain was also presented in the form of a Bode plot that gives both the power gain and the phase shift in the signal as a function of frequency. The bandwidth was defined as the frequency difference between the upper and lower 3 dB points on the Bode plot, and it was shown that if the gain was high enough, the gain-bandwidth product of the amplifier was a constant. If the gain was not sufficiently high, the gain decreased superlinearly with increasing bandwidth. This was highly reminiscent of the conclusions in our primitive switching analysis in the last chapter.

We then turned to the important subject of measuring the small-signal parameters and of scaling the parameters from measured bias points to other bias points or to other temperatures. The measurements are listed in Table 8.1. They are all quite simple in principle, although the proper instrumentation is frequently elaborate and subtle. Two new terms were introduced in the determination of the capacitive terms: these were $f_\beta$—the frequency at which the forward current gain has fallen to $1/\sqrt{2}$ of its low-frequency value—and $f_T$—the frequency at which the extrapolated plot of $|h_{fe}|$ versus $f$ goes to $|h_{fe}| = 1$.

As long as $\beta$ remained constant, we found that most of the small-signal conductances scaled with collector current. The principal exception was our old nemesis, $r_{bb'}$. The diffusion capacities also scaled linearly with $I_C$, but the depletion capacities were voltage rather than current dependent.

Variation of the parameters with temperature provided a bit more of a problem because $I_C$ is normally a function of temperature, the dependence being determined in large measure by the particular biasing arrangement employed. To avoid this indeterminacy, we examined the temperature effects under conditions that stabilized the collector current. This corresponds to a design goal in many circuit applications, so taking $I_C$ as a constant is a particularly suitable choice. Some terms, such as $g_m$, had very explicit temperature dependence, but even those that did not would be temperature dependent in silicon devices if they depended in any way on $\beta$. This was the result of the fact that in silicon, $\beta$ depended to an important degree on the minority-carrier diffusion length in the base. This coefficient goes up reasonably rapidly in most materials because the lifetime of the minority carriers generally increases much more rapidly than the diffusion constant decreases with increasing temperature. The net results of both the explicit and implicit temperature dependence for the $\pi$-model parameters is given in Table 8.2.

Part of the reason for learning to scale small-signal parameters is to be able to use the spec sheets published by the manufacturers. To see what one can and cannot get from a good set of spec sheets, we did several problems based on Fairchild's 2N708. The data sheets provided us with enough information to do a complete determination of the $\pi$ parameters and to observe the variation in $\beta$ that changes in $I_C$ and $T$ induced. The need to provide some form of bias stabilization was quite clear.

We took a brief look at the bias-stability problem by examining an amplifier's performance at room temperature and at 100°C with and without some emitter degeneration. Without the stabilizing resistor in the emitter circuit, the amplifier was in saturation at 100°C. With emitter degeneration, the collector current went up by a relatively small amount. After we had converted to a simpler equivalent circuit, we discovered what this DC stability had cost us—gain! The room-temperature gain had dropped from 45 dB to a measly 12 dB. Although we could get our AC gain back by bypassing the emitter resistor, DC stability and DC gain are antithetical.

Having developed electrical and physical models based on a structure that had appeared deus ex machina, in the beginning of the previous chapter, we returned to the topic of fabrication to see if the structure we had used was really what a transistor looked like. The double-diffusion and postalloy diffusion methods were discussed for a reasonably realistic problem: putting two isolated transistors on a single chip. This brought up such items as buried layers, mesas, and surface passivation. After the descriptions of the fabrication techniques were finished, it became quite clear that one of our model assumptions was incorrect. We had assumed that the base doping profile was essentially flat, but the diffusion profiles that give proper base and emitter widths would lead to strongly graded profiles. This brought up the final topic of the chapter—the drift transistor and high-frequency operation.

Handling the nonuniform doping turned out to be relatively simple, at least as long as low-level injection could be assumed. The drift field produced by the impurity-concentration gradient effectively shortens the base and lengthens the emitter. Both of these are very desirable results. The base width is shortened by a factor equal to the ratio of the average doping to the doping at the edge of the depletion region. The emitter is effectively lenthened by a similar factor.

The very short base and low collector capacity that can be achieved by combining the diffusion and epitaxy techniques suggested that extremely high frequencies were within the possible range of transistor technology. Closer inspection, however, showed that several substantial and fundamental problems arose in efforts to push $f_T$ well up into the microwave range. To push $f_T$ up, one wants to make $g_m$ as large as possible compared to $c_{c'e} + c_{b'c}$. This involves increasing the current to the point where noticeable saturation begins to set in. Unfortunately, the maximum useful current is decreased by just those steps that decrease $c_{b'c}$. The two factors that limit the current are collector debiasing resulting from ohmic drops in the lightly doped collector material and base stretching that arises from the space charge of the carriers in transit across the collector depletion region. Both of these would be ameliorated by doping the collector more heavily, but that step, unfortunately, would give not only a larger $c_{b'c}$ but also a lower collector breakdown voltage. Pushing the frequency limits into the microwave range becomes a very difficult exercise in making proper trade-offs between a large number of interacting structure and processing variables.

We are now ready to return to the "original" transistor—the FET. These are, for all practical purposes, completely different devices, although much of the physics and all of the manufacturing techniques of bipolar devices are employed in their analysis and realization. One unfinished topic is carried over from this chapter to the last: the rather mysterious fact that the $p^+n$ junction on a silicon surface breaks down in a rather sloppy fashion at anomalously low voltages. The reasons for this will become clear when we see how a MOSFET operates.

**Problems**

1. An $n$-$p$-$n$ silicon bipolar transistor is being operated at an average collector current of 1 mA. The voltage applied to the base (emitter grounded) is $V_{be} = 0.64 + 0.0075 \cos{(\omega t)}$ ($\omega = 10^4 \sec^{-1}$). Find the ratio of the r.m.s. values of the collector current at the second and first harmonics (i.e. $2\omega$ and $\omega$). (Let $kT = 0.026$ meV.)

*Answer:* $v(2\omega)/v(\omega) = 0.069$. Some caution must be exercised in obtaining this answer. The approach is simple enough. One writes

$$I_c = I_{es}e^{qV_{be}/kT} = I_Ce^{qv_{be}/kT}$$

Then, by expanding the exponential in a Taylor series, the answer is at hand. The caution is to note that in expanding terms such as

$$\cos^n{(\omega t)}$$

if $n$ is even (or odd), the expansion will contain all of the even (or odd) harmonics less than or equal to the $n$'th. Thus, expanding to the fourth, we get

coefficient of $\cos(\omega t)$:    $a + a^3/8 + \ldots$

coefficient of $\cos(2\omega t)$:    $a^2/4 + a^4/24 + \ldots$

where $a = qv/kT$. If $a$ gets reasonably large (i.e., if the input signal voltage becomes comparable to 26 mV) the higher-order terms can make an important contribution. For this problem, they make about 1 % difference.

2. If you are in the business of building integrated circuits that somebody has to sell, you may be very loath to specify too narrow a range of bias voltage. You might like a narrow range, your circuit might like a narrow range, but your customer will not like a narrow range. The usual out is to give a wide range for the customer, but design the circuit so that it stabilizes the voltage on the actual bias buss. You might think that a Zener diode would do the trick, but it is quite difficult to control the breakdown voltage on the integrated Zener. The circuit shown below behaves very much like a Zener diode. Furthermore, it is surprisingly independent of the hazards of processing. It is very easy to hold the *ratio* of the base circuit resistors to very close tolerances, and as you will see, that assures a reasonably stable voltage, $V_1$.

(*a*) The open-circuit output voltage $V_1$ of this circuit is expected to be 4 V if the input voltage is 5 V. What will this voltage be if the input voltage rose to 9 V?

(*b*) If $V_1 = 4$ V for $V_S = 5$ V, what would the value of $I_{es}$ be?

(*c*) If the value of $I_{es}$ were to increase in another unit to 5 times the value of the one under consideration above, give the sign of and a reasonable upper bound to the change in $V_1$ that would be observed at $V_S = 5$ V.

(*d*) By how much would $I_{es}$ vary if the temperature went from 300°K to 350°K?

(*e*) Find the small-signal resistance looking into the terminals $A$-$A'$ if $V_S = 5$ V.

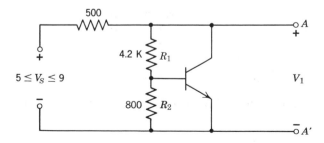

3. An interesting question arises as to whether one wants to operate a narrow epitaxial collector—such as one might find in a microwave transistor—just above or just below the epitaxial layer "punch-through" voltage. Consider a transistor with a uniform base layer doping of $10^{17}$ cm$^{-3}$, a collector epitaxial layer doping of $N_{D_1} = 10^{15}$ cm$^{-3}$, and a substrate doping of $N_{D_2} = 10^{19}$ cm$^{-3}$. Let the width of the collector epi layer be 4 $\mu$m. Find the ratio of the common-emitter output conductance just below punch-through to the conductance just above punch-through.

*Answer:* The ratio of the conductances is the same as the ratio of $\partial W_b/\partial V_{CE}$ evaluated just above the punch-through to $\partial W_b/\partial V_{CE}$ evaluated just below punch-through. That ratio is given by $N_{D_1}(N_{D_2} + N_A)/N_{D_2}(N_{D_1} + N_A)$. In view of the fact that $N_{D_1} \gg N_A \gg N_{D_2}$, the ratio becomes $N_A/N_{D_2} = 10^2$.

4. Some properties of a silicon *n-p-n* transistor are measured at 300°K DC bias point: $I_c = 1$ mA, $V_{CE} = 15$ V. They are

$$h_{FE} = 100$$
$$V_{B'E} = 0.62 \text{ V}$$
$$C_{ob} = 10 \text{ pF}$$
$$\omega_T = 3.64 \times 10^8/\text{sec}$$

(*a*) If the collector current were increased to 3 mA, which of these values will change and by how much?

(*b*) If $V_{CE}$ were decreased to 10 V, what changes would you expect?

(*c*) If the temperature were increased to 325°K with the collector current and voltage held constant, what changes would you expect?

5. An *n-p-n* silicon transistor has a 12 $\mu$m wide uniformly doped collector region containing $5 \times 10^{15}$ Sb atoms/cm$^3$; a graded base whose width is 1.4 $\mu$m and whose acceptor concentration is exponentially graded from $2 \times 10^{18}$ to $5 \times 10^{16}$ B atoms/cm$^3$ across that width; and an emitter containing $5 \times 10^{19}$ P atoms/cm$^3$ that is 2.4 $\mu$m wide. At 300°K, the relevant properties in each region are:

| | Excess-carrier lifetime | Minority-carrier mobility |
|---|---|---|
| collector | $10^{-6}$ sec | 400 cm$^2$/Vsec |
| base | $5 \times 10^{-7}$ sec | 1000 cm$^2$/Vsec |
| emitter | $5 \times 10^{-8}$ sec | 100 cm$^2$/Vsec |

The emitter area is $3 \times 10^{-3}$ cm$^2$.

(*a*) Determine the maximum value of $h_{FE}$ for this transistor at 6 V of collector bias.

*Answer:* The *effective* base width is only 0.37 $\mu$m, so the current gain is limited almost entirely by reverse injection. That term gives an $h_{FE}$ of 80.

**(b)** Would high-level injection set in before $h_{FE}$ reached 90% of this value?

*Answer:* No. The extremely narrow depletion region gives very little recombination current. With the heavy doping at the emitter side of the base, high-level injection will not set in until the voltage reaches about 0.85 V. The depletion current falls below the 10% level at about 0.52 V.

## CHAPTER 9

# DEVICES THAT WORK PRINCIPALLY BY DRIFT: (A) THE JFET

The bipolar transistor of the last chapter is certainly an extremely un-obvious device. So obscure are the principles that underlie its action that it is almost impossible to give a meaningful "simple" explanation of its operation. The fact is that the bipolar transistor was not invented; it was discovered. The recognition of what was going on earned three men a Nobel Prize.

Most prior efforts to invent a solid-state current valve had been directed to methods of modulating the conductance of a channel. The resistance of a conducting channel in an electron conductor can usually be written:

$$R = L/\sigma A = L/q\mu nA \qquad (9.1)$$

where $L$ is the channel's length, $A$ is its cross-section, and $\sigma$ is the conductivity of the material. With the exception of $q$, all of the parameters in (9.1) are potentially variable. The task in inventing a useful current valve is to find a convenient method of modulating $R$ that gives a great deal of modulation with a small expenditure of power. There are several esoteric devices that have made very successful use of variations in $\mu$ (e.g., strain gauges and Gunn and Read diodes), but for the most part, variable-resistance devices have been based on inducing changes in $n$ and/or $A$. Although one may induce such changes optically or thermally, the most useful devices are those that vary the resistance of the channel by means of an electric field. Such devices are called *field-effect transistors* (FET's). They are also called *unipolar* transistors since, in contrast to their distant cousin, the bipolar transistor, conductance through the FET requires the flow of carriers of only one polarity.

FET's are currently the "latest thing" in solid-state electronics, and you might get the idea that they are a very recent development. Actually, their history dates back a remarkably long time. It is only success that is recent. We could argue that the traditional vacuum tube (or vacuum valve, as the

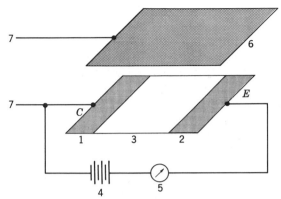

**Figure 9.1** An early conception of a field-effect transistor. This is a drawing from British Patent 439,457 of O. Heil. Heil described the device as comprising a thin layer (3) of semiconducting material (such as tellurium, iodine, or one of several oxides) with ohmic contacts (1 and 2) at either end. An insulated metallic layer (6) is placed over the semiconducting layer, serving to modulate (by an unspecified mechanism) the conductivity of the underlying semiconducting layer.

British persist in calling it) is an FET. That would place the invention of the FET not long after the turn of the century.

The first efforts to develop a solid-state FET led to patents in the 1930's [Lilienfeld, U.S. patents, 1,745,175 (1930), 1,877,140 (1932), and 1,900,018 (1933), and Heil, British patent 439,457 (1935) ] but no useful devices. These first solid-state FET's were strips of semiconducting material with an insulated metal layer on top of them. Figure 9.1 shows Heil's conception. The idea was that one could attract or repel majority carriers from the semiconducting strip by applying a voltage between the strip and the semiconducting substrate. An effect was observable, but it was much less than one would have expected from simple electrostatic theory. It appeared as if the charge that moved in and out of the semiconductor under the influence of the metal-to-substrate voltage was not free to participate in conductance across the strip.

To explain this anomalous lack of conductivity, John Bardeen postulated the existence of a very high density of localized surface states that trapped the surface carriers and prevented their participation in the conductance of the channel [J. Bardeen, *Phys. Rev.* **71**, 717 (1947)]. A year later, Shockley and Pearson demonstrated the existence of these states [*Phys. Rev.* **74**, 232 (1948) ], but a chance discovery in those investigations almost ended progress on the FET. It was in an examination of the surface states that Bardeen and Brattain placed two diodes side by side on the same germanium sample, saw some very unexpected behavior, realized what was going on, and announced the bipolar transistor to the world [J. Bardeen and W. H. Brattain, *Phys. Rev.* **74**, 230 (1948) ]. With the bipolar transistor off and running, work on unipolar transistors faded into the background.

Then, in 1952, Shockley went against the tide a bit and invented a very practical FET [*Proc. IRE* **40**, 1365 (1952) ]. It was reduced to practice within a year [Dacey and Ross, *Proc. IRE* **41**, 970 (1953) ]. However, even though this new FET was based on the same technology as the bipolar transistor, it required a higher degree of processing sophistication than was available in the 1950's. It could not be produced competetively with the bipolar devices that were readily available. Once again the FET languished in obscurity.

In the early 1960's two factors led to the reemergence of the FET. The first was the development of the planar and epitaxial technologies, which permitted the inexpensive manufacture of high-quality FET's. The second, which really derived from the first, was the realization that whole circuits rather than just components could be produced by the planar process. The economics of such integrated circuits strongly favor devices that require less wafer area and fewer processing steps. The original FET—the one pursued by Heil and the solid-state group at Bell Telephone Laboratories— offered the most promise, although high yields and high reliability were to take almost a decade of research and development. Today, both bipolar and unipolar transistors enjoy much favor, and it seems very unlikely that either will be eclipsed for some time to come.

Two forms of FET enjoy current vogue. The *junction field effect-transistor* (JFET) of Shockley obtains current control by moving a depletion region into and out of the conducting channel. Thus, the JFET depends on the variation of $A$ in (9.1). The other form, variously called the IGFET (insulated-*g*ate FET), MOSFET (*m*etal-*o*xide-*s*emiconductor FET), MISFET (*m*etal-*i*nsulator-*s*emiconductor FET), or MOST (*m*etal-*o*xide-*s*emiconductor *t*ransistor), modulates both $n$ and $A$ in (9.1). The modulation is obtained through capacitive coupling not unlike that envisaged by its early inventors. In operation, the two devices bear strong resemblance to each other, although there are some very important differences. We shall begin our analysis with the JFET because it is the easier of the two to understand. Then, after a close look at the nature of the semiconductor-insulator interface, we shall turn to the IGFET, which is commercially the more important of the two.

## THE JUNCTION FET

Figure 9.2 shows the structure of a typical modern planar-junction FET. It would be made in much the same fashion as a bipolar transistor with the principal difference being in the mask shapes. One would start with an "epi"

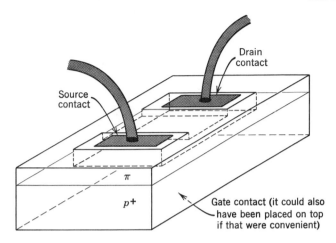

**Figure 9.2** A planar epitaxial junction FET. The structure comprises a $p^+$ substrate with a $\pi$ epitaxial layer on top. Into this has been diffused an $n$-type box with a $p$-type crossbar diffused on top of that. The resulting structure can be thought of as a tube of $n$-type material surrounded by $p^+$ material. The dimensions have been grossly distorted to improve the drawing's clarity. The channel is normally short, shallow and very broad.

wafer having a $\pi p^+$ profile, and then diffuse in a rectangular $n$-type region, driving the junction all the way to the $p^+$ layer. This is followed by a $p$-type diffusion that extends out beyond two of the sides of the $n$ region. This extension connects the upper and lower $p$ layers. The resulting structure is essentially a rectangular tube of $n$-type material running through a sea of $p$-type material. Leads (labeled *source* and *drain*) are connected to the two ends of the tube and a third connection (called the *gate*) is made to the $p$-type material.

Figure 9.3 shows a section of the resulting structure with the notation and coordinates that we are going to use in the analysis. Keep in mind that, compared to any of the dimensions shown in the figure, the breadth ($z$-direction) of the $n$-type tube (the *channel*) is enormous. The breadth may be fully two and a half orders of magnitude larger than the width. However, unlike the bipolar transistor's analysis, the planar structure in this case does not generally allow a completely one-dimensional approximation. Our primary concern in the FET is with the *transverse* flow of majority current from source to drain. This, like the transverse flow in the base, usually must be analyzed in at least two dimensions.

Thinking of the $n$-type region as a tube suggests how this device is going to operate. If one applies reverse bias to the junction between the channel

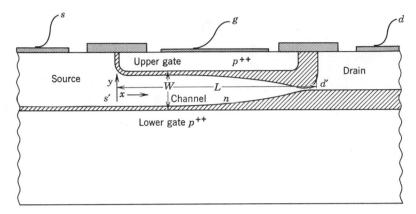

**Figure 9.3** A section drawing of an $n$-channel junction FET. The channel is shown prior to pinch-off but with substantial drain current flowing. The channel length ($L$) and maximum width ($W$) are shown. The breadth of the channel ($B$) is measured in the $z$-direction. The hatched areas are the depletion regions.

and its surroundings, the depletion region moves into the $n$-type region and reduces its cross-section. This is quite analogous to pinching a water hose; pinch hard enough and the flow through the hose stops completely. Thus, by varying the bias between the gate and the channel, we can vary the conductance between the source and drain. Note that the gate draws almost no current, since the junction is reverse biased. At low frequency it will take very little power to control the current in the channel, so junction FETs should show large power gains.

The analysis of the current flow through the channel is conveniently divided into three regions of operation. The division is based on the magnitude of the voltage across the channel, $V_{ds}$. If that voltage is very small compared to the voltage necessary to pinch the channel significantly, the channel's resistance is independent of the drain current, $I_d$. This situation is called the *linear mode* of operation. When operated in this mode, the FET serves as a very useful component—a voltage-variable linear resistor. At the other extreme, if the voltage between gate and drain is large enough to pinch the channel completely closed at the drain end, the drain current becomes almost entirely independent of $V_{ds}$. (This should NOT be intuitively obvious! If you pinch a hose shut, no water flows. But as you shall see, our hose analogy is not complete.) In this mode of operation, the FET acts like a voltage-controlled current source. The current-source mode is frequently called *pentode operation* because vacuum-tube pentodes have somewhat similar characteristics. It is also called the *saturated* mode, since $\partial I_d / \partial V_{ds} \simeq 0$. The

third mode of operation—called the *transition region* or the *triode mode*—
is intermediate between these first two. In the transition region, the drain
current is dependent on the drain voltage, but in a nonlinear way. The nonlinearity results from the interaction between the ohmic drop in the channel
and the channel width; the effort to drive more current through the channel
causes the channel to pinch off more. Thus, the current increases less and
less rapidly with increasing drain voltage until the channel pinches off entirely
at the drain end. The linear and transition modes permit a reasonably direct
analysis; the simple but thorough saturated-mode analysis has yet to see
the light of day. Our plan of attack is to do the simple analysis that describes
the drain characteristics prior to pinch-off. With only a little sophistication,
we will get very accurate results from this analysis. Then we will patch together a solution for the pentode mode that will do a pretty good job of showing what to expect, although its analytical completeness will leave something
to be desired.

## DRAIN CHARACTERISTICS OF THE JUNCTION FET

Figure 9.3 shows a vertical section through a junction FET operating
in the transition mode (i.e., the drain end is much more pinched than the source
end). The aspect ratio of the channel, with $B \gg L > W$, assures us that this
section would look the same no matter where along the $z$-axis it was taken.
In other words, the variation in the current density is two dimensional.
Carriers enter the channel from the source, drift through the channel under
the influence of the electric field in the channel, and emerge into the drain
region. The voltage drop along the channel causes the channel to pinch down
progressively from source to drain.

Our primary objective is to obtain the relationship of the drain current,
$I_d$, to the drain-to-source voltage, $V_{ds}$. If we make two reasonable approximations, our task is quite easy. The first is one of arithmetic convenience: Let
us assume that the entire volume of the channel is uniformly doped. The
second assumption looks innocent enough, but it will come back to haunt
us very shortly: Let us assume that the channel is pinched sufficiently gradually from left to right in the figure that the field lines can be considered to
be almost parallel with the $x$-axis. This *gradual-channel* assumption permits
us to calculate the resistance of the channel very simply, but it does not hold
even approximately at the narrow end of a pinched-off channel. More of
this will appear at the conclusion of our preliminary analysis.

Our first task is to compute the drain characteristics of the JFET. These
will serve the same useful functions that the collector characteristics did
for the bipolar transistor. The voltage from drain to source can be divided

into three reasonably distinct components. The principal drop is across the thin, high-resistivity channel—that is, from $d'$ to $s'$ in Figure 9.3. The section from the ends of the channel to the respective ohmic contacts also comprises resistive material with associated potential drops. Although these drops are small and readily represented by simple ohmic resistances in series with the channel, they are important enough to require their inclusion in even a first-order analysis. What we will do is to determine $V_{d's'}$ versus $I_d$ for constant $V_{gs'}$ and then add the resistive drop a posteriori.

Having assumed that the channel width changes slowly along the $x$-axis and not at all along the $z$-axis, and that the channel is uniformly doped, we may write the voltage drop, $dV$, across a differential length of channel, $dx$, as

$$dV = I_d \, dR = I_d \, \frac{dx}{2yB\sigma} \tag{9.2}$$

where $y$ is the half width of the channel at the point $x$, $B$ is the breadth of the channel ($z$-direction), and $\sigma$ is the conductivity of the channel material. To obtain the channel width, $2y$, we must subtract the depletion regions on either side of the channel from the metallurgical-channel width, $W$. Since the $p$ side of the junction is very heavily doped, the depletion region is almost entirely on the $n$ side. Then, from (3.51), noting that $N_A \gg N_D$, we have the channel width as

$$2y = W - 2 \sqrt{\frac{2\varepsilon}{qN_D}[V(x) + V_D - V_{gs'}]} \tag{9.3}$$

where $V(x)$ is the voltage from the point $x$ to the source end of the channel ($s'$) and $V_D$ the diffusion (or "built-in" or contact) potential. To find $I_d$ versus $V_{d's'}$, we put (9.3) into (9.2), rearrange a bit, and then integrate from $s'$ to $d'$:

$$\int_0^{V_{d's'}} \left\{ 1 - \left( \frac{8\varepsilon}{qN_D W^2} \right)^{1/2} (V(x) + V_D - V_{gs'})^{1/2} \right\} dV = \frac{I_d}{\sigma BW} \int_0^L dx \tag{9.4}$$

which, upon integration, gives

$$I_d = \frac{\sigma BW}{L} \left\{ V_{d's'} - \frac{2}{3} \left( \frac{8\varepsilon}{qN_D W^2} \right)^{1/2} [(V_{d's'} + V_D - V_{gs'})^{3/2} - (V_D - V_{gs'})^{3/2}] \right\} \tag{9.5}$$

Noting that the conductance of the metallurgical channel is $G_0 = \sigma BW/L$ and that the *total* potential that just pinches the channel shut is [from (9.3)]: $V_p = qN_D W^2/8\varepsilon$, we get a neater version of (9.5) and one more closely related to terminal measurements:

$$I_d = G_0 \left\{ V_{d's'} - \frac{2}{3} \frac{1}{V_p^{1/2}} \left[ (V_{d's'} + V_D - V_{gs'})^{3/2} - (V_D - V_{gs'})^{3/2} \right] \right\} \qquad (9.5')$$

Keep in mind that the junction is reverse biased, so $-V_{gs'}$ is a positive number. A look at (9.5') shows that what we expected to happen does. Increasing $V_{d's'}$ increases $I_d$ at a rate that is generally less than linear; decreasing $V_{gs'}$ (increasing the reverse bias) reduces the drain current.

Although (9.5') is a perfectly reasonable function, it is easier to see what is going on by looking at the differential conductance:

$$g_{d's'} = \partial I_d / \partial V_{d's'} = G_0 \left[ 1 - \left( \frac{V_{d's'} + V_D - V_{gs'}}{V_p} \right)^{1/2} \right] \qquad (9.6)$$

For small drain voltages ($\ll V_p$), the channel conductance is independent of drain voltage and we have a linear resistor that we may adjust by varying $V_{gs'}$. As the drain voltage rises, it begins to cause some pinching at the drain end of the channel, so the conductance starts to decrease. This state of affairs continues until the total voltage from drain to gate reaches the pinch-off voltage. When that point is reached, $g_{d's'}$ goes to zero. However, although the *differential* conductance vanishes, the drain current does not. Figure 9.4 shows a plot of equation 9.5' using typical values for the three constants. The drain curves are drawn out to the point where $g_{d's'}$ vanishes, and, as you can see, the drain current at pinch-off can assume a wide variety of values depending on the value of gate bias, $V_{gs'}$.

This brings us to the obvious question: What happens to the drain current if the drain voltage is increased beyond the point where the drain end of the channel pinches off? The question may be obvious, but the answer is not. Briefly, the answer is that the drain current becomes almost independent of the drain voltage. Before we tackle the problem of explaining why this should be, note that if $I_d$ were completely independent of $V_{d's'}$, our first job would be done. To complete the characteristic drain curves, all we would have to do is take the curves drawn in Figure 9.4 and extend them horizontally to the right. The JFET would be acting like a voltage-controlled current source, a most useful device. This is not far from the truth, although there will always be some finite output conductance. Since the ultimate limit on the gain that the JFET can deliver is determined by its output conductance, it behooves us to get some rudimentary concepts on what to expect.

## OPERATION IN THE SATURATION REGION

A large number of very clever people have taken a crack at the JFET output conductance problem, and yet the problem still remains quite obdurate.

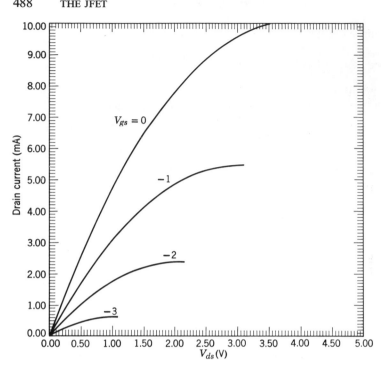

**Figure 9.4** Calculated drain characteristics of a JFET for the linear and transition regions. This is a plot of equation 9.5' with zero source and drain resistance and $G_0 = 0.01$ mho, $V_D = 0.9$ V, and $V_p = 5$ V.

In a recent rather uncompromising numerical analysis of JFET behavior beyond "pinch-off," Kennedy and O'Brien [*IBM J. of Res. Develop.* **14**, 95 (1970)] found that they could make no all-inclusive generalizations. There simply are no neat little packages to tie the whole problem up in. What we are going to do here is to examine some of the more important phenomena that are taking place and then follow through to a numerical solution that should approximate long channels.

We begin with one of the implications of high conductivity. In regions of high conductivity, electric fields are supported by current flow ($\mathbf{E} = \sigma \mathbf{J}$) with negligible field support by charge accumulation or polarization of the medium ($\nabla \cdot \mathbf{E} = 0$). Conversely, in regions of low conductivity, fields are associated with charge accumulation and polarization. Obviously, in intermediate situations, intermediate conclusions hold.

It is important to convince yourself that these statements center on the ability of mobile charge to respond vigorously to an applied field. If the nor-

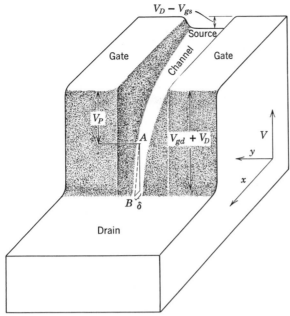

**Figure 9.5***a* Potential-energy diagram for electrons in *n*-channel JFET operating in the saturation (pentode) mode. The very steep regions are depleted of free carriers. Electrons enter the channel from the source end, accelerate as they move toward the drain, and reach the saturated-drift velocity in the vicinity of the "pinched-off" end of the channel.

mally mobile charges in a particular region have achieved their saturated drift velocity, they do not respond to increases in the field, and that portion of the material is, at least incrementally, an insulator. Such "incremental insulators" can have substantial current flowing in them, but the current is field independent. In fact, since the electrons (or holes) are incrementally immobile, they may serve as origin or termination points for field lines. This they certainly cannot do in regions of substantial differential conductivity. (We have seen a similar role for normally mobile charges in the bipolar transistor when they served to debias and stretch the base during their transit across the collector depletion region.)

Consider then the following description of a saturated JFET as illustrated by Figure 9.5*a* and *b*. Let us assume that the JFET is being operated well into the saturation region so that $V_{ds}$ may be quite large while $I_d$ is not very large at all. Our task is to account for the voltage drop from drain to source and to see how the current must respond to variations in that voltage. Both figures illustrate the potential distribution to be expected; one is a per-

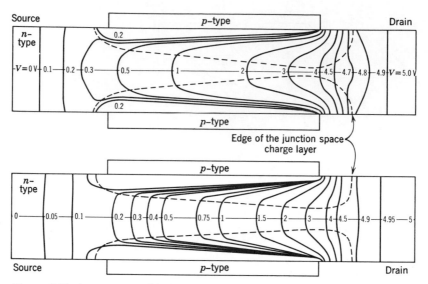

**Figure 9.5***b* A computer-aided plot of the voltage distribution in a JFET. This is a "contour map" of Figure 8.5*a*. The upper plot assumes constant mobility. The lower plot includes the effects of field-dependent mobility. [From D. P. Kennedy and R. R. O'Brian, *IBM J. of Res. Develop.*, **14**, 95 (1970).]

spective drawing, the other a topograph. Proceeding from source to drain, electrons enter into the relatively horizontal source region that is effectively a pool of electrons. This pool is drained through the channel, which is quite wide at the source end, but narrows progressively as it goes toward the drain. (If you try a water analogy here, put a top on the channel to make its conducting cross-section proportional only to its width.) The flow density must get progressively larger to account for the total current, so the slope increases toward the drain. This ohmic situation can continue for a while, but by the time the channel has fallen a few volts below the source, the channel is so steep that the electrons can no longer accelerate to follow it. In this steep region, the flow has saturated. We may now draw a steep cliff to connect the saturated channel end with the lower pool of the drain. The question is, how steep? The saturated region will support as large a field as there is charge to create.

The steepness of the cliff is limited only by avalanche breakdown, but achieving a given steepness requires a sufficiently dense supply of both positive and negative charge to form a dipole layer of the proper field strength. Just as in the depletion region of a *p-n* junction, a diffuse layer of charge leads to a relatively weak field and long depletion region. The supply of positive charge in our *n*-channel device is quite apparent; it is the ionized donor atoms. If the saturation point occurs some distance from the heavily doped region of the drain—say, 1 or 2 $\mu$m—then the density of donor atoms is that of the channel itself.

The source of negative charge is not so obvious. One source is the acceptors in the gate material. Since these are on the side of the channel, only a portion of each field line contributes to the voltage drop in the channel. The field lines must curve up through the channel and spread out to get to the gate charge. The upper topograph in Figure 9.5b shows such a situation. That figure was calculated with constant mobility, so there is no velocity saturation and no charge accumulation within the channel itself.

If we allow velocity saturation, the charge in transit through the saturated-drift region can provide a substantial portion of the required negative charge. The local charge density is simply the number of electrons in excess of the number of donors. This situation is illustrated in the lower topograph of Figure 9.5b. Note that the "cliff" is much steeper in this figure and that the field lines tend to go straight up the channel. In the upper figure, where the field lines get out of the channel, much more of the drain voltage drop occurs in the gate depletion region and not in the narrow portions of the channel. In the lower figure, the drop is principally in the saturated-drift region. Judging from the drop in the ohmic region in each figure, the case that properly accounts for velocity saturation shows only about half of the current that is seen in the other figure. It should be clear that we are apt to get very wrong numbers unless we include some of the features of velocity saturation in our calculations. On the other hand, unless we adopt some simplification, a large computer will have to be included as an appendix to this book. Consider the following model of the saturated portion of the channel.

Let us assume that the transition to saturated-drift velocity is aburpt (i.e., that the differential mobility has only two values, $\mu$ or 0). With this assumption, the output current saturates when the channel has pinched down to the point where, with that particular current, the carriers have reached the saturation velocity, $U_s$. It is easy enough to specify the channel width at that point. It is

$$2y_s = \frac{I_{d\,sat}}{U_s B q N_D} \tag{9.7}$$

At the beginning of saturation, the "cliff" has just begun to form. Further increases in the drain voltage will be absorbed principally by this cliff, which will spread into the drain and the channel to acquire the charge to support the field. The charge in the channel arises from the increase in current traveling through the saturated-drift region. The increase in current comes about because the unsaturated portion of the channel has been shortened by the expansion of the saturated section. Since these two events are coupled, we may obtain the length of the saturated section, $\delta$, in terms of the drain voltage in excess of the saturation voltage. $\delta$ is defined in Figure 9.5a. The channel is saturated for the fields (slopes) between points $A$ and $B$ in the figure.

First note that if the voltage drop across the unsaturated portion of the channel remains relatively unaffected by the increases in drain voltage, the integrals in (9.4) may be performed over the new length $L - \delta$. The result is

$$I_d = I_{d\text{sat}} / (1 - \delta/L) \tag{9.8}$$

If we may now assume that the channel width remains constant at the value given by (9.7), we may write the excess charge in terms of the current in excess of $I_{d\text{sat}}$:

$$n - N_D = \frac{I_d - I_{d\text{sat}}}{U_s B q 2 y_s} = \frac{I_{d\text{sat}}}{2 y_s} \left[ \frac{(I_d/I_{d\text{sat}}) - 1}{U_s B q} \right] = \left( \frac{I_d}{I_{d\text{sat}}} - 1 \right) N_D$$

The last step used (9.7). By combining the above equation with (9.8),

$$\frac{n - N_D}{N_D} = \frac{\delta/L}{1 - \delta/L}$$

From our standard-depletion-region formula (3.52), we may express $\delta$ in terms of the voltage across the depleted-drain and saturated-channel region as

$$\delta^2 = \frac{2\varepsilon (V_d - V_{d\text{sat}})}{q} \frac{N_D}{N_D^2 \left( \dfrac{\delta/L}{1 - \delta/L} \right) \left( \dfrac{1}{1 - \delta/L} \right)}$$

or, upon solving for $\delta$:

$$(\delta/L)^3 = \frac{2\varepsilon (V_d - V_{d\text{sat}})}{q N_D L^2} (1 - \delta/L)^2 \tag{9.9}$$

For long channels, (9.9) says the saturated-channel length will behave like the depletion region in a graded junction, growing as the cube root of the applied voltage. For shorter channels, (9.9) predicts an even slower rate of growth. Notice that $\delta < L$ for all $V_d$; no "punch-through" takes place. Equation 9.9 also shows the intuitively satisfying result that $\delta$ decreases with increasing doping. Now let us see what (9.9) says about the output conductance, $g_{ds} = \partial I_d / \partial V_{ds}$.

By combining (9.8) and (9.9), we may obtain

$$g_{ds} = \frac{I_{d\text{sat}}}{(1 - \delta/L)^2} \frac{\partial \delta/L}{\partial V_{ds}} = I_{d\text{sat}} \left( \frac{2\varepsilon}{q N_D} \right) \frac{(1 - \delta/L)}{3\delta^2 - \delta^3/L} \tag{9.10}$$

For the common case where $\delta \ll L$, the simpler form directly from (9.9) is

$$g_{ds} \simeq \left( \frac{2\varepsilon L}{qN_D} \right)^{1/3} \frac{I_{d\,sat}/3L}{(V_d - V_{d\,sat})^{2/3}} \tag{9.10'}$$

Equation 9.10′ shows the expected result that the output conductance decreases with increasing doping, channel length, and bias.

It should be very carefully noted that (9.10′) will give a somewhat pessimistic estimate (too large) of $g_{ds}$. In deriving (9.10′), the entire voltage drop was assigned to the drifting space charge. The field between the gate-depletion region and the drain will make up a portion of the field strength parallel to the axis of the channel. This will decrease the required amount of charge in the saturated portion of the channel, reducing $\delta$ and $g_{ds}$ accordingly. There is no reason to assume that this gate field is negligible. In fact, a crude guess would put its contribution to the drain-source voltage drop at about 20% to 50% of the total voltage from gate to drain. We will cover this sort of "sharing the burden" in greater detail for the MOSFET, so for the moment let us content ourselves with a look at the pessimistic predictions of (9.10′).

Since there are four relatively independent variables in (9.10′), it is easiest to see what sort of numbers to expect by picking some values that are typical in practice and seeing what the various combinations of the four variables yield. The results for a silicon JFET are presented in Table 9.1. If you compare these predictions with Figure 9.6, you will see that the values from (9.10′) are plausible. The three different members of the same "family" of JFET's presented in Figure 9.6 appear to differ principally in channel thickness.

**Table 9.1.** Values of $g_{ds}$ and its Reciprocal, $r_d$, for Several Values of Drain Current, Drain Voltage, Channel Length, and Channel Doping[a]

| $N_D$ (cm$^{-3}$) | | | \multicolumn{4}{c}{$10^{15}$} | | | | \multicolumn{4}{c}{$10^{16}$} | | |
|---|---|---|---|---|---|---|---|---|---|---|
| $L$ ($\mu$m) | | | 10 | | 20 | | 10 | | 20 | |
| $I_{d\,sat}$(mA) | | | 1 | 5 | 1 | 5 | 1 | 5 | 1 | 5 |
| $V_d - V_{d\,sat}$ | 1 V | $g_{ds}$ | 79 | 395 | 49.7 | 248 | 36.7 | 184 | 23.1 | 116 $\mu$mhos |
| | | $r_d$ | 12.7 | 2.5 | 20.1 | 4.02 | 27.2 | 5.45 | 43.3 | 8.66 K ohms |
| | 15 V | $g_{ds}$ | 12.8 | 64 | 8.02 | 40.1 | 5.92 | 29.6 | 3.73 | 18.6 $\mu$mhos |
| | | $r_d$ | 78.4 | 15.6 | 125 | 25 | 169 | 33.8 | 268 | 53.7 K ohms |

[a] These values are calculated with equation 9.10. Since the gate-to-drain field is not included, the conductances so calculated are somewhat high.

# GENERAL PURPOSE AMPLIFIER
## SILICON EPITAXIAL JUNCTION
### N-CHANNEL FIELD EFFECT TRANSISTOR

| C680 | C681 |
| C682 | C683 |
| C684 | C685 |

The C680 through C685 are general purpose FET's which combine a high GM/I$_{DSS}$ ratio with low gate capacitance and leakage currents. These devices are particularly well suited for use in high impedance amplifiers from Sub Audio to Low R.F. frequencies. These devices can also be used as bilateral resistive elements for switching and voltage controlled resistance applications. They are manufactured by the epitaxial junction process which combines the advantages of alloy, epitaxial, and planar techniques. Both devices are bed mounted, oxide passivated, and utilize a unique gold bonding process which eliminates "purple plague", resulting in extreme ruggedness and parameter stability.

## AVAILABLE IN TO-18 ONLY

### ELECTRICAL DATA ABSOLUTE MAXIMUM RATING

| | | |
|---|---|---|
| Drain to Source Voltage | BV$_{DSO}$ | 30 Volts |
| Drain to Gate Voltage | BV$_{DGO}$ | 30 Volts |
| Gate to Source Voltage | BV$_{GSO}$ | −10 Volts |
| Gate Current | I$_G$ | 50 mA |
| Junction Temp (operating and storage) | T$_J$ | −65°C to +200°C |
| Power Dissipation (free air) | P$_D$ | 200 mW |
| Lead Temp (@ ⅟₁₆" ± ⅟₃₂" from case) | T$_L$ | 240°C for 10 sec. |
| Derating Factor (free air) | D$_F$ | 1.14 mW/c° |

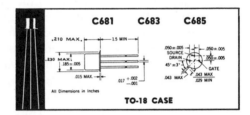

C681    C683    C685

TO-18 CASE

## ELECTRICAL CHARACTERISTICS: T$_A$ = 25°C (UNLESS OTHERWISE STATED)

| Parameter | Symbol | Condition | C680 – C681 MIN. | C680 – C681 MAX. | C682 – C683 MIN. | C682 – C683 MAX. | C684 – C685 MIN. | C684 – C685 MAX. | All Typ. | Units |
|---|---|---|---|---|---|---|---|---|---|---|
| Gate Leakage Current | I$_{GDO}$ | V$_{GD}$ = −15 V | — | 10.0* | — | 10.0* | — | 10.0* | 0.2 | nA |
| Zero Gate Voltage Drain Current | I$_{DSS}$ | V$_{DS}$ = 15 V, V$_{GS}$ = 0 | 0.08 | 0.4 | 0.4 | 1.60 | 1.50 | 6.0 | — | mA |
| Pinch-Off Drain Current | I$_{D(off)}$ | V$_{DS}$ = 10 V, V$_{GS}$ = V$_{PO(max.)}$ | — | 10.0 | — | 10.0 | — | 10.0 | — | μA |
| Transconductance | Gm | V$_{DS}$ = 10 V, V$_{GS}$ = 0, f = 1 KC | 200 | 500 | 400 | 1000 | 600 | 1500 | — | μmho |
| Pinch-off voltage | V$_{PO}$ | V$_{DS}$ = 10 V | 0.5 | 2.5 | 1.0 | 5.0 | 2.0 | 10.0 | — | Volts |
| Gate to Source Cap. | C$_{GS}$ | V$_{GS}$ = −10 V, f = 140 KC | — | — | — | — | — | — | 3 | pfd |
| Gate to Drain Cap. | C$_{GD}$ | V$_{GD}$ = −10 V, f = 140 KC | — | — | — | — | — | — | 1.5 | pfd |
| Gain-Bandwidth Product | Ft | V$_{DS}$ = 15 V, V$_{GS}$ = 0 | — | — | — | — | — | — | 50 | Mc |
| Output Admittance | Y$_{OS}$ | V$_{DS}$ = 15 V, V$_{GS}$ = 0 | — | 10 | — | 20 | — | 60 | — | μmho |
| Spot Noise Figure | N.F. | R$_G$ = 1 MΩ, f = 1 KC | — | — | — | — | — | — | 2.0 | db |

*For 1 nA max. I$_{GDO}$ add A suffix to part number

## CRYSTALONICS
### A TELEDYNE COMPANY

147 Sherman St. • Cambridge, Mass. 02140

Tel: (617) 491-1670 • TWX: 710-320-1196

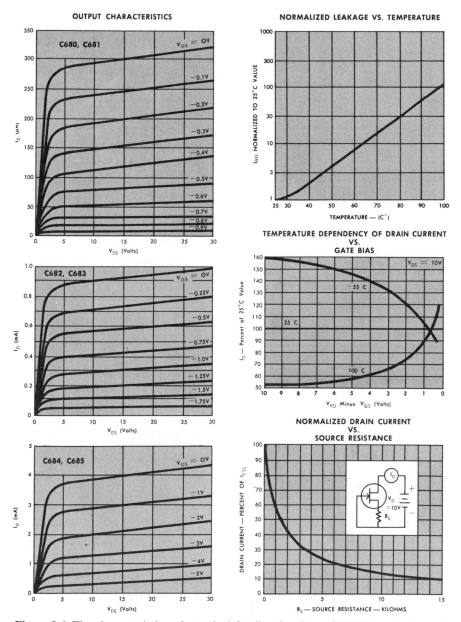

**Figure 9.6** The characteristics of a typical family of *n*-channel JFET's—the Crystalonics C680 series.

(That presumption is based on the identical gate capacities and the increase in $V_p$ with increases in $I_{DSS}$.) Since the channel width, $W$, enters (9.10′) only through $I_{dsat}$, the ratio of the currents should give the ratio of the $g_{ds}$'s. At least crudely, it does. The error is attributable most probably to the gate-drain voltage, which does depend on $W$.

Note that the heavier doping gives a distinct improvement in $g_{ds}$. It would also give a larger $g_m$ and higher current capacity. Are there any parameters that would suffer with increasing $N_D$? (Answer: Yes! The gate capacities would increase, the breakdown voltage would decrease, and the pinch-off voltage would increase.) As usual, compromise is the name of the game.

## THE SOURCE AND DRAIN PARASITIC RESISTANCES

Up to this point, we have dealt solely with the channel characteristics. To connect this channel to the outside world, one is obliged to proceed through the resistive source and drain regions. In designing a JFET, one may not make these resistances arbitrarily small because the doping of the drain and source regions determines in large measure the breakdown voltage of the gate junctions. You might wonder why these small parasitic resistances would be important in a device that typically finds itself in high-impedance circuitry. After all, if all your impedance levels are of the order of $10^5$ ohms, what difference can 50 to 100 ohms make?

The answer is shown graphically in Figure 9.7. In that figure, the channel characteristics of Figure 9.4 are shown as dashed lines. These would be the output-characteristic curves for a JFET with negligible source and drain resistance. The solid curves are for a device with the same channel characteristics, but with the addition of a source resistance ($R_s$) and a drain resistance ($R_d$) each equal to $1/2G_0$ (50 ohms in this case). The effect on the characteristic curves is dramatic, especially for low gate bias. Although both resistors have some effect, the principal villain is $R_s$. It provides degenerative feedback to the gate-source voltage in proportion to the drain current. Thus, for any given gate bias, the drain current and its rates of change with gate and drain biases are reduced by the presence of $R_s$. For large drain currents, the reduction is quite substantial.

The solid curves in Figure 9.7 were obtained from equation 9.5′ by substituting for the channel bias terms the appropriate expressions in terms of the terminal variables (e.g., $V_{d's'} = V_{ds} + I_d R_s + I_d R_d$). The resulting expression, although soluble, is impractical for analytic purposes. Figure 9.7 shows a numerical evaluation for one particular example. Some more general results are shown in subsequent figures.

Since FET's are usually operated in the saturation mode, one of the most important results of the analysis performed above is the relationship between

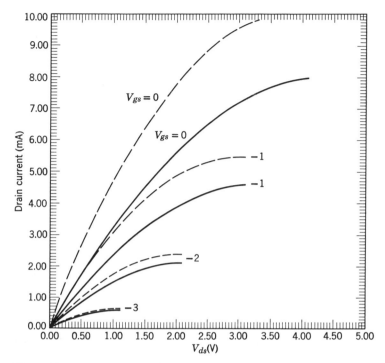

**Figure 9.7** JFET characteristic curves calculated from equation 8.5′ with source and drain resistances included (solid lines) and without them (dashed lines). The assumed values for the JFET are the same as in Figure 9.4. The two series resistors are each taken to be 50 ohms. $G_0 = 0.01$ mhos for this device.

$I_{dsat}$ and $V_{gs}$. If there were no source resistance, the appropriate relationship would be equation 9.5, with $V_{d's'} - V_{gs'}$ set equal to $V_p$. With slight rearrangement, this becomes

$$I_{d\,sat} = (G_0 V_p / 3)\left[ 1 - 3\left( \frac{V_D - V_{gs'}}{V_p} \right) + 2 \left( \frac{V_D - V_{gs'}}{V_p} \right)^{3/2} \right] \quad (9.11)$$

Equation 9.11 is a good description only for JFET's with very long channels. Such devices are poor designs in many respects, so very few commercially viable JFET's behave as (9.11) predicts.

A well-engineered JFET typically has a source conductance that is one to two times $G_0$. It is not that the designer would not like to get a smaller value of $R_s$, but rather that the various constraints on the design seem to conspire to make $R_s G_0$ lie between 0.5 and 1. For these typical devices, the saturation

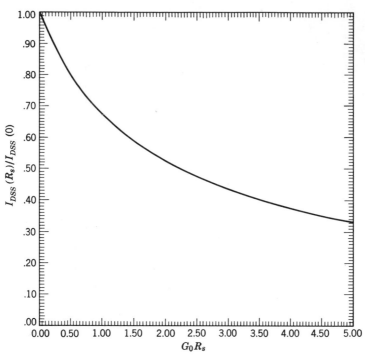

**Figure 9.8** $I_{DSS}$ as a function of the source resistance. The plot is normalized to the maximum drain current for $R_s = 0$. The horizontal axis gives the source resistance in terms of the reciprocal of the channel conductance. Thus, the plot is useful for all devices whose channels behave as predicted by (9.11).

current is found to follow a square law of the form:

$$I_{d\,sat} = I_{DSS}\left[1 - \left(\frac{V_D - V_{gs}}{V_p}\right)\right]^2 \Big/ \left[1 - \left(\frac{V_D}{V_p}\right)\right]^2 \qquad (9.12)$$

$I_{DSS}$ is the saturated-drain current with the gate shorted to the source. It is thus the maximum current that the JFET will carry and a most readily measured parameter. In fact, all of the variables on the right side of (9.12) are easily measured. They are the terminal variables. That makes (9.12) a very utilitarian expression. Figure 9.8 shows how $I_{DSS}$ depends on $R_s G_0$ and Figure 9.9 shows how well $I_{d\,sat}$ follows (9.12). For source resistances such that $R_s G_0 = 0.5$, the fit is remarkably good. Since this is a very typical value, equation 9.12 proves quite accurate for most commercial JFET's.

One caution is in order in using (9.12) or any other expression based on a fixed value of $R_s$. The biggest part of the parasitic source resistance is to

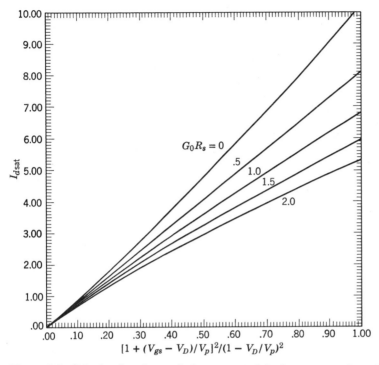

**Figure 9.9** Calculated values of the saturated-drain current plotted against the right-hand side of equation 9.12. Each curve represents a different ratio of channel conductance to source conductance. Athough none of the curves show much curvature, it is only the curve for $G_0R_s = 0.5$ that gives a really straight line for equation 9.12.

be found in the high-resistivity material that forms the transition region between source and channel. This is not a sharply defined region, nor is it entirely independent of gate bias and drain current. Although these changes in $R_s$ may not be very large, Figure 9.8 shows that a small change in source resistance can produce a noteworthy variation in $I_{DSS}$. We must therefore conclude that (9.12) is only a first-order approximation even if $R_sG_0$ happens to come out at the magic number 0.5.

## A SHORT EXERCISE

For both the JFET and the bipolar transistor, the maximum useful current that a given device can handle is determined to a large extent by the magnitude of the parasitic lead-in resistance. There are also fundamental limitations such as the onset of high-injection conditions in the bipolar transistor and the finite conductance of the channel in the JFET, but in practice, it is usually the lateral voltage drop in the base of the bipolar transistor or the voltage drop across the source in the JFET that limits the current. To get some idea of the dimensions of a JFET required to handle moderate currents, consider the following problem.

Find the maximum current that can be carried by the JFET given below and find the minimum $V_{ds}$ for that current. Do the problem first, assuming that $R_s = R_d = 0$; then determine what the maximum current capability would be if $R_s = R_d = 1/2G_0$. The channel is made of 3 ohm-cm $n$-type silicon (doping about $2 \times 10^{15}/cm^3$) and has a width of 3 $\mu$m, a length of 20 $\mu$m, and a breadth of 1 mm. (Devices with much shorter channels are routine today, but if the channel is not long compared to its width, one cannot justify the plane, parallel-current-flow assumption that leads to the one-dimensional analysis that we employed.)

[*Answer:* You are asked to find $I_{DSS}$ under two assumptions. If $R_s = 0$, we may go directly to (9.11) to find the saturation current with $V_{gs'} = 0$. To use (9.11), we must find $V_D$ and $V_p$. The first is a direct application of (3.49), but that requires a guess for the doping on the $p$-type side of the gates. The exact value is not too essential because an order-of-magnitude error in $N_A$ will produce only about 60 mV error in $V_D$. Since the gates are always heavily doped, a guess of about $10^{18}/cm^3$ could not be too far off the mark. That immediately gives us $V_D = 0.775$ V at 300°K. Solving for $V_p$ using (9.3) gives the total pinch-off potential as 3.4 V. The full-channel conductance is $G_0 = = 0.005$ mhos. Plugging these sundry constants into (9.11) yields the first answer:

$$I_{DSS}(R_s = 0) = 3.3 \text{ mA}$$

That is not a particularly impressive current capability, especially when you consider that a 1 mm² bipolar device would easily handle an ampere or more. This comparison is a little unfair because the only reason to make a device square is that scribing and breaking (or almost any way of separating the wafer into individual components) is much easier if square dice are employed. However, it is generally true that if current capacity is your object in life, the bipolar transistor is your solution. On the other hand, it is pretty hard to beat the current *gain* of an FET. For most applications, either a bipolar or unipolar transistor will do the job required. It is only in a few cases—high current capacity or high input impedance, for example —that one or the other device enjoys a distinct intrinsic advantage.

The value of $V_{ds}$ needed to achieve saturation (given that the source and drain regions contribute no drops) is simply $V_{d\,sat} = V_p - V_D = 2.65$ V. To compute the saturation voltage for the case that includes the source and drain resistances, note that $I_{DSS}$ is the saturation current for $V_{gs} = 0$. Thus, $V_{d\,sat} = = V_p - V_D + I_{DSS}R_d$. To get $I_{DSS}$, we may use Figure 9.8. That figure shows that a source resistance of $1/2G_0$ (100 ohms in this case) reduces $I_{DSS}$ to 80.5 % of its zero source resistance value. Thus the new $I_{DSS}$ is 2.66 mA and $V_{d\,sat}$ is 2.92 V.]

## A SMALL-SIGNAL MODEL OF THE JFET

For most design purposes and for direct comparison with the bipolar transistor, the $\pi$ model offers many advantages. I shall stick to this one model, although you should not construe this choice to be much more than personal preference. The other models all have their uses and advantages. Figure 9.10 shows a reasonably complete $\pi$ representation of the JFET. Figure 9.11 shows an approximation that is adequate for the vast majority of applications.

The circuit elements of Figure 9.10 each represent charge storage, energy dissipation, or current transport in reasonably distinct portions of the JFET's structure. The separation is possible only in the saturated mode of operation because the pinched-off end of the channel effectively delineates the input from the output circuits in a physically small portion of the whole structure. For example, we may divide the gate leakage current into two components corresponding to currents flowing in the drain and the source leads. These leakage currents in a good junction are exceedingly small and not very voltage dependent, so on an incremental basis they may be represented by two extremely small conductances, $g_{gs}$ and $g_{gd}$. Since values less than $10^{-9}$ mhos are typical for these conductances at room temperature, only in rather special circumstances (so-called *electrometer* applications) will they influence the circuit performance of the JFET.

The capacities, $c_{gd}$ and $c_{gs}$, that shunt these leakage conductances represent the depletion layer capacities NOT associated with the channel itself. An examination of Figure 9.2, noting that the source and drain regions must be at least as long as the channel to accomodate the ohmic contacts, should convince you that these two capacities should be at least of the same order as the channel capacity itself. Typical values for these capacities in small JFET's are a few picofarads.

The channel itself is really a distributed tapered RC transmission line. As a first approximation, such a line can be represented by a single $r$ and $c$ as shown in Figure 9.10. Such an approximation should be reasonably accurate up to frequencies of the order of $1/rc$. Since $(1/rc)$ is generally well in excess of $10^9$ sec$^{-1}$, the approximation shown in Figure 9.10 is good over just about the entire range of JFET operation. In fact, over most of that range, $r_c$ is so much less than $1/\omega c_c$ that it can be comfortably ignored. Thus, for most frequencies and most signal-source impedances, the input of the JFET can be represented by two capacitors, as shown in Figure 9.11. When this approximation is valid, the channel voltage and the input voltage are identical.

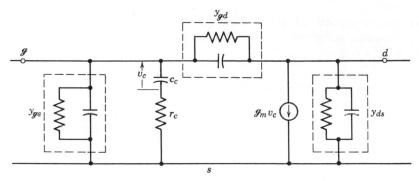

**Figure 9.10** The π representation of the JFET operating in the saturated mode. The degenerative effects of the parasitic source resistance have been taken into account in the evaluation of the components. Accordingly, no $R_s$ appears explicitly in the model. The symbol $(g_m)$ for transconductance is the most popular, although $(g_{fs})$ would be more correct in "standard" nomenclature.

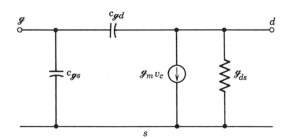

**Figure 9.11** The simplified π representation of the JFET. This model is adequate for almost all circuits. Only at very high frequencies or for electrometer applications are the resistive components of the input important.

The remaining three components—the current source, $g_{ds}$, and $c_{ds}$—characterize the incremental behavior of the pinched-off channel region. The current that gets to the drain depends only weakly on the drain voltage. Thus, we may usefully represent the output as a current source whose current strength is determined by the gate-source voltage (independent of the drain-source voltage) and a shunt conductance that stands for the incremental effects of channel-length modulation. Associated with the channel-length modulation is some charge storage in the depleted channel. This is represented by the capacity, $c_{ds}$. Since $c_{ds}$—including the associated packaging capacity—is frequently insignificant compared to the capacity of the load (typical measured values of $c_{ds}$ being of the order of 0.5 to 1 pF for a small device), it is often ignored in modeling the JFET. With this final approximation, we obtain the simple but quite functional representation shown in Figure 9.11. Let us take a close look at the four components in Figure 9.11 to see how they depend on the device, the bias point, and the temperature.

The two capacities, $c_{gs}$ and $c_{gd}$, are quite conventional depletion-layer capacities. They depend on $V$ but not much on $T$. A number of circumstances tend to restrict their range of variation. Most JFET's are made with epitaxially deposited, relatively uniform channel layers, so the capacities tend to be inversely proportional to the square root of the applied reverse gate bias. Since the channel is usually kept quite thin to make $V_p$ reasonably small, the upper gate must be a shallow layer. One also usually makes an $n$-channel device, so the upper layer is $p$-type. A thin layer, and a $p$-type layer at that (in silicon), will have a relatively low breakdown voltage because of the high field concentration at the edge of the layer. Guard rings help to prevent too low a breakdown, but the typical $n$-channel device is limited to a reverse bias on the order of 25 to 35 V or less. Since the pinch-off voltage is of the order of 2 to 5 V, the variation in capacity with bias is limited to a factor of about 2 to 4. Capacity values are usually quoted at their minimum value (It certainly sounds like a better JFET that way!), and for typical small devices, all capacities turn out to be a few picofarads. (See Figure 9.6.) Such small values inevitably include, as an important fraction of their total value, the packaging capacity associated with getting the current from the die to the outside world. This immutable capacity turns out to be about 0.5 to 1 pF for the common forms of package, so the possible variation in the capacity is still further reduced. For example, for the device described in Figure 9.6, $c_{gd}$ is given as 1.5 pF. About half of that is the package.

There are few variables available to the designer who would minimize the capacities. To some extent, he may minimize one capacity or the other by going to a nonrectangular geometry (concentric circles being the obvious and reasonably popular choice). If you are given this choice, which of the two capacities would you minimize and why? [Answer: The proper choice is $c_{gd}$. That capacity, multiplied by the gain of the stage, appears across the input terminals (i.e., the Miller effect discussed in the last chapter).] Thus, in any circuit of moderate gain, the critical capacity is $c_{gd}$. Note that the designers of the C680 series (Figure 9.6) managed to get a factor-of-2 difference between $c_{gd}$ and $c_{gs}$. There is also one other advantage in choosing to make the source area the larger of the two. If skillfully used, the increase in source area can be used to minimize the parasitic source resistance. As Figures 9.7 to 9.9 show, this is highly desirable.

The transconductance of the JFET (called either $g_m$ or $g_{fs}$) contains terms that depend on bias and on temperature. Unfortunately, just what the dependence is is determined in no small measure by the relative magnitude of the source resistance. Figures 9.7 and 9.9 should convince you of that fact. Given any particular relationship for $I_{d\,\text{sat}}$ versus $V_{GS}$, it is easy enough to differentiate to obtain $g_m$. For example, equation 9.12, which describes most JFET's quite well, gives

$$g_m = \frac{\partial I_{d\,\text{sat}}}{\partial V_{GS}} = (I_{DSS}/V_p) \left[ 1 - \frac{(V_D - V_{GS})}{V_p} \right] \Big/ \left[ 1 - \frac{V_D}{V_p} \right]^2 \quad (9.13)$$

The linear relationship between $g_m$ and $V_{GS}$ is typical of many JFET's, but only because the product $R_s G_0$ tends to be between 0.5 and 1 for most JFET structures. Figure 9.12 shows how $g_m$ varies with $V_{GS}$ for a number of different source resistances. These curves are the derivatives of the curves in Figure 9.9. Note that the straight-line relationship between $g_m$ and $V_{GS}$ is characteristic of JFET's with $G_0 R_s = 0.5$, a likely value in practical devices.

The temperature dependence of $g_m$ is not easy to predict because $R_s G_0$ depends on $T$. Thus, as the temperature shifts, not only do the parameters in (9.12) or (9.13) change but also the relationship itself. On a qualitative basis, it is possible to discuss what will happen, but it is quite difficult to be very quantitative. As the temperature increases, the diffusion potential and the carrier mobility both decrease. The percentage change in the mobility is much larger in the lightly doped material that comprises the channel than it is in the heavily doped material of the source. Accordingly, the product $R_s G_0$ will tend to decrease with increasing temperature. From Figure 9.8, you can see that the decrease in $R_s G_0$ will make $I_{DSS}(R_s)$ a larger *fraction* of $I_{DSS}(0)$, but the decrease in $G_0$ means that $I_{DSS}(0)$ itself decreases. It is not at all obvious whether $I_{d\,\text{sat}}$ for any given gate bias will increase with temperature. In fact, although most devices have $I_{DSS}$ decreasing with increasing temperature, positive temperature coefficients are not unknown.

A complete analysis would also have to include the variation of $V_D$ with temperature. $V_D$ is effectively the gate bias for zero applied voltage, so one would expect the variation of $V_D$ to have the maximum influence at low gate bias. Since $V_D$ decreases with increasing temperature, it tends to produce a positive temperature coefficient for $I_{DSS}$ in (9.12).

What emerges from this plethora of counterbalancing effects is the general and important result that the JFET is not very temperature sensitive. $I_{d\,\text{sat}}$ for a given gate voltage shifts only slightly with temperature, generally decreasing with increasing temperature but not very rapidly. This means that thermal runaway, the perennial plague of high-power bipolar transistors, is not a problem with FET's.

It is important to step back for a moment and compare the bipolar and JFET transistors. What is the fundamental difference between these two devices that renders one so exceptionally dependent on temperature and the other relatively insensitive to its thermal environment? [*Answer:* The

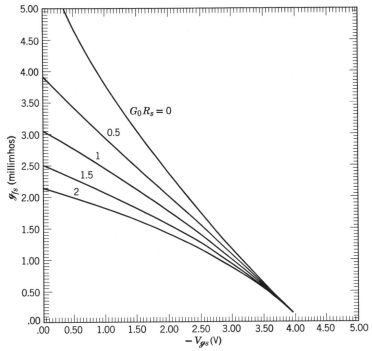

**Figure 9.12** Transconductance versus gate voltage for a JFET. Each curve represents the behavior of a device with a particular ratio of source-to-channel resistance. The channel characteristics are the same as in Figures 9.4 and 9.7 (i.e., $V_p = 5$ V, $V_o = = 0.9$ V, and $G_0 = 0.01$ mho).

bipolar transistor depends on the transport of minority carriers across a depletion region; the JFET is strictly a majority-carrier device. The fraction of carriers that can traverse the depletion-region barrier is an exponential function of temperature. Furthermore, the number that can get across without recombining is also exponentially dependent on temperature. Thus, the residual collector current ($I_{co}$) at a given bias and the $\beta$ of the bipolar transistor are extraordinarily temperature dependent. Unfortunately, the temperature dependence is regenerative, leading to "thermal runaway." In the JFET, on the other hand, the principal current is dependent on temperature only through the variation of the carrier mobility. This is a weak dependence. Accordingly, one would not expect majority-carrier devices to suffer all of the problems at low and high temperatures that beset the minority-carrier controlled devices. This expectation is borne out in practice.]

JFET's have only one fundamental problem with elevated temperatures: the gate current increases extremely rapidly with temperature (see Figure

9.6). Although these currents are generally pretty low over the entire operating range of the JFET, it must be remembered that FET's usually operate in circuits designed for high gate impedances. Thus, relatively small currents flowing in the gate circuit can result in large shifts in the gate bias. A microamp might readily clobber a circuit if it were designed to operate in the megohm impedance range. If this increase in gate current does not affect the circuit, however, you can use JFET's right up to the temperatures that result in mechanical or chemical damage to the device. (Note in Figure 9.6 that the maximum use temperature [200°C] is only 40° below the maximum temperature permitted for a ten-second soldering operation!)

JFET's may also be used in very-low-temperature environments without any loss of effectiveness. The only problem you would be likely to run into would be mechanical. At very low temperatures, the difference in expansion coefficients between the device, the package, and the metallization is quite likely to result in a fracture or an open circuit. [Very low being a relative word, a caveat is in order. If you REALLY lower the temperature (e.g., liquid He temperatures), you'll freeze out all of the carriers. At that point your JFET is little more than an insulator with three leads.]

## SMALL-SIGNAL CIRCUIT PERFORMANCE OF THE JFET

In "normal" applications, the JFET finds itself being driven by a moderate impedance source at a moderate frequency; it, in turn, drives a load of moderate impedance. By moderate I mean less than one megohm for impedances and less than ten megahertz for frequency. When used under these circumstances, the model of Figure 9.11 is quite adequate. This model is quite simple to analyze, and its use will give us some interesting data to compare with bipolar transistor performance.

A one-stage amplifier of elementary design is shown in Figure 9.13. In it, the JFET is represented by the circuit of Figure 9.11. The load comprises the real load shunted by $g_{ds}$. The signal-source resistance includes the bias network. (Note that there is some ambiguity in the meaning of *source* when discussing JFET circuits. In general, you can tell if the writer means the source of the FET or the signal source, but when he says "source resistance," it is not at all obvious what is meant. In this section, we are using the small-signal model of a JFET in which the effects of $R_s$ are contained in $g_m$ and the gate capacities. Thus, for this section, the source resistance, $R_S$, is the effective resistance of the signal source. The capitalized subscript indicates that it is an external circuit element rather than a device parameter, but in a subscript, that is a subtle distinction.)

The gain of the amplifier could be found by exactly the same method employed in the last chapter, but for a reason that will shortly appear, let us

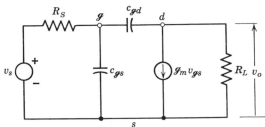

**Figure 9.13** A small-signal representation of a single-stage JFET amplifier. The JFET is represented by the circuit of Figure 8.11 with $g_{ds}$ absorbed into $R_L$ and any load from the bias network included in the effective source resistance $R_S$.

not make the unilateralization approximation for the JFET circuit. Making no approximations, the circuit of Figure 9.13 yields the following relationships:

$$v_{ds} = -\frac{R_L(g_m - sc_{gd})}{1 + sc_{gd}R_L}v_{gs}$$

$$v_{gs} = \frac{(1 + sc_{gd}R_L)}{1 + sR_S c_T + sR_L c_{gd} + s^2(c_{gs} + c_{gd})R_L R_S c_{gd}}v_s$$

where $c_T = c_{gs} + c_{gd}[1 + (g_m - sc_{gd})R_L]$ is the usual unilateralized input capacity. These two expressions may be combined to give the gain as

$$A = v_{ds}/v_s = -\frac{R_L(g_m - sc_{gd})}{1 + sR_S c_T + sR_L c_{gd} + s^2(c_{gs} + c_{gd})R_L R_S c_{gd}} \qquad (9.14)$$

Under almost all circumstances of real interest, $g_m \gg |sc_{gd}|$ and $R_S c_T \gg R_L c_{gd}$. When these two strong inequalities hold, we are back to the unilateralized solution:

$$A \cong -\frac{R_L g_m}{1 + sR_S c_T} \qquad \text{with the 3 db frequency,} \qquad \omega_{3db} = 1/R_S c_T$$

$$(9.15)$$

For the bipolar transistor, with its high $g_m$ and with an input resistance in excess of $r_{bb'}$, it is almost impossible to violate the conditions that lead to (9.15). For the JFET, on the other hand, it is possible at least in principle to reduce the signal-source impedance to negligible proportions. Such a situation is most unlikely, but if we may turn the second inequality around so that $R_L c_{gd} \gg R_S c_T$, (9.14) can be represented as

$$A \cong -\frac{R_L g_m}{1 + sR_L c_{gd}} \qquad (9.16)$$

If we extrapolate (9.16) out to frequencies where the magnitude of the gain approaches unity, we obtain a number that is reminiscent of $\omega_T$, the bipolar transistor's gain-bandwidth product. For the JFET (9.16) gives this unity-gain point as

$$\omega_T = g_m/c_{gd} \qquad (9.17)$$

This frequency is well in excess of the unity-gain point for (9.15) and represents a sort of upper limit for device performance. As such, it enjoys some popularity as a figure of merit for JFET's. (E.g., see Figure 9.6.)

From both (9.17) and (9.15) it is clear that high-frequency performance demands the smallest possible $c_{gd}$ and the highest possible $g_m$. Maximizing $g_m$ calls for the largest possible $I_D$ and that in turn calls for maximizing $G_0$ and minimizing $R_s$. For a given structure, these values are determined by the mobility of the majority carriers. Since electrons generally have substantially higher mobilities than holes, the $n$-channel device enjoys a decided advantage.

If you compare the gain-bandwidth product given in Figure 9.6 with that shown for the 2N699 (Figure 8.21), you will see that the two devices operate over similar bandwidths. Neither of these devices has outstanding frequency performance per se. There are commercial JFET's delivering useful gain to 0.5 GHz and bipolar transistors operating in excess of 5 GHz. Although gain-bandwidth products are not the whole story, it is generally the case that the bipolar transistor gives more gain at high frequencies than a comparable JFET. The principal difference is that the critical dimension in a bipolar transistor—the base width—may be held to much smaller values than its counterpart—the channel length—in the JFET. Such a statement presumes a diffused upper gate. If instead a Schottky-barrier junction forms the upper gate, some really startling high-frequency performance can be realized. At the end of the next chapter, we will examine a Schottky-barrier JFET that appears to have useful gain from DC to 30 GHz!

If we ignore the somewhat esoteric subject of the ultimate upper frequency for a given class of devices, how well does the JFET compare with the bipolar transistor? The answer, not surprisingly, depends on what you're doing with them. In general, FET's are employed where they enjoy a cost advantage (i.e., that's mostly MOSFET's in large integrated circuits) or where their extremely high input impedance offers some circuit advantage. Since bipolar devices have $c_T$'s that are within an order of magnitude of typical JFET values and $g_m$'s that are usually much larger, (9.15) suggests that the use of a JFET at a given gain will give less bandwidth than a comparable bipolar device. On the other hand, if the signal source has a very high impedance (e.g., most optical detectors have impedances in the megohm range or above), a JFET can yield substantially more *realized* signal gain.

You could not be faulted if you now concluded that bipolar devices are superior except for these really high-impedance applications. However, although you might not be faulted, you would be wrong in two rather important respects. FET's are frequently found in low-impedance amplifiers because they lead to improved *stability* and greater *linearity*. The question of stability arises in high-gain, tuned amplifiers. The net capacitive coupling between input and output sets an upper limit on the gain that may be achieved. If that upper limit is exceeded, the amplifier will break into spontaneous oscillation. (See Chapter 17 of P. E. Gray and C. L. Searle, *Electronic Principles*, Wiley, 1969.) Thus, the amount of RF gain that can be achieved is determined in large measure by $c_{b'c}$ in the bipolar transistor and by $c_{gd}$ in the FET. If you look at the data for the 2N708 (Figure 8.11a) and the C680 (Figure 9.6), you will see that a rather impressive RF bipolar transistor has five times the coupling capacity of a really run-of-the-mill JFET. A genuine RF JFET will double that difference; a MOSFET in a proper package may double it again. This inherently greater stability of the FET makes the design and manufacture of tuned RF amplifiers substantially simpler.

The linearity problem is particularly severe in the tuned RF amplifier of sensitive radio receivers. In principle, the tuned first stage should discriminate between the wanted signal and others that may be much stronger but at different frequencies. However, if the amplifying element is not strictly linear—and none of them are—the nearby spurious signals may modulate the true signal. If you have an all-transistor radio in your car and spend any time listening to remote stations, you are probably acutely aware of this obnoxious effect. To explore this particular failing for any of our devices, we need only expand $\partial I_{out}/\partial V_{in}$ beyond the first term (which is represented by $g_m$ in our linear model). As it turns out, it is the odd terms in the series expansion that give troubles; the even terms are all filtered out by the tuned circuit in the output. To see this, consider a signal of the form:

$$v_{in} = \alpha(t) \sin(\omega_1 t) + \beta(t) \sin(\omega_2 t) \qquad (9.18)$$

$\alpha$ and $\beta$ represent the modulation on the two signals. $\omega_1$ is the angular frequency of the desired signal; $\omega_2$ is that of the strong nearby signal. The tuned stage has gain only in the vicinity of $\omega_1$. Its output can be represented by a Taylor series of the form:

$$i_{out} = \sum_n A_n v_{in}^n \qquad (9.19)$$

$A_1$ is simply $g_m$. If we now put (9.18) into (9.19), we will get terms of the general form: $\sin^j(\omega_1 t) \sin^k(\omega_2 t)$. Each of these may be expanded to single sinusoids containing either harmonics of the two frequencies or sums and/or differences of the two frequencies. The tuned filter discriminates against all of these that are not at $\omega_1$. Contributions at $\omega_1$ come only from terms of

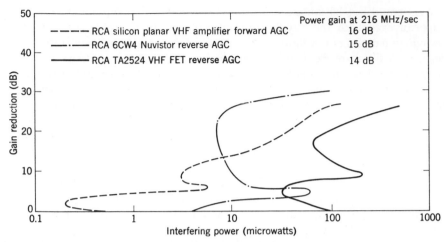

**Figure 9.14** Comparison of the cross-modulation performance of a MOSFET, a bipolar transistor and a vacuum tube. The abscissa is the power to produce 1% distortion. (From P. E. Kolk and H. Johnson, in J. T. Wallmark and H. Johnson, Eds., *Field-Effect Transistors,* Chapter 5, Prentice Hall, 1966.)

the form $\sin(\omega_1 t)\sin^{2k}(\omega_2 t)$. These terms occur only in the odd powers of (9.19). For example, the chief offender is likely to be from the $n = 3$ terms of (9.19):

$$\alpha(t)\beta^2(t)\sin(\omega_1 t)\sin^2(\omega_2 t) = \beta^2\alpha\sin(\omega_1 t)\left[1 - \cos(2\omega_2 t)\right]/2 \quad (9.20)$$

[There will also be a term in $\omega_1$ from the expansion of $\alpha^3\sin^3(\omega_1 t)$ from the $n = 3$ components. That term represents a distortion of the desired signal by itself. The sum of such self-distortion terms is called the *automodulation distortion*.] The amount of *cross-modulation* from (9.20) depends on the square of the spurious signal voltage at the input of the amplifier. Thus, the spurious audio-signal voltage will be badly distorted itself. This undesirable signal is proportional to the spurious RF power.

Since the bipolar transistor's collector current is exponentially dependent on the input voltage, there is a nonvanishing $A_n$ for all $n$. Furthermore, the terms fall off only as $1/n!$. That's pretty fast for large $n$ but rather slow for the first few terms. The third term is $1/6$th of the first ($g_m$).

The JFET, on the other hand, tends to be pretty well represented by (9.12) —a square-law relationship with no terms for (9.19) above $n = 2$. That means that for a JFET that really follows (9.12), there would be NO cross- or automodulation distortion at all! JFET's afen't really quite THAT good, but they are enormously better than bipolar devices. MOSFET's enjoy similar immunity to the various types of modulation distortion. Figure 9.14

shows the great superiority of FET's over bipolar transistors and even vacuum tubes. The vertical axis refers to the shift in gain due to a shift in bias point produced by an AGC (*automatic gain control*) circuit. Since this shift in bias point affects both the input impedance of the amplifier and the expansion of (9.19), the cross-modulation sensitivity is a function of the "gain reduction." The curve for each device represents the input power to achieve 1% cross-modulation. The advantage of using an FET is clearly evident.

## BIASING THE JFET AND THE PROBLEM OF CASCADED DC STAGES

Routine biasing of a single-stage JFET amplifier is rather easy. The only problem is to avoid having the gate bias voltage shift excessively with changes in temperature. This problem is most evident in the "fixed-bias" circuit shown in Figure 9.15a. The gate voltage in that circuit is given by

$$V_{GS} = V_{GG} - I_{GSS}R_G$$

Note that these are all DC quantities and that $I_{GSS}$ is the current flowing through the reverse-biased gate diode. Both $V_{GG}$ and $I_{GSS}$ are negative for the $n$-channel device shown. This fixed-bias arrangement becomes very temperature dependent if $I_{GSS}R_G$ becomes significant compared to $V_{GG}$. In general, this means that if you need even a modicum of temperature stability, the value of $R_G$ must be less than you probably want it to be.

A much improved scheme is shown in Figure 9.15b, with still better versions in $c$ and $d$. These are all circuits that employ "self-bias" in that the source current flowing through $R_S$ provides the principal gate bias. $R_S$ also provides some thermal stability, since the loss in gate bias resulting from an increase in $I_{GSS}$ is partially compensated by the increase in the source voltage, $I_DR_S$, that follows the gate debiasing. This may be seen by writing the loop equation for the gate circuit in Figure 9.15b:

$$V_{GS} = I_DR_S - I_{GSS}R_G$$

The degeneration provided by the source resistance is frequently sufficient for the limited temperature swings met with in most commercial electronic equipment. However, the simple self-bias circuit of Figure 9.15b is limited in the amount of degeneration that can be obtained, since one is obliged to keep $V_{GS}$ well away from the pinch-off voltage if much output power is to be obtained.

The circuit of Figure 9.15c allows substantially more degeneration to be employed by providing a DC level shift for the gate circuit. The gate-to-source voltage may then be set independently of the source-to-ground voltage.

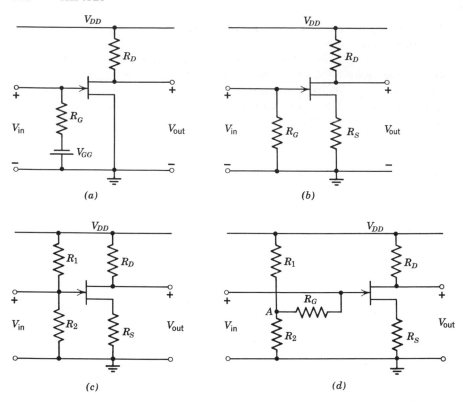

**Figure 9.15** Some bias arrangements for single-stage JFET amplifiers. (*a*) Fixed bias; (*b*) simple self-bias; (*c*) self-bias with DC level shift for the gate; and (*d*) an improved version of the circuit in (*c*) that uses smaller resistance values in the gate circuit.

In this circuit, the amount of degeneration is limited only by the power-supply voltage, $V_{DD}$.

The circuit of Figure 9.15$d$ takes care of a practical problem that certainly isn't obvious in the preceding circuit. The difficulty with 9.15$c$ is that if one wants to have a very large input resistance, one is obliged to use very large values for $R_1$ and $R_2$. Resistors with values in the megohm range and above tend to be rather unstable, noisy, and temperature dependent. Considering how resistors are made, this is not very surprising, but it is highly undesirable. The last circuit provides a partial solution by separating the voltage-division function from the input-resistance function. Using moderate values for $R_1$ and $R_2$ provides a stable, reasonably noise-free voltage at point $A$. $R_G$ can then be selected to provide sufficient input resistance. Since $R_G$ will

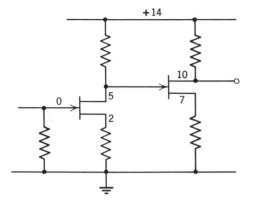

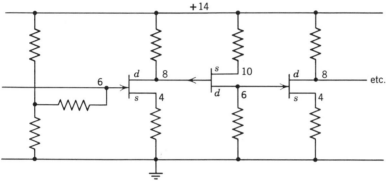

**Figure 9.16a** A cascaded JFET DC amplifier, showing how moderate amounts of bias on each stage very quickly saturate the drain supply. The numbers give the voltage with respect to ground.

**Figure 9.16b** A cascaded JFET DC amplifier using both $p$ and $n$ channel units. Each transistor has $|V_{DS}| = 4$ and $|V_{GS}| = 2$. The same amount of source degeneration could be used on each stage, and the cascade could obviously be extended indefinitely.

probably be substantially less than the maximum of $R_1$ or $R_2$ in circuit 9.15c, a substantial improvement in both stability and noise can be obtained. (Note that we are discussing noise and instability that are in excess of the thermodynamically inherent values. The $kT\Delta f$ noise power is there whether you use one resistor or ten. On the other hand, the low-frequency noise associated with large resistors grows very rapidly with increasing $R$. Thus, going from very large to moderate resistors can result in a decided improvement.)

For most purposes, a single amplifying stage is insufficient. A cascaded, multistage circuit is normally called for. At the same time, in today's modern stress on miniaturization, the economics of integrated circuits puts a very high value on coupling the stages together directly without isolation capacitors. Thus, even if we have no intention of amplifying DC, we often must resort to a cascaded DC amplifier. Devices that require a reverse-biased gate (with respect to the drain), such as JFET's, vacuum tubes, and some MOS-FET's, are at a distinct disadvantage in such circuits. The problem is clearly evident in Figure 9.16a. With 14 V from the supply, if we would have 3 V across each transistor and $-2$ V of gate bias, we may cascade only two stages. By using two bias supplies (one $+$, one $-$) and throwing away part of the gain of each stage, one can circumvent this problem, but such a solution is unsatisfactory for a number of reasons. The immediate and accurate conclusion is that by themselves, $n$-channel JFET's are not particularly suitable for DC cascaded amplifiers.

An interesting solution is possible if $p$-channel JFET's are also available. By alternating $p$- and $n$-channel JFET's, a "sawtooth" biasing arrangement results that can be extended indefinitely. Figure 9.16b illustrates this method. Note that with the same supply voltage (14 V) as in Figure 9.16a and the same gate bias, the circuit of Figure 9.16b permits more voltage across each transistor and also more degeneration on each source. Such a circuit is very appealing, but the lack of a wide selection of $p$-channel JFET's and their generally inferior performance make this circuit less practical than it seems on paper. Furthermore, in integrated circuitry, getting both $p$- and $n$-channel JFET's on the same substrate is difficult and expensive. This cascading problem pretty well eliminates the all-JFET amplifier from competition in the monolithic integrated-circuit market.

## SUMMARY OF RESULTS FOR THE JFET

We have begun our study of majority-carrier current valves with the JFET, the first successful majority-carrier transistor. We first examined the relationship of the channel current to the gate and drain bias, ignoring for the moment the parasitic source and drain resistances that appear to be inherent in the JFET structure. For fixed-gate bias, the drain current was found to increase with $V_{d's'}$, linearly a first and then less and less rapidly until it became almost independent of the drain bias. The three regions of operation were dubbed linear, transition, and saturation, respectively. The addition of the source and drain resistances made the analysis more complicated. Although they did not change the qualitative description above, they had a significant effect on both the large- and small-signal performance of the JFET. For operation in the saturation mode, the typical values of $R_s G_0$ make the JFET into a square-law amplifier (equation 9.12). The inclusion of the parasitic resistances in the analysis permitted us to use the true terminal variables but rendered somewhat obscure the variation of the device parameters with temperature. In spite of this analytic handicap, we were able to conclude that the JFET was relatively insensitive to its thermal environment, being equally useful at both high and low temperature. It did not suffer from the thermal run-away that afflicts bipolar devices. The only term that presented any problem was the gate current which, although always small, tended to increase very rapidly with temperature. In the very-high-impedance circuits that are frequently employed with FET's, special precautions must be taken to prevent this leakage current from debiasing the circuit.

A small-signal model was developed that was adequate for most JFET applications (with passing reference to a somewhat more complete representation). In a comparison between the small-signal models of the JFET and the bipolar transistor, it appeared at first that the bipolar transistor, with its higher gain-bandwidth product, was generally superior except for certain inherently high-impedance applications. However, the small coupling capacity between gate and drain makes the JFET more stable in tuned RF amplifiers, and its greatly superior linearity makes it much less subject to cross- and automodulation distortion. We must conclude that the bipolar and field-effect transistors are competitive in most applications, with each having some special domains in which they are obviously superior.

We now turn our attention to the "other" FET, the IGFET. Here we will find a device that competes with the bipolar trsnsistor not only electrically but also economically. JFET's will never revolutionize electronics, but the IGFET might well.

Our first efforts must be devoted to the oxide-semiconductor surface, which constitutes the dominant element in IGFET behavior. Then we may return to device analysis. Although the IGFET bears a strong resemblance to the JFET, it is a much more variegated beast with fascinating possibilities. We begin with a rather odd two-terminal device, the *MOS capacitor.*

### Problems

The problems below all concern the same high-power JFET. In order to get high output current, it is necessary to make the gate extremely broad. In this case, the design called for a gate stripe that is 1 cm by 0.001 cm. Obviously, such a shape would be almost impossible to handle, so the design is folded to fit onto a convenient rectancular die. The source and drain metallization patterns are shown below as well as the source-drain-channel regions for the resulting device. The channel doping is phosphorus at $10^{16}$ atoms/cm$^3$. The electron mobility in the channel is 1000 cm$^2$/Vsec.

1. Why have 21 source stripes and only 20 drain stripes? (I.e., why not have 21 drain stripes and 20 source stripes?)

    *Answer:* To minimize the parasitic source resistance and the drain-to-gate capacity for the given channel breadth.

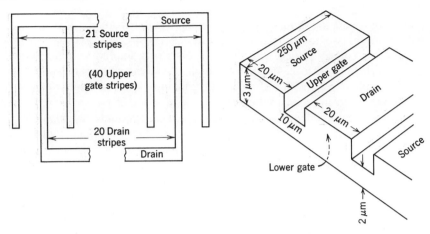

2. What is the pinch-off voltage for this transistor?

   *Answer:* 7.55 V including $V_0$.

3. What is the approximate value of $I_{DSS}$?

   *Answer:* Assuming That $R_s G_0 = 1$, $I_{DSS} = 0.41$ A.

4. Assuming a 1-microsecond excess-carrier lifetime and $T = 300°K$, what is the approximate value of the gate leakage current under the bias conditions: $V_{GS} = -3$ V, $V_{DS} = 0$? Does the inactive interface between the gate and the source and drain make an important contribution?

   *Answer:* The area of the gate-channel interface is $2 \times 10^{-3}$ cm$^2$. The total area of the junction between the gate and the rest of the device is approximately $4.3 \times 10^{-3}$ cm$^2$. Thus, the nonutilitarian area exceeds the useful area by about 15%. The reverse current under the bias conditions listed is 0.69 nA.

5. If the lifetime of the minority carriers were unaffected by temperature, what would the reverse current be at 400°K?

   *Answer:* The current is proportional to $n_i$, which would increase by a factor of 1720 over that range. Accordingly, the gate leakage current will be about 1.2 μA at 400°K.

6. Under the bias conditions of problem 4, estimate the small-signal input conductance of the gate.

   *Answer:* $0.92 \times 10^{-10}$ mhos.

7. Under bias conditions: $V_{GS} = 0$, $V_{DS} = 10$ V, estimate the gate-to-drain capacity.

*Answer:* The capacity for the area *not* at the channel end is about 12 pF. Most of this is at the epitaxial-layer interface. There would also be a few pF from the saturated channel region. A total of between 14 to 16 pF is a good estimate.

8. What is "$\omega_\tau$" for this JFET for $V_{DS} = 10$ V?

*Answer:* The maximum $g_m$ is approximately 0.23 mhos. The approximate value of $c_{gd}$ is 15 pF. Thus, $\omega_\tau = g_m/c_{gd} = 1.5 \times 10^{10}$ or $f_T = = 2.4 \times 10^9$ Hz.

## CHAPTER 10

# DEVICES THAT WORK PRINCIPALLY BY DRIFT: (B) THE MOSFET

In the JFET, the carriers in the channel are there as the direct consequence of the impurities introduced in processing. The source of the carriers, however, had no influence on the behavior of the transistor. Any method by which the carriers can be induced to lie in a channel connecting the source and drain can result in a viable FET. The simplest way to draw charged carriers into a selected region is by application of an electric field. That certainly sounds a great deal easier than the many steps needed to create the structure of Figure 9.2. In fact, field-induced channels are so easy to generate that they are difficult to avoid in some circumstances. The problems with the original FET's of Lilienfeld and Heil lay not with the concept of drawing carriers in under the "field plate" but rather with the difficulties in keeping these carriers mobile at the surface. The difference between failure in the 1930's and enormous success in the 1960's is the technology of surface preparation. It is interesting to note that difficulties in achieving a workable IGFET inspired the research that produced the bipolar transistor, and that it was in pursuit of better bipolar devices that enough surface technology was developed to make possible the eventual realization of excellent IGFET's. Thus, at the end, we go back to the beginning.

## CHARGE DISTRIBUTION IN AN MOS CAPACITOR

We return now to the central concept in the original FET—the idea that an electric field can modulate the conductivity of a semiconductor by repelling or drawing in free carriers. We have been dealing indirectly with this idea ever since Chapter 3, when the depletion approximation was introduced. The volume of the conductive region (the undepleted region) can be modified by the application of external potentials. We just finished making extensive use of that effect in the last chapter.

Nothing in the previous paragraph is dependent on the presence of a p-n junction; it is solely the result of an electric field. Couldn't the same thing be accomplished by applying the field capacitively? The answer is a most emphatic yes! Consider the solution of the junction problem illustrated in Figure 3.9. If the left-hand side ($x < 0$) were replaced with an insulator terminated at $x = -a$ by a metal layer, the solution might well be expected to look like Figure 10.1. This would be the solution in the depletion approximation if the insulator-semiconductor (IS) heterojunction did not alter the nature of the semiconductor in its vicinity. Although we have already seen (Chapter 6) that the surface is quite different from the bulk, let us pursue this somewhat simplistic "no-surface" solution a bit further. The effects of surface states and charge in the insulator can be easily added a posteriori.

The object of this exercise is to obtain the band diagram as a function of applied potential. From the band diagram we may readily determine the local free-carrier density and thus (in principle) the conductivity. One curious observation must be accepted before we may proceed. In Chapter 3 it was stated with much emphasis that a system had achieved thermal equilibrium if and only if all net flows that were energetically possible had ceased. It does not conflict with that description of thermal equilibrium to state that a system in a box sitting on a table can come to equilibrium within itself even through the box could in principle "flow" through the table. In any practical sense, the table is an impassible barrier.

In the same way, we may conclude that the semiconductor in the MOS structure of Figure 10.1a can be in thermal equilibrium even with an applied potential between the metal and the semiconductor. The insulator prevents any flow just as the table does. Changing the potential is like turning gravity up and down; as long as the flow through the barrier is quite inconsequential, the system on each side of the barrier can reach its own thermal equilibrium. This permits us to employ the useful concept of a Fermi level. However, the Fermi levels in the metal and in the semiconductor need not be the same, since these two systems are effectively isolated from each other.

As long as the depletion approximation holds, our problem is easily managed. The situation, shown in Figure 10.1a, is nothing more than the parallel-plate capacitor with one plate being a p-type semiconductor. If a charge, $+ Q$, per unit area is distributed on the surface of the metal plate, the field in the oxide is $q/\varepsilon_i$ (Gauss's law) where $\varepsilon_i$ is the dielectric constant of the insulator.* Since $D_\perp$ must be continuous across the boundary at $x = 0$ (we are assuming no surface charge), $Q/\varepsilon_s$ is the field at the surface of the semiconductor ($x = 0 +$). $\varepsilon_s$ is the dielectric constant of the semiconductor.

---

* The words *insulator* and *oxide* are used interchangeably here, although all oxides are not insulators and the insulators used in MOS devices need not be $SiO_2$.

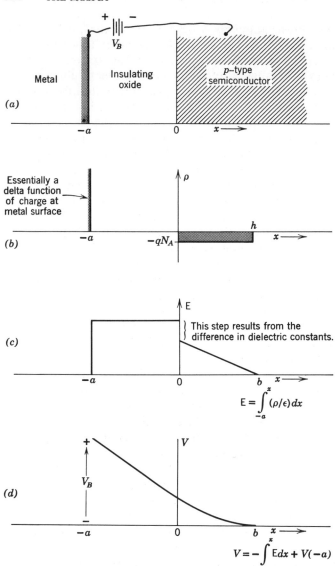

**Figure 10.1** The solution of Poisson's equation for the MOS capacitor in the depletion approximation. This should be compared with Figure 3.9, which portrays the $p$-$n$ junction. Surface states on the semiconductor and charge in the oxide are *not* included, so the electric displacement is constant in oxide and continuous across the oxide-semiconductor junction. (*a*) MOS structure in one dimension; (*b*) charge density in the depletion approximation; (*c*) electric field as a function of position; and (*d*) the electric potential as a function of position.

(Note that the typical semiconductor has a much higher $\varepsilon$ than the typical oxide; accordingly, the electric field will drop by a factor of 3 to 6 on entering the semiconductor.) The solution in the semiconductor is quite identical to the *p-n* junction solution (equations 3.47 and 3.48). However, the total depletion length, $l_p$, is simply the length necessary to give a total charge of $-Q$ per unit area on the semiconductor:

$$Q = -qN_Al_p \tag{10.1}$$

The total voltage and charge on the capacitor are then related by

$$V = V_i + V_s = \frac{Qa}{\varepsilon_i} + \frac{Q^2}{2\varepsilon_sqN_A} \tag{10.2}$$

The incremental capacity of our MOS sandwich ought to be the inverse of the derivative of (10.2):

$$1/c = \frac{\partial V}{\partial Q} = \frac{1}{c_i} + \frac{1}{c_s} = \frac{a}{\varepsilon_i} + \frac{Q}{\varepsilon_sqN_A} = \frac{a}{\varepsilon_i}\sqrt{1 + \frac{2\varepsilon_i^2 V}{\varepsilon_sqN_Aa^2}} \tag{10.3}$$

Equation 10.3 states that we have two capacitors in series, one of which (the insulator capacity) is constant and one of which falls off with increasing charge (the depletion-layer capacity). For large enough $V$'s, (10.3) suggests that the capacity should decrease in inverse proportion to the square root of the voltage. Does it?

Well, if you were to make the measurement, you would get a very curious result. At moderate-to-high frequency (for the small AC signal used to measure the incremental capacity), the capacity would follow (10.3) for a while, and then rather abruptly become *constant*. As if that isn't peculiar enough, if you were to lower the measurement frequency to the low- or subaudio range, the capacity would follow (10.3) for a somewhat shorter range, then rather abruptly swing back toward its initial ($Q = 0$) value, $c_i$. Figure 10.2 shows some actual data. (Ignore for the moment the displacement of the curves along the voltage axis. That is the result of surface states, charge in the oxide, and work-function differences, all of which we will examine shortly. Let us pretend, for the next few paragraphs, that the upper knee occurred at $V = 0$.)

The reasonable explanation for this bizarre $c$ versus $V$ curve is obtained from Figure 10.3, which shows the band diagram for varying amounts of positive bias (as defined in Figure 10.1a). A useful way of looking at the experiment that gives Figure 10.2 is to think of putting a charge $dQ$ on the metal plate and observing where its opposite, $-dQ$, appears within the volume of the semiconductor. The farther away the compensating charge is, the larger the voltage ($\Delta V = -\int \Delta E dx$) and the smaller the capacity. If a demand is made for negative charge, it may be answered by increasing

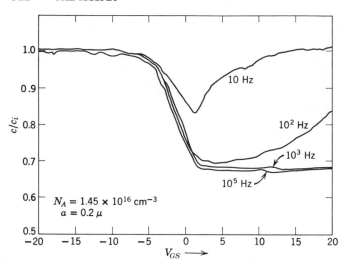

**Figure 10.2** The bias dependence of the capacity of an isolated MOS capacitor as a function of the measurement frequency. The capacity is plotted as a fraction of the oxide layer capacity, $c_i$. [From A. S. Grove, B. E. Deal, E. H. Snow, and C. T. Sah, *J. Appl. Phys.* **35**, 2458 (1964).]

the length of the depletion region (getting $dQ = -qN_A dx$) or by acquiring some more electrons. If the first method is operative, the capacity should be given by (10.3) where $c_s = \varepsilon_s/l_p$. On the other hand, if the charge acquired is mobile (in contrast to the acceptors, which are not), the excess carriers will move to the oxide-semiconductor interface and the capacity measured will be $c_i$, the capacity of the oxide (insulator) layer alone. In going from the flat-band condition of Figure 10.3a to the bent bands of Figure 10.3b, it is evident that the depletion layer has grown as the intrinsic level has been bent down toward the Fermi level. If more bias than what is illustrated in Figure 10.3b is applied, the intrinsic level near the surface will pass below the Fermi level. This means that the number of electrons in that region will exceed the number of holes—the material at the surface has been *inverted*; it is *n*-type. However, not until the surface-electron population becomes significant compared to the acceptor population can we expect to see much effect. Remembering that motions in the band diagram are reflected exponentially in the carrier density, it does not take a great deal of voltage across the semiconductor to obtain a significant population inversion as illustrated in Figure 10.3c. Once the inversion is achieved, every 60 mV of surface potential gives a factor-of-ten increase in the charge density.

At this point, our depletion-model calculation is obviously no longer functioning properly. Increases in the field at the surface of the semiconductor

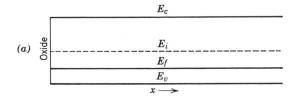

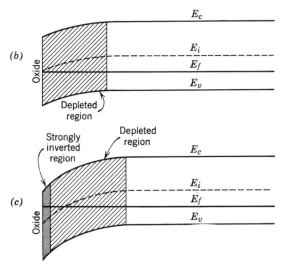

**Figure 10.3** The band diagram for the MOS capacitor of Figure 10.1$a$ for varying amounts of total potential. ($a$) No field at the OS interface; ($b$) a moderate positive field at the OS interface; and ($c$) a positive field at the interface sufficiently large to invert the layer near the surface from $p$- to $n$-type.

can now be compensated by acquiring mobile charge essentially on the surface. Very little extension of the depletion region is required. This explains why the depletion region stops growing with increasing bias, as indicated by the higher frequency measurements of Figure 10.2, but it does not explain the variation of $c$ with $\omega$. That requires a look at the dynamics of achieving equilibrium.

Consider the situation shown in Figure 10.3$c$. If I apply a charge of $dQ$ to the metal plate as a step function, I must instantly find $-dQ$ in the $p$-type semiconductor. Why not the inverted region? Well, the inverted region has no available electron population. Every electron is already accounted for by a positive charge on the metal plate. The nearest source of *new* negative charge is at the edge of the depletion region. The edge of the depletion region moves to the right, to provide $- dQ = qN_A dx$, but the resulting potential distribution is not an equilibrium situation. As electrons are generated within the depleted volume, they will drift under the influence of the nonequilibrium field toward the surface. Awaiting equilibration through the generation process may be a lengthy procedure, but eventually the system will relax into its equilibrium state with the extra charge, $- dQ$, comfortably ensconced on the semiconductor's surface and the depletion layer back essentially to its former state. Thus, if I make my capacity measurements at a frequency that permits equilibration, I will be putting charge on opposite sides of the insulator and observing $c = c_i$. On the other hand, if I make my measurements at high frequency, $c_i$ appears in series with a *fixed* value of $c_s$. Why fixed? (Answer: By definition, the DC bias always has time to achieve equilibrium, so once substantial inversion is achieved, increases in bias simply increase the surface charge and do not materially affect the length of depleted material underneath. Since the high-frequency AC voltage samples the now-constant depletion-layer length, the observed value of $c_s$ is essentially independent of $V$.)

## APPLICATIONS FOR THE MOS STRUCTURE

You might well ask why there is so much interest in another peculiar nonlinear capacity. Bias-tunable capacitors are of great interest in themselves. Under the name of *varactors* (variable reactor), they find application in tuners, frequency and phase modulators, and AFC (*automatic frequency control*) circuits, and they are the essential element in almost all parametric amplifiers. The temperature stability, minute size, and low cost of the solid-state varactors make them highly competitive with the traditional mechanical tuning capacitor in even the most mundane of applications. (Almost all European TV sets use them for tuning, and they are likely to make inroads into the American sets soon.) In parametric amplifiers, where the capacity must follow RF bias signals in the microwave region, solid-state varactors are sine qua non. The traditional varactor is simply a *p-n* junction diode with very light doping to give the maximum change of $c$ with $V$. (Frequently it is PIN.)

Although *p-n* junction varactors can provide large $c$ variation and moderately high $Q$, there is distinct room for improvement. For one thing, it would be nice to have a capacitor that did not suddenly start conducting under forward bias, and for another, it would be nice to have a capacitor that did not have such a high leakage at elevated temperatures. To overcome these shortcomings, J. L. Moll [*Wescon Convention Record*, Part 3, p. 32 (1959)] and Pfann and Garrett [*Proc. IRE*, **47**, 2011 (1959)] proposed the MOS capacitor in essentially the form we have been examining.

Had the MOS structure simply added another variable capacitor to the repertory, it would not have had a very profound impact on electronic technology. What makes the MOS structure so interesting is the possibility of using the inverted region as a voltage-controlled channel in a MOSFET. Although there are many problems involved in achieving a good MOSFET, insulated-gate FET's have so many endearing qualities that they threaten to displace the bipolar transistor from its position of dominance.

To get a good picture of the MOSFET and the problems involved in making one, we must take a somewhat more realistic look at the system one would obtain if one grew an oxide on a silicon wafer and then evaporated a layer of metal on it. What we must do is account for charges, fields, and states that are "built in." These will include various types of surface states, charged and somewhat mobile ions within the insulator, and work-function differences between the layers. We will obtain not only the background needed for analyzing MOSFET's but also some insight into the breakdown behavior of oxide-passivated *p-n* junctions.

## THE METAL-OXIDE-SEMICONDUCTOR SYSTEM

In the discussion of the ohmic contact (Chapter 6, Figures 6.11 through 6.13), we found that the equilibrium-band diagram was determined in part by the alignment of the Fermi levels from the bulk of the metal to the bulk of the semiconductor (i.e., the contact potential or work-function difference between the two) and in part by the localized surface states with energies in the forbidden band. If we now add an intermediate dielectric layer—the $SiO_2$ layer—those conclusions still hold, but we also have the possibility of charges and states within the insulator. Since some of the charges can move and many of the states can exchange charge with the semiconductor (or even the metal), and since we can apply substantial voltages to the MOS sandwich, there is the possibility of a moderately complicated interaction between the applied potential and the character of the semiconductor near the surface. Let us begin with the contact potential.

Several pages back, I took some pains to defend the idea that the Fermi levels within the metal and semiconductor were independent of each other because they were isolated from each other. That's true with respect to the short term, but even quartz has some conductivity. Left alone for long enough, the system will equilibrate. However, unlike the ohmic-contact or Schottky-barrier situations in which the full contact potential had to be absorbed by the shift in the semiconductor bands, for the MOS sandwich, a significant fraction of the voltage will appear across the dielectric layer. In the limit of zero oxide thickness, the full contact potential appears across the semi-conductor depletion region; for the very thick oxide, the full contact potential appears across the insulating layer.

Figure 10.4 shows the effect of the contact potential on the band diagram for an aluminum-quartz-silicon sandwich. Since the Fermi level for an isolated sample of aluminum is about 1 V higher than the Fermi level for an isolated sample of $p$-type silicon, the bands must be bent to align the Fermi levels in the M, O, and S layers. This diagram inevitably raises certain questions. Consider, for example, that we have ignored the problem of making the Fermi level in the oxide align with either side. Why doesn't the contact potential between the insulating oxide and the semiconducting silicon lead to significant band bending? [*Answer:* The junction of O and S may well lead to a significant shift in the O bands; it is the semiconductor that hardly changes at all. What is required is enough charge displacement on either side of the junction to account for the contact-potential difference. A small shift in the S bands and a completely trivial shift in the M bands will produce the same amount of charge as a shift of many volts in the O bands. In fact, it takes such an immeasurably small amount of charge to swing the Fermi level 7 or 8 eV (the band gap is about 9 eV for quartz) in the O layer that on a practical basis, the Fermi level is indeterminate within a $\pm$ 4 eV range. Since the alignment of the oxide Fermi level has no observable influence on the rest of the system, we are fully justified in simply ignoring it.]

Figure 10.4a shows the situation when the entire system is in equilibrium. The surface of the aluminum has a positive charge, $Q$; the insulating layer is assumed to be charge-free; and the depleted layer of silicon provides a charge, $- Q$. The Fermi level is constant throughout the system.

Figure 10.4b represents a situation known as the *flat-band condition*. To flatten the bands in the semiconductor, an external voltage must be applied that puts $+ Q$ in the semiconductor and $- Q$ on the metal plates. The voltage that accomplishes this is called the *flat-band voltage*, $V_{FB}$. The application of an external voltage results in a nonequilibrium state, but the M and S layers are so well insulated from each other that they are effectively separate systems, each of which may achieve a local thermal equilibrium. Accordingly, the bands are flat and a Fermi level exists in the M and S layers, but the Fermi levels are no longer aligned. For this particular system, the flat-band voltage would be the contact potential.

Defining a flat-band voltage would not be very utilitarian if it were not physically observable. One can, in fact, measure it quite directly. In Figure 10.2, the upper-left-hand knee occurs at the flat-band voltage. Once the flat-band condition is achieved, the application of still more negative bias creates an accumulation layer that, being completely mobile, collects at the surface of the semiconductor. Accordingly, for bias in excess of $V_{FB}$,

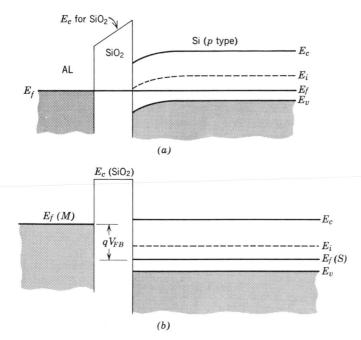

(a)

(b)

**Figure 10.4** The effect of the contact potential on the energy-band diagram for the MOS capacitor. (a) gives the thermal-equilibrium case. The Fermi levels for the M, O, and S layers are aligned. A portion of the contact potential appears across the oxide; the remainder is absorbed by the semiconductor depletion layer. (b) shows the flat-band case, the bands having been flattened by the application of an external voltage, $V_{FB}$. There is now no longer an equilibrium situation within the oxide.

the capacity of the MOS sandwich is that of the oxide layer alone, $c_i$. Notice in Figure 10.2 that $V_{FB} = -5$ V. That is rather obviously too large to be solely the contact potential, which by itself is seldom more than a volt. The sources of this extra component of $V_{FB}$ form our next topic.

## CHARGE AND CHARGEABLE STATES AT THE SURFACE AND IN THE OXIDE

As we saw in the discussion of the ohmic contact and Schottky barrier, there are states characteristic of the semiconductor surface that tend to bend the bands at the surface. Although it is hardly surprising that the surface is different from the interior, the surface states that are found do not seem to correspond to those that would be expected from the transition from crystal to vacuum. Not only are there fewer states than theory predicts, but their distribution within the gap is also highly sensitive to variations in the processing of the surface. The number of such states on a well-prepared surface can be less than $10^{10}/cm^2$. On poorly prepared surfaces, they may exceed $10^{13}/cm^2$. An extremely important property of these states is their ability to exchange charge very readily with the bulk of the semiconductor. Thus, as the Fermi level within the semiconductor is moved up and down by the application of external potentials, these states can fill or empty. Such states are called *fast-surface states*. Since these states are localized, carriers occupying the fast states are not free to transport charge. It was the high density-of-surface states in the early IGFET's (e.g., Figure 9.1) that prevented them from operating well; the charge collected on the surface was trapped and could not affect the conductivity.

To see how these fast-surface states will affect the c-V characteristics, go back to equation 10.2. The first term in that equation gives the voltage drop across the oxide for a charge, $Q$, on the metal surface. That will be unaffected by the presence of surface states in the semiconductor. The second term gives the voltage drop across the semiconductor. Since that is the voltage necessary to acquire $-Q$ by depleting the semiconductor, the presence of charge on the surface will change that term. If the surface charge due to fast-surface states is $-Q_{fs}$, then the relationship becomes

$$V = \frac{Qa}{\varepsilon_i} + \frac{(Q - Q_{fs})^2}{2\varepsilon_s q N_A} \tag{10.4}$$

Equation 10.4 may not look much worse than (10.2), but it is. Since $Q_{fs}$ represents the filling of surface states, (10.4) takes us back to the problem of solving for the Fermi level in the presence of significant numbers of deep states. We had to resort to graphical techniques then, and the situation here is somewhat worse. However, if we want to see only the sort of influence the surface states will have on the c-V curve, (10.4) will suffice.

Let us consider the simplest arrangement of surface-state energies that we can. Let us say that they all lie at the middle of the gap. Let the temperature be low enough so that the Fermi distribution function changes rapidly from 0 to 1. Finally, let the material be p-type. Now consider how the derivative of (10.4) $[1/c = dV/dQ]$ varies with $V$ as $V$ goes from the flat-band voltage to the voltage required to achieve significant inversion of the surface. Since the fast-surface states have $E_{fs} = E_i$ (by assumption), they will remain empty until the intrinsic level is bent down to the Fermi level. Thus, initially $dQ_{fs}/dV = 0$; $1/c$ is given by (10.3); and the c-V curve shows no evidence of our fast-surface states. As $E_i \Rightarrow E_f$, however, the fast states begin to fill. If the Fermi function were a step, we would have to charge the surface states before we could raise the surface potential. This would mean that the charge would be changing only at the surface; $dQ/dV = dQ_{fs}/dV$; and the capacity would be given by $1/c = 1/c_i$. Until the surface states filled, it would look as if the surface had been inverted. Since the Fermi function is not abrupt and the surface states are not all at one energy, in a realistic situation a portion of $dQ$ comes from an increase in the depletion length so that $c < c_i$.

Eventually, we will have supplied enough $V$ and $Q$ to have completely filled the surface states, $dQ_{fs}/dV \Rightarrow 0$ once more, and the $c$-$V$ curve will once again behave as if there were no surface states. Note, however, that to get over "the lump in the middle" required some extra voltage. If there were $10^{12}$ fast-surface states per $cm^2$ and the oxide thickness were 0.2 $\mu$m, how much extra voltage does it require to go from the flat-band to the in-inverted-surface condition? [Answer: $Q_{fs}$ appears on the surface of the semiconductor, so the capacity that is being charged is the oxide capacity. Since that capacity is invariant with voltage, we may write the total voltage to charge surface states (regardless of whether the charging was abrupt or gradual) as $V_{fs} = Q_{fs}/c_i$. For $c_i$ we will need the dielectric constant of quartz (about 3.78 $\varepsilon_0$). Thus, $V_{fs} = qaN_{fs}/\varepsilon_i = (1.6 \times 10^{-19} \times 2 \times 10^{-5} \times 10^{12})/(3.34 \times 10^{-13}) = 9.59$ V. That is certainly a most obvious displacement of the $c$-$V$ curve.] Since really well-prepared surfaces can have $N_{fs} < 10^{10}/cm^2$, it is possible to have essentially no displacement; on the other hand, a badly prepared surface could have such a large displacement that it would be impossible to fully charge the surface states without puncturing the oxide with the large field. (Note that for the example above, the field in the oxide required to charge the surface states is $0.48 \times 10^6$ V/cm, a pretty substantial field in anybody's oxide.) It is worthwhile to remember that breakdown in an insulating oxide is almost always a destructive process. This is because the discharge is almost always filamentary and thus associated with very high thermal dissipation. Unlike a junction device, MOS devices seldom get more than one avalanche in the insulating layer.

For the typical broad distribution of surface states within the forbidden gap and a Fermi function that takes about 0.1 V to go from 0 to 1, you should expect the charging of the surface states to result in a gradual displacement of the $c$-$V$ curve rather than anything very startling or abrupt. Figure 10.5 shows a typical example of $c$-$V$ data with and without significant numbers of surface states.

Notice that even without the presence of fast-surface states, the $c$-$V$ curve shows a displacement that is too large to be entirely due to the contact potential. This extra shift must indicate the presence of some charge that is not subject to variation with the applied voltage. The obvious place to find such charge is in the insulating layer. Two principal sources of this fixed charge have been identified. Both are evidence of charged ions within the oxide.

The first of these fixed charges to be identified turned out to be alkali

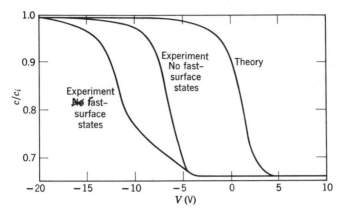

**Figure 10.5** Capacity-versus-bias voltage for an MOS capacitor with and without a significant number of surface states. (From A. S. Grove, *Physics and Technology of Semiconductor Devices*, Wiley, 1967.)

elements—principally Na—in the quartz. That it should be there is not too surprising; sodium is extraordinarily plentiful. It is an important fraction of everything in the laboratory from fingerprints to laboratory glassware. It is also quite soluble in quartz. The sodium ions in quartz turn out to be an insidious source of instability in MOS-device characteristics. Calling it a *fixed* charge is not really correct. Although it retains its charge of $+q$ regardless of the local field, it is mobile enough at room temperature and above to move around under the influence of the applied field. Since its influence on the flat-band voltage is dependent on its location in the oxide, its mobility means that the flat-band voltage may drift with time. Some pretty extraordinary (and proprietary) steps are taken to keep $Na^+$ out of the oxide in MOS devices. A common solution uses a layer of phosphorous glass on the quartz. The phosphorous glass apparently leaches the Na from the quartz.

The other source of fixed charge in quartz that has been identified is associated with the presence of Si in the $SiO_2$. [See B. E. Deal, M. Sklar, A. S. Grove, and E. H. Snow, *J. Electrochem. Soc.* **114**, 266 (1967).] In the formation of $SiO_2$, the wafer of silicon is usually placed in an $O_2$ or $O_2 + H_2O$ atmosphere at 1000°C to 1300°C. After a very short while, the reaction rate for $Si + O_2 \Rightarrow SiO_2$ is limited by the rate at which the reactants can diffuse toward each other through the $SiO_2$ layer already formed. The oxidizing agent is soluble and much more mobile than Si, so the reaction takes place very, very close to the silicon surface. However, the silicon atoms do have some solubility and mobility; there is an observable density of elemental Si in the $SiO_2$ at the interface. A fraction of this elemental silicon is ionized, the fraction being strongly dependent on the final heat treatment that the $SiO_2$ receives. (See Deal et al.'s paper referenced above.) The result is a layer of charge that is fixed in magnitude and position. It lies essentially on the Si surface (i.e., within 0.02 $\mu$m of the surface), contributing a fixed charge of about 1 to 10 $\times$ $10^{10}$ electron charges per $cm^2$. Since this rather stable charge lies so close to the surface, it is frequently referred to as a *slow-surface charge* to distinguish it from the fast-surface states that can readily exchange charge with the bulk of the semiconductor.

The effect of the charge in the insulating layer on the flat-band voltage may be obtained directly from the observation that when the flat-band condition has been achieved, the field normal to the surface of the semiconductor must be zero. (This is just another way of saying that if the bands are flat, there is no accumulation of charge in the semiconductor; no charge, no field.) Considering the surface of the semiconductor to be $V = 0$, the potential at the metal plate is then obtained by integrating Poisson's equation twice for the charge in the insulator, $dQ_i = \rho_i dx$:

$$dV(dQ_i) = \frac{xdQ_i}{\varepsilon_i} = \frac{\rho_i xdx}{\varepsilon_i}$$

or, on integration:

$$V_{FB} = -\int_0^a \frac{\rho_i xdx}{\varepsilon_i} = \frac{-1}{c_i} \int_0^a \frac{\rho_i xdx}{a} \tag{10.5}$$

The last form of (10.5) states that the charge is weighted by the fraction of its distance across the oxide. Thus, for the slow-surface charge, $Q_{ss}$, the shift in the flat-band voltage is $Q_{ss}/c_i$. For $10^{11}$ slow states per $cm^2$ and an oxide 0.2 $\mu$m thick, the flat-band voltage is shifted by 0.97 V. An equivalent voltage for a uniformly distributed charge throughout the oxide layer requires about $10^{16}$ ions/$cm^3$ (i.e., $2 \times 10^{11}/cm^2$ in a layer $2 \times 10^{-5}$ cm thick). Note that both of these shifts depend on the oxide thickness, so that a voltage shift without an oxide thickness tells you nothing about the effective surface charge. The data in Figure 10.5 show a shift of 7 to 8 V without fast-surface states and a shift in excess of 15 V for the case with substantial numbers of fast-surface states. Considering the numbers given above for typical values of $Q_{ss}$, these relatively large shifts are an indication of a fairly thick oxide layer. For a 0.2 $\mu$m layer, a shift of about 2 V might be considered more typical.

It is important to note that both the Na and Si ions will tend to make the silicon surface become more *n*-type. That the silicon surface is *naturally* *n*-type has some important ramifications for both bipolar and FET technology. For example, if you measure $c$ versus $V$ for an MOS sandwich in the middle of a lightly doped $p$-type wafer, you will *always* get the "low-frequency" behavior such as that labeled 10 Hz in Figure 10.2. ("Always" in the previous sentence assumes that you don't have a 100 MHz capacitance bridge.) The reason for this is that with no bias applied, the entire surface is already inverted. The bias applied to the metal dot affects the region immediately under the dot. The remainder of the surface is always thoroughly loaded with electrons. Thus, when the surface under the metal dot begins to invert, that region is readily supplied with electrons from the surrounding inversion layer. Not only does this yield anomalous $c$ versus $V$ data, it also means that $n$-type regions—either diffused or metallurgically generated—can extend beyond the regions intended. [For a complete discussion of the frequency response of the $c$ versus $V$ data, see E. H. Nicollian and A. Goetzberger, *IEEE Trans. on Electron Devices*, ED-12, 108 (1965).]

The extension of the $n$-type region is chiefly responsible for the difference in breakdown voltage for diffused $n$-on-$p$ and $p$-on-$n$ junctions. For the $n$-on-$p$ junction, the extension of the $n$-region into the $p$ reduces the junction curvature, thus raising the breakdown voltage. For the $p$-on-$n$ junctions, on the other hand, the $p$-type region is bent back at the surface, increasing the curvature of the junction. This increase in curvature is sufficient in itself to lower the breakdown voltage, but if the $n$-type region extends far enough, it implies that a heavily inverted region is adjacent to a very heavily doped region. This can easily result in tunneling between the two regions at the surface. This lateral current flow would be bad enough in a diode; in a $p$-$n$-$p$ transistor it is a disaster. The lateral current flow adds directly to the base current without contributing anything to the collector current. This makes it relatively difficult to get a high-$\beta$ $p$-$n$-$p$ transistor in silicon. Effective designs require $p$-type guard rings to terminate any induced $n$-type channels.

(A series of very elegant experiments to show that this surface-channel formation is indeed responsible for many of the problems with the planar technology are described by A. S. Grove, *Physics and Technology of Semiconductor Devices*, Chapter 10, Wiley, 1967.)

## A COMMENT ON THE EXTENT OF OUR COVERAGE OF SURFACE STATES

Probably no other single part of device technology has received more attention than the characteristics of the semiconductor surface. There is a great deal of both science and black magic in getting surfaces to behave themselves. The intense interest in good surface control is a direct result of the fact that most junctions terminate at a surface. If that surface is not properly prepared and protected, the junction will leak and break down at the surface long before it would within the bulk. Thus, from the earliest days of the bipolar transistor (which itself was a by-product of the study of the germanium surface), the demand for semiconductor devices has stimulated an extraordinary interest in semiconductor surfaces.

The literature on the subject of surfaces is voluminous and still growing rapidly. We have covered only the rudiments of the MOS system, developing only those details that are needed for a simple analysis of the MOSFET. For those interested in obtaining a detailed introduction to semiconductor-surface physics, a clear presentation of the knowledge up to 1965 can be found in A. Many, Y. Goldstein, and N. B. Grover, *Semiconductor Surfaces*, Wiley, 1965. As mentioned above, a reasonably thorough summary of the influence of the surface on *p-n* junction behavior can be found in A. S. Grove, *Physics and Technology of Semiconductor Devices*, Chapter 10, Wiley, 1967. The $SiO_2:Si$ system receives some detailed examination in several of the chapters of J. T. Wallmark and H. Johnson, Eds., *Field-Effect Transistors*, Prentice-Hall, 1966. All three of these references contain detailed and frequently nonoverlapping bibliographies. Some of the more recent work is found in S. M. Sze, *Physics of Semiconductor Devices*, Chapter 9, Wiley, 1969.

For our purposes, we have what we need: the concepts of surface depletion and inversion, and the several factors that determine the flat-band voltage. With these simple but worthy tools in hand, we may now proceed to "discover" the MOSFET.

## MOSFET STRUCTURES

Although effective MOSFET's may be realized in a number of different ways, each with its benefits to device performance or manufacturing success, the structure most commonly used is that shown in Figure 10.6. In this MOSFET, the source and drain regions are diffused into a moderately doped substrate and a thin, clean oxide layer grown over the intervening *p*-type material. Aluminum is then evaporated over the surface and etched to leave the gate and the contacts for the source and drain. Note that only one diffusion is required. This makes the MOSFET simpler (in principle) to process than the bipolar transistor. However, hidden in the phrase "clean oxide" are enough difficulties to keep even the most sanguine process engineer well entertained.

In the structure shown in Figure 10.6, the channel is induced by the combined action of the applied and built-in bias. Since the bias is used to turn the conductivity on, such structures are called *enhancement-mode* transistors. Note that the sign of the gate bias is the same as that of the drain bias.

It is also possible to build the channel in just as in the JFET. For example, if a thin *n*-type layer had been grown epitaxially on the *p*-type substrate in Figure 10.6, a metallurgical channel would exist with no applied bias. The

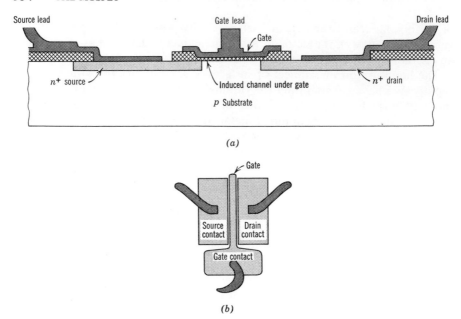

**Figure 10.6** Structure of a simple rectilinear insulated gate FET. The dark regions in the cross-section represent aluminum; the hatched regions are quartz. As usual, scale is sacrificed to clarity. Typical channel lengths are 2 to 10 $\mu$m; the gate oxide layer is usually about 0.1 $\mu$m thick; the wafer (substrate) is typically 200 $\mu$m thick. (*a*) Cross-section; (*b*) plan view showing the aluminization pattern that would be seen when looking down on the finished device.

gate bias would then be used to deplete the *n*-type layer just as it is in the JFET. Such transistors are said to operate in the *depletion* mode. Depletion transistors require a gate bias of opposite sign from the drain bias. This means that they are not easily cascaded in direct-coupled amplifiers.

It is important to remember that both types of MOSFET's will have their characteristics shifted somewhat by the built-in gate bias. Thus, either kind can be manufactured to be on or off at zero applied gate bias. A *normally on* (on for zero gate bias) transistor has the advantage that the gate electrode need not cover the entire channel. Since the gate-to-drain capacity is so critical in establishing the upper frequency of operation for a transistor, leaving the drain end of the channel uncovered can result in a substantial improvement in performance. Such gates are called *partial* gates.

Whether the channel is induced or built in, its behavior is almost the same as in the JFET. There are two important differences, however. First, since the oxide layer is a true insulator (in contrast to the *p-n* junction, which is an insulator for one polarity of bias only), the channel conductance may be varied by the application of bias of either sign. For normally on devices, this means that one can have DC amplification at zero bias—occasionally useful—and one need not worry about circuit or component damage or a short circuit if the gate voltage changes sign.

The second difference is the relative inefficiency of the substrate as a gate. Five volts applied to the substrate will usually have less effect than one volt applied to the gate. Since the gates in the JFET behave reasonably symmetrically, why is the substrate so inefficient in the MOSFET? (*Answer:* In the JFET, both gates are heavily doped with respect to the channel. Thus, the junction depletion region is almost entirely in the channel and the applied voltage almost entirely devoted to modulating the channel conductance. For the MOSFET structure of Figure 10.6, most of the substrate bias appears across the wide-substrate depletion region. Accordingly, very little voltage remains to modulate the channel conductance. Another way to look at the same thing: Since the depletion region is typically almost an order of magnitude wider than the insulating oxide, the gate electrode is much more strongly coupled to the channel than the substrate is.) However, even though the gate is the dominant electrode, the substrate is far from unobtrusive. That it should increase the Miller capacity is fairly obvious. However, it also influences the output conductance in the saturated mode and provides the highly temperature-dependent leakage currents typical of junction diodes. Whether any of these effects will bother you depends rather obviously on the application you have in mind. In general, unfortunately, the substrate is too important to ignore.

It is possible to eliminate the substrate entirely by making a MOSFET along the lines originally conceived by Lilienfeld and Heil (Figure 9.1). Although technology has still to catch up with concept in these *thin-film* transistors (TFT), they open whole new vistas to the transistor engineer. In principle, all you need for a depletion-mode MOSFET is a thin film of semiconducting material with an insulated gate in the middle flanked by two ohmic contacts. Since no junctions are required, polycrystalline material may be used. (I.e., there is no problem with diffusion along grain boundaries, since no diffusions are performed.) This means that any semiconductor that can be deposited in a convenient fashion (e.g., evaporation, pyrolytic decomposition, precipitation, etc.) on an inert insulating substrate may be used regardless of whether it can be doped both *p* and *n* type, grown as a single crystal, or processed at elevated temperatures. Since the manufacture of thin film passive components such as resistors and capacitors is quite compatible with TFT production, it would seem that thin-film microcircuits would have captured the market. They probably would have, were it not for the inconvenient fact that they don't generally work as well as their single-crystal cousins. The polycrystalline film seldom yields a mobility even close to that of single-crystal material, and the critical problem of controlling the surface-state density is much more severe in thin films than single-crystal surfaces. For example, the orientation of the surface is found to have a large effect on the density of slow-surface states. Single-crystal wafers with any crystalline orientation are readily obtainable. [ (100) surfaces seem to be best and are most widely used.] Getting a thin, polycrystalline film to orient itself, however, is exceedingly difficult.

Generally speaking, TFT's have not yet become commercially competitive with monolithic circuitry, although there does not seem to be any fundamental reason why they cannot. Whether TFT's ever become very important is more likely to hinge on economic advantages than on device physics.

## CALCULATION OF THE CHANNEL CONDUCTANCE FOR AN ENHANCEMENT-MODE MOSFET

As the first step in analyzing the drain characteristics of a MOSFET such as that in Figure 10.6, we must determine the conductance of a small length of channel as a function of its voltage with respect to the gate and substrate. If we assume that the free carriers in the channel have a well-defined mobility, the conductivity is directly proportional to the amount of free charge in the channel. This reduces the problem of finding the conductivity to one of determining the fraction of the field lines emanating from the gate that terminate on mobile charge. The remainder will terminate on the fixed charge within the depletion region. Although this problem is easily stated as a one-dimensional version of Poisson's equation, an easy solution is barred by the same difficulties that led us to make the depletion approximation in Chapter 3. The problem is that the density of mobile charge is exponentially related to the voltage. Although the depletion approximation will still stand us in good stead where there is a negligible density of mobile charge, that is insufficient for the MOSFET problem, where we are vitally interested in the solution in the highly conductive channel region. Here we require a second clever approximation first proposed (I believe) by H. K. J. Ihantola in 1961.

Ihantola's method was to break the problem into two parts. He first considers the case in which no inversion layer exists. For that case, the variation of gate-to-substrate bias is accompanied by variation of the depletion width in the substrate. The analysis of this variation is what we just did for the MOS capacitor; it is developed neatly and simply using the depletion approximation. The second part of the analysis begins after the formation of the conducting channel. Since the free-carrier concentration is exponentially dependent on the surface potential, once a channel forms we may assume

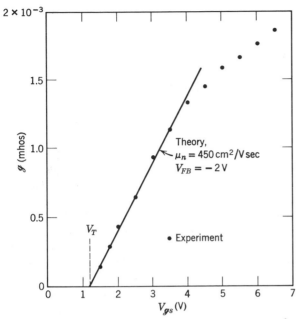

**Figure 10.7** Channel conductance versus gate-to-substrate bias in a MOSFET. These measurements were taken for very small drain voltages, so the channel conductance is undisturbed by the drain current. (From A. S. Grove, *Physics and Technology of Semiconductor Devices*, Wiley, 1967.)

that the depletion width remains fixed and that any additional gate bias results in an increased number of mobile carriers in the conducting channel. Since the channel is behaving like a conductor, this second problem is nothing more than the parallel-plate capacitor.

There is one arbitrary step in this analysis: We must assume some specific potential at which the channel is formed. It is obvious that the channel does not form abruptly at some specified voltage, but like the transition from bulk to depletion region, the transition from little conductance to much conductance takes place with little enough voltage change to be considered abrupt.

Once the channel is formed, parallel-plate capacitor analysis tells us that the amount of mobile charge on the "lower plate" is given by

$$Q_e = c_i(V - V_T) \tag{10.6}$$

where $c_i$ is the capacity of the insulator layer and $V_T$ the substrate-to-gate voltage at turn-on. If this charge is really mobile with a mobility that does not depend on the surface field, (10.6) states that the conductance of the channel should be proportional to the voltage in excess of $V_T$. Figure 10.7

shows that this is pretty much the case. Ihantola's assumption of an abrupt turn-on and a linear $g$ versus $V$ curve looks quite reasonable.

What just happened may seem so simple that you missed its value. Restating it: The assumption that the transition from depletion to inversion at the surface takes place abruptly at a particular surface potential makes it very easy to calculate the conductance of the channel as a function of the local surface potential. When the MOSFET is used as a transistor, the local surface potential will be a function of the current and the various biases. To relate the terminal voltages to the drain current, we must be able to integrate the potential drop along the channel. Having a ready expression for the incremental channel conductance, as we did for the JFET, is the essence of obtaining a solution.

The only thing standing between us and a direct solution for the terminal characteristics of the MOSFET is making a suitable choice for the turn-on voltage, $V_T$. Let us begin with an examination of the electron potential as a function of position in the device of Figure 10.6. If no bias is applied within the substrate itself (including the two $n$ regions), the substrate will be in equilibrium. Two such equilibrium diagrams are shown in Figure 10.8. The upper figure shows the electron potential as a function of position with the flat-band voltage applied to the gate. Since electrons tend to "fall down" in such a diagram, the electrons will be found in the low spots corresponding to the $n$-type regions. These are separated from the $p$-type ohmic regions by steep cliffs representing the equilibrium depletion regions.

The second figure shows the equilibrium case with the gate sufficiently positively biased to invert the surface. This gate-to-substrate bias must be enough to lower the surface potential energy for electrons in the channel region to the point at which a noticeable number of states are filled. It seems reasonable to say that the number of electrons is "noticeable" when the surface concentration is typical of a lightly to moderately doped $n$-type material. Since the substrate, with its Fermi potential $\phi_p$, is lightly to moderately doped $p$-type material, the surface will meet our noticeability criterion when the Fermi potential at the surface, $\phi_s$, is given by

$$\phi_s = -\phi_p \tag{10.7}$$

(N.B. $\phi_p \leq 0$.) According to (10.2), to go from the flat-band condition of Figure 10.8$a$ to the inverted surface of Figure 10.8$b$ will require a change in surface potential of at least

$$\Delta\phi_s = -2\phi_p \tag{10.8}$$

Using (10.8) and our assumption about the abrupt change from depletion to inversion, we may now calculate the change in voltage to go from the flat-band condition to the formation of a conducting channel between the two diffused $n$-type regions.

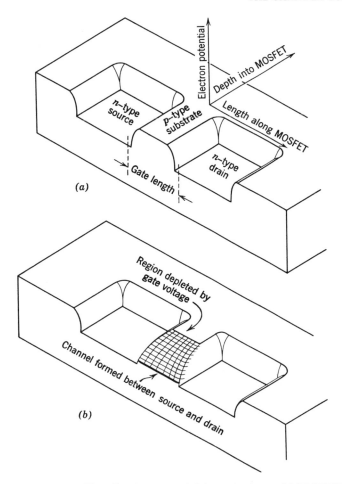

**Figure 10.8** The electron potential in an $n$-channel MOSFET ($z$-axis) as a function of position in the device. For both figures, the source, drain, and substrate are at the same voltage. ($a$) shows the flat-band condition and ($b$) shows the potential at "turn-on," with sufficient gate bias to form a channel between source and drain.

At turn-on, the surface potential differs from the bulk substrate potential by the amount, $-2\phi_p$. Since this potential difference is the result of a depletion region, we may use (3.47) and (3.48) to give us the total charge stored in the depletion region, $Q_D$. The voltage across the insulator layer is then simply $-Q_D/c_i = -Q_D a/\varepsilon_i$ from (10.2). From the depletion approximation we have

$$Q_D = -qN_A l_p = -\sqrt{-4\varepsilon_s\phi_p q N_A} \qquad (10.9)$$

Measuring the turn-on voltage, $V_T$, from the flat-band point, (10.9) gives

$$(V_{gb} - V_{FB})|_{\text{turn-on}} = V_T = -Q_D/c_i - 2\phi_p \qquad (10.10)$$

[The subscript, $b$, refers to the substrate (or *base*). $s$ will be used for the source.] Typical numbers for the terms in (10.10) are $Q_D = -10^{-8}$ coulombs/cm$^2$, $c_i = 10^{-8}$ F/cm$^2$, and $\phi_p = -0.3$ V. Thus, the difference between Figures 10.8a and 10.8b corresponds to about 1.5 to 2 V between gate and substrate.

Since any additional gate voltage will, according to our assumption, be accommodated by an increase in mobile charge on the surface with little change in $Q_D$, we may generalize (10.10) to include a substantial amount of surface charge, $Q_s$:

$$V_{gb} - V_{FB} = -(Q_s + Q_D)/c_i - 2\phi_p \qquad (10.11)$$

Generally, we are most interested in the mobile surface charge, so we may rewrite (10.11) as

$$Q_s = -Q_D - c_i(V_{gb} - V_{FB} + 2\phi_p) = -c_i(V_{gb} - V_T) \qquad (10.12)$$

It should be possible to go directly from the free charge to the channel conductance, although you might well have at least a twinge of doubt about whether the mobility in a surface layer is well behaved. Such doubts are well founded, but, for a moment, let us assume that the concept of mobility is as applicable to the surface as it is to the bulk. Under that assumption, the conductance of a channel with electron charge per unit area, $Q$, length, $L$, breadth, $B$, and mobility, $\mu_e$, is given by

$$g(V_{gb}) = -\mu_e QB/L = \mu_e c_i(V_{GS} - V_T)B/L \qquad (10.13)$$

Figure 10.7 shows how well (10.13) stands up. Pretty well indeed! However, note that the measured mobility is noticeably less than might be expected for the bulk and that at sufficiently large values of $Q$, the conductance does begin to fall off from that predicted. These are both indications that the bulk and the surface are not entirely similar. We shall return to this anon. In the meantime, Figure 10.7 indicates that we may use (10.13) with the expectation of reasonable accuracy.

## THE CHANNEL CONDUCTANCE FOR THE NONEQUILIBRIUM CASE

Until this point, we have assumed that the substrate, the source, and the drain are in thermal equilibrium with each other. Such an assumption is based on the fact that we applied no bias between any of these terminals; with it we were able to establish the turn-on surface potential as $-2\phi_p$.

In normal operation, the source and drain will be at substantially different voltages and the substrate at still another. Thus, we must extend our analysis to the nonequilibrium case in which currents are flowing and the channel conductance is a function of the potential differences between all four MOS-FET terminals. It is likely that you will find this analysis somewhat less "intuitive" than the JFET's because of the addition of the substrate terminal. Hopefully, Figure 10.9 will help to elucidate what is going on as the bias at each terminal is varied.

If we began with the equilibrium case shown in Figure 10.8b and then applied some reverse (positive) bias to both the source and the drain, we would obtain the potential-energy diagram shown in Figure 10.9a. The substrate is taken as the reference point for all voltages. Thus, the application of reverse bias to the source and drain lowers the potential energy of the electrons in these two regions. For the most part, this simply results in an increase of the depletion-region width (mostly by motion into the lightly doped substrate) with its accompanying reverse current. At the surface under the gate, however, something more interesting goes on. The electrons that were initially occupying the channel can now "fall" into the source and drain, and they promptly do so. This decreases the conductivity of the channel. When the positive bias on the source and drain becomes great enough, there are so few carriers in the channel that it no longer can be considered as conducting. The field lines from the positively biased regions must terminate on negative charges. The only source of these is the substrate-depletion region, so the substrate-depletion region expands. If the channel is long enough, the middle will still be inverted, but this inverted region no longer connects the source and drain, since the middle represents a potential barrier that is too high for the carriers in the source or drain to surmount. In fact, if there is to be an inverted region in the center, there must be some leakage of electrons out the edges, a loss of charge that must be replaced by generation within the depletion region. Since this source of charge is quite limited, the inverted region must be well beyond the depletion region caused by the reverse bias between the substrate and the n-type material.

If we want to reestablish the channel in the presence of the reverse bias on source and drain, we must "pull" the surface potential toward the level of the n regions. There is no reason why we couldn't do this by applying more positive bias to the gate. That will push the depletion region back into the substrate until the channel forms all the way from source to drain. It should be evident in the drawing that if the bias from substrate to source and drain is − 1 V, the gate must supply + 1 V of *surface potential* (requiring even more gate bias) before the channel reforms.

There is, of course, no reason why the source and drain should be at the same bias; they usually are not. The most typical case has the source tied to

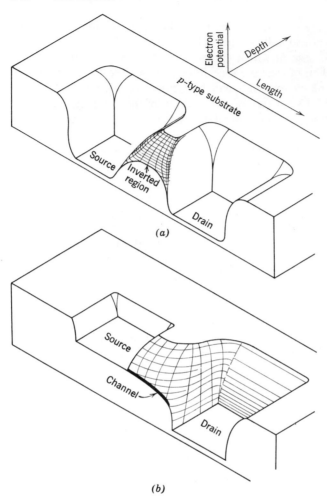

**Figure 10.9** The electron potential in an *n*-channel MOSFET. For both cases substantial bias is applied between the substrate and the drain. (*a*) In this figure the gate-substrate voltage is the same as in Figure 10.8*b*. However, here the source and drain are strongly reverse biased with respect to the substrate. Thus, although an inverted region exists, no continuous channel is formed. (*b*) The MOSFET operating in its normal saturation mode. The source and substrate are at the same voltage. The gate and drain are both quite positive.

the substrate and the drain at substantial positive bias. This situation is illustrated in Figure 10.9*b*. Such a bias arrangement introduces a very interesting new feature: Even if the drain is reverse biased to the point that the channel is completely depleted at the drain end, the drain current is no longer limited by generation in the substrate-depletion region. There is a ready supply of electrons at the source. The operation in the condition illustrated is very similar to the saturation mode of the JFET. The gradient in voltage from source to drain is accompanied by a gradient in the conductivity. To maintain the same current, the carriers are obliged to travel faster and faster as they go from source to drain. If this is carried far enough, the electrons reach their saturated-drift velocity, and the current becomes essentially independent of the drain voltage. This is the situation illustrated in the last figure.

To calculate the drain current for various bias arrangements, we proceed in much the same fashion as for the JFET. The channel conductance is determined for biases that give a continuous channel from source to drain. The calculation is presumed to hold up to the point where the channel just vanishes at the drain end. Beyond that point, the drain current is expected to be almost independent of drain voltage.

## GENERAL SOLUTION FOR THE UNSATURATED CHANNEL CONDUCTANCE

Figure 10.10 shows the situation we wish to analyze. A channel has been induced between the source and drain. All four terminals may be at different potentials. For convenience, let us ground the substrate terminal ($B$) and reference the channel voltage, $V_{cb}(x)$, at any point in the channel to this ground potential. Our purpose is to relate the drain current to the drain-to-source voltage. If we can determine the resistance, $dR$, of the element of channel length $dx$, we may obtain the voltage from source to drain by integrating $I_d dR = dV$ from source to drain.

To make the determination of $dR$ simple, two reasonably plausible assumptions are in order. As with the JFET, we first assume that the current flows sufficiently parallel to the surface to obviate the need for doing a two- or three-dimensional field-theory problem. The thinness of the conducting channel makes this a pretty safe assumption as long as the drift velocity has not saturated. Our next one, however, is not so obvious. We must assume that the carrier mobility throughout the channel is a constant. For the JFET case, we ignored the gradualness of the saturation of the carrier drift velocity near the drain on the premise that the saturation was effectively the same as channel pinch-off. Since both effects would produce current saturation, they are functionally similar. For a first-order analysis it is much, much easier to pretend that the saturation is the result of a true pinch-off than it is to do a legitimate calculation of $I_d$ versus $V_{ds}$, including drift saturation.

The MOSFET case is somewhat less obvious, since the mobility of carriers in the surface layer need not be as well behaved as bulk mobility. There is evidence in Figure 10.7 that large-surface carrier densities adversely affect the carrier mobility, but that same figure suggests that over a pretty wide range of carrier density, a reasonably constant mobility can be defined. We might hope that the drain behavior could be well represented by the pinch-off myth of the JFET. Such a hope will not be badly disappointed.

Having assumed constant mobility and plane flow, we may write $dR$ from (10.13) as

$$dR = - \frac{dx}{\mu Q_s(x)B} \qquad [Q_s \leqq 0] \qquad (10.14)$$

where $B$ is the channel breadth and $Q_s$ the surface mobile-charge density in coulombs/cm$^2$. The total charge density in the element $dx$ in Figure 10.10 comprises the mobile charge, $Q_s$, indicated by the dark-gray area, plus the depletion-layer charge, $Q_D$, indicated by the light-gray area. The gate and substrate biases are not functions of position, but since the channel bias is, both the depletion-layer charge and the mobile charge are functions of position along the channel. The voltage across the gate insulating layer is related to the surface charge by logical extension of (10.12):

$$Q_s(x) = - Q_D(x) - c_i \left[ V_{gb} - V_{cb}(x) - V_{FB} + 2\phi_p \right] \qquad (10.15)$$

Similarly, $Q_D$ may be obtained from (10.9) as

$$Q_D(x) = - \sqrt{2\varepsilon_s \left[ V_{cb}(x) - 2\phi_p \right] qN_A} \qquad (10.16)$$

Putting (10.15) and (10.16) into (10.14) yields

$$dV_{cb} = \frac{I_d dx}{\mu B} \left\{ \sqrt{2\varepsilon_s qN_A(V_{cb} - 2\phi_p)} + c_i(V_{gb} - V_{cb} - V_{FB} + 2\phi_p) \right\}^{-1} \qquad (10.17)$$

Rearranging (10.17) and integrating from the source ($V_{cb} = V_{sb}$) to the drain ($V_{cb} = V_{db}$) yields the mildly obnoxious relationship between drain current and drain voltage:

$$I_d = (\mu B/L) \left\{ - \tfrac{2}{3} (2\varepsilon_s qN_A)^{1/2} \left[ (V_{db} - 2\phi_p)^{3/2} - (V_{sb} - 2\phi_p)^{3/2} \right] + \right.$$
$$\left. + c_i \left[ (V_{gb} - V_{FB} + 2\phi_p - V_{db}/2) V_{db} - (V_{gb} - V_{FB} + 2\phi_p - V_{sb}/2) V_{sb} \right] \right\} \qquad (10.18)$$

Equation 10.18 is the proper expression for the drain current from the point where the gate voltage turns on the channel ($V_{gb} = V_T$) to the point where the channel pinches off at the drain end ($V_{db} = V_{d\,sat}$).

To determine these "end points" for equation 10.18, we must find the appropriate gate bias to open the channel at the source end for a given $V_{sb}$ and

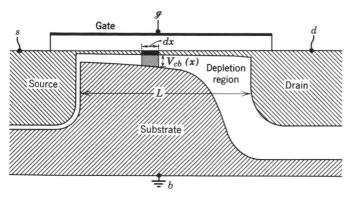

**Figure 10.10** Diagram for the calculation of the channel conductance. The depletion region separating the substrate from the source, channel, and drain regions is shown as a clear area. The channel length is $L$, its breadth $B$, and its depth (not used in the calculation) is a function of the surface potential.

the drain voltage necessary to shut off the channel at the drain end for a given $V_{gb}$. Both of these may be obtained from (10.15) and (10.16), since the mobile charge, $Q_s$, is zero at turn-on or turn-off. The turn-on gate-to-substrate voltage is given by

$$V_T = V_{FB} + V_{sb} - 2\phi_p + \sqrt{\frac{2\varepsilon_s q N_A}{c_i^2}(V_{sb} - 2\phi_p)} \qquad (10.19)$$

If you compare this with (10.10) with (10.9) inserted, you will see that $V_{sb}$ behaves just like an additional inversion potential (i.e., just like $-2\phi_p$). The pinch-off voltage at which the drain current saturates is found in a similar way to be

$$V_{d\,sat} = V_{gb} - V_{FB} + 2\phi_p + \frac{\varepsilon_s q N_A}{c_i^2}\left[1 - \sqrt{1 + \frac{2c_i^2}{\varepsilon_s q N_A}(V_{gb} - V_{FB})}\right]$$

$$(10.20)$$

When the substrate is lightly doped and the gate oxide layer is thin (i.e., $c_i$ large and $N_A$ small) for sensible values of $(V_{gb} - V_{FB})$, (10.20) takes on the particularly simple form:

$$V_{d\,sat} \simeq V_{gb} - V_{FB} + 2\phi_p \qquad (10.20')$$

Quantitatively, lightly doped means of the order of $10^{14}/\text{cm}^3$ and thin means of the order of $10^{-5}$ cm. These are roughly the lower bounds for practical processing, so the simple form of (10.20') can be encountered in practice.

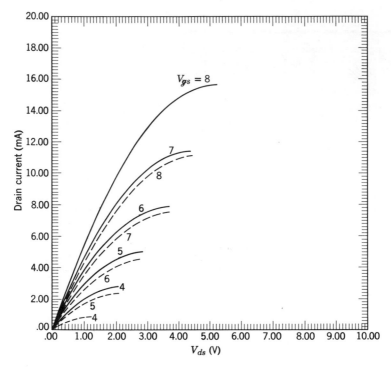

**Figure 10.11** The drain characteristics of a large *n*-channel enhancement MOSFET. These curves are plots of equation 10.18 with the following choices of parameters:

$$N_A = 10^{15}/cm^3 \qquad V_{FB} = 0 \qquad L = 10\ \mu m$$

$$B = 1\ mm \qquad a = 0.2\ \mu m$$

The remainder of the parameters are values typical for Si and $SiO_2$. The solid lines are for $V_{sb} = 0$; the dashed lines for $V_{sb} = 4$ V.

You would not be considered unduly pessimistic if you expressed some doubt about the analytic usefulness of (10.18) and chortled a bit over putting (10.20) into (10.18). They are not really very appetizing. To use (10.18) directly, a digital computer with plotting capability is a most helpful companion. It yields curves such as those shown in Figures 10.11 through 10.13. Plots of (10.18) are the drain characteristics for an *n*-channel enhancement mode FET. These are shown in Figure 10.11 for a large device with a substrate of moderate resistivity (about 10 ohm-cm material). Note that reverse-biasing the substrate by 4 V has about the same effect as lowering the gate bias by 1 V. How would this control ratio of 1:4 change if the substrate resistivity were increased? (*Answer:* The difference between the effectiveness

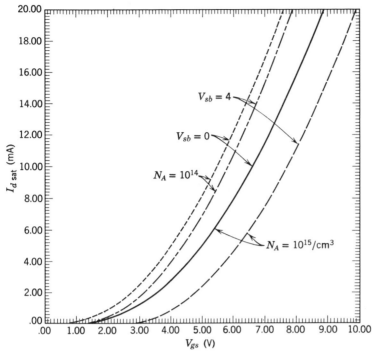

**Figure 10.12** The drain current at saturation versus the gate bias. The dimensions of the *n*-channel enhancement MOSFET are the same as for Figure 10.11. The curves show how the drain current depends on gate bias for two different substrate dopings, each with two examples of substrate bias.

of the two gates increases with increasing substrate resistivity. This is because the depletion region is wider in lightly doped material.) Figure 10.11 should be compared with the JFET curves in Figure 9.7. Both devices show a linear gate-controlled resistance for small values of $V_{ds}$ and a saturated-drain current at large values of $V_{ds}$. Although it need not be the case, the gate voltages for typical JFET's are lower than the gate voltages employed with the MOSFET.

The saturated-drain characteristics are shown in Figure 10.12. Two values of substrate doping are illustrated. Note that the more lightly doped substrate has a higher saturation current at any given gate bias. The more lightly doped one also shows less response to the substrate bias. This is no mere coincidence; the current at the pinch-off point is determined in part *by the effectiveness of the substrate gate.*

It is important to note that the curves in Figure 10.12 cannot be extended indefinitely. Not only do the unmodulated parasitic source-and-drain resistances enter the calculations at high current densities, but as Figure 10.7 clearly shows, the average mobility of the carriers begins to fall when the carrier density at the surface gets too large. In much the same way as $\beta$ falls off at high current densities in bipolar transistors, the gate loses effectiveness at high drain current in a MOSFET. This fall off in $\mu$ is blamed on the surface scattering. As $I_d$ increases, the fraction traveling near the surface also increases.

For small-signal analysis in the saturated mode, we want the derivatives of the curves in Figure 10.12. These are given in Figure 10.13. For the lightly doped substrate, it is pretty clear that the transconductance is essentially linearly proportional to the gate voltage. If $g_{fs}$ is linearly dependent on $V_{gs}$, $I_{dsat}$ should be proportional to $V_{gs}^2$. That certainly sounds a lot more pleasant than anything (10.18) suggests to the casual observer. However, since Figure 10.13 comes directly from (10.18), it would appear that some very useful simplification might be possible. That worthy goal we now pursue.

The approximation that rapidly reduces (10.18) to useful proportions is the same one that yields (10.20′): If the gate insulating layer is sufficiently thin and the substrate doped lightly enough, then

$$\sqrt{\varepsilon_s q N_A} \ll c_i \sqrt{V} \tag{10.21}$$

for any of the $V$'s in (10.18). If (10.21) holds, we may use (10.20′) and also drop the terms that are raised to the three-halves' power in (10.18). Observing that the first three terms in the inner brackets of the remaining portion of (10.18) sum to $V_{dsat}$ as given by (10.20′), we may write (10.18) as

$$I_d = (\mu B c_i/L)\left[\,(V_{dsat}V_{db} - V_{db}^2/2) - (V_{dsat}V_{sb} - V_{sb}^2/2)\,\right]$$

which may be readily rearranged to yield

$$I_d = (\mu B c_i/2L)\left[\,(V_{d\,sat} - V_{sb})^2 - (V_{d\,sat} - V_{db})^2\,\right] \tag{10.22}$$

This is certainly a vast improvement over (10.18).

If we make the usual assumption that the drain current for $V_{db} > V_{d\,sat}$ is independent of the drain voltage, (10.22) gives us

$$I_{d\,sat} = (\mu B c_i/2L)(V_{d\,sat} - V_{sb})^2 \tag{10.23}$$

Then, differentiating (10.23) to get the transconductance (called either $g_m$ or $g_{fs}$) gives

$$g_m = (\mu B c_i/L)(V_{d\,sat} - V_{sb}) = (\mu B c_i/L)(V_{gb} - V_{FB} + 2\phi_p - V_{sb}) \tag{10.24}$$

These last two equations should be compared to those for the JFET (Equations 9.12 and 9.13). In both cases, we find that the FET's behave as square-law devices. Although such a result is only approximate for either device, the relatively low content of higher-order terms in (9.12) and (10.23)

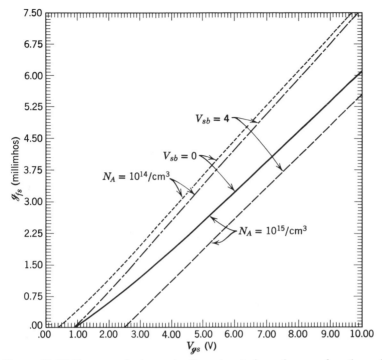

**Figure 10.13** Transconductance in the saturated mode as a function of gate bias. These curves are the derivatives of the curves in Figure 10.12. They should be compared with the straight lines predicted by (10.24).

implies a very high resistance to the deleterious effects of intermodulation distortion.

## THE MODEL COMPARED TO A REAL MOSFET

Since the results we have obtained so far are all based on a number of simplifying assumptions, an example from the real world is certainly in order. Figure 10.14 presents the manufacturer's spec sheets for a pair of enhancement-mode, $n$-channel MOSFET's. Since the manufacturer indicates that this transistor family is designed for low-power audio applications, it is probably safe to guess that he has attempted to minimize the noise figure (something we have not yet discussed) while relaxing the requirements on the gate capacities and the various breakdown voltages. To see how well such devices fit our analytical model, let us begin with the large-signal characteristics in Figure 10.14$c$.

First, note that this $n$-channel FET is quite distinctly "on" with no applied gate bias. Our discussion of the natural tendency of the silicon surface to turn $n$-type certainly makes this a plausible result. This offset of the turn-on voltage makes the operational difference between enhancement- and depletion-mode devices somewhat obscure, since both may be operated with gate biases of either sign. For the somewhat more common $p$-channel silicon MOSFET, on the other hand, the difference is usually quite clear: The enhancement-mode device is "off" at zero gate bias; the depletion-mode device is normally "on."

From the data given in Figure 10.14c, estimate the flat-band voltage for the 2N3796. [Answer: The turn-on voltage is evidently about $-3$ V. From (10.19), making the usual simplifying assumption that the term within the radical is negligible, we have $-3 = V_{FB} - 2\phi_p$. In moderately lightly doped silicon, $\phi_p$ is of the order of $-0.3$ V. Accordingly, $V_{FB} \cong -3.6$ V, a not implausible value.]

Now consider the saturated-drain characteristics. How well does our simple model of equation (10.23) fit the given data? Before making that comparison, it is important to observe that the manufacturer is not really plotting $I_{d\,\text{sat}}$; his measurements are not made at the pinch-off point. At best, his data represent $I_{d\text{sat}}$ plus the current resulting from channel-length modulation. This makes it a bit pointless to try to fit the more complete equation 10.18 to the data. As it turns out, (10.23) gives a reasonably encouraging fit with the parameters $V_T = -3$ V and $(\mu B c_i/2L) = 0.18$ mA/V$^2$. Some points calculated from (10.23) are shown on the transfer characteristics of the 2N3796. The fit is certainly better than needed to characterize the "typical" device.

It should follow that if (10.23) fits the transfer characteristics, (10.24) should give the transconductance. The result of trying that experiment is shown as the dashed line in the upper-left-hand graph in Figure 10.14d. Note that the curves are plotted with drain current rather than gate voltage as the abscissa. From (10.23) and (10.24), our model predicts that the transconductance should be proportional to the square root of the drain current. The manufacturer's data show $g_{fs}$ increasing slightly less rapidly than $I_d^{1/2}$, falling off at the highest currents from one of several causes. (A most likely cause is the falloff of mobility in the channel at high bias voltages as shown in Figure 10.7. However, it is also possible that the FET is not operating in the saturated mode at the highest currents. For the 2N3796 curves given, $V_{ds} = 10$ V is insufficient to reach saturation for currents above about 17 mA.) Whether you consider the fit we have obtained good or bad depends on what you want to do with it. For routine amplifier analysis it is more than adequate. The variation from one transistor to another will be a more important source of error. On the other hand, if you wanted to use the square-law characteristic of equation 10.23 for an amplifier-linearity

# 2N 3796 (SILICON)
# 2N 3797

Silicon $N$-channel MOS field-effect transistor designed for low-power applications in the audio frequency range.

**CASE 22(2)**

(TO-18)

MAXIMUM RATINGS ($T_A = 25°C$ unless otherwise noted)

| Rating | Symbol | Value | Unit |
|---|---|---|---|
| Drain-source voltage | $V_{DS}$ | | Vdc |
|         2N3796 | | 25 | |
|         2N3797 | | 20 | |
| Gate-source voltage | $V_{GS}$ | $\pm 30$ | Vdc |
| Drain current | $I_D$ | 20 | mAdc |
| Power dissipation at $T_A = 25°C$ | $P_D$ | 300 | mW |
|   Derate above 25°C | | 1.7 | mW/°C |
| Operating junction temperature | $T_J$ | $+ 200$ | °C |
| Storage temperature range | $T_{stg}$ | $- 65$ to $+ 200$ | °C |

HANDLING PRECAUTIONS:

MOS field-effect transistors have extremely high input resistance. They can be damaged by the accumulation of excess static charge. Avoid possible damage to the devices while handling, testing, or in actual operation, by following the procedures outlined below:

1. To avoid the build-up of static charge, the leads of the devices should remain shorted together with a metal ring except when being tested or used.
2. Avoid unnecessary handling. Pick up devices by the case instead of the leads.
3. Do not insert or remove devices from circuits with the power on because transient voltages may cause permanent damage to the devices.

**Figure 10.14***a.* Maximum ratings for the Motorola 2N3796 and 2N3797 silicon enhancement MOSFET's.

ELECTRICAL CHARACTERISTICS ($T_A = 25°C$ unless otherwise noted)

| Characteristic | Symbol | | Min | Typ | Max | Unit |
|---|---|---|---|---|---|---|
| Drain-source breakdown voltage | $BV_{DSX}$ | | | | | Vdc |
| ($V_{GS} = -4.0\,V$, $I_D = 5.0\,\mu A$) | | 2N3796 | 25 | | — | |
| ($V_{GS} = -7.0\,V$, $I_D = 5.0\,\mu A$) | | 2N3797 | 20 | 25 | — | |
| Zero-gate-voltage drain current | $I_{DSS}$ | | | | | mAdc |
| ($V_{DS} = 10\,V$, $V_{GS} = 0$) | | 2N3796 | 0.5 | 1.5 | 3.0 | |
| | | 2N3797 | 2.0 | 2.9 | 6.0 | |
| Gate-source voltage cutoff | $V_{G(off)}$ | | | | | Vdc |
| ($I_D = 0.5\,\mu A$, $V_{DS} = 10\,V$) | | 2N3796 | — | 3.0 | 4.0 | |
| ($I_D = 2.0\,\mu A$, $V_{DS} = 10\,V$) | | 2N3797 | — | 5.0 | 7.0 | |
| "On" drain current | $I_{D(on)}$ | | | | | mAdc |
| $V_{DS} = 10\,V$, $V_{GS} = +3.5\,V$ | | 2N3796 | 7.0 | 8.3 | 14 | |
| | | 2N3797 | 9.0 | 14 | 18 | |
| Drain-gate reverse current* | $I_{DGO}$* | | | | | pAdc |
| ($V_{DG} = 10\,V$, $I_S = 0$) | | | — | — | 1.0 | |
| Gate-reverse current* | $I_{GSS}$* | | | | | pAdc |
| ($V_{GS} = 10\,V$, $V_{DS} = 0$) | | | — | — | 1.0 | |
| ($V_{GS} = -10\,V$, $V_{DS} = 0$, $T_A = 150°C$) | | | — | — | 200 | |

| Characteristic | Symbol | Min | Typ | Max | Unit |
|---|---|---|---|---|---|
| Small-signal, common-source forward transfer admittance | $|y_{fs}|$ | | | | $\mu$ mhos |
| $(V_{DS} = 10\,\text{V},\ V_{GS} = 0, f = 1.0\,\text{KHz})$    2N3796 | | 900 | 1200 | 1800 | |
|      2N3797 | | 1500 | 2300 | 3000 | |
| $(V_{DS} = 10\,\text{V},\ V_{GS} = 0, f = 1.0\,\text{MHz})$    2N3796 | | 900 | — | — | |
|      2N3797 | | 1500 | — | — | |
| Small-signal, common-source output admittance | $|y_{os}|$ | | | | $\mu$ mhos |
| $(V_{DS} = 10\,\text{V},\ V_{GS} = 0, f = 1.0\,\text{KHz})$    2N3796 | | — | 12 | 25 | |
|      2N3797 | | — | 27 | 60 | |
| Small-signal, common-source input capacitance | $c_{iss}$ | | | | pF |
| $(V_{DS} = 10\,\text{V},\ V_{GS} = 0, f = 1.0\,\text{MHz})$    2N3796 | | — | 5.0 | 7.0 | |
|      2N3797 | | — | 6.0 | 8.0 | |
| Small-signal, common-source reverse transfer capacitance | $c_{rss}$ | | | | pF |
| $(V_{DS} = 10\,\text{V},\ V_{GS} = 0, f = 1.0\,\text{MHz})$ | | — | 0.5 | 0.8 | |
| Noise figure | $NF$ | | | | dB |
| $(V_{DS} = 10\,\text{V},\ V_{GS} = 0, f = 1.0\,\text{KHz},\ R_S = 3\,\text{megohms})$ | | — | 3.8 | — | |

* This value of current includes both the FET leakage current as well as the leakage current associated with the test socket and fixture when measured under best attainable conditions.

**Figure 10.14b.** Specifications for some of the electrical parameters for the Motorola 2N3796 and 2N3797 silicon enhancement MOSFET's.

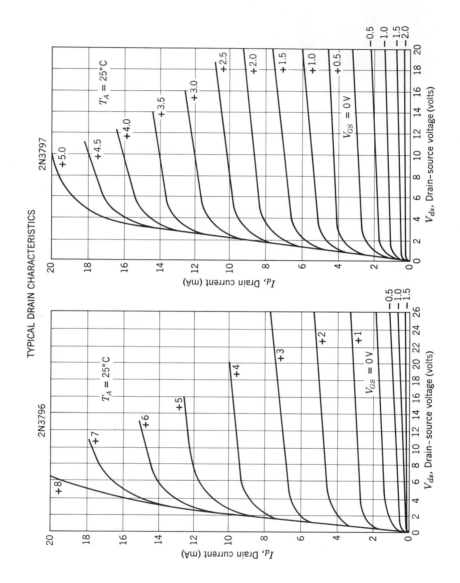

TYPICAL DRAIN CHARACTERISTICS

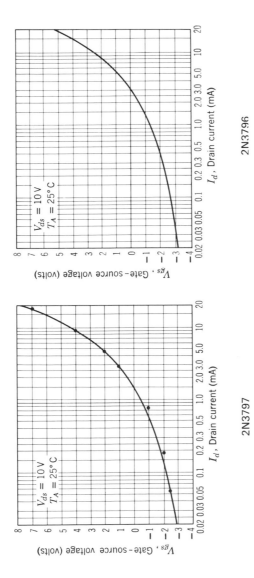

**Figure 10.14c** Large-signal low frequency terminal characteristics for the 2N3796 and 2N3797 silicon enhancement MOSFET's. The points shown on the common-source transfer characteristics for the 2N3796 are a fit to equation 10.23, with $(\mu B c_i / 2L) = 0.18$ mA/V$^2$ and $V_{d\,sat} = V_{gb} + 3$.

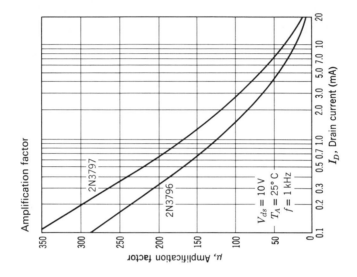

Amplification factor

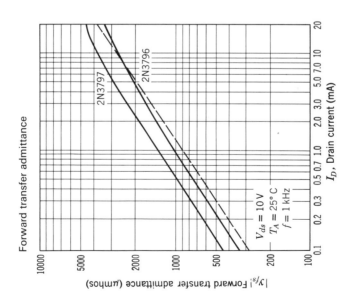

Forward transfer admittance

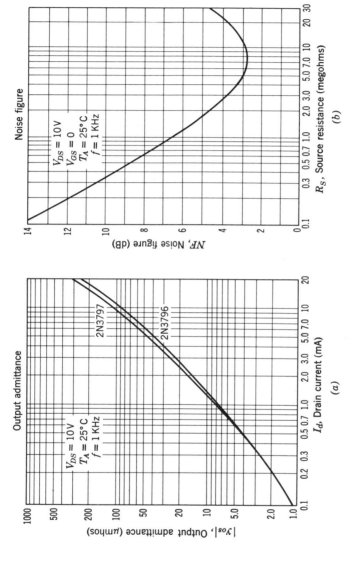

**Figure 10.14d** Small-signal parameters for the Motorola 2N3796 and 2N3797 silicon enhancement MOSFET's. The dashed line on the transconductance curve is the derivative of the transfer characteristics represented by the points in the previous figure.

analysis (as you well might if this FET were appearing on a hi-fi or telephone repeater amplifier), you would want a much more accurate representation. For our purposes here, it certainly seems adequate.

What we have accomplished at this point is an analysis that is reasonably accurate in the linear and transition regions. We still must examine the saturation mode of operation, since it is in this mode that the FET is most often employed as an amplifier.

## THE OUTPUT CONDUCTANCE OF THE SATURATED MOSFET

In obtaining the output conductance for the saturated JFET, we reduced the formidable two-dimensional field theory problem to an easy one-dimensional problem by resorting to three assumptions. We assumed that the saturated channel maintained a constant cross-section, that the differential mobility changed abruptly from $\mu_0$ to 0, and that the field between the gate and drain did not make a significant contribution to the source-drain potential drop. None of these assumptions was really accurate, but they were not so very bad that you would expect the answer to be meaningless.

Unfortunately, if you tried the same thing with the IGFET, your results would be likely to be very bad. It is not that the same phenomena of drift saturation and channel length modulation do not occur, but rather that the simple symmetry of Figure 9.5 is missing in the IGFET. The IGFET equivalent of Figure 9.5 is Figure 10.9b. Figure 10.9b doesn't encourage you to make a one-dimensional approximation, and it isn't even the full story. It doesn't show the gate electrode, which is certainly close enough to the channel and drain to contribute a substantial amount of field. The field is distinctly two dimensional. With the several media, the nonlinear mobility, and currents that depend on the transverse field, it is a decidedly unattractive problem to attack analytically. Indeed, for many years, most people employed a one-dimensional plane-diode approximation even though it gave results that enjoyed only limited success in describing real devices.

Recently, however, a new approach has been suggested that smacks something of Alexander the Great's solution of the Gordian-knot problem. Frohman-Bentchkowsky and Grove [*IEEE Trans. on Electron Devices*, ED-16, 108 (1969)] used their axe to disassemble the impossible problem into three simple, soluble subproblems. They begin their solution by observing that there are three principal places for field lines to terminate:

1. They may terminate on the fixed acceptor charges anywhere within the substrate. (Frohman-Bentchkowsky and Grove ignored the possibility of field lines terminating on the electrons drifting at the saturated velocity.)

2. They may terminate (or originate) on the metal gate electrode.

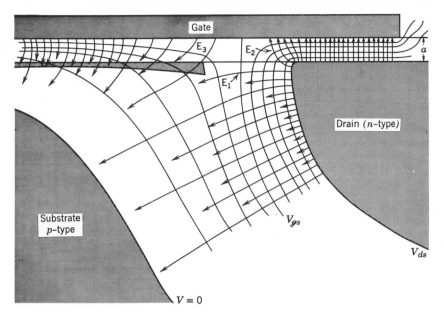

**Figure 10.15** A field map of the drain end of an *n*-channel enhancement MOSFET operating in the saturated node. The equipotentials are a volt apart. The distance from the channel end to the drain is of the order of a micrometer.

**3.** They may terminate on mobile electrons within the channel (e.g., field lines from the gate to the channel).

If we were proceeding along traditional lines, we would determine the amount of channel-length modulation by integrating the electric field from the drain to the point where the channel pinches off. Since the voltage at those two end points is known, the integral would then determine the length of the depletion region. Unfortunately, it is quite clear at the outset that no simple algebraic representation of the local field is going to appear to permit that integration to be performed. What Frohman-Bentchkowsky and Grove did was to reverse the classical method of solving such problems: They found the average field from the voltage differences that they knew rather than finding voltages by averaging the fields that they didn't know. Getting the right answer with their approach, however, took both some genuine cleverness and a bit of luck.

To see what they did, consider the "best-guess" field map of the drain end of a saturated *n*-channel enhancement MOSFET (Figure 10.15). Besides being an excellent illustration of the fact that field-mapping is a dead art, Figure 10.15 shows how the fields from each of the various charge distributions contribute to the voltage $(V_{ds} - V_{d\,\text{sat}})$ from drain to channel end.

The situation illustrated is reasonably representative of real devices. The oxide is 0.2 $\mu$m thick; the distance from the end of the unsaturated channel to the drain is a bit less than 1 $\mu$m. The difference between equipotentials represents 1 V. The substrate and source are assumed to be at the same voltage.

One of the equipotentials—the one representing the gate potential—is particularly revealing. To the right of it, field lines from the drain terminate on the gate electrode; to its left, field lines originate on the gate and terminate on the channel or in the depleted substrate. Note that both of these sets of field lines contribute to the transverse field between the drain and the channel end. The remainder of the transverse field comprises field lines that originate in the drain and end on acceptor sites (or the electrons drifting at their saturated velocity) in the "pinched-off" channel region. Following the notation of Frohman-Bentchkowsky, I have numbered these three fields as shown in Figure 10.15.

Now, to find the drain conductance, we begin, as we did with the JFET, with the drain current at "pinch-off" (equation 10.23) and modify the equation to include the possibility of channel-length modulation. Thus, (10.23) becomes

$$I_d = I_{d\,\text{sat}}\left[1/(1 - \delta/L)\right] \cong \frac{1}{1 - \delta/L}(\mu Bc_i/2L)(V_{gb} - V_{FB} + 2\phi_p - V_{sb})^2$$

$$(10.25)$$

from which we obtain

$$g_{ds} = \partial I_d/\partial V_{ds} = I_{d\,\text{sat}}\left[1/(1 - \delta/L)\right]^2 \partial(\delta/L)/\partial V_{ds} \qquad (10.26)$$

Our task is thus to find $\delta$.

To achieve this goal, Frohman-Bentchkowsky and Grove begin with the fairly obvious fact that the average field is the voltage difference divided by the depletion length from drain to channel end. From this one has

$$\delta = (V_{db} - V_{d\,\text{sat}})/\bar{E}_t \qquad (10.27)$$

They then proceed to assign a portion of the three fields of Figure 10.15 to make up the total transverse field, $E_t$. For the lines labeled $E_1$ in the figure, if the field is not very strongly curved at the surface, and if the drifting electrons' charge can be ignored, we may use the simple planar-junction theory [i.e., the one-sided step-junction version of (3.50)] to get the average transverse field as

$$\bar{E}_{1t} = (qN_A/2\varepsilon_s)^{1/2}\left[V_{db} - V_{d\,\text{sat}}\right]^{1/2} \qquad (10.28)$$

Up to this point we have followed the traditional approach. Now comes the coup: We add two more terms to account for the fringing of the fields from drain to gate and from gate to channel and substrate. The simplest approach—and one that proves effective experimentally—is to assume that the average transverse field is proportional to the average normal field across the oxide. Frohman-Bentchkowsky and Grove chose to present the proportionality constant as the product of two components: a coefficient ($\alpha$ or $\beta$) called the *fringing factor*, which is multiplied by a second coefficient, the ratio of the dielectric constants of quartz and silicon. The latter factor takes care of the fact that the field is reduced in a medium of high dielectric constant. On that basis, they write the remaining two field contributions as

$$\bar{E}_{2t} = \alpha \, (\varepsilon_i/\varepsilon_s) \left[ (V_{db} - V'_{gb})/a \right] \tag{10.29}$$

and

$$\bar{E}_{3t} = \beta \, (\varepsilon_i/\varepsilon_s) \left[ (V'_{gb} - V_{d\,\text{sat}})/a \right] \tag{10.30}$$

$V'_{gb}$ is the gate bias measured from the flat-band voltage (to include the *total* field). They state, without discussion, that $\alpha$ and $\beta$ ought to be bias and structure independent. This is certainly not an obvious conclusion, so consider the following argument.

To find that the constants, $\alpha$ and $\beta$, are insensible to bias and structure, we must establish that the average fraction of the field that is directed along the channel axis should be insensible to these same two variables. Two arguments support such a conclusion. First, the "flow" of equipotentials shown in Figure 10.15 will be roughly the same regardless of the bias (beyond pinch-off), oxide thickness, or substrate doping. Thus, regardless of what we do, the equipotential lines "come in" running horizontally through the oxide and "flow out" to the lower right following the contour of the drain. Since the flow of such free equipotentials must be smooth and regular and "locally parallel," the field map for the saturated mode of operation will always look pretty much like Figure 10.15. Changes in bias may expand or contract the depletion region and add or substract equipotentials, but the pattern will not change very much. This first argument strongly suggests that the coefficients, $\alpha$ and $\beta$, will be reasonably independent of bias.

That the pattern should be independent of structure is true only up to a point. It should be clear that if one makes the oxide sufficiently thick, the field lines will run vertically in the upper portions of the oxide. When that situation obtains, the variations in oxide thickness affect $\alpha$ and $\beta$. However, as we shall now show, when the oxide thickness is substantial, $\bar{E}_{2t}$ and $\bar{E}_{3t}$ no longer have much influence on the output conductance. Thus, the second argument is that the variations in $\alpha$ and $\beta$ that result from oxide-thickness differences occur over an oxide-thickness range that makes them irrelevant.

Having argued for the constancy of $\alpha$ and $\beta$, we may now write the drain

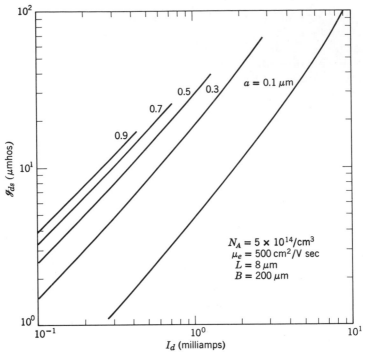

**Figure 10.16** Calculated values of $g_{ds}$ versus $I_d$ for various values of oxide thickness. $V_{ds} = 10$ V for all the curves. The curves run from turn-on until $V_{gs} = 8$ V.

to channel-end depletion length from (10.27)–(10.30):

$$\frac{1}{\delta} = \left(\frac{qN_A}{2\varepsilon_s}\right)^{1/2} \left[V_{db} - V_{dsat}\right]^{-1/2} + \left(\frac{\varepsilon_i}{a\varepsilon_s}\right) \times$$

$$\times \left[\frac{\alpha(V_{db} - V'_{gb}) + \beta(V'_{gb} - V_{dsat})}{V_{db} - V_{d\,sat}}\right] \tag{10.31}$$

The drain conductance may now be obtained by differentiating the inverse of (10.31) and inserting the result into (10.26). Differentiation yields

$$\frac{\partial \delta}{\partial V_{ds}} = \frac{\sqrt{K_1(V_{db} - V_{d\,sat})/4} + K_2(\beta - \alpha)(V'_{gb} - V_{d\,sat})}{K_1(V_{db} - V_{d\,sat}) + 2K_2\sqrt{K_1(V_{db} - V_{d\,sat})}(V^*) + K_2^2 V^{*2}} \tag{10.32}$$

where

$$V^* = \alpha(V_{db} - V'_{gb}) + \beta(V'_{gb} - V_{d\,sat})$$

$$K_1 = qN_A/2\varepsilon_s$$

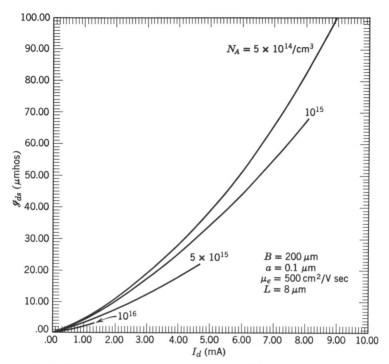

**Figure 10.17** Calculated values of $g_{ds}$ for several different values of substrate doping. $V_{ds} = 10$ V for all the curves. The curves run from turn-on until $V_{gs} = 8$ V. Compare with Figure 10.14$d$.

$$K_2 = \varepsilon_i / a\varepsilon_s$$

Note that $K_1$ is the coefficient affected by doping, while $K_2$ is the coefficient that reflects the oxide thickness.

Although the management disclaims any responsibility for the accuracy of the field map in Figure 10.15, it does give a pretty good picture of what to expect for $\alpha$ and $\beta$. Each is to give the average horizontal field per unit of associated vertical field. Inspection of the figure shows that the gate-to-channel field is a good deal more horizontal than the drain-to-gate field. Using a wide variety of dopings, channel lengths, and oxide thicknesses, Frohman-Bentchkowsky and Grove found that they got consistently accurate predictions of the drain conductance using $\alpha = 0.2$ and $\beta = 0.6$. These values are certainly consistent with Figure 10.15.

Figures 10.16 and 10.17 show some plots of the calculated drain conductance versus drain current for several different oxide thicknesses and substrate dopings. As can be clearly seen, thin oxides and heavy substrate dopings

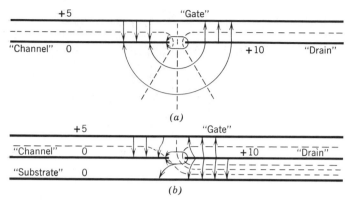

**Figure 10.18** A metal-plate analog of the saturated MOSFET that illustrates how the substrate can decouple the drain from the channel. The upper figure is for a thin-film transistor or a conventional MOSFET with substantial substrate bias. The lower figure is for the conventional MOSFET with the source connected to the substrate.

reduce the output conductance. Since a thin oxide also contributes to a high $g_m$, a thin oxide is usually extremely desirable. Heavier doping is generally undesirable. It leads to a lower $g_m$, lower drain-breakdown voltage, and a higher substrate capacity.

The advantages obtained by using thin oxide layers mean that most devices will be manufactured with a configuration that requires including the gate field in the calculation of $g_{ds}$. Equation 10.31 shows that if the oxide thickness is small enough to make the second term in that equation dominant, the depletion length tends to remain constant. Thus, the output conductance should be satisfactorily low.

Physically, what equation 10.31 does is divide the field into the portion that terminates on the charges in the channel-depletion region and another portion that does not. Only the first class contributes strongly to channel-length modulation, but both portions contribute to supporting the voltage from the drain to the source. Thus, those steps that minimize the depletion width (e.g., moderate to heavy substrate doping) and maximize the field in the insulating layer (e.g., using a thin oxide) will produce the highest output impedance at a given current level.

## THE INFLUENCE OF THE SUBSTRATE BIAS ON THE OUTPUT CONDUCTANCE

Up to this point, we have analyzed the output conductance under the assumption that $V_{sb} = 0$. Although you undoubtedly bore up well under the suspense, it would be inappropriate to ignore $V_{sb}$ indefinitely, since the vast

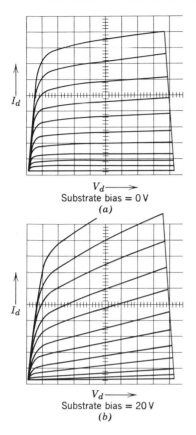

$I_d$

$V_d \longrightarrow$
Substrate bias = 0 V
(a)

$I_d$

$V_d \longrightarrow$
Substrate bias = 20 V
(b)

**Figure 10.19** The drain characteristic curves for an $n$-channel enhancement MOSFET. The upper family has $V_{sb} = 0$; the lower family has $V_{sb} = 20$ V. The gate increment is 0.2 V/step. The vertical scale is 0.1 mA/div and the horizontal scale is 2.0 V/div. (From S. R. Hofstein, in J. T. Wallmark and H. Johnson, Eds., *Field-Effect Transistors*, Prentice Hall, 1966.)

majority of MOSFET's find their way into monolithic integrated circuits, where many of them must operate with substantial bias between the source and substrate. This reverse bias has the desirable effect of maximizing capacitive isolation, but as we shall see, it tends to reduce the output impedance.

In principle, to analyze the effect of substrate bias, we should determine what effect that bias had on Figure 10.15. Unfortunately, not only is my field mapping inadequate to that considerable task, but the result is also sufficiently counterintuitive that you probably wouldn't believe it even if I could produce the figures. Instead, I will adapt an analysis first presented by Hofstein and Warfield, *IEEE Trans. on Electron Devices*, ED-12, 129 (1965). [See also J. T. Wallmark and H. Johnson, *Field-Effect Transistors*, Prentice-Hall, 1966, Chapter 5.] Consider the field distribution surrounding the three metal plates shown in Figure 10.18a. This is a highly idealized model of the field distribution that one would obtain with a truly insulating substrate (i.e., a *thin-film* transistor or TFT). Note that the field lines from the drain can terminate on the channel. Accordingly, for this configuration, we should expect the drain to act as a rather inefficient gate. An increase in drain voltage will increase the channel conductance and result in an increase in drain current. It would lead to a substantial increase in $g_{ds}$.

Now look at the field distribution with a fourth plate added on the lower side. This is illustrated in Figure 10.18b. The lower plate is a crude analogy to the conducting substrate of a conventional MOSFET. With the lower plate close to the middle two, very few field lines go from the "drain" to the "channel." In effect, the lower plate has shorted out most of the capacity between the two middle plates.

Although the parallel-plate analogy is a bit strained here, it illustrates pretty clearly what one might expect in a MOSFET on a lightly doped substrate. For relatively low substrate bias, the substrate effectively shorts out the capacitive coupling between drain and channel. For large reverse bias, on the other hand, the substrate-depletion region becomes so large that substantial coupling will exist between the drain and the channel. Accordingly, for MOSFET's on lightly doped substrates, substantial substrate bias implies low output impedance. This is illustrated very clearly in the drain characteristics shown in Figure 10.19. Although there are very few applications in which one would want 20 V of bias on the substrate, there are many applications in which a loss of output impedance much less spectacular than that illustrated would be intolerable.

## AN ILLUSTRATIVE PROBLEM

Let us say that you want to produce a good high-frequency silicon MOSFET. You have good mask alignment equipment, so you think you can hold alignments to $\pm 1$ $\mu$m. Accordingly, you design your FET with a channel of only 6 $\mu$m. To make sure that you cover the entire channel, you use a gate electrode of 8 $\mu$m width. The source and drain are to be diffused in to a depth of 2 $\mu$m. They have the same breadth as the channel. They extend on either side of the channel axis 25 $\mu$m. (Note that this makes the whole device only 0.002 in. across. In an integrated circuit where surface area is a precious commodity, this very narrow dimension is most useful.) The channel characteristics are listed in tabular form below. If $V_{gs} = 5$ V, $V_{sb} = 0$, and $V_{ds} = 12$ V, find $g_m$, $\delta$, $I_d$, and $g_{ds}$.

*Channel Data*

$L = 6$ $\mu$m          $B = 200$ $\mu$m
$a = 0.15$ $\mu$m
$\mu_e = 400$ cm$^2$/Vsec
$V_{FB} = -3$ V          $N_A = 5 \times 10^{14}$/cm$^3$

[*Solution*: $g_m$ may be obtained using (10.24). To obtain the coefficients for that equation, we go back to Chapter 3 to get

$$2\phi_p = 2\,(kT/q)\log(n_i/N_A) = -\,0.54\text{ V}$$

Then, from (10.20′), we get $V_{d\,\text{sat}} = 7.46$ V, and from (10.23), $I_{d\,\text{sat}} = 8.25$ mA. Putting these into (10.24) yields

$$g_m = 2210\ \mu\text{mhos}$$

$\delta$ is obtained directly from (10.31) as

$$\delta = 1.35\ \mu\text{m}$$

In obtaining $\delta$, note that the two terms in (10.31) are of roughly comparable weight. This shows that a substantial portion of the drain-to-channel potential drop is supported by the gate fields that do not cause channel length modulation.

The next step is to get $\partial\delta/\partial V_{ds}$ from (10.32) This comes out to be 0.149 $\mu$m/V. The last two answers are then obtained directly from (10.25) and (10.26):

$$I_d = 10.6\text{ mA}$$

and

$$g_{ds} = 339\ \mu\text{mhos}$$

Note that if the next stage is another FET, this low output resistance (less than 3 kohms) may cost us much of the advantage of the second FET's high input impedance.]

So far, things look mediocre when compared to a typical high-frequency bipolar transistor. At 8 mA, the bipolar transistor would yield a $g_m$ more than 150 times that of the MOSFET we have just analyzed. The collector resistance would probably be better than 10 Kohms. However, raw gain and high output resistance are not the whole story. First there is the freedom from intermodulation distortion that characterizes the FET's. Seond, there is the question of interlead capacity. To get some feeling for the capacitive loading, estimate the direct gate-to-drain and gate-to-channel capacity for the MOSFET we have just discussed:

[*Discussion*: The biggest part of $c_{gd}$ will probably be the overlap capacity. The overlap area is of the order of $200 \times 10^{-8}$ cm$^2$, not exactly an unbounded expanse. The capacity per unit area, $c_i$, is $\varepsilon_i/a = 2.2 \times 10^{-8}$ F/cm$^2$. Multiplying these two numbers together yields $c_{gd} = 0.044$ pF (or 44 femtofarads). You might well argue that that is *no capacity at all*! Each electron arriving on such a capacitor would raise the potential by 3.6 $\mu$V! It is certainly true that getting leads into the gate and drain will probably add a good bit more than this tiny capacity, but even the audio FET of Figure 10.14 gives only 0.5

pF as its typical $c_{gd}$. Since some of that capacity is associated with the package and the overlap area is probably larger than in our example, the value we have just obtained is not too unreasonable. In a monolithic integrated circuit, where the "packaging capacity" can be much, much lower because of the proximity of the successive stages, the MOSFET offers some truly remarkable low capacities. Thus, in spite of the relatively low $g_m$, amplifiers using MOSFET's can have very competitive gain-bandwidth products. For example, if we want to compare devices using the gain-bandwidth product of equation 9.12, the MOSFET we have just examined gives

$$f_t = \omega_t/2\pi = g_m/2\pi c_{gd} = 0.8\text{ GHz}$$

That's certainly a respectable number. Still higher values are possible by using partial gates to reduce $c_{gd}$ or shorter channels to increase $g_m$. In the low-impedance circuits that must be used at these high frequencies, the shorter channel's lower $r_{ds}$ is no handicap. MOSFET circuits operating well above 1 GHz are in active use today.

Although one can often ignore $c_{gs}$ in a JFET because the Miller capacity $(g_m R_L c_{gd})$ is much larger, with $c_{gd}$'s as small as we have just found, that may be an unreasonable approximation. For example, in the device under discussion, there should be about seven times more gate area over the channel and source than over the drain. Accordingly, the gate-source capacity should be about 0.3 pF.* Since the maximum gain is limited by $r_{ds}$ to $g_m r_{ds} = 60$, $c_{gs}$ must be at least 10% of the input capacity.]

In cascaded stages, one frequently ignores the driving stage's output capacity because it normally is dwarfed by the next stage's input capacity. Is this a justifiable conclusion here? [*Discussion*: The principal capacitive load at the drain itself will be the drain-to-substrate junction. Its area is given as $25 \times 200\ \mu\text{m}^2 = 5 \times 10^{-5}\text{ cm}^2$. Assuming that the drain side of the junction is heavily doped, the depletion width from (3.51) would be about 2.24 $\mu$m, giving a drain-junction capacity at this bias of 0.24 pF. This is about 10% of the Miller capacity for an identical next stage. With somewhat heavier doping for the substrate, $c_{db}$ would become a more substantial fraction of the total load capacity.]

## A VERY BRIEF LOOK AT NOISE

Back in Chapter 3, we took a look at Johnson noise. This is only one of the several sources of noise that limit the lowest useful signal level for any given active device. When weak signals must be amplified, the overriding

---

* Actually, the channel capacity is less than $c_i BL$ (see problem 5 at the end of this chapter), but there is also the overlap capacity to be added. The 0.3 pF is only an approximation.

criterion for evaluating an amplifier's performance is usually the amount of noise that the amplifier adds to the signal.

The sundry different sources for unwanted random signals are usually divided into three classifications. The first includes all the contributors to Johnson noise. These random signals are simply the low-frequency ($h\nu \ll kT$) blackbody radiation that comes from any resistive (i.e., absorptive) circuit element. The only way to reduce this sort of noise in an intrinsicly resistive device is to lower the temperature. Thus, such noise is often called *thermal noise.*

The second classification is *quantum* or *shot* noise. This is quite literally the noise one hears from a tin roof in a rainstorm. The quantization of the rain into drops is precisely what causes the noise. (Tin roofs don't make such noises in continuous flows. If you don't believe me, listen to a submerged tin roof.) In a similar way, the quantization of the electron flow implies that even direct current is really a hail of "bullets," with all the noise attendant to their random and very discrete arrival. This kind of noise can be reduced in many cases by lowering the bias current, although it should be pretty obvious that such a move usually cannot be made without compromising some other parameters. However, it does tell you that if you are amplifying a nanoampere, don't use a transistor that must carry a milliamp to function well.

The third noise classification is a huge grab bag called *flicker* noise or $1/f$ noise. It includes almost anything that is not either Johnson or quantum noise. The second name makes specific reference to the fact that most of these noise sources contribute noise power that is roughly inversely proportional to the frequency at which this power is measured. This type of noise is usually a serious plague in emerging device technologies and becomes much more manageable as the technology matures. For example, the introduction of planar processing into transistor fabrication substantially reduced the $1/f$ noise in bipolar transistors by controlling the buildup of large numbers of unstable surface states. However, you should not get the idea that proper processing is all that is needed to make this sort of noise vanish entirely. Good technique helps, but some of the sources of $1/f$ noise are the result of using real rather than ideal materials. The ideal-material market being what it is, these noise sources are likely to be with us for some time.

A thorough examination of the fundamentals of noise theory is the proper subject of another book. What will be done here is to discuss the most common figure of merit, the *noise figure,* and then do a very brief examination of the noise performance of our three transistors. Let us begin with the noise figure.

An ideal noise-free amplifier is defined as one whose output noise power is equal to the amplified noise at its input. For all real amplifiers, the noise power available at the output will exceed this ideal lower limit. The *noise figure* (*NF*) for the amplifier is defined as the ratio of the total noise power to the ideal noise power. Normally, the noise figure represents the total broadband output of the amplifier, but measurements may also be taken within a narrow bandwidth. These narrow-band measurements are called *spot-noise figures.*

Measuring noise figures at moderate frequencies is not too difficult, but it does require reasonably strict adherence to some definitions to make the measurements meaningful. First, let us define the power gain ($A_p$) of an amplifier as the ratio of the output power to the *available signal power*. In most cases, the amplifier will not be perfectly matched to the source, so all of the available power will not be utilized. In fact, we will shortly derive a rather curious result that the optimum noise figure is not obtained in a matched condition. Let us use the symbol $NP_S$ for the signal source noise power, $NP_o$ for the output noise power with no input, and $\Delta f$ for the amplifier bandwidth. Our definition of noise figure then becomes

$$NF = \frac{A_p NP_S + NP_o}{A_p NP_S} = 1 + \frac{NP_o}{A_p NP_S} \qquad (10.33)$$

If the source is clean of other noise sources, $NP_S$ is limited by Johnson noise [see (3.60)] to $kT_S\Delta f$, where $T_S$ need not be associated with the amplifier itself. For example, looking at the sky with an antenna can give temperatures from a few degrees K to the temperature of the sun. $NP_S$ may also have components of other man-made or natural noise sources. This definition makes the noise figure depend on the signal source. One may also use a *standard noise figure*, which simply sets $NP_S = kT\Delta f$ where $T$ is taken as a standard temperature. Room temperature (290°K) is usually employed.

It is self-evident from (10.33) that one wants to minimize $NP_o$ and maximize $A_p$. However, at this point in our analysis, we have no way of determining how adjusting one will affect the other. We will get to this in a few pages. For the moment let us see how one would make a noise-figure determination.

The circuit employed at moderate frequencies is shown in Figure 10.20. The noise source is a random-signal generator with a flat spectrum. (The usual noise source gets its random signals from shot noise.) The noise source is well represented as an adjustable, calibrated current source with either internal or external provision for setting its admittance. The noise power available from the source is related to the conductance and to the mean-square noise current (which is adjustable and calibrated) by

$$NP_S = \overline{i_S^2}/4G_S \qquad (10.34)$$

(N.B. Since the power spectrum of the noise source is flat, the value of $\overline{i_S^2}$

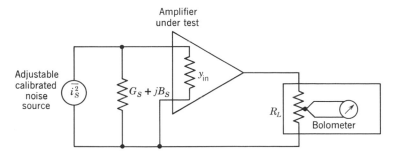

**Figure 10.20** Schematic of the circuit for determining the noise figure for an amplifier.

is proportional to the bandwidth, $\Delta f$. We are presuming here that you know both $A_p$ and $\Delta f$ for the amplifier.)

The remainder of the measurement is quite direct. One connects a wide-band power meter at the output (typically a bolometer, which is the equivalent of the chemist's bomb calorimeter — power is absorbed by a resistor and the temperature rise measured) and measures the power with and without a signal from the noise source. Plugging into (10.33) gives the noise figure. To get a "standard" noise figure, one should set $NP_S = kT\Delta f$. More often than not, the noise figure is given as the logarithm of (10.33) in dB.

## SOME SOURCES OF NOISE IN A TRANSISTOR

If you look at the noise-figure plot in Figure 10.14*d*, you will see that the noise figure has a very distinct minimum for a source (signal source, that is!) resistance that doesn't correspond to any of the MOSFET parameters that we have found so far. Why should the noise figure be best at 8 Mohms? To make some sense out of that curve we need a circuit model of the noise sources in the transistor itself. When we have these, we will be able to see how the gate loading affects the noise power at the output.

Since we will want to combine the effects of a number of noise sources represented by electrical current or voltage generators, we must have a rule for addition of two noise signals. The rule is quite clearly evident if you shine two random, incoherent flashlights (that's the usual kind of flashlight) on the same spot. Since there is no stable phase relationship between the two signals, there can be no stable interference pattern. The average light intensity *at any point* is simply the sum of the two intensities. Thus, although the instantaneous pattern (which you can't see) would reflect the sum of the electric vectors, the average pattern is the sum of the average powers. Accordingly, the r.m.s. voltage from the sum of two random, independent voltage sources,

$e_1$ and $e_2$, is

$$\bar{e}_{\text{sum}} = \sqrt{\bar{e_1^2} + \bar{e_2^2}} \qquad (10.37)$$

This combinatorial rule* is the reason why noise sources are most often labeled by their mean square value.

The sources of noise within a transistor include the lossy elements that give Johnson noise, the currents themselves (which contribute shot noise), and all sorts of low-frequency surface and junction fluctuations that give $1/f$ noise. The usual noise limit comes either from shot noise or Johnson noise unless you are working in the lower audio portions of the spectrum. Over most of the spectrum of interest in everyday electronics, shot noise has a flat power spectrum, with the power being proportional to the current. The same spectral flatness is characteristic of Johnson noise, but in this case the noise is proportional to the absolute temperature. Thus, if one plotted the noise power, $NP_o$, per unit spectral width as a function of frequency, the curve would fall off with increasing frequency roughly as $1/f$ until the sum of the shot and Johnson noise began to dominate. Beyond that point, the curve should be flat out to very high frequencies, where it would once again begin to fall off. The break point between $1/f$ and the flat portion of the curve is found in the low-to-middle audio range in most of today's solid-state devices. In the "good old days," it frequently was found up at MHz frequencies.

Modeling the noise sources is reasonably obvious once the source of the noise is identified. Most of the identification problems have to do with $1/f$ noise sources, a subject of continuing research and interest. Here let us limit ourselves to the shot and Johnson noise sources to keep the discussion clear and within reasonable bounds.

The model for Johnson noise is one that we have already employed in Chapter 3 in the Nyquist derivation. (See Figure 3.11.) Once a lossy element has been identified and its incremental conductance determined, one replaces the real conductance with an ideal noise-free conductance in parallel with a noise current generator whose abailable output power is $kT\Delta f$. (A Thevenin equivalent is obviously equally good.) Thus the mean-square current from the source in parallel with ideal conductance $g$ is

$$\bar{i_n^2} = 4kT\Delta f g \qquad (10.38)$$

Shot noise arises from the quantization of the current. The mean-square noise current for an average current $I$ is given by the Schottky formula (which we will make plausible in a moment):

$$\bar{i_n^2} = 2\Delta f g I \qquad (10.39)$$

---

* Formally, one states that, since by presumption $\bar{e_1 e_2} = 0$, $\overline{(e_1 + e_2)^2} = e_1^2 + e_2^2 = \sum_i \bar{e_i^2}$.

Equation 10.39 suggests a model that is simply a current generator. Note that the power obtained from such a model depends on the admittance that the generator drives.

Once one has cataloged the sundry sources of noise within the transistor, it is usually advantageous to combine them into as few entities as possible. Circuit theory and equation 10.37 suffice for that. The usual place to put these composite sources is at the input. Before proceeding to do that, however, we should establish the plausibility of (10.39).

## SHOT NOISE IN A QUANTIZED FLOW

To do a really legitimate proof of (10.39) starting from where we are here would be a bit lengthy. (A very clear presentation can be found in W. B. Davenport, Jr. and W. L. Root, *Introduction to Random Signals and Noise*, McGraw-Hill, 1958.) However, the principal properties of shot noise are quite pervasive. They can be found in any quantized flow in which the quanta arrive in random order. Let us take a particularly simple example of such a system: the game of coin flipping.

The "arrival time" of a head is completely indeterminate a priori in a sequence of flips of any length. Although the expectation value in 500 flips is 250 heads, the probability that you won't get exactly 250 is much greater than the probability that you will. How far from 250 heads are we likely to get? The average deviation from the mean is obviously zero, but not so for the mean-square deviation. Finding that number leads quite directly to (10.39). We begin by finding the mean because the arithmetic is of use in finding the mean-square deviation.

The probability of getting exactly $M$ heads in $N$ flips is

$$P_{M,N} = \frac{1}{2^N}\left[\frac{N!}{(N-M)!M!}\right] \tag{10.40}$$

The average number of heads in $N$ flips is given by

$$\overline{M(N)} = \sum_{M=0}^{N} P_{M,N}M = \sum_{M=0}^{N} \frac{1}{2^N}\left[\frac{N!M}{(N-M)!M!}\right] \tag{10.41}$$

$$= \frac{1}{2^N}\sum_{M=1}^{N}\frac{N!}{(N-M)!(M-1)!} \tag{10.41'}$$

Letting $M - 1 = K$:

$$\overline{M(N)} = \frac{N}{2^N}\sum_{K=0}^{N-1}\frac{(N-1)!}{(N-1-K)!K!} \tag{10.42}$$

Since the summation in (10.42) is simply the total number of possible arrangements of $N - 1$ flips, which is $2^{N-1}$, (10.42) becomes

$$\overline{M(N)} = \frac{N}{2} \tag{10.43}$$

which can hardly rate as much of surprise. However, our next result is fa. less intuitively obvious.

What we want is the average value of the square of the deviation from the mean. This is obtained from

$$\overline{\Delta M^2} = \overline{(M - N/2)^2} = \sum_0^N \frac{1}{2^N} \left[ \frac{N!}{(N-M)!M!} \right] \left( \frac{N^2}{4} - MN + M^2 \right) \tag{10.44}$$

Since the sum of the probabilities is unity, the first term in the round bracket contributes $N^2/4$ upon summation. The second term is $-N$ times the right-hand side of (10.41), which in view of (10.43) yields $-N^2/2$ after summation. The third term may be reduced in the following steps:

$$T\#3 = \sum_{M=0}^N \frac{1}{2N} \left[ \frac{N!M^2}{(N-M)!M!} \right] = \sum_{M=1}^N \frac{1}{2^N} \left[ \frac{N!M}{(N-M)!(M-1)!} \right] =$$

$$= N \sum_{K=0}^{N-1} \frac{1}{2^N} \left[ \frac{(N-1)!(K+1)}{(N-1-K)!K!} \right] \tag{10.45}$$

$$= \frac{N}{2} \sum_{M=1}^{N-1} \frac{1}{2^{N-1}} \left[ \frac{(N-1)!}{(N-1-M)!(M-1)!} \right] +$$

$$+ \frac{N}{2} \sum_0^{N-1} \frac{1}{2^{N-1}} \left[ \frac{(N-1)!}{(N-1-M)!M!} \right]$$

The first summation is like (10.41), while the second sums to 1 since it is the sum of all the probabilities for $N - 1$ flips. Thus, the third term becomes

$$T\#3 = \frac{N(N-1)}{4} + \frac{N}{2}$$

Putting all three terms together to evaluate (10.44) gives us the mean-square deviation:

$$\overline{\Delta M^2} = N/4 = \overline{M}/2 \tag{10.46}$$

Had we used a system that does not obey the exclusion principle of only 1 head per flip, the result would have differed by a factor of 2:

$$\overline{\Delta M^2} = N/2 = \overline{M} \tag{10.46'}$$

Both forms state that the r.m.s. deviation increases as the square root of the mean number of events (i.e., heads, electrons, etc.). This means that the percentage error in determining the mean experimentally will decrease as one over the square root of the number of events.

The relationship between (10.46′) and (10.39) is quite direct. Measurement of modulation on a current requires measurement of the deviations from the mean. One way or another, these deviations must be determined by counting the number of electrons that arrive in a half cycle at the modulation frequency and comparing this number with the expectation number in the carrier current. Keep in mind that deliberate and random deviations look the same. An example will take us from (10.46′) to (10.39).

Let us say that we had 1 mA flowing through a transistor that is driving a 5 Kohm load. Let the bandwidth be 5 MHz (roughly the bandwidth of a TV channel). A half cycle is $10^{-7}$ seconds, and in that time $6.25 \times 10^8$ bias electrons are expected. This same number, according to (10.46′), is the expected mean-square deviation. Taking its square root gives an r.m.s. noise current of $2.5 \times 10^4 \times 1.6 \times 10^{-19}/10^{-7} = 4 \times 10^{-8}$ amperes. What we have just done there is nothing more nor less than to apply (10.39), since $\overline{M} = \overline{I}/2fq = \overline{\Delta M^2}$ and $\overline{I_n^2} = q^2 \overline{\Delta M^2}/(2f)^2 = 2qf\overline{I}$. With $f = \Delta f$, we have (10.39).

Let us pursue the example a bit further. The noise output power and r.m.s. output voltage into the 5 Kohm load are

$$NP_0 = 8 \times 10^{-12} \text{ watts}$$

$$v_{\text{r.m.s.}} = 0.2 \text{ mV}$$

Neither of these may sound like very much to you, but they are really quite substantial. For example, the noise power is roughly equivalent to the amount of power your eye receives from a flashlight at a distance of 4 miles. That is an easily seen signal at night. If the voltage gain of the stage were 10, the input voltage would have to be 20 μV to give a signal-to-noise ratio of 1. This is about the right value when you consider that most good FM receivers with a bandwidth of about 5% of our 5 MHz operate well at a few μV. Though one can get broadband gains of 100 or so at lower frequencies, you can see that devices that must carry a milliamp of bias current require signals in excess of a few microvolts/MHz.

## A REASONABLY GENERAL NOISE MODEL AND THE MINIMUM NOISE FIGURE

What we need now is a good but simple model of a transistor with noise generators included. We may divide the localized sources into two classes:

those that are and those that are not affected by the input loading. As an example of each, let us consider shot noise from the gate current and shot noise from the drain current of a JFET. We would like to represent each by an appropriate generator located in the gate circuit of the transistor. Let us start with the drain current.

Current in the drain is related to voltage on the gate by the transconductance. As long as the capacitive coupling between drain and gate is not too large, there is insignificant feedback, and the voltage on the gate is pretty independent of the current in the drain. When this simple situation obtains, we may represent the shot noise current in the drain by a voltage generator that adds $\bar{v}_i^2 = 2\Delta f q \bar{I}/g_m^2$ to the square of the gate voltage. Putting a voltage generator between the gate lead and the gate itself will do just that. This is the $v_i^2$ generator shown in Figure 10.21. Properly scaled, all of the noise sources that are independent of gate loading may be added into $v_i^2$.

Shot noise in the gate current, on the other hand, leads to noise only to the extent that it can develop some power. If you were to short out the gate lead, the fluctuations in gate current would not correspond to any change in the gate voltage and would not yield any noise in the output. Such a situation is well portrayed by the generator $i_i^2$ in Figure 10.21. Once again, all of the sources of noise that depend on the gate loading can be incorporated into this one noise-current generator.

With the circuit of Figure 10.21, we may now ask the question: Is there an optimum value of source admittance for minimum noise figure? The question is readily answered by finding $NF$ from (10.33). First we need $A_p$. The minimum power† available from the source is

$$P_S = kT\Delta f = \bar{i}_S^2/4G_S \tag{10.47}$$

The output power to $G_L$ with the given circuit is

$$P_o = \frac{\bar{i}_S^2 g_m^2/4g_{ds}}{(G_S + g_i)^2 + (B_S + b_i)^2} \tag{10.48}$$

The gain is then

$$A_p = \frac{g_m^2}{g_{ds}} \left[ \frac{G_S}{(G_S + g_i)^2 + (B_S + b_i)^2} \right] \tag{10.49}$$

In a fashion similar to (10.48), we obtain the output noise power due to the

---

† N.B. By our definition of $NF$, it is always possible to improve the amplifier's performance by increasing the signal power. That's only the observation that to be heard in a noisy room, you have to shout. However, to find out how the amplifier performs with weak signals, we make available the minimum signal possible—just the Johnson noise of the signal source.

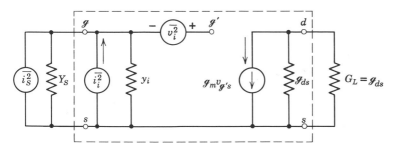

**Figure 10.21** A reasonably general model of a transistor amplifier, including noise sources. The output is matched for maximum power transfer, but no assumptions are made on the relationship of $g_s$ to the amplifier's input admittance, $y_i$.

two internal noise generators as

$$NP_o = \frac{g_m^2}{4g_{ds}}\left[\frac{\overline{i_i^2}}{(G_S + g_i)^2 + (B_S + b_i)^2} + \overline{v_i^2}\right] \qquad (10.50)$$

Putting the various component parts into (10.33) gives the noise figure:

$$NF = 1 + \frac{\overline{i_i^2} + [(G_S + g_i)^2 + (b_i + B_S)^2]\,\overline{v_i^2}}{4kT\Delta f G_S} \qquad (10.51)$$

(Notice that if $i_i$ is in the usual units [amperes], $kT$ must be in joules, not eV.) It is clear there there is a minimum value of $NF$ and that the noise figure is generally reduced by tuning out the input susceptance, $b_i$ (i.e., setting $B_S = -b_i$). It would appear that maximizing the bandwidth would improve the noise figure, but that is illusory. Both $\overline{i_i^2}$ and $\overline{v_i^2}$ are proportional to $\Delta f$ (above the $1/f$ region), so the noise figure is really independent of bandwidth as long as the gain is.

Having set $B_S = -b_i$, the minimum value for $NF$ is obtained by differentiating (10.51) w.r.t. $G_S$. The minimum occurs when

$$G_S = \sqrt{g_i^2 + \frac{\overline{i_i^2}}{\overline{v_i^2}}} \qquad (10.52)$$

The first term is what you would expect—the best performance calls for a matched input. However, that result is true only when the input conductance is much greater than the ratio of the noise sources. For FET's with their negligible input conductances, the second term dominates completely. Thus, even though the MOSFET of Figure 10.14$d$ has an input conductance that is probably less than $10^{-13}$ mhos, the minimum noise figure is found at $R_S = = 8$ Mohms. Similarly, for the JFET in Figure 9.6, the noise figure is specified at 1 Mohm.

The existence of a distinct minimum suggests a rather important conclusion. If you are after optimum noise performance, you must select a device that has a minimum noise figure in the impedance range that you are going to use. Although JFET's have remarkably low noise figures—values such as 0.1 dB being flaunted about these days—if you cannot use them with megohm-gate-resistance levels, you may be substantially better off with a bipolar transistor.

Generally speaking, the optimum noise figures for JFET's are superior to both MOSFET's and bipolar transistors. Low-noise versions of the latter two usually run around 2 to 4 dB. JFET's NF's run from 0.1 to 0.5 dB. For the run-of-the-mill transistor, noise figures are substantially higher. Values from 10 to 20 dB are not uncommon. Since low-noise devices are inevitably employed in very-low-signal work, these devices get their low-noise figures by operating at low currents and low bias voltages. Devices that can perform in more trying circumstances usually come with much larger noise figures.

## UPSMANSHIP IN THE FREQUENCY DOMAIN

Progress in microwave transistors is usually measured in fractions of a gigahertz or a few dB. Thus, a few waves are created when a device leap-frogs over all the competition and increases its performance by better than an order of magnitude. Such a giant stride was taken rather recently by the JFET. By using a thin epitaxial layer of moderately heavily doped GaAs with a 1 $\mu$m wide Schottky-barrier gate, Drangeid, Sommerhalder, and Walter achieved the remarkable unity-gain frequency of 30 GHz [Electronics Letters, 6, 228 (1970)]. The device itself is shown in Figure 10.22 and its DC and gain versus frequency characteristics in Figure 10.23. The device is particularly interesting because it brings into very sharp focus the requirements on the material and the structure that must be met if one would operate transistors in a frequency range in which everything is a distributed component.

If you will go back for a moment and review the criteria for high-frequency performance [(9.17), (9.15), and (10.24)], you will see that maximizing $g_m$ and minimizing the gate capacities seems to be the way to go. The capacity of greatest interest is $c_{gd}$. That is generally minimized by operating at maximum drain voltage and minimizing the overlap area between gate and drain. $g_m$ is maximized by getting a material with the highest possible mobility, making the transistor breadth as large as possible and the channel length as short as possible, and doping the channel reasonably heavily. This last step compromises the gate breakdown-voltage and the gate-to-drain capacity, but up to a point it is a winning play.

There is nothing really wrong with these conclusions, but we should proceed with some caution. All of the results that we derived were based on a gradually

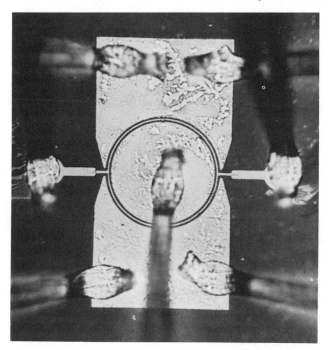

**Figure 10.22** The high-speed GaAs Schottky-barrier FET. The drain is at the center. The gate is the 1 $\mu$m wide circle surrounding the drain. The gate is driven from two 15 $\mu$m diameter pads insulated from the substrate. The two source pads are driven with four leads to minimize both resistance and inductance in the source. The active GaAs layer is 0.27 $\mu$m thick and doped to 6 $\times$ $10^{10}$/cm$^3$ with Se ($n$ type). The active layer was deposited epitaxially on semiinsulating GaAs. The mobility of the deposited material was 2600 cm$^2$/Vsec. The data are from the same source as Figure 10.23. (This photograph courtesy of International Business Machines Corporation.)

varying channel, negligible displacement currents, and a rather diffident attitude on whether pinch-off or velocity saturation marked the transition to saturated performance. When we go to very high frequencies and very short channels, a new and more careful look is required.

Consider a JFET in a common gate circuit delivering gain at some elevated frequency like 10 GHz. At such frequencies it is obviously unwise to ignore the displacement current necessary to charge the gate. The proper way to describe the progress of the signal through the transistor is in terms of a transmission line. The DC bias establishes the dimensions of the channel and space-charge layers so that at any given point along the channel there is a capacity per unit length and a resistance per unit length. This sounds like an RC transmission line—an obnoxiously tapered one at that—but watch

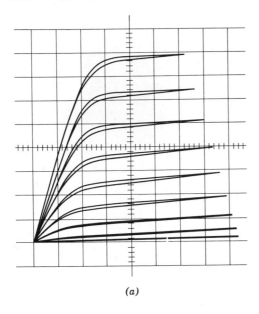

(a)

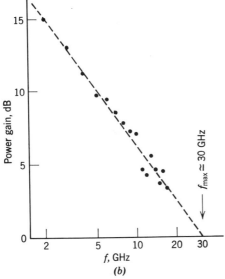

(b)

**Figure 10.23** Characteristics of the FET shown in Figure 10.22. (*a*) Drain characteristics: 5 mA per vertical division, 1 V per horizontal division. 1 V gate steps starting
from 0 V. [From Drangeid, Sommerhalder, and Walter, *Electronics Letters* **6**, 228
(1970).] (*b*) Power gain versus frequency. $V_{ds} = 2.5$ V, $V_{gs} = 0$ V. (Courtesy of International Business Machines Corp.)

what happens as we get near the drain end where the electron drift velocity saturates because of the high field.

We begin by defining each of the appropriate variables in terms of its local DC level plus a small AC (sinusoidal) term:

$V(x) = V_0(x) + v(x)$     voltage from channel to gate

$I(x) = I_0(x) + i(x)$     channel current

$Q(x) = Q_0(x) + q(x)$     depletion-layer charge

$U(x) = U_0(x) + u(x)$     local electron drift velocity

The channel itself may be described in terms of the differential resistance per unit length, $R_0$, and differential capacity per unit length, $C_0$, for each differential length. To greatly simplify the analysis, let us assume that the drift velocity increases linearly with field (constant mobility) until it reaches an abrupt saturation at $U = U_s$. With this assumption, $\mu Q_0 = 1/R_0$ where $\mu$ is either the zero field mobility or zero. With these definitions, the analysis proceeds readily.

The AC current at any point $x$ is given by

$$i = U_0 q + Q_0 u \qquad (10.53)$$

Continuity of current (charge conservation) requires that

$$\frac{\partial i}{\partial x} = -\frac{\partial q}{\partial t} = -j\omega q \qquad (10.54)$$

Putting (10.54) into (10.53) gives

$$i = -(U_0/j\omega)\frac{\partial i}{\partial x} + Q_0 u \qquad (10.55)$$

The AC voltage and the increment in velocity therefrom are

$$v = q/C_0 = -(1/j\omega C_0)\frac{\partial i}{\partial x} \qquad (10.56)$$

$$u = -\mu\frac{\partial v}{\partial x} \qquad (10.57)$$

Putting the last two equations into (10.54) and taking advantage of the relationship between $R_0$ and $Q_0$:

$$i = -(U_0/j\omega)\frac{\partial i}{\partial x} - (1/R_0)\frac{\partial v}{\partial x}$$

$$= -(U_0/j\omega)\frac{\partial i}{\partial x} + (1/R_0)\frac{\partial i}{\partial x}\left[(1/j\omega C_0)\frac{\partial i}{\partial x}\right] \qquad (10.58)$$

If we take the ratio of (10.58) to (10.56), we obtain the characteristic impedance of this transmission line as

$$1/Z = i/v = U_0 C_0 - \frac{C_0 \dfrac{\partial}{\partial x}\left(\dfrac{1}{C_0}\dfrac{\partial i}{\partial x}\right)}{R_0 \dfrac{\partial i}{\partial x}} \tag{10.59}$$

We now come to the surprising result. As the current proceeds from source to drain (left to right), the velocity increases continually until it reaches the saturation velocity, $U_s$. At that point $R_0$ goes to infinity, and (10.58) and (10.59) reduce to a particularly simple and unexpected form:

$$i = -(U_s/j\omega)\frac{\partial i}{\partial x}$$

which upon integration gives

$$i = i_o \exp\left(-j\omega x/U_s\right) \tag{10.60}$$

and the impedance becomes

$$Z = 1/U_s C_0 \tag{10.61}$$

The current thus becomes an unattenuated wave *traveling only to the right*! The wave solution shows very explicitly the isolation between source and drain that saturation of the channel produces. Note also that the magnitude of the current at the drain end of the saturated channel is the same as the magnitude of the current on the other side. This means that if the channel is short enough so that the attenuation of the AC gate-to-channel voltage is insignificant, we may obtain the transconductance in terms of the impedance at the saturation point (sp):

$$g_m = i_d/v_{sg} = i_{sp}/v_{sp} = 1/Z_{sp} = U_s C_0 \big|_{x=sp} \tag{10.62}$$

This is certainly a very different $g_m$ than the one that we got in (9.13) or the one that you would get directly from the Shockley analysis (9.11) that ignores the source resistance. Which result is better?

To answer that question, note that to get (10.62) we had to assume that the signal voltage was unattenuated—that is, that the channel was reasonably short—while for our previous analysis we had to assume that the saturation point was essentially the same whether one invoked pinch-off or velocity saturation. "Essentially the same" means that the difference is small compared to the channel length, that the voltage drop across the channel is about the same for either analysis. Broadly speaking, the Shockley theory is good for long channels; the Drangeid-Sommerhalder theory, which we have just

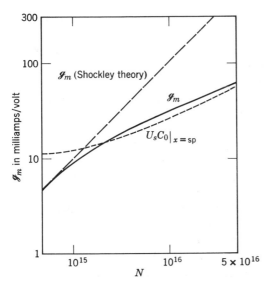

**Figure 10.24** JFET transconductance as a function of channel doping. The parameters of the channel in question are $L = 4$ $\mu m$, $b = 1$ mm, $w = 1$ $\mu m$, $\mu = 2500$ cm$^2$/ Vsec (GaAs), and $V_{sg} = 0$. [From K. E. Drangeid and R. Sommerhalder, *IBM J. Res. Devel.* **14**, 82 (1970).]

derived, is good for short channels. For a given channel length, whether the channel is "long" or "short" is really a measure of just how wide the channel is at the saturated end. This we may readily obtain if we know the doping density, $N_D$:

$$y_{sp} = I/N_D U_s B \qquad (10.63)$$

From (9.11) we have that $I_{DSS} \propto G_0 V_p$, and since each of the terms on the right is proportional to $N_D$, the numerator in (10.63) is proportional to the square of $N_D$ (and also, incidentally, to the channel breadth). This means that (10.63) can be written:

$$y_{sp} \propto N_D/U_s \qquad (10.64)$$

Thus, in a sense, the channel becomes "shorter" as it is more heavily doped. Drangeid and Sommerhalder have calculated $g_m$ over the transition from "long" to "short" for a particular example. Their results are shown in Figure 10.24.

The reason why knowing about (10.62) is so important is that it points out very clearly what you must do to get very-high-frequency performance. You must select a material with the highest possible saturation velocity (the authors chose GaAs); you should make the channel short, thin, and as heavily doped as you can without compromising the breakdown voltage or the saturation velocity; and you should use a gate that has as small a depletion region on the gate side as possible. The choice of a Schottky barrier for the gate means that there is no depletion region on the gate side. It further permits one to process at relatively low temperatures so that the extremely thin active layer will not be diffused out into the substrate.

The need for the thin channel layer is not entirely self-evident. It is required to prevent excessive DC power dissipation. The power dissipated increases with channel depth, but the areas of top and bottom surfaces through which heat can be removed do not change with depth. The proper way to increase $C_0$ is to increase the breadth of the channel, $B$. The thin channel also helps to reduce the drain-saturation voltage, which reduces both the breakdown requirements and the DC power dissipation.

The device shown in Figure 10.22 represents a remarkable combination of remarkable achievements. Growing very thin layers of very-high-quality epitaxial material is quite difficult (0.27 $\mu$m is roughly 500 atoms thick!). Superimposing a sequence of masks (at least 3 were required, although 4 or 5 were probably used) while holding misalignment to a small fraction of the 1 $\mu$m base width is practically impossible. Now if only they could get *another* factor of 10. . . . The people at Westinghouse who coined the phrase "molecular electronics" may have had something. How many atoms would you like in your gate?

## SUMMARY OF RESULTS FOR THE IGFET

This chapter began with an analysis of the principal building block of the IGFET, the MOS capacitor. Following a discussion of the independence of the Fermi levels in the semiconductor and gate metal, we examined the $c$ versus $V$ characteristics for such a sandwich and developed a physical model to explain the results. The flat-band voltage was introduced. Shifts in this voltage were ascribed to the usual contact potential, to chargeable states at the silicon-oxide interface (fast-surface states), and to ions within the oxide layer. The two dominant sources for ions in the oxide are Na and Si. Since these both contribute positive charge to the oxide, there is a natural tendency for the surface of the silicon under the oxide to be $n$-type regardless of the bulk doping. The $n$-type surface on $p$-type substrates prevented the observation of a "high-frequency" $c$ versus $V$ curve and tended to make $p$-on-$n$ junctions break down or leak at lower voltage than $n$-on-$p$ junctions.

Several different structures suitable for use as an IGFET were examined. Those in which the channel conductivity was induced were called "enhancement-mode" devices; those with a built-in metallurgical channel were called "depletion-mode" devices. Since the source and drain regions were created by a noncritical diffusion, it was possible to use polycrystalline materials deposited in thin films on insulating substrates. Such thin-film transistors (TFT's) have permitted the use of many materials that would otherwise be unavailable to the transistor manufacturer, but they have not yet challenged single-crystal FET's for a share of the market.

With a structure to analyze, we set about obtaining the drain characteristics for the enhancement-mode MOSFET. Taking the substrate as our reference point, we established the relationship between the drain current and the potentials of the source, gate, and drain. To achieve this goal, we had to use a somewhat simpler model than the real device suggests. First we followed Ihantola's assumption that the free charge at any point in the channel was linearly proportional to the difference between the local surface potential and the potential for "significant" inversion. Then we made the same assumptions about the gradual channel and constant mobility that served us well for the JFET. These permitted us to obtain a somewhat cumbersome representation for the drain characteristics in the transition region (equation 10.18). When the substrate was lightly doped and the oxide thin, the drain characteristics were shown to have a much simpler form. The drain current was found to follow a square-law relationship (equation 10.22) that at saturation gave a transconductance which increased linearly with the gate-to-source voltage (equation 10.24). A comparison of these results with a real MOSFET showed that at least in this one case, the simplified model was reasonably accurate.

We then turned to the problem of calculating $g_{ds}$ in the saturated mode. We employed the method of Frohman-Bentchkowsky and Grove to reduce this very complicated two-dimensional field-theory problem to workable proportions. The results (equations 10.21, 10.31 and 10.32) were hardly a handy formula to commit to memory, but they showed very clearly the importance of having a thin oxide layer to minimize $g_{ds}$. Heavier doping also helped to minimize $g_{ds}$, but not without having adverse effects on other parameters.

A very crude metal-plate model was used to examine the influence of the substrate bias on $g_{ds}$. No numerical results were obtained, but it was clear that for lightly doped substrates, a large amount of reverse bias on the substrate-to-drain junction substantially increased the output conductance. From the model, it was clear that the TFT would suffer from this problem more than most other structures.

We next examined some of the general features of the noise performance of active devices. Three types of noise were described: Johnson noise is the blackbody radiation emitted by any dissipative element; shot noise is the result of randomness in the flow of discrete (quantized) particles; and $1/f$ or flicker noise is the result of the random fluctuations that characterize the charging or discharging of surface or bulk states. The first two have flat power spectra; the last has a $1/f$ spectrum that makes it important only at low frequencies.

The noise figure for an amplifying stage was defined and its measurement discussed. The definition (equation 10.33) makes the noise figure depend on the character of the source. "Standard" noise figures are measured with a noise source equivalent to a resistor at some "standard" temperature.

With this definition, we set about finding how the noise figure would depend on the impedance of the signal source. A circuit model of an amplifying stage was developed for shot noise in the gate and drain currents. To properly characterize these noise sources, a plausibility argument to justify the Schottky formula (equation 10.39) was developed and the proper combinatorial rule for noise sources was stated (equation 10.37). Then, with a small amount of circuit manipulation, we found that indeed there was a source admittance for minimum noise figure (equation 10.52). Unlike most other "optimum" admittances, this one did not prove to be the complex conjugate of the input admittance. The imaginary part was conjugate, but the real part was greater than the real part of the input admittance. For FET's, the optimum conductance generally lies between $10^{-6}$ and $10^{-7}$ mhos. An important corollary to this result is that the best noise figure in a given application will be achieved with the device that has the minimum noise figure at the signal-source admittance.

The final topic was a brief analysis of a Schottky-barrier JFET that appears to give useful gain out to centimeter wavelengths. This JFET was an example of some of the best of the state of the art in epitaxy and pattern resolution with a channel only 1 $\mu$m long and 0.27 $\mu$m thick. With such a short channel, we had to reexamine some of the premises on which our earlier analyses had been based. The current traveling through the channel was found to propagate as if it were in an RC transmission line. As the current entered the saturated portion of the channel, the solution changed to an unattenuated wave that propagated *only* in the direction of the drain. This result, plus the shortness of the channel, led directly to the surprising expression for $g_m$ (equation 10.62), which showed that the transconductance for such structures was determined almost solely by the saturation velocity and the gate capacity per unit length at the saturation point. It became evident at that point that one wanted a thin, heavily doped channel with maximum saturation velocity. The remarkable unity gain point of 30 GHz was the end product of the implementation of these results.

## Problems

1. An MOS capacitor is to be used to tune the RF end of a TV receiver. The capacitor is part of an integrated circuit, so it must be compatible with the rest of the manufacturing steps for the other devices. The substrate

is $n$-type with $N_D = 10^{15}/cm^3$. The oxide may be any thickness, although any value less than $10^{-5}$ cm is apt to have pinholes.

(*a*) If an area of $10^{-3}$ cm$^2$ is set aside for the capacitor, how thick should the oxide be to have $c_{max} = 30$ pF?

(*b*) What will the minimum capacity?

(*c*) Assuming $10^{11}$-fast-surface states/cm$^2$, what will the bias voltage difference be between $c_{max}$ and $c_{min}$?

2. A piece of $n$-type Ge with a resistivity of 0.1 ohm-cm and an electron mobility of 3000 cm$^2$/V sec is coated with a 1/2-micron thick layer of SiO$_2$ and then with a layer of Al. This forms an MOS sandwich. The energy it takes to get an electron out of Al (the so-called "electron affinity") is 3.38 eV, out of Ge (from the valence band) 4.5 eV, and out of SiO$_2$ ~ 11 eV. The band gap for Ge is 0.66 eV; for SiO$_2$ it is 9 eV. Sketch the band edges with respect to the Fermi level throughout the sandwich, giving exact values where possible. (Ignore surface states in doing this problem.)

3. One may readily observe the equilibration in an MOS capacitor which shows a $c$-$V$ characteristic such as Figure 10.2. One can charge the capacitor extremely rapidly—say in $10^{-7}$ sec—and then watch the open-circuit voltage fall as the capacitor relaxes to its thermal-equilibrium state by generation of minority charge within the semiconductor. The change is quite large, as this problem will show. Consider a simple MOS sandwich with an oxide layer $10^{-5}$ cm thick on an $n$-type silicon wafer with $5 \times 10^{15}$ donors/cm$^3$. Let the capacitor's area be 1 cm$^2$.

(*a*) If the capacitor is initially charged to 20 V, how much charge is placed on the plates?

*Answer:* From (10.2), $Q_0 = 1.57 \times 10^{-7}$ C.

(*b*) If the capacitor remains open circuited while it relaxes, so that the stored charge remains the same, what is the voltage across the capacitor at equilibrium?

*Answer:* In the equilibrium state, most of the charge ends up on the surface of the semiconductor, and the voltage drops to 5.35 V.

(*c*) How much energy is stored initially by the capacitor and how much is stored after thermal equilibrium has been achieved? (Be careful here to proceed from the appropriate definition of incremental work, $dW = = V dQ$. This is not a linear capacitor!) Where does the extra energy go?

*Answer:* Immediately after being charged, the capacitor has 1.13 microjoules stored as recoverable electrical energy. After relaxation to the new thermal-equilibrium state, the capacitor has only 0.37 microjoules available as electrical energy. The remaining 0.76 microjoules has been degraded to heat.

4. Four-point-probe measurements of the resistivity of lightly doped $p$-type wafers are very difficult to make. The same measurements on $n$-type wafers offer no problems. Why should this be true?

*Answer:* The charge in the thin layer of native oxide on the surface is enough to invert the lightly doped surface. One thus reads an anomalous value that really represents only the surface layer.

5. The calculation of the gate-channel capacity for the arbitrary MOSFET requires the integration of the root of a fourth-order polynomial. Possible though this may be, it is unattractive at best. However, when the approximation that led to equation 10.20' is valid (thin oxide and lightly doped substrate), the calculation reduces to very sensible proportions. Show that, under that approximation, the channel portion of the capacity, $c_{gs}$, at saturation is independent of the gate voltage and equal to 2/3 of the intrinsic oxide capacity ($c_i BL$). For simplicity, let $V_{sb} = 0$.

*Solution.* Equation 10.18 may be used to write down the channel voltage as a function of position. Under the approximation, this is a quadratic equation. Using (10.15) and (10.16), one may write the gate charge density as a function of the local gate-to-channel voltage. Then with (10.20') and (10.23), one may integrate along the channel to get the total charge on the gate. Differentiation then gives the desired result.

6. If one uses polycrystalline silicon or molybdenum as the upper gate in a silicon MOSFET, one may use the gate itself to define the apertures for the source-and-drain diffusions. Such a structure is said to have a *self-aligned* gate. Having deposited the gate, and then having etched away the oxide right up to the edge of the gate, one may diffuse in the source and drain with the assurance that they will align perfectly with the gate and with very little overlap. The refractory nature of these gate materials allows one to continue processing at normal furnace temperatures and without too many new problems. The self-alignment is substantially better than one could hope to obtain by superposition of successive masks. This assures a minimum overlap-diode capacity and a very low parasitic source resistance. Furthermore, if the gate is silicon, the work-function difference between the substrate and the gate will be very small. Such a structure is shown on the next page. Let us use it for some problems on device performance.

*Pertinent data:*

Substrate is $n$-type with $5 \times 10^{15}$ donors/cm$^3$

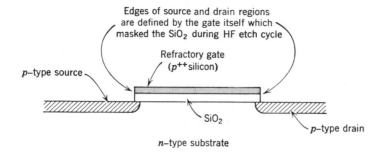

Gate oxide is 0.15 $\mu$m thick and contains $10^{16}$ Na$^+$ ions/cm$^3$
Surface hole mobility is 200 cm$^2$/V sec
Gate material is polycrystalline $p^{++}$ silicon with about $10^{19}$ acceptors/cm$^3$
The gate overlaps the source and drain regions by 0.5 $\mu$m
The gate is 10 $\mu$m long and 200 $\mu$m broad

(a) Determine $V_{FB}$.

*Answer:* $V_{FB} = -0.24\ V$. Note that the gate doping partially compensates for the positive ions in the oxide.

(b) Find the gate potential needed to get $I_d = -1$ mA. Assume negligible surface-state density and use the approximation of (10.20′) and (10.23).

*Answer:* $V_{gs} = -5.65$ V.

7. The Crystalonics Company puts out a "low-noise" silicon *p-n-p* bipolar transistor on which they publish the following data:

Transistor type 2N2372
$h_{FE}$ ($V_{CE} = -4V, I_C = -25\ \mu A$) = 20
Equivalent input noise voltage (same bias conditions) = 1.3 $\mu$V
Equivalent input noise current (same bias conditions) = 75 pA
Spot noise figure (same bias conditions; signal-source resistance of 10 Kohns) = 2.5 db

The manufacturer is not so kind as to give you his measurement bandwidth, so for problem (a) find the approximate bandwidth that makes the noise figure and the two source values agree with one another.

*Answer:* From (10.51), $\Delta f = 34$ KHz. This rather improbable value makes one wonder whether the data are correct, but let us proceed as if they were.

(b) Has the manufacturer chosen the optimum source resistance for his noise measurement?

*Answer:* From (10.52), he is not too far astray. The optimum value is 13.1 Kohm.

(*c*) In Figure 9.6, the same manufacturer gives a noise figure for one of his general-purpose JFET's. He gets a 2 db noise figure with a source resistance of $10^6$ ohms. Assuming that this figure is determined at the optimum point, find the equivalent noise-source values for 34 KHz bandwidth.

*Answer:* $\overline{i_i^2} = 4.5 \times 10^{-22} \, A^2$; $\overline{v_i^2} = 10^{12} \, \overline{i_i^2}$.

(*d*) You now wish to choose between the JFET and the bipolar transistor for an application requiring a source resistance of $10^5$ ohms. Which device will have the lowest noise figure in this circuit?

*Answer:* For the JFET, $NF = 9.5$ db. For the bipolar transistor, $NF = 3.2$ db. The bipolar device is better by a factor of 4.

8. From the data in Figure 10.14*c*, determine $\overline{i_i^2}$ and $\overline{v_i^2}$ for the 2N3797 transistor for a bandwidth of $10^4$ Hz.

*Answer:* From the fact that the optimum noise figure is 2.8 db at 7 Mohms, (10.51) and (10.52) give

$$\overline{i_i^2} = 10.7 \times 10^{-24} \, A^2$$
$$\overline{v_i^2} = 5.25 \times 10^{-10} \, V^2$$

9. More often than not, the Johnson noise is the principal noise in a system. Shot noise should not be ignored, but it is dominant mostly in systems that put large amounts of energy (compared to $kT$) into each quantum. A comparison is called for. Consider a germanium *p-n* junction used to detect near infrared light quanta by collecting the carrier pairs swept across the reverse biased junction. Let the reverse bias be $-5$ V, the load resistor 1 Kohm, and the system bandwidth $10^7$ Hz. The current through the diode is well described by the equation:

$$I = I_{sat} \left[ e^{qV/kT} - 1 \right] - I_l$$

where $I_{sat} = 10^{-7}$ amperes at $300°K$ and $I_l$ is the current generated by the light quanta.

(*a*) Find $I_l$ such that the ratio of the signal power to the noise power, $S/N = 1$. (Note that the signal is the power $I_l$ delivers to the load.) Which is dominant, Johnson or shot noise?

*Answer:* $S/N = I_l^2/[2q\Delta f(I_{sat} + I_l) + 4kT\Delta f] = 1$. By far the dominant term is the Johnson noise, so we obtain:

$$I_l = 4kT\Delta f = 4.1 \times 10^{-7} \text{ amperes}$$

(*b*) Is there an optimum value of $R_L$ as far as $S/N$ is concerned?

*Answer:* No. $S/N$ does not contain $R_L$.

(c) Noting that $I_{sat}$ is proportional to $T^3 e^{-E_g/kT}$, find $S/N$ as $T \Rightarrow 0$, with $I_l$, $R_L$ and $E_g$ being independent of $T$. Is shot noise or Johnson noise the limiting noise here? [*Hint:* The answer is NOT infinity.]

*Answer:* $S/N \Rightarrow I_l/2q\Delta f = 1.28 \times 10^5$ as $T \Rightarrow 0$. Note that the limiting noise is the shot noise *in the signal itself.*

# LIST OF SYMBOLS

Though most of the symbols that appear in the text are listed below, some were excluded, either because they were deemed to be of such common usage as to be obvious (e.g., the cartesian coordinates) or because they differ from similar listed symbols only by the exchange of a subscript (e.g., $D_e$ and $D_h$ with $D$ and the subscripts defined below). Symbols are generally unique within a chapter, but not necessarily throughout the book. (E.g., from Chapter 1 to Chapter 8, subscript $e$ is "electron"; in Chapter 8, it is "emitter.") In the device chapters, a lower case letter is an incremental (small signal) function, an upper case letter with lower case subscript is the total (large signal) function, and one with an upper case subscript is the quiescent (DC) value. A subscript 0 is a thermal equilibrium value unless some other initial value is explicitly stated.

| Symbol | Name of the Function | Page | Equation |
|--------|----------------------|------|----------|
| $A$ | Gain | 417 | 8.11 |
| $a$ | A fixed length or spatial coordinate | — | — |
| $B$ | Magnetic flux density | — | — |
|  | Channel breadth | 486 | 9.2 |
| $BV$ | Breakdown voltage | 385 | — |
| $b$ | A fixed length or spatial coordinate | — | — |
|  | Small signal susceptance | — | — |
|  | (subscript) base or substrate | — | — |
| $b_m$ | Mean separation of scattering centers | 161 | 4.13 |
| $C$ | External capacity | — | — |
| $C_s$ | Surface concentration of impurities | 308 | 6.20 |
| $c$ | Speed of light | — | — |
|  | A fixed length or spatial coordinate | — | — |
|  | (subscript) collector | — | — |
|  | Differential (small signal) capacity | 323 | 6.27 |
|  | (subscript) capture | — | — |

| Symbol | Name of the Function | Page | Equation |
|---|---|---|---|
| $D$ | Diffusion coefficient | 192 | 4.60 |
| $D^*$ | Ambipolar diffusion coefficient | 252 | 5.37 |
| $d$ | Plane separation in a crystal | 41 | 2.2 |
|  | (subscript) drain | — | — |
| $E$ | Energy | — | — |
| $E_f$ | Fermi energy level | 112 | — |
| $E_i$ | Intrinsic Fermi level | 117 | 3.27 |
| E | Electric field intensity | — | — |
| $e$ | (subscript) electron | — | — |
|  | (sgbscript) emission | — | — |
|  | (subscript) emitter | After Chapter 7 | |
| $F$ | Force | — | — |
| $f$ | Frequency | — | — |
|  | Distribution function | 111 | 3.11 |
|  | (subscript) forward transfer | 411 | — |
| $f_B$ | Bose distribution function | 144 | 3.57 |
| $f'$ | Fermi distribution function for impurity states | 124 | 3.36 |
| $f_\beta$ | Beta cut-off frequency | 435 | 8.26 |
| $f_T$ | Unity-current-gain cut-off frequency | 436 | 8.27 |
| $G$ | External conductance | — | — |
| $G_0$ | Conductance of the metallurgical channel (JFET) | 487 | 9.5' |
| $g$ | Density of states | 79 | 2.32' |
|  | Incremental conductance | — | — |
|  | (subscript) gap | — | — |
|  | (subscript) gate | — | — |
| $g_m, g_{fs}$ | Transconductance | 414 | 8.2 |
| $H$ | Magnetic field intensity | — | — |
| $h$ | Planck's constant | flyleaf | — |
|  | Hybrid parameter | 411 | — |
|  | (subscript) hole | — | — |
| $\hbar$ | Planck's constant divided by $2\pi$ | flyleaf | |
| $I$ | Current | — | — |
|  | Optical intensity | — | — |
| $I_{es}$ | Ebers-Moll parameter | 370 | 7.6' |
| $I_{cs}$ | Ebers-Moll parameter | 370 | 7.7 |

| Symbol | Name of the Function | Page | Equation |
|--------|---------------------|------|----------|
| $I_{\epsilon o}$ | Ebers-Moll parameter | 372 | 7.12 |
| $I_{co}$ | Ebers-Moll parameter | 372 | 7.13 |
| $I_{DSS}$ | JFET saturation current with $V_{gs} = 0$ | 498 | 9.12 |
| $i$ | Differential (small signal) current (subscript) input | — — | — — |
| $J$ | Large-signal current density | — | — |
| $j$ | $\sqrt{-1}$ Small-signal (differential) current density | — 324 | — 6.31 |
| $K$ | Spring constant Constant of concentration in a linear-graded junction | 163 328 | 4.16 6.41 |
| $k$ | Wave number Boltzmann's constant | 6 flyleaf | 1.2 — |
| $L$ | Diffusion length Channel length (FET) | 245 480 | 5.28 9.1 |
| $L^*$ | AC diffusion length | 323 | 6.29 |
| $l$ | Mean free path | 263 | 5.49 |
| $l_n, l_p$ | Depletion length on the $n$ (or $p$) side | 136 | 3.47 |
| $M$ | Mass | — | — |
| $m$ | Mass | — | — |
| $m^*$ | Effective mass | 67 | 2.21 |
| $m^{**}$ | Density-of-states effective mass | 114 | 3.19 |
| $m_\perp, m_\parallel$ | Effective mass $\perp$ to or $\parallel$ with the major axis of the constant energy ellipsoid | 81 | — |
| $NF$ | Noise figure | 570 | 10.33 |
| $NP$ | Noise power | 570 | 10.33 |
| $N_c, N_v$ | Effective density of states at the conduction (valence) band edge | 116 | 3.23 |
| $n$ | Electron density (subscript) $n$ side | — — | — — |
| $n_i$ | Intrinsic number of free carrier pairs | 117 | 3.25 |
| $n_T$ | Electron density if $E_f = E_T$ | 235 | 5.10' |
| $n'$ | Excess electron density | 237 | 5.17' |
| $o$ | (subscript) output | — | — |
| $p'$ | Excess hole density | — | — |

| Symbol | Name of the Function | Page | Equation |
|---|---|---|---|
| $p$ | Momentum | 10 | 1.6 |
| | Hole density | — | — |
| | (subscript) $p$ side | — | — |
| $p_T$ | Hole density if $E_f = E_T$ | 235 | 5.11 |
| $Q$ | Impurity atoms per unit surface area | 310 | 6.21 |
| | Charge | — | — |
| $Q_D$ | Charge per unit area in the depletion region | 539 | 10.9 |
| $Q_{fs}$ | Fast-surface-state charge per unit area | 528 | 10.4 |
| $Q_s$ | Mobile surface charge per unit area | 540 | 10.11 |
| $q$ | Charge of the electron (absolute value) | flyleaf | — |
| | Small-signal (incremental) charge | 581 | — |
| $R$ | External resistor | — | — |
| | Radius | — | — |
| | Reflectivity | 52 | — |
| | Rate | 232 | 5.1 |
| $R_s$ | Parasitic source resistance (JFET) | 496 | — |
| $R_S$ | Signal-source series resistor | — | — |
| $r$ | Differential (small signal) resistance | — | — |
| | (subscript) reverse transfer | 411 | — |
| | Radius vector | — | — |
| $S$ | (subscript) signal-source | — | — |
| $s$ | Complex angular frequency | — | — |
| | Surface recombination velocity | 258 | 5.43 |
| | (subscript) source (FET) | — | — |
| $T$ | Temperature | 111 | — |
| | (subscript) trap | — | — |
| $T_L$ | Lattice temperature | 214 | 4.101 |
| $t$ | time | — | — |
| | (subscript) thermal | — | — |
| $U$ | Gain reduction factor | 456 | 8.30 |
| | Electron drift velocity | 581 | — |
| $U_s$ | Saturated drift velocity | 581 | — |
| $u$ | Incremental drift velocity | 581 | 10.53 |
| $V$ | Volume | — | — |
| | Large-signal voltage | — | — |
| | Potential energy | 11 | 1.7 |

| Symbol | Name of the Function | Page | Equation |
|--------|---------------------|------|----------|
| $V_D$ | Diffusion potential | 136 | 3.49 |
| $V_{d\,sat}$ | Saturation voltage | 545 | 10.20 |
| $V_{FB}$ | Flat-band voltage | 531 | 10.5 |
| $V_p$ | Pinch-off voltage | 486 | — |
| $V_T$ | Turn-on voltage | 537 | 10.6, 10.19 |
| $V'$ | Complex velocity | 202 | 4.73 |
| $v$ | Velocity | — | — |
|  | Small-signal (differential) voltage | — | — |
| $v_p$ | Phase velocity | 64 | — |
| $v_g$ | Group velocity | 65 | 2.17 |
| $W$ | Metallurgical channel width (i.e. depth) | 486 | 9.3 |
| $X$ | External reactance | — | — |
| $x$ | Differential reactance | — | — |
| $Y$ | External admittance | — | — |
| $y$ | Differential reactance | — | — |
| $Z$ | External impedance | — | — |
| $z$ | Differential impedance | — | — |
| 0 | (subscript) equilibrium value | — | — |
|  | (superscript) neutral | — | — |

Greek and Special Symbols

| Symbol | Name of the Function | Page | Equation |
|--------|---------------------|------|----------|
| $\alpha$ | $\sqrt{(V_0 - E_i)(2m/\hbar^2)}$ | 24 | — |
|  | A density variable in the distribution function | 188 | — |
|  | Common-base current transfer ratio | 371 | 7.8 |
|  | Ionization coefficient | 264 | 5.53 |
| $\beta$ | $\sqrt{E_i(2m/\hbar^2)}$ | 24 | — |
|  | Common-emitter current transfer ratio | 381 | 7.17 |
| $\gamma$ | Asymmetric displacement in distribution function | 189 | 4.51 |
| $\delta$ | Symmetric displacement in the distribution function | 189 | 4.51 |
|  | Length of the saturated portion of the channel | 492 | 9.8 |
| $\Delta$ | Increment | — | — |
| $\epsilon$ | Dielectric constant (permittivity) | flyleaf | — |

| Symbol | Name of the Function | Page | Equation |
|---|---|---|---|
| ε | (subscript) emitter | before Chapter 8 | |
| μ | Mobility | 192 | 4.60 |
| ν | Frequency | — | — |
| ρ | Resistivity | — | — |
| σ | Conductivity | 64 | — |
| | Cross-section | 232 | 5.1 |
| $\tau_R$ | Randomization (relaxation) time | 162 | 4.15′ |
| τ | Lifetime | 233 | — |
| $\tau_c$ | Mean time between collisions | 160 | — |
| φ | Eigenfunction of the wave equation | 18 | — |
| ψ | General wave function | 10 | 1.2 |
| | Electrostatic potential | 134 | 3.45 |

# INDEX